Reengineering IBM Networks

Anura Gurugé

WILEY COMPUTER PUBLISHING

John Wiley & Sons, Inc.

New York • Chichester • Brisbane • Toronto • Singapore

Publisher: Katherine Schowalter
Editor: Robert Elliott
Managing Editor: Mark Hayden
Text Design & Composition: Publishers' Design and Production Services, Inc.

Library of Congress Cataloging-in-Publication Data:
Gurugé, Anura.
Reengineering IBM networks / Anura Gurugé.
p. cm.
Includes index.
ISBN 0–471–14274–3 (pbk. : alk. paper)
1. Local area networks (Computer networks) 2. Wide area networks (Computer networks). 3. IBM computers. 4. Asynchronous transfer mode. I. Title.
TK5105.7.G88 1996
004.6—dc20 96–10799
CIP

Printed in the United States of America
10 9 8 7 6 5 4 3 2 1

To Linda, Danielle, Matthew Gordon, & My Mother
For All the Good Times . . .

About the Author

Anura Gurugé is an Independent Strategic Technical Consultant who specializes in LAN/WAN internetworking in IBM environments, SNA/APPN/HPR, Frame Relay, ATM and System Management.

Over the last five years he has worked with most of the leading bridge/router, intelligent hub, FRAD, Token-Ring Switching, and Gateway vendors over the last few years and has designed many of the SNA-related features now found on bridge/routers and "gateways."

He is the author of the best-selling *SNA: Theory and Practice* as well as several other books on SNA, APPN, and SAA.

He has published over 200 articles and is the author of the Business Communications Review supplements *BCR's Guide to SNA Internetworking* and *Beyond SNA Internetworking.*

He is a regular contributor to *Network World, Business Communications Review,* and *InSight IS.*

He conducts technical and sales training seminars worldwide on a regular basis.

In a career spanning 22 years he has held senior technical and marketing roles in IBM, ITT, Northern Telecom, Wang, and BBN.

Contents

Foreword

"What is every major institution in the world focusing on?" asks Gerstner. "It's called reengineering. It's called getting competitive. It's called reducing cycle time and cost, flattening organizations, increasing customer responsiveness. All of these require a collaboration with the customer, and with suppliers, and with vendors." All of these efforts, in short, depend on networking.

"The View from IBM," *Business Week*, October 30, 1995

IBM Chief Executive Officer Louis Gerstner, whose vision of network-centric computing is driving the IBM of the 1990s, should know a thing or two about reengineering.

Big Blue and its customers are currently involved in the biggest collective reengineering project in the history of computing—trying to rebuild the Systems Network Architecture networks that run their companies to support new applications, new kinds of traffic, new end user expectations, and new technologies.

Talk about a challenge. This rearchitecting of SNA networks is akin to changing the wheels on a train while it's still running.

Simply put, SNA defined corporate networking as we know it. Sure, you've heard plenty about LANs and internetworks and client/server computing. But those technologies are only starting to come up to speed in the stay-up-or-go-broke world of business computing.

SNA is the foundation network technology that keeps the bank ATM machines running, makes the airlines fly, and keeps Wall Street in the money. Still.

But SNA—like the B-17s and B-24s that liberated Europe—was designed for a different war. The battlefield conditions have changed, the business world is more mobile, more multimedia, more real-time. So the SNA networks must change to meet those needs or fade away.

IBM can't afford to let SNA fade away. SNA represents a form of once-in-a-business-lifetime account control that Big Blue needs to maintain if Gerstner and company are going to have a major place in the next phase of computing. And despite all the hype about companies dumping mainframes, the fact is most customers don't want to scrap their SNA networks. Network managers from major corporations have told me on countless occasions that nothing in the client/server world compares with SNA in terms of reliability. The mean-time between front-end processor outages is measured in months. Sorry, internetwork vendors, there's no comparison.

Simply put, few companies have the budget or the stomach to rebuild their SNA networks from scratch. They need transition strategies and technologies. They need ways to tie mainframes more effectively into their LAN-based networks, to make it easier for end users to get at high-level corporate information and business-proven applications.

So how do we get from the old SNA to the new SNA? How, readers ask, can we:

- Carry SNA and multiprotocol traffic on a common network?
- Support new multimedia applications that assume a whole new set of network conditions?
- Make it possible for end users and customers to collaborate across the network in real-time?
- Cut the cost and complexity of managing our corporate networks?

I wish I could tell you. I'd make a killing in consulting fees.

IBM has lots of answers—some shipping, some still on the drawing board—in the form of new products and technologies like ATM, APPN, HPR (these acronyms will become second nature to you after reading this book). They're all designed to smooth the transition into your future network.

And IBM isn't alone. Lots of other suppliers are seizing on this massive network reengineering in the hopes of cashing in on one of the biggest technology opportunities ever. Why do you think Cisco Systems, Inc. has set up shop jeek-by-jowl with IBM's networking folks in Research Triangle Park, N.C.?

There will be many choices, many difficult decisions. So who do you turn to for answers?

Relax, you've already taken a big step in the right direction by buying this book (at least I hope you've purchased this copy and aren't just schlepping around with someone else's).

Anura Gurugé is a real answer man when it comes to SNA and SNA migration issues. He's been writing about these topics for *Network World* for five years now, treating our 150,000 readers to honest opinions and valuable insights about the state of the large-systems network market and the evolution of IBM.

One thing about Anura: You don't have to wonder where he stands on an issue. He calls them straight, even if it means pointing out where IBM and customers are wrong in their choices about SNA networking. I know he drives IBM crazy on occasion, because I have to field the calls from their flaks.

But Anura isn't an SNA basher—a field that's gotten quite crowded of late. That's too easy. Anyone can say that SNA can't cut the mustard in today's distributed computing world, that its time has come and gone. It's much tougher to assess the various means of moving SNA forward, to provide analysis of what will and won't work for SNA customers trying to deal with new business challenges.

That's where Anura excels. He's generous with his time and ideas. More important, he has the experience to put technologies and strategies into perspective, to keep the dialogue down to earth when so many other so-called experts are only issuing sound bites on the death of Big Blue networking.

Our readers run big SNA networks, as well as all sorts of other networks. And Anura helps us provide the best coverage we can. It's as simple as that.

So read and learn. Take the plunge. Because the companies that successfully migrate their enterprises into the next generation of networking will be the market share leaders of tomorrow.

John Gallant
Editor-in-Chief
Network World

Introduction

These are turbulent but nonetheless curiously exhilarating times in commercial sector networking. An era, one could even say a dynasty, is coming to an end; a new and beguiling era beckons. The benevolent but unswaying rule of Systems Network Architecture (SNA) over the majority of the networks used by large commercial enterprises is rapidly coming to a close. IBM is no longer a predominant force in networking. Nobody talks about Open System Interconnection (OSI). Transmission Control Protocol/Internet Protocol (TCP/IP) is ubiquitous and does appear poised to inherit the earth. SNA as an end-to-end protocol will be around well into the 21st century but only in the context of multiprotocol networks. SNA-only networks are dead.

Cisco, Bay Networks, 3Com, Motorola, Hypercom, and Fore Systems et al. are the new high rollers of networking. ATM, referring not to automated teller machines but to the broadband, multimedia technology of Asynchronous Transfer Mode, has become the holy grail of contemporary networking. Though the technology is already available and the hype abounds, the IBM networking community is unlikely to be ready to embrace ATM-based networking for another few years. There are likely to be a few other intermediary Rubicons to cross. Multiprotocol networking is inevitably going to be one. Local Area Network (LAN) switching and 100Mbps LAN technology could be another. Advanced Peer-to-Peer Networking (APPN) and High Performance Routing (HPR) may also require evaluation. Boredom will be one thing that the IBM networking community will not have to concern itself about over the next few years.

This book is intended to serve as an in-depth, no-holds-barred, nonpartisan reference guide cum tutorial to this new era of SNA/APPN-capable multimedia, multiprotocol networking. It covers all pertinent technologies, concepts, methodologies, and issues from Transparent Bridging to ATM, Token-Ring Switching to HPR, and Data Link Switching to AnyNet Gateways. This book could be used in two distinct modes. It could be read sequentially, to serve as a systematic step-by-step guide to the entire network reengineering process. It could also, however, be profitably used as a reference guide to a particular topic, for example, the pros and cons of replacing channel-attached 37xx's with a channel-attached bridge/router, or a comparison of Data Link Switching (DLSw) against Request for Comment (RFC) 1490-based encapsulation. If the book is to be used in this mode, the Index is likely to be the best means of locating the pertinent topic(s). The Index, however, is not the only means of navigating through the book. Each chapter of this book is prefixed by a one-page quick guide that enumerates all the key topics covered in that chapter.

Reengineering their existing networks into SNA/APPN-capable multiprotocol networks is now the most profound and urgent challenge confronting networking professionals in the so-called IBM shops. These new generation networks will invariably be built around bridge/routers, RFC 1490-based Frame Relay Access Devices (FRADs) or in some instances even AnyNet Gateways. Technologies such as DLSw, Synchronous Data Link Control (SDLC) to Logical Link Control 2 (LLC:2) conversion, frame relay encapsulation, and APPN/HPR Network Node (NN) routing are all likely to play a role in such integrated networks. LAN switches, in particular token-ring switches, will be liberally deployed as a means of cost-effectively and nondisruptively increasing campus-level LAN bandwidth. Most of these integrated, multiprotocol networks, particularly in North America, will initially use Frame Relay as their Wide Area Network (WAN) fabric of choice.

Chapters 5 to 8 of this book cover every aspect of implementing multiprotocol networks. Chapter 3, in addition, describes a unique and objective scheme for choosing between six or so viable technologies that can be used to realize this type of network. This scheme relies on the WAN traffic mix, in the 1997–1998 period, as the exclusive metric for selecting the optimum technology. Chapter 4 serves as a detailed treatise on LAN switching—in particular Token-Ring Switching. SNA/APPN-specific techniques relevant to accommodating SNA/APPN traffic within a multiprotocol network are described in Chapter 2. Managing multiprotocol networks and the notion of Total System Management are the focus of Chapter 10.

Ironically, multiprotocol networking will be but an interim step in what will in the end prove to be a much larger networking reengineering process. When it comes to ATM the primary question is no longer "if" but "when." ATM will start to make its presence felt in the campus arena by mid-1996. In many such instances ATM's introduction into these environments will be in the form of 155 Megabits per second (Mbps) fat pipes between hubs, LAN switches, or bridge/routers. Widescale multimedia ATM, WAN ATM, and ATM to the desktop is unlikely to prevail in IBM shops until after 1997.

Chapter 3, in addition to providing an introduction to ATM technology, describes a realistic migration strategy to ATM-centric networking and points out all the major milestones that will be encountered during this migration. Chapter 9 deals with IBM's comprehensive repertoire of ATM offerings, including its Networking BroadBand Services (NBBS) architecture for ATM-centric networks. This chapter, however, is not in any way an attempt to propagandize IBM products; far from it. If anything, it points out the shortcomings of some of IBM's current offerings. This chapter concentrates on IBM for two very good reasons.

IBM at present is unique in having a full-spectrum product offering that addresses every facet of ATM networking from 25Mbps PC adapters all the way to 52 Gigabits per second (Gbps) WAN switches. Thus this product range serves as a useful reference point to describe the types of offerings that are available. In each instance competing product offerings are mentioned. The second reason for picking IBM is that it is IBM. This book, as its title proclaims, is expressly targeted at the IBM networking community. Most members of this community are likely to have some degree of familiarity with IBM's ATM story. If nothing else they probably have all heard the term *Nways*. Ironically, one of the things described in this chapter is that *Nways* per se does not refer just to IBM's ATM initiatives.

Over the last five years I have been heavily involved in bridge/router and FRAD-based SNA/APPN-capable multiprotocol networking. During that time I worked with nearly all leading bridge/router, SNA LAN gateway, and FRAD vendors. I was instrumental in designing some of the SNA/APPN-related facilities now on the market. These included: SDLC-to-LLC:2 conversion, SNA traffic prioritization, Binary Synchronous Communications (BSC) support, LAN-over-SNA solutions, and even an LAN-to-SNA/APPN Gateway. In addition I have worked closely with a few token-ring switch vendors. During this period I had over 150 articles published in trade magazines and conducted at least 100 training-related seminars. This book is an attempt to condense what I learned in this process into a single reference text. My goal was to essentially provide a training seminar on reengineering IBM networks in book form. I have had repeated requests for this type of book, and I am glad to be able to deliver it at long last.

This book can be read by anybody involved with networking. It will, however, only truly be of interest to those dealing with networks that have some level of IBM content—whether it be SNA/APPN traffic or IBM-supplied ATM campus switches. While it does not rely on any specific knowledge of networking it does however assume readers do have at least a cursory understanding of what SNA and TCP/IP are. A glossary and a list of acronyms appear at the end of the book to guide users through any unfamiliar terms. This book should be of particular value to networking professionals, on both the demand and supply sides of the market, who are actively involved with trying to reengineer what were SNA-only networks. Though it is a technical book, it is not aimed exclusively at "technocrats." It is structured in such a way that readers can easily identify the overtly technical "bits-and-bytes" sections and skip over them.

The organization of the book is as follows:

Chapter 1: The Future of Networking in IBM Environments. This chapter examines all the various business, cultural, and technological forces that are shaping the future of commercial networks and acting as the catalyst for the network reengineering process. It introduces all of the key technologies described in the remainder of the book and highlights the key issues that need to be addressed when building a new SNA/APPN-capable multiprotocol network.

Chapter 2: State-of-the-Art SNA Networking. This chapter concentrates on the implications of having to support SNA/APPN traffic across a multiprotocol LAN/WAN network. It includes exhaustive descriptions of all the various types of SNA LAN gateway, SNA Subarea-to-Subarea Routing, traffic prioritization, and the potential role of 37xx Communications Controllers in multiprotocol networks.

Chapter 3: The Blue Brick Road to ATM. This chapter provides an introduction to ATM. It then lays out a detailed migration strategy for realizing multimedia ATM-centric LAN/WAN networks. It addresses all the issues related to campus level ATM and desktop ATM. It also includes a section that deals with how to choose the optimum technology for parallel network consolidation.

Chapter 4: Token-Ring Switching: The On-Ramp to ATM. This chapter provides an in-depth tutorial into all aspects of token-ring switching including cut-through, store-and-forward, adaptive cut-through, and transparent switching.

Chapter 5: The First Steps in Building a Multiprotocol Network. This chapter covers Source-Route Bridging, Transparent Bridging, Source-Route Transparent Bridging, Translation Bridging, TCP/IP encapsulation of SNA/APPN à la DLSw(+), Protocol Independent Routing, Synch Pass-Through, Remote Link Polling, SDLC-to-LLC:2 conversion, and Remote SNA Switching.

Chapter 6: Frame Relay: A Springboard to ATM. This chapter introduces frame relay, and then examines the exciting possibilities of using the RFC 1490 native mode encapsulation standard as a means of realizing SNA/APPN-capable multiprotocol networks. It includes a detailed comparison of RFC 1490 vis-à-vis DLSw.

Chapter 7: High Performance Routing: The Latest Incarnation of SNA. HPR is the successor to both SNA and APPN. This chapter examines all aspects of HPR including the three new protocols associated with it. These protocols are: Automatic Network Routing, Rapid Transport Protocol, and Adaptive Rate-Based Congestion Control. It also includes a comparison of HPR NN-based routing vis-à-vis DLSw and RFC 1490.

Chapter 8: Multiprotocol LAN/WAN Networking over a Single-Protocol IP or HPR WAN. It is possible to build IP-traffic-only or APPN/HPR-traffic-only WANs that can nonetheless serve as the backbone for multiprotocol networking. This chapter looks at all the technologies that can be used to realize this type of network, including AnyNet protocol conversion.

Chapter 9: IBM's Nways ATM Solutions. This chapter looks at IBM's entire ATM story including the product repertoire, the Packet Routing Integrated Zurich Modular Architecture (PRIZMA) ATM chip technology, the NBBS architecture, and the Switched Virtual Networking framework.

Chapter 10: Total Enterprise Management: An Elusive Goal. Multiprotocol networks supporting a large population of SNA users typically need to be managed using both a Simple Network Management Protocol (SNMP) manager and the mainframe resident NetView/390. This chapter examines how to realize this type of cooperative management. It also uses IBM's SystemView framework to introduce the possibilities and promise of Total System Management.

Chapter 11: Putting It All Together. This last chapter sets out to synthesize and reinforce all of the key concepts, methodologies, and technologies discussed in the previous ten chapters. It includes many real-life, before-and-after examples of various network reengineering initiatives in what were hitherto very IBM-centric networking environments.

Appendix A: Essential SNA in a Nutshell. This is a succinct but comprehensive tutorial on the SNA architecture and covers all the key concepts and constructs including Logical Units (LUs), Physical Units (PUs), System Services Control Points (SSCPs), Nodes, Sessions Types, and the Path Control Network.

Appendix B: Thumbnail Sketch of APPN. This is a brief overview of APPN that augments the description of HPR provided in Chapter 7. By the end of 1996 most of today's APPN NN implementations will conform to HPR.

Glossary

List of Acronyms and Abbreviations

Bibliography

With the era of SNA-only networks now coming to an end, it it appropriate to describe my initial contact with SNA. I had been with IBM a few weeks—joining as a junior programmer in the Systems Support Group at the Research Laboratory in Hursley, in the summer of 1974—when I received a cryptic and curt memo to the effect that "Henceforth SNA will be known as Systems Network Architecture rather than Single Network Architecture." Since I had never heard of SNA, and since both titles were equally baffling, I sought clarification from my manager—the head of System Support. I was told quite categorically to forget about it as SNA would never catch on! Twenty-two years later we are getting around to building SNA/APPN-capable multiprotocol networks and thinking about supporting SNA/APPN across ATM.

Anura Gurugé
Shaws Pond, 1995

Acknowledgments

I am indebted to many, on both the supply and demand side of the market, for providing me with the opportunities to explore at first hand the various technologies and methodologies that I discuss in this book. They are too innumerable to name individually—but I thank you all. Each in their own way provided me with some insight that helped me get a better perspective of this field.

The contributions of some, however, do warrant explicit gratitude. Alison Conliffe, formerly of *Network World*, helped me tremendously by doing the first-cut edit on the manuscript while Susan Marples did a near miraculous job on the copyedit. John Gallant, Editor in Chief of *Network World,* despite the arduous demands on his time, kindly undertook the task of writing the Foreword to this book without me even having to twist his arm. Speaking of editors I also need to thank Susan Collins, Charlie Bruno, and Paul Desmond of *Network World* who over the last few years have tried gallantly to influence my writing. This book would not have been possible without the help of Bob Elliott and Mark Hayden of John Wiley.

Mark Lillycrop, of Xephon PLC and the Editor of *InSight IBM,* has been a good friend and a collaborator for over a decade and as a result has had the misfortune of having had to edit the equivalent of a few of these books over the years. Bill Kwan, the CTO of Jupiter Technology, has been my trusted source and sounding board on technology while Michael Cooney of *Network World* has been my other ear to the market for a very long time.

Special thanks are also due to: Greg and Bill Koss, Larry Samberg and Nigel Machin who were leading lights at CrossComm; Nick Grewal and Bob Rosenbaum of Nashoba Networks who pandered to my interest in token-ring switching; Nick Francis, Selby Wellman, Betsy Huber, Scott W. Anderson, Donna Cronheim, Wayne Clark, and Don Listwin et al. of Cisco; Vicki Noreen and Brian D. Clark of Unisys Network Enable; Richard Tobacco, Lark Allen, Don Haile, Ron Bond, etc. of IBM; Jim O'Connor of Bus-Tech; Maks Wulcan of Eicon; Bob Koch of Proteon; Nina Saberi, the president of Netlinmk; the Wallner brothers of Hypercom; and Isaac "Itzik" Nosatzki, my alter ego in the Holy Land.

Linda helped in multiple ways and made sure that this book did get done while Danielle and Matthew Gordon provided me with enough distractions to ensure that my perspectives were not getting warped.

CHAPTER 1

The Future of Networking in IBM Environments

A QUICK GUIDE TO CHAPTER 1

Commercial-sector networks are now in the throes of what promises to be a spectacular metamorphosis. SNA-only networks that were the stalwart backbone of business-critical networking for close onto two decades are rapidly becoming anachronisms. SNA is no longer the only protocol of import vis-à-vis mission-critical applications. By mid-1994, over 40% of IBM mainframe sites were augmenting their classic SNA-centric mainframe applications with Unix-based, TCP/IP applications. Other protocols such as Internetwork Packet Exchange/Sequenced Packet Exchange (IPX/SPX), NetBIOS, and Banyan Vines are being widely used for file-server, database, print-server, and new-wave client/server applications.

In order to accommodate this potent cocktail of protocols, enterprises are beginning to build multiprotocol, multivendor LAN/WAN networks. Many of these will initially resort to LAN switching as a cost-effective means of gaining instant gratification when it comes to enhancing the productive bandwidth available for campus-level LAN applications. The need for 56 Kilobits per second (Kbps) to 4Mbps WAN bandwidth for these multiprotocol networks will be sated by frame relay.

Such multiprotocol networks, however, are but the first step in a much larger network reengineering process. The siren call of

ATM-centric, multimedia networks running in the megabit to gigabit can already be heard—loud and clear. The technology, if not the multimedia applications, for ATM is already at hand. Networking professionals within the IBM community no longer have any choice but to start laying down concrete plans for migrating their existing parallel networks into a multimedia, multiprotocol ATM-centric network over the next five to ten years. This chapter is a detailed roadmap for this migration.

This chapter starts off by examining all the various business (e.g., electronic funds transfer), cultural (e.g., the Internet), and technological (e.g., ATM) forces that are shaping the future of commercial networks. It articulates all the issues such as System Management and multivendor interoperability that have to be confronted and tackled. Section 1.2 describes how today's networks have to reflect and adapt to the dramatic changes taking place in the application arena. SNA applications are now being routinely augmented or even usurped by non-SNA applications.

Section 1.3 onwards concentrates on the concepts, issues, and technologies that pertain to consolidating SNA and non-SNA parallel networks. This section provides an initial introduction to technologies such as DLSw, HPR, AnyNet, Frame Relay RFC 1490, and SNA Session Switching. It dwells on the importance of these integrated networks maintaining the high-availability characteristics of today's awesomely reliable SNA networks. A mere 1% drop in the availability of such a network actually translates into an additional 88 hours of revenue-sapping downtime per year! The chapter concludes by looking at techniques for congestion control, minimizing packet discard, and multivendor interoperability relative to this new generation of integrated, multiprotocol LAN/WAN networks.

A tantalizing new era in commercial-sector networking around the world is at hand; of that there can be no shadow of a doubt. Networking in the so-called IBM shops, in particular, will never be the same again. IBM's remarkable three-decade-long domination of networking, particularly so in the case of large commercial enterprises, is nearly at an end. The networking requirements of today are more dynamic and diverse than they have ever been before, and invariably revolve around the need for reliable, high-performance, and at least partially intuitive interoperability between disparate and often dispersed computing systems. The impetus for much of these new networking requirements has arisen as a direct consequence of the initiatives under way at **enterprises**, worldwide, to reengineer their business process in a bid to further enhance their competitive edge.

The hardware and software technology required to adequately satisfy much of today's networking needs is becoming increasingly complex, and inevitably either emerging or rapidly evolving. No longer can a single vendor claim with any conviction or credibility to be able to provide even the overall architectural framework, à la SNA of old, for the entire network, let alone the myriad of necessary products. This applies to IBM as well as all other vendors.

In terms of the various devices a network must support and the equipment required to actually realize the network, multivendor networking has already become an inescapable fact of life. These multivendor networks will have to be able to dexterously support a wide spectrum of assorted protocols including SNA. Single-protocol networks, including SNA-only networks, are now a severely endangered species. The ultimate onus, in most cases, of sourcing, integrating, configuring, and managing these multifaceted networks will now rest, caveat emptor so to speak, with the customer. System integrators, anxious to please, will do their very best to provide some much needed relief in these circumstances.

The new generation of networks, in addition to being multiprotocol, will also be markedly different from existing computer networks. They are unlikely to be restricted to data traffic. They will also not rely on Time Division Multiplexing (TDM), a technology from the early 1970s, as the only means of sharing bandwidth between the data network and a voice network. Instead, many, if not all, of these networks, reacting to the availability of compellingly cost-effective broadband (i.e., 100Mbps+) bandwidth and multimedia integration techniques such as ATM, will set out to implement a new generation of integrated networks. These integrated networks will simultaneously handle digitized, full-motion video (as in videoconferencing), fax, voice, and data traffic.

To accommodate these impending changes, networking professionals in IBM shops no longer have any option but to start taking decisive steps toward reengineering their current data-only, IBM-biased, and SNA-centric networks. *The overriding goal of this reengineering process is to nondisruptively transition the existing networks into multiprotocol, multimedia, multivendor, brodband networks.*

Already the 35,000 or so blue-tinged networks, worldwide, that were built to ensure resilient and painless access to application programs and data resident on IBM mainframes or minicomputers (e.g., IBM AS/400s) are in the throes of a dramatic and far-reaching metamorphosis. Though still prevalent, slow-speed (19.2Kbps or less) long-haul SDLC links in North America and X.25 connection in Europe are no longer the primary means of device interconnection. Nonprogrammable 3270 terminals have been usurped by the seemingly ubiquitous PCs and to a lesser extent by Unix/RISC workstations. LANs, both token-ring and Ethernet, laden with PCs, workstations, and servers (e.g., file or print servers) now crisscross buildings and campuses, and serve as relatively high-speed conduits for a potent cocktail of protocols. At a minimum this cocktail is likely to consist of the protocols associated with SNA, TCP/IP, Novell's IPX/SPX, or IBM's NetBIOS traffic.

The preponderance and preeminence of these multiprotocol, PC-centric LANs is by far the most profound change affecting all aspects of networking in IBM environments and a major spur for a systematic network reengineering initiative. LAN-centricity has to be a key element of the new networking paradigm.

Contrary to the misguided thinking in the late 1980s, mainframes are not going to emulate dinosaurs and become extinct—at least not for awhile. Mainframes, as demonstrated by IBM's financial results of late, will still be around well into the 21st century. They will act as highly secure and relatively cost-effective repositories for business critical data and will run very high-volume, transaction-processing applications. They will, however, not be the foci around which all the processing and data storage are centered, as was the case even just a few years ago. Instead, mainframes will complement the processing and data storage being performed by LAN-interconnected PCs, workstations, and RISC/Unix systems (e.g., IBM RS/6000s or SP/2s). In this context, the mainframe will become the server of servers.

1.1 THE GENERAL BEARINGS FOR SYSTEMATIC REENGINEERING

From a data networking perspective, the key goal of reengineering today's SNA-only networks is to provide a high-performance, multiprotocol transport mechanism for interactions occurring between dispersed LANs, LANs and mainframes, and multiple mainframes. This new transport mechanism obviously needs to be flexible, reliable, and extensible. Many (at least 50%) of today's non-PC SNA devices are still link-attached (e.g., SDLC links). Hence, the new networks, despite their LAN-centricity, will still have to provide mechanisms to effectively accommodate link-attached devices alongside the LAN-attached devices. In the case of SDLC devices, the integration with LAN environments will more than likely be realized via SDLC-to-LLC:2 conversion. This a technique widely available as an integrated function in many of today's internetworking devices such as bridge/routers and FRADs. SDLC-to-LLC:2 conversion makes an SDLC-attached device look like it is LAN-attached. Figure 1.1 illustrates the overall structure of a consolidated, multivendor, multiprotocol, next generation data-only network.

Though noteworthy, realizing an effective multivendor, multiprotocol network over the next couple of years, is, however, only going to be the first hurdle for most networking professionals. The next step would be to heed the beckoning of cell-switching technology à la ATM and the broadband bandwidth that invariably comes with it. With ATM technology on tap, the next logical and fiscally attractive step in the reengineering effort is to extend the data-only network to embrace all in-house (i.e., intraenterprise) voice, fax, and video traffic. Figure 1.2 demonstrates the concept of an all-in-one, multimedia network that is supporting voice, fax, video, and data. It is unlikely, though, that many would want to go all the way toward such an all-in-one network in one giant step. Instead, they would first want to put the technology and the networking equipment, as well as the network administration and management processes, through a rigorous regime of real-life testing with representative volumes of traffic.

Any large-scale disruption to a mission-critical data network is often traumatic and can result in the enterprise suffering some heavy financial losses due to lost business. The business criticality of an all-in-one network is even greater in that it carries both the data and in-house voice traffic. Consequently many enterprises may opt not to put both voice and data on one network until the technology and the infrastructure of that network has been adequately tried, tested, and proven. Hence, the transition to an all-in-one multimedia network is likely to occur in two stages, probably spread over three to four years.

The first stage would be to integrate the traffic from nascent video applications with the data traffic. Possible video applications include desktop videoconferencing, live CNN-type news feeds to PCs/workstations, and video-based training programs piped to the desktop. This integration will be realized using a multimedia networking technology—most likely ATM. But it may also involve some forms of high-speed LAN-switching (e.g., a token-ring switch) solutions as an interim step prior to the full-scale adoption of ATM. Once the multimedia aspects of the network have been proven, voice and fax traffic can be gradually integrated into what would then become an all-in-one network capable of sustaining all the electronic data of an enterprise for the next two decades or more.

Whether fully multimedia, or data-only to begin with, it is unlikely

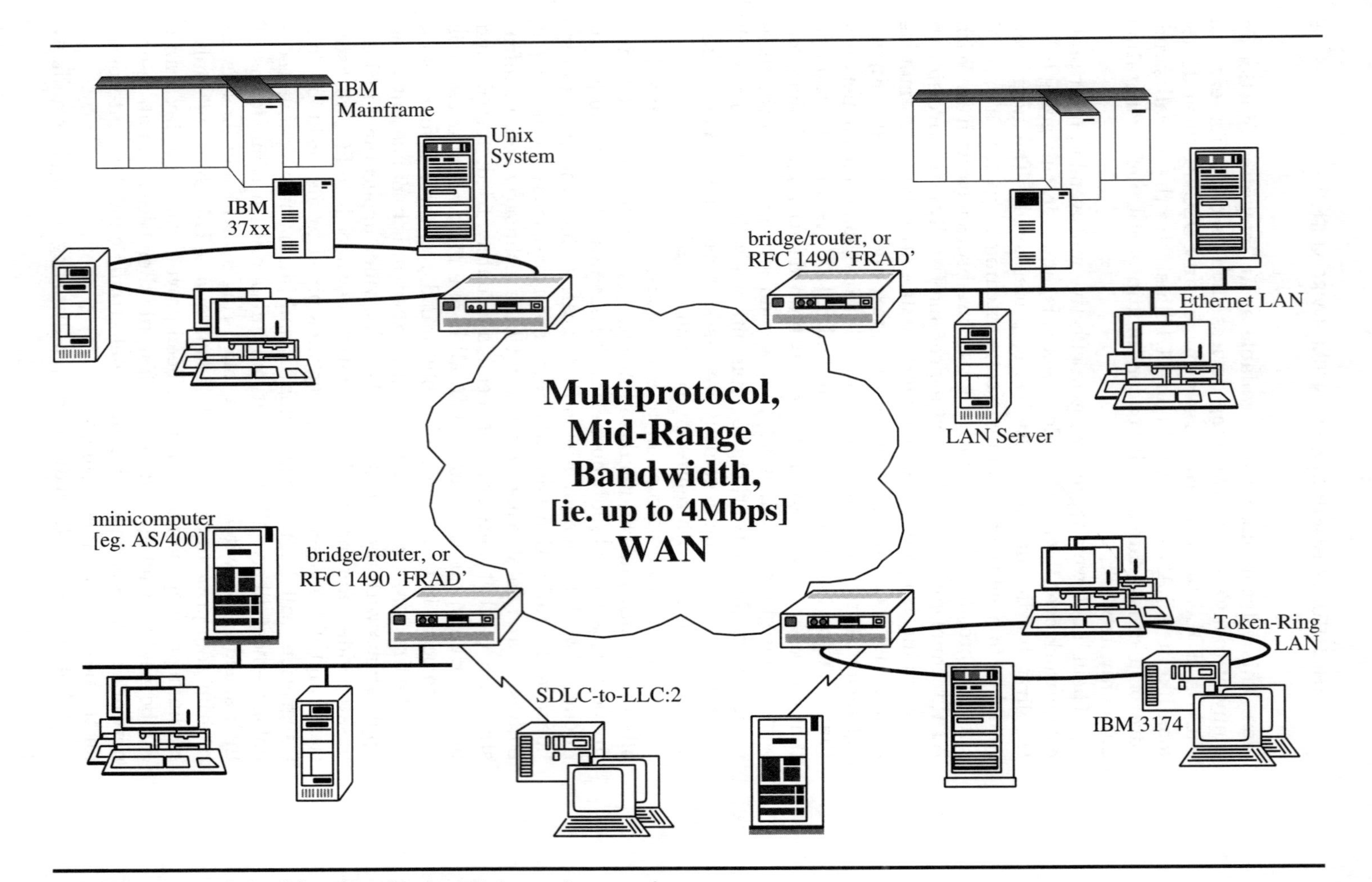

FIGURE 1.1 Consolidated, multiprotocol data-only network.

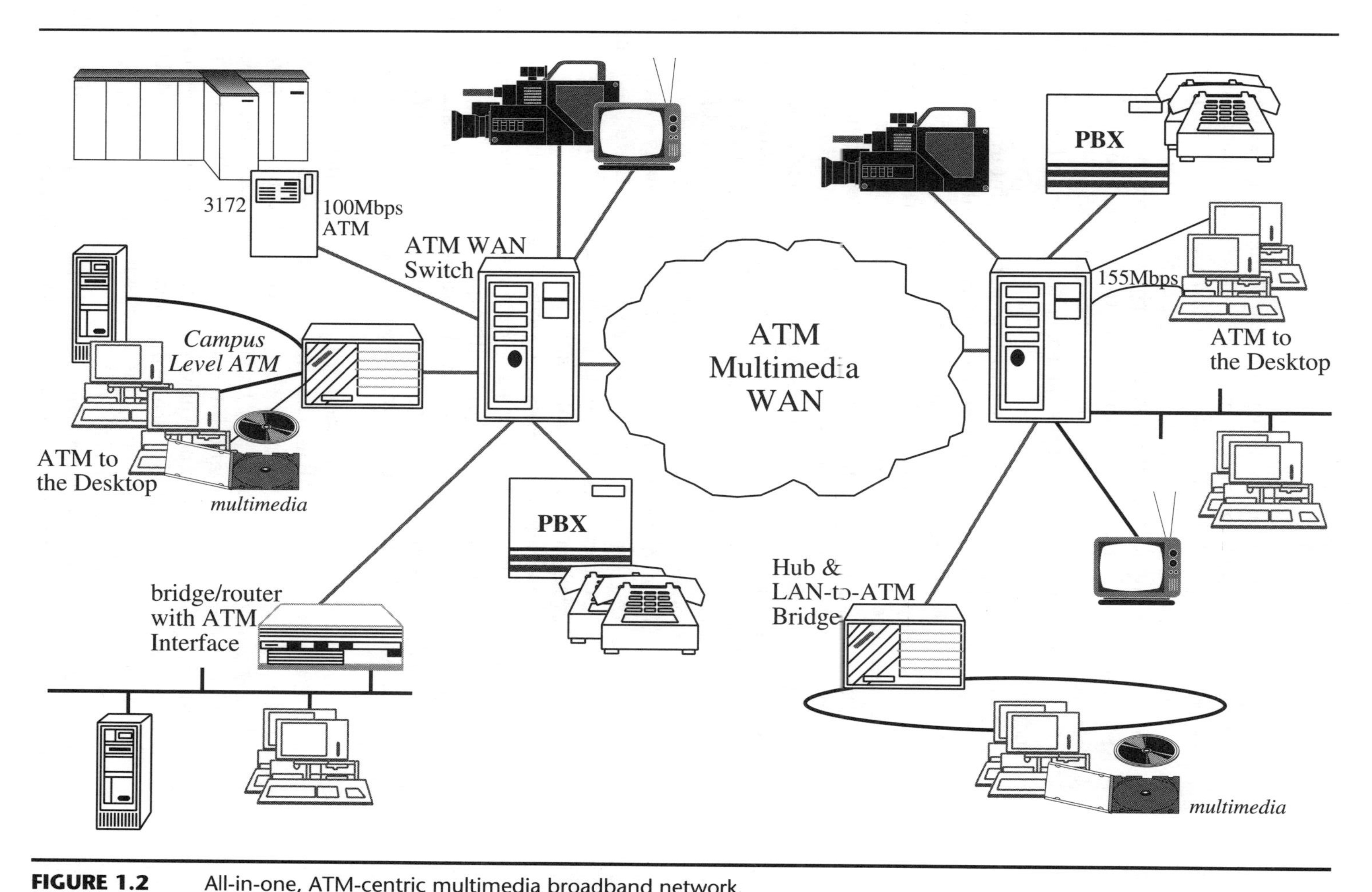

FIGURE 1.2 All-in-one, ATM-centric multimedia broadband network.

that this new breed of network will end up being totally self-contained and constrained just for intraenterprise interactions. Electronic Funds Transfer (EFT) continues to grow in popularity as a very efficient, fast, cost-effective, and secure means for taking care of accounts receivable. It is increasingly ousting the need for printing and mailing checks. EFT is obviously contingent on the ability to establish networking connections, in most instances just for the duration of the transfer, between either the payer and the payee, or the payer and a third-party clearinghouse such as a bank.

Permanent intercompany, network-to-network interconnection is also relatively common. Such interconnections could be used to facilitate on-line, real-time access to data required to execute a collaborative venture (e.g., parts of an aircraft being built by different companies). In other instances such interconnections may be the result of company mergers or acquisitions. Since 1984, SNA has provided explicit support for this type of interconnection via the SNA Network Interconnection (SNI) gateway facility. SNI permits authorized sessions to be dynamically and transparently established between autonomous SNA networks that have their own naming (e.g., LU and application names) and addressing (e.g., SNA subarea numbers) schemes. Interoperability between autonomous networks is also a quintessential feature of IP. In parallel to EFT and intercompany interconnection, many enterprises, especially in the financial sector, now rely heavily on up-to-the-minute information obtained via electronic feeds (e.g., stock market ticker prices) from third-party organizations.

The freewheeling Internet is, however, indubitably destined to become the biggest and most pervasive data feed that most enterprises will have to contend with and hopefully capitalize upon. The Internet vividly demonstrates the true potential of today's PC/workstation and mid-range bandwidth networking technology when it is innovatively synthesized by human ingenuity unrestrained by the needs of commerce.

Few would want their mission-critical enterprise network to be as dynamic and unregulated as the Internet. However, an increasing number of commercial enterprises, particularly computer and networking vendors, are using the Internet as a very low-cost and far-reaching media for intra- and intercompany e-mail. Many also use it for advertising in the form of World Wide Web home-page showcases. One can already shop and place orders over the Internet. With the right authentication and encryption software it is likely to soon become a major conduit for EFT.

In addition to its mesmerizing possibilities as a public data dissemination utility, the Internet is also likely to become the showcase and proving ground for new networking technology. The Internet Engineering Task Force (IETF), which oversees the technical infrastructure of the Internet, has already become the de facto arbitrators of multiprotocol, multivendor interoperability standards. The IETF uses the Request for Comment (RFC) process as the basis for establishing standards. One such standard is RFC 1490, which has become a nearly ubiquitous standard for encapsulating other protocols such as SNA, IP, IPX, and NetBIOS within Frame Relay frames.

Given these forces dictating the nature of future networks, the key issues that have to be tackled, head on, over the next few years, in order to realize an all-in-one multimedia network, with the appropriate external interfaces, are as follows:

- Ensuring that multiprotocol LAN environments, built around PCs, workstations, and LAN servers, become an integral and essential component of an enterprise's information processing and information dissemination mechanism. This must apply not only to data applications but also for certain voice and video applications such as desktop videoconferencing and live television feeds to workstations.
- Integrating the dispersed LAN environments with IBM mainframe resident, mission-critical applications in a seamless synergistic manner. This must be done such that there will always be unrestricted data interchange possibility between these applications and any applicable LAN-attached device (e.g., a Unix workstation).
- Implementing a high-performance, multiprotocol WAN that would interconnect all of the distributed LAN environments as well as the mainframes. This WAN would initially provide up to 4Mbps of bandwidth (e.g., with a public frame relay service) for data-only applications, but would in time offer bandwidth in excess of 45Mbps for multimedia applications.
- Adopting a full-spectrum and powerful network management system capable of incisively controlling and monitoring a multiprotocol, multimedia network. This system should be extensible to embrace total system management (e.g., management of LAN servers, workstations, applications, databases, and mainframe hardware).
- Incorporating SNA traffic associated with the business-critical applications into the multiprotocol, and in time multimedia, LAN and WAN environments. The sacrosanct reliability and response times of these applications can at no time or in no way be jeopardized. In some instances, maintaining two parallel networks, i.e.,

SNA and non-SNA, across a frame relay or ATM backbone might prove to be the most prudent and optimal solution.

- Providing controlled and protected access to external networks and data feeds—paying particular attention to the inevitable need to have some kind of bidirectional interface to the Internet.
- Determining that all of the various preferred technologies and products comply, whenever possible and applicable, to the relevant industry standards. This will facilitate, with luck, minimum-hassle multivendor interoperability—always bearing in mind that in the IBM world many of the widely endorsed and proven interoperability schemes, such as SNA, BSC, APPN, and Data Link Switching (DLSw), may not be bona fide industry standards.
- Ensuring that the reliability and resilience of the network, at every stage during its evolution, lives up to the stringent high-availability requirements of revenue-generating, mission critical applications. Always keep in mind that a 1% drop in network uptime over a year is equal to an additional 88 hours of revenue disrupting network outages!
- Keeping the networking overheads (e.g., length of data encapsulation headers or broadcast search traffic) to a minimum to optimize the network's overall efficiency. Unnecessary overhead just saps bandwidth, increases overall network congestion, undermines network performance, and impacts application response times. Minimizing overhead increases the total useful bandwidth available for actual data transfer. Despite becoming increasingly cost-effective, bandwidth is not, and is unlikely to ever be, free. Optimizing efficiency will permit more work (e.g., processing of more transactions or transfer of more data) with a given amount of bandwidth. It will also reduce the frequency at which additional bandwidth may need to be acquired as network usage grows.
- Exploring all possibilities on an ongoing basis to consolidate parallel networks (e.g., SNA and non-SNA, data and voice, data and video), eliminate redundant equipment and processes, streamline operations and obtain the most economical service to contain network costs.

1.1.1 A Technological Potpourri

In addition to the now commodity items such as LAN adapters and intelligent LAN hubs, the major technologies, protocols, and products that are likely to figure prominently in the network reengineering process are as follows.

- Conventional, multiprotocol bridge/routers (from vendors such as Cisco, Bay Networks, and 3Com), with SNA traffic integration facilities such as DLSw and Advanced Peer to Peer Networking (APPN) Network Node routing.
- Frame relay routers (from vendors such as Hypercom, Motorola, NetLink, and Sync Research), that use the RFC 1490 alien protocol encapsulation standard to support SNA and non-SNA protocols.
- IBM's AnyNet protocol conversion Gateways (e.g., 2217-200) that can be used to realize multiprotocol LAN/WAN solutions, with a uniquely SNA heritage, that are direct alternatives to bridge/router or frame relay router-based solutions.
- LAN switches, particularly token-ring LAN switches (from Bay Networks, Nashoba Networks, Cisco, 3Com, etc.), that provide a cost-effective means of enhancing LAN throughput. These switches are particularly useful in configurations that have large LAN segments or where multiple LANs are interconnected to each other across a backbone LAN.
- Frame relay (public or private), Switched Multimegabit Data Service (SMDS), and Integrated Services Digital Network (ISDN) services as means for obtaining mid-range (i.e., 56Kbps up to 4Mbps) bandwidth for LAN-interconnection, WANs, or WAN access.
- ATM as the strategic, long-term basis for realizing broadband, multimedia networks along with ATM WAN switches (e.g., IBM 2220, Cisco LightStream, Northern Telecom Magellan), ATM capable Hubs (e.g., IBM 8260, BayNetworks, Cabletron, etc.) and ATM adapters for PCs and workstations.
- Simple Network Management Protocol (SNMP) as well as IBM's NetView and SystemView offerings for network and, in time, total system management.
- APPN and its eventual successor High Performance Routing (HPR) as means for providing some degree of SNA-oriented, session-level routing within multiprotocol LAN/WAN networks.
- TCP/IP, Novell NetWare IPX/SPX (and in time NetWare Link Services Protocol [NLSP]), NetBIOS, Banyan Vines, AppleTalk, DECnet, OSI and Xerox Network System (XNS)—the major non-SNA protocols now likely to be found alongside SNA in many of today's enterprises.
- Cost-effective, rugged, and IP-capable LAN-to-mainframe gateways such as IBM 3172s for interconnecting LAN/WAN networks to mainframes.
- Techniques for integrating traffic from link-attached devices (i.e., devices with RS-232 type serial ports as opposed to LAN adapters) with those from LAN-attached devices such as SDLC-to-LLC:2

conversion, BSC remote polling, or the encapsulation with frame relay of Async traffic.

Figures 1.3 and 1.4 expand upon Figures 1.1 and 1.2, respectively, to show where and how the aforementioned technologies, protocols, and products come together within a network.

1.2 APPLICATIONS CALL THE TUNE AND UNIX SETS THE TEMPO

Amazingly, the overall structure and methodology of major applications, in particular the mission-critical ones, within the core IBM mainframe world have undergone very little change over the last 20 years. These applications invariably adhered to a "master-slave" paradigm that revolved around the notion of mainframe-resident applications serving a population of nonprogrammable (i.e., "dumb") terminals. The application was responsible for all of the data processing and data management. This despite the fact that PCs started displacing the dumb 3270 terminals, at a significant rate, starting in the mid-1980s. Fortunately, things did at last started to change, with a vengeance, over the last couple of years.

Mainframe-resident, 3270-terminal-oriented, mission-critical applications are now being routinely complemented by a new wave of client/server applications. In some cases, hitherto mission-critical applications are even being usurped by client/server applications. File- and print-server applications that use protocols such as Novell NetWare IPX/SPX, NetBIOS, Banyan Vines, AppleTalk, or even TCP/IP are ubiquitous. E-mail is pervasive.

client/server, program-to-program, and database server applications, at long last, having ridden out the global economic downturn of the late 1980s, which precluded the necessary investment in this new wave of applications, are now in vogue. These applications use protocols such as TCP/IP, SNA LU 6.2, OSI, DECnet, and IPX/SPX for their interprogram communications. LAN-oriented, workgroup collaboration software, such as IBM/Lotus Notes, which permit multiple dispersed users to access and update a common set of documents, are gaining increasing corporate attention as tangible means for improving productivity and further facilitating communications within project/program teams.

Unix is enjoying a remarkable renaissance of popularity right across the IBM world—most notably within enterprises that hitherto had been beholden to mainframes for nearly all of their computing power. There are two overriding reasons for this Unix ground swell: a

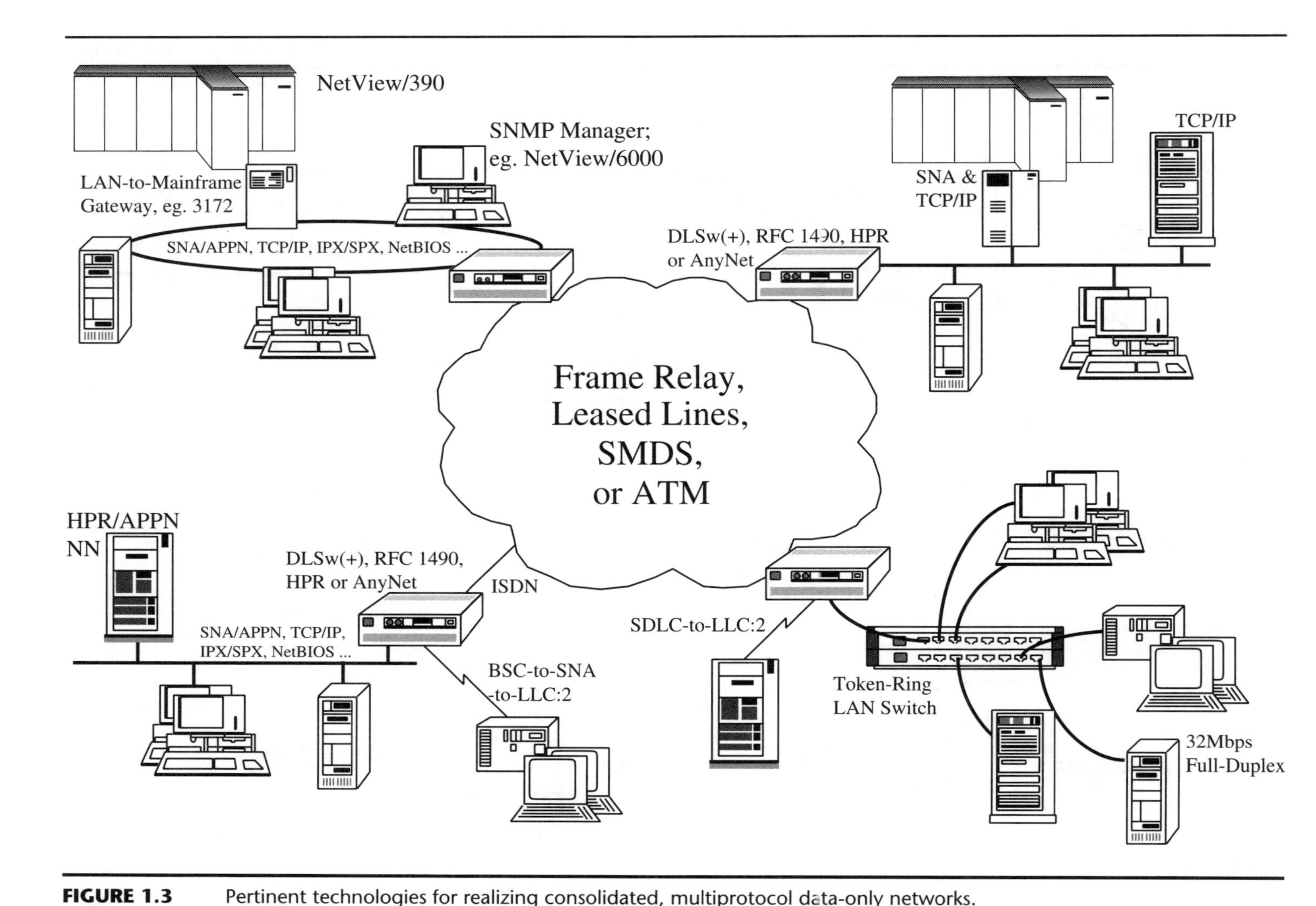

FIGURE 1.3 Pertinent technologies for realizing consolidated, multiprotocol data-only networks.

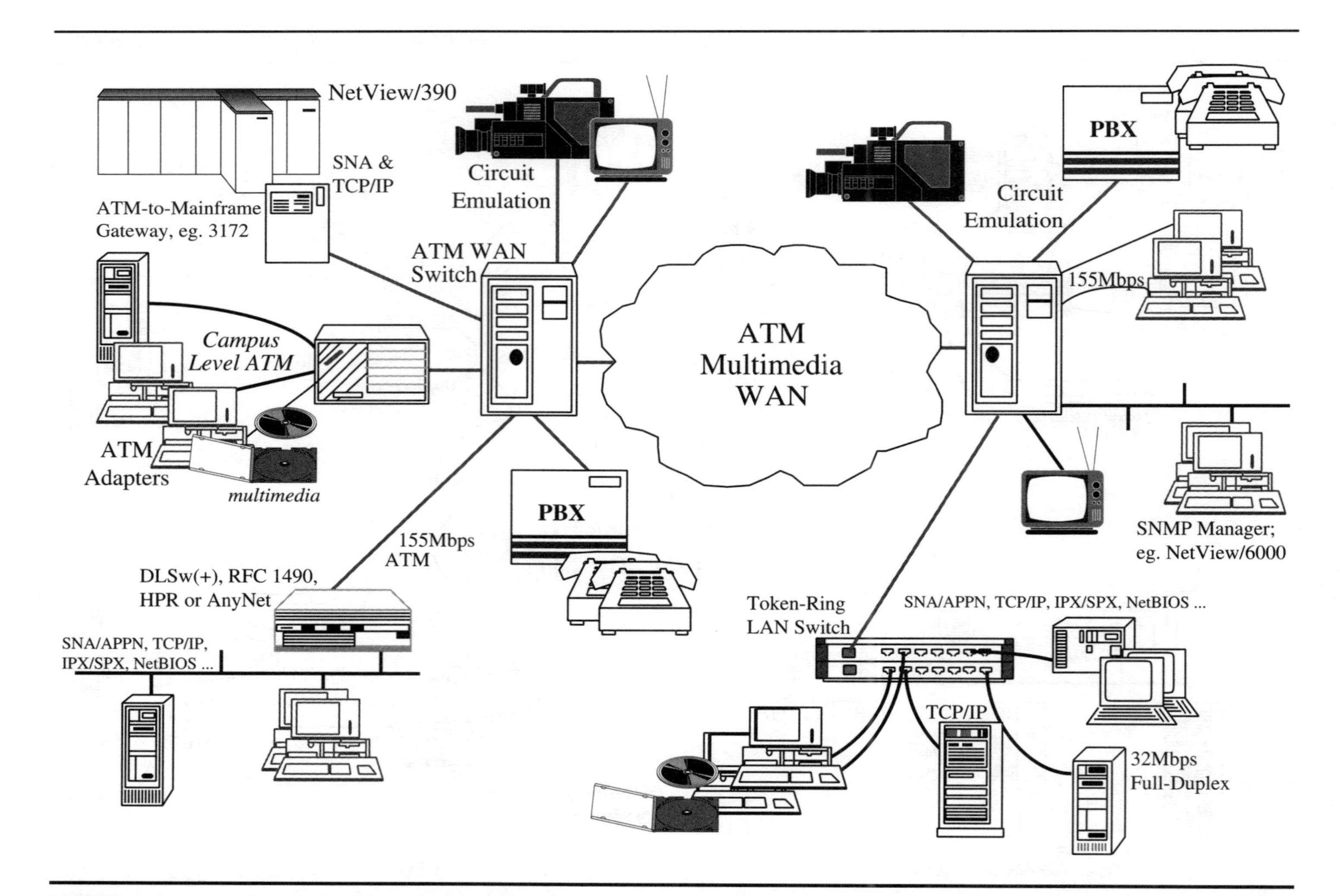

FIGURE 1.4 Pertinent technologies for building ATM-Centric multimedia broadband networks.

large base of Unix programmers, and powerful but very attractively priced Unix/RISC workstations. Thanks largely to the largess of AT&T, which donated many systems, most North American universities and colleges have been using Unix systems since the late 1970s to induct students to the joys of computer science. Since then, the notion of using Unix as an educational platform has been adopted by most international higher education institutions to an extent motivated by the fact that these systems are now relatively inexpensive. Unix systems, the perceived precursor to open systems, now appear to embody the very essence of the academic aspect of the Internet.

As a direct result of this academia-based on-ramp to Unix, all enterprises now have an ever-growing battalion of Unix-literate programmers who, given a choice, would prefer to develop new applications on Unix systems. What makes this choice even more compelling is that these programmers are also likely to be noticeably more productive when using Unix. This preference of Unix among the programming community has been complemented somewhat serendipitously by the dramatic enhancements in the price/performance characteristics of RISC workstations, from Sun, IBM RS/6000 and SP/2, HP, and DEC among others, that happen to use Unix as their native operating system. Consequently, Unix has become the new platform of choice for most new business applications that are being developed, as well as for applications that are being moved away from mainframes.

A worldwide survey of nearly 900 mainframe sites conducted by Xephon PLC in mid-1994 showed that close to 40% of these sites were already using one or more Unix-based applications. This represented a 50% increase in such applications within this base from just a year before. This trend in the growing role of Unix within what were mainly mainframe shops is going to continue. IP is the native networking protocol of Unix systems. The growth in the use of Unix will invariably result in a proliferation of IP. Thus, IP will, in time, join SNA to become one of the key applications protocols being used by commercial enterprises.

Alongside this Goth in non-SNA applications, even existing SNA-based mission-critical applications are being given a much needed face-lift and torqued into becoming client/server by the introduction of graphical and iconic front-end applications running on PCs/workstations. (See sidebar on "PC-Based Face-Lifts for Mission-Critical Applications.") Independent of the changing structure and dynamics of in-house applications, the interactions with and across the TCP/IP based Internet is burgeoning. Consequently, most enterprises today have to contend with a potent cocktail of protocols that invariably include at least 3270-SNA, SNA/LU 6.2, TCP/IP, IPX/SPX, and NetBIOS.

PC-BASED FACE-LIFTS FOR MISSION-CRITICAL APPLICATIONS

The bulk of today's mainframe-resident, mission-critical applications was originally developed in the 1970s and early 1980s. Fixed-function, alphanumeric, record-by-record, data-field (e.g., highlighted or protected fields) oriented 3270 terminals were the state-of-the-art and prevalent terminals of that era. Consequently, they became the universal terminal of choice for mission-critical applications. The interactions processed by these applications are molded around 3270 commands and data streams. This hard-coded 3270 specificity is unfortunate given that PCs and workstations have replaced 3270 terminals in most enterprises.

The processing, local storage, and display capabilities of these PC/workstations provide the ideal basis for developing user interfaces that are much more user-friendly, less demanding (e.g., fewer key strokes), and more intuitive than was ever possible with 3270s. In addition, some local processing could be done on the input data such as verifing that all the necessary data is present. This local processing can both ease the amount of processing that has to be done at the mainframe and preclude the unnecessary transmission of invalid or incomplete transactions across the network. Modifying mission-critical applications is, however, a risky and costly proposition. Hence, many MIS managers are understandably reluctant to modify these applications, especially if it entails major surgery to the core terminal I/O handling routines.

Instead, some MIS managers are using customized datastream-reshaping software, interspersed between mainframe applications and the PC/workstations users. Reshaping datastreams in this manner permits them to provide a contemporary look and feel and appropriate local processing capabilities to old applications. Using techniques specifically oriented for PC/workstations, the new face-lift software not only generates and manages the new user interface but also performs any preprocessing (e.g., error or completeness checking). In addition, the face-lift software diligently maintains a 3270 appearance (i.e., sends and accepts standard 3270 commands and datastreams) vis-à-vis the mainframe applications. This ensures that they still think that they are interacting directly with 3270 terminal users. Thus, while the mainframe applications remain unchanged, their users get the benefits of a new and improved user interface.

The face-lift software could be deployed in four places, these being:

1. On each PC/workstation
2. On the mainframe, alongside the mission-critical applications
3. On an intermediary PC or Unix based server
4. Distributed across any two of the above locations (e.g., central component on the mainframe and smaller components in each PC/workstation)

The face-lift software has the option of using any protocol/datastream between itself and the PC/workstation. It only has to use SNA/3270, at the back end, on its interface with the mission-critical applications. Figure S1.1 illustrates how different combinations of protocols may be used between the face-lift software and the mission-critical applications, depending on where the face-lift software is located.

1.2.1 SNA Is Not Dead—Just SNA-Only Networks

The burgeoning of non-SNA applications should, however, not be construed in any way as portending the imminent demise of SNA; far from it. SNA as a high-level (i.e., Layer 4 to 7), application-to-PC/workstation protocol is destined to be around—as a major component of the traffic mix of most commercial enterprises—for at least another 10 years! This is not an idle and unsubstantiated prediction. The very nature of mission-critical applications, combined with the monumental investment in these applications, will, if nothing else, sustain the use of SNA for another decade. Enterprises have invested about 20 trillion dollars over the last three decades in developing today's base of mission-critical applications. Such investments are not written off lightly.

While most enterprises are now committed to eventually reengineering all their current mission-critical applications to be client/server–based, and possibly in some cases even to be mainframe-independent, this radical conversion is not going to happen overnight. In addition to the immense costs of such conversions, there is also the vexing issue of application veracity and integrity. Even the slightest discrepancy between the new and the old applications, such as the way the third decimal point in a numerical field is rounded up or down during an arithmetic calculation, could wreak havoc to the successful functioning of an enterprise.

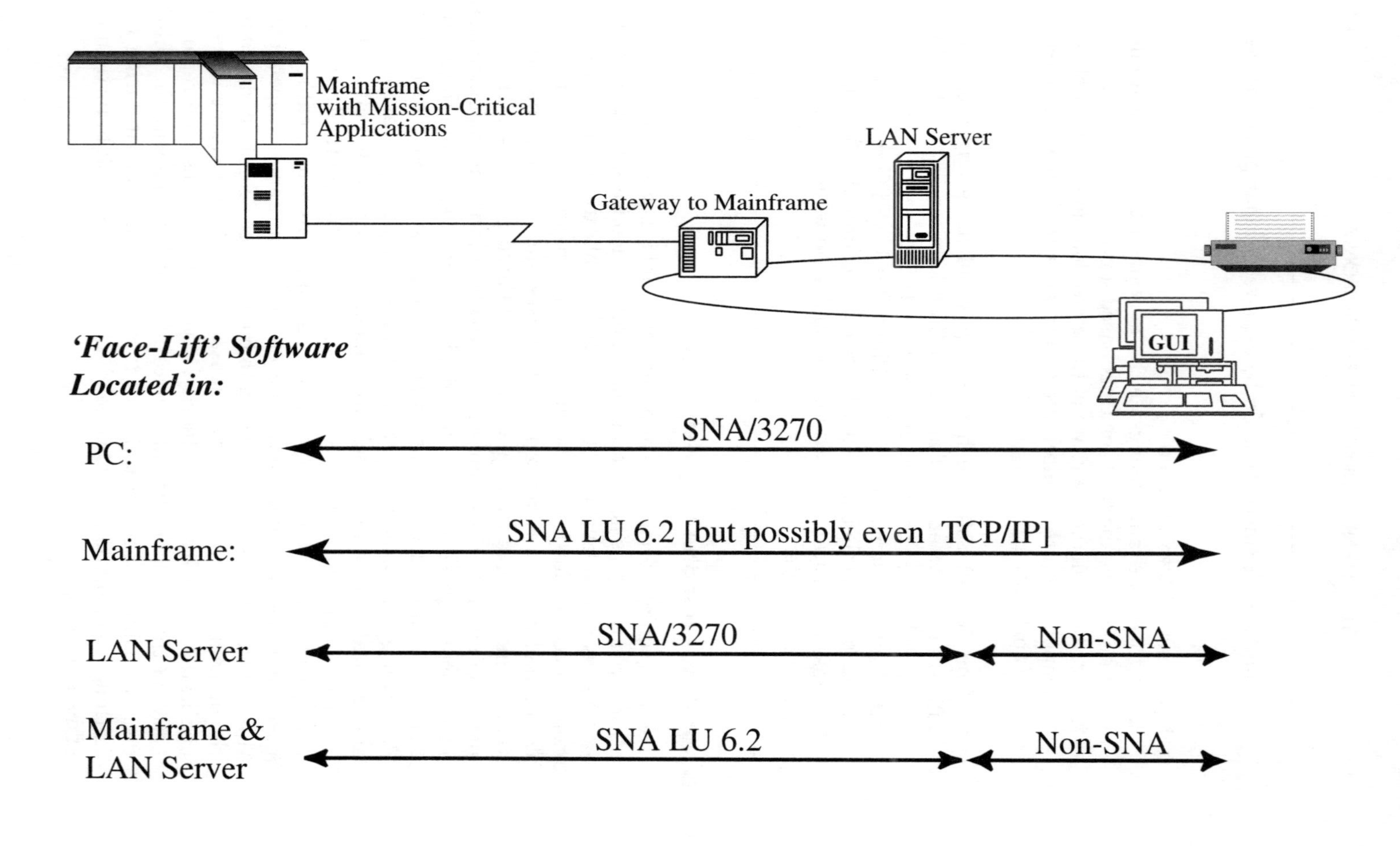

FIGURE S1.1 Some protocol permutations for providing 3270-based mission-critical applications with a face-life when dealing with PCs.

The successful and consistent operation of mission-critical applications, as explicitly testified to by their very name, is the lifeblood that sustains the day-to-day operations of most commercial enterprises. The considerable and embarrassing delays and cost overruns incurred in trying to get the new Denver International Airport to operate smoothly, so that it could take the place of the still-functioning old airport, have been widely chronicled. This cut-over from a functioning old airport to a brand new one can easily be compared to the trials and tribulations of cutting over from a revenue-producing mission-critical application to a brand new and unproven version.

Disrupting the smooth operation of mission-critical applications could be very hazardous to one's career as well as health. Hence, there will be no great stampede to get rid of today's SNA-based mission-critical applications. Conversion will be slow, systematic, and spread over a fairly long time. With over 20,000 enterprises around the world using SNA-based mission-critical applications today, it will be a very long time before SNA, in its current 3270 datastream-centric form, ceases to be a major networking scheme. In this context, it is salutary to remember that nearly 30 years after it first saw the light of day in 1967, BSC is still in use today, and that multiprotocol bridge/router and frame relay router vendors are being pressured into offering support for BSC traffic.

The recrafting of today's mission-critical mainframe applications to make them client/server and more user-friendly does not automatically mean that none of the new applications will be using SNA. None, it is true, is likely to be using traditional, 3270-based SNA. Some, however, are likely to opt for SNA's rather sophisticated, and now mature, peer-to-peer, program-to-program protocol suite—the so-called LU 6.2 facility. LU 6.2 was developed, circa 1982, for client/server applications. It has powerful error recovery capabilities, including the ability to undo (i.e., roll back) the effects of a transaction (e.g., the update of a database) in the event of an error. LU 6.2 is thus extremely well-suited for the demands of applications that have to access and update multiply distributed databases (e.g., debit funds from one account in a database in New York and credit that amount to an account on a database in Chicago)—a scenario that will gain considerable momentum with the spread of EFT and electronic shopping applications. SNA, therefore, is going to be around, in the coming decade, in two forms: in "born-again" LU 6.2 form from new client/server applications, and in "legacy" form from old, unconverted applications.

SNA-based mission-critical applications and SNA as an end-to-end high-level protocol are, thus, going to remain prominent fixtures of enterprise networking for yet another decade. What will be markedly different from the current scenario is that SNA will no longer rule the network. Other protocols, most significantly IP, will soon be as important as SNA, even in mainframe shops. Non-SNA data traffic will start to account for more and more of the overall networking traffic mix. *This will lead to the eventual death, over the next few years, of the SNA-only networks that have been the cornerstone of commercial-sector networking for the last two decades.* They will be replaced by multiprotocol networks that will be capable of transporting SNA traffic alongside non-SNA traffic.

A clear distinction should be made between the death of SNA-only networks, and the continued longevity of SNA within the realms of multiprotocol networks. A few years ago, there was much discussion about the death of SNA—at the hands of the rampaging non-SNA protocols. This, as is now clear, was the garbled outcome of attempts to articulate the endangered status of SNA-only networks. The indubitable fact that SNA-only networks were being usurped by multiprotocol networks somehow got interpreted as also meaning that SNA was dying—which is not the case. One of the key tasks involved in reengineering today's IBM networks is to elegantly and safely transition SNA traffic from SNA-only networks to a new breed of multiprotocol backbones.

1.2.2 The Ingredients for a Potent Protocol Cocktail

Despite their innate diversity, the application repertoire that is likely to be found in any enterprise today could be divided into three generic categories. These three categories are: file- or print-server applications, mainframe or minicomputer resident applications, and the new client/server, program-to-program applications that include e-mail, distributed database management, and Notes-like workgroup collaboration software. Each of these categories has a native or predominant set of protocols associated with it. These protocol sets are as follows:

- File/Print Server Applications: Novell NetWare IPX/SPX, NetBIOS, TCP/IP, Banyan Vines, AppleTalk, and XNS
- Mainframe/minicomputer Resident Applications: SNA LU-LU Session Type 2 (i.e., 3270 datastream) or Type 7 (i.e., 5250 datastream), TCP/IP (e.g., tn3270)
- client/server/Program-to-Program Applications: TCP/IP, SNA LU Type 6.2 (including SNA Distribution Services [SNADS], Document Interchange Architecture [DIA], Distributed Data Management [DDM], SNA/FS), IPX/SPX, OSI, DECnet, NetBIOS, XNS

This multiplicity of protocols per se does not pose a major networking challenge provided they are restricted to LANs. LANs can support different streams of traffic, each based on a different protocol. Thus, it is not uncommon to see SNA, IPX/SPX, TCP/IP, and NetBIOS all being used on the same token-ring LAN by different applications.

Unfortunately, LAN-based interactions, despite their local denotation, are unlikely to be restricted to physically adjacent LANs within one building or one campus. The growing trend is for LAN-to-LAN and LAN-to-mainframe interactions across a WAN, for example, a PC on a LAN in Boston requiring either access to data on a file-server attached to a LAN in San Francisco, or needing to log on to applications resident on a mainframe in Chicago.

1.3 THE DREADED PARALLEL NETWORK SYNDROME

The current networking challenges in IBM environments begin to surface as the scope of multiprotocol LAN-based applications need to be extended across a WAN. Today, nearly all of these enterprises have a proven and trusted WAN that reaches out to all of the remote sites containing LANs. This WAN happens to be SNA-based. Existing SNA WAN backbones do not have the generalized multiprotocol handling capability nor the necessary bandwidth to totally address the demands of multiprotocol LAN/WAN networking. For all their maturity and apparent sophistication, not to mention cost, SNA backbones are but single-protocol networks—really only capable of handling SNA traffic with panache.

Over the last few years, IBM has gradually and grudgingly added the capability to natively route IP traffic across SNA backbones, albeit with some caveats (e.g., the first release of the IBM 3746 Model 950 Nways MultiNetwork Controller that is positioned to replace the aging 3745 does not have IP support). Support for LAN protocols, even the more prevalent ones within IBM environments such as NetBIOS and IPX/SPX, can only be accommodated across an SNA backbone using auxiliary products such as the IBM 2217 Nways MultiProtocol Concentrator. Figure 1.5 summarizes the problem besetting today's SNA backbones.

Due to SNA's frustrating inability to deal with the other now popular protocols, many enterprises have already been forced to implement and maintain two parallel WANs, these being the existing, typically IBM 37xx Communications Controller-based SNA backbone for mission-critical application access, and alongside it a bridge, bridge/router, or frame relay router (or FRAD) multiprotocol WAN for non-SNA, inter-LAN traffic. Figures 1.6 and 1.7 depict the dreaded

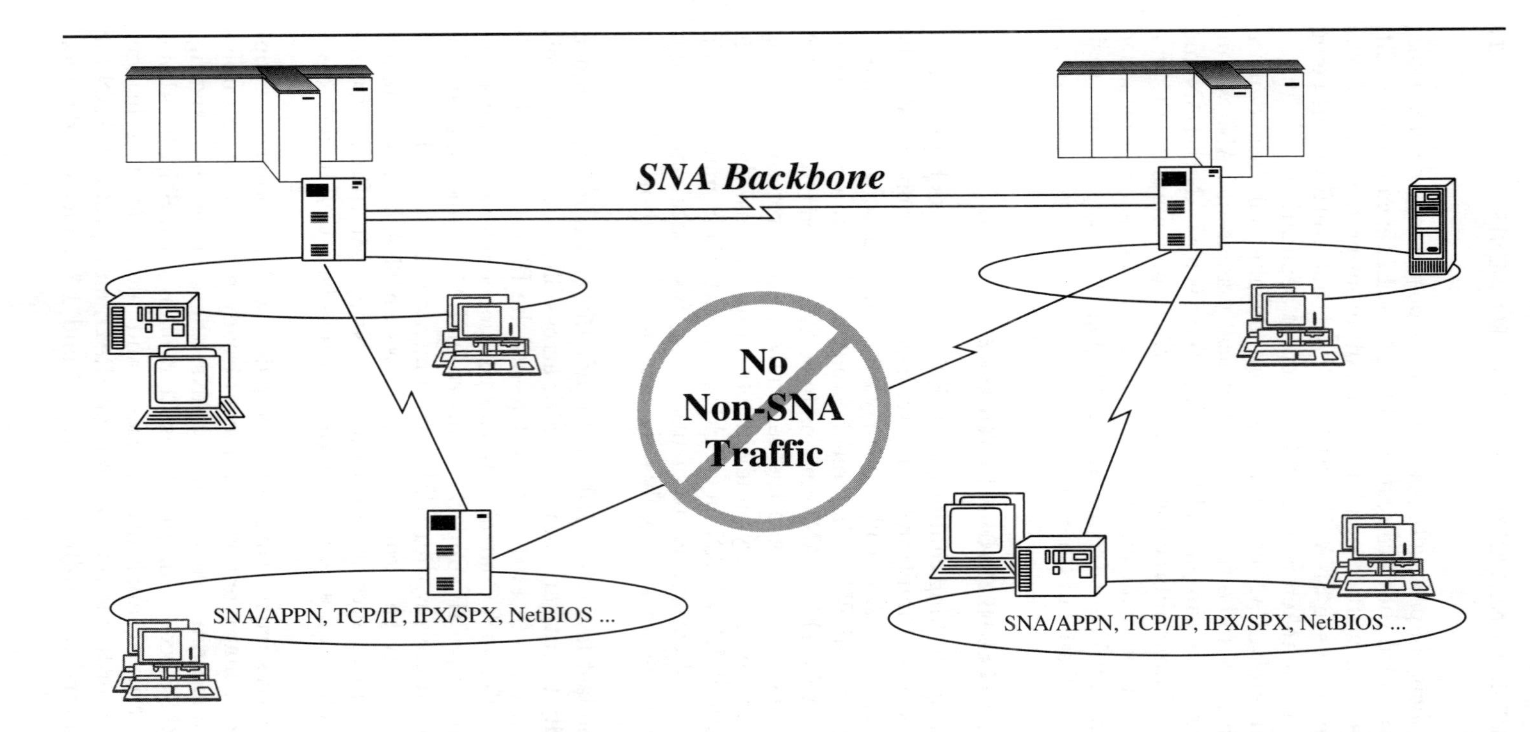
SNA Backbone
No
Non-SNA
Traffic
SNA/APPN, TCP/IP, IPX/SPX, NetBIOS ...
SNA/APPN, TCP/IP, IPX/SPX, NetBIOS ...
Protocols being used by different Application Categories:
❖ File/Print Server: IPX, NetBIOS, TCP/IP, Vines, AppleTalk ...
❖ Host Appl. Access: SNA LU-LU Session Type 2, 7 ...
❖ Client/Server: TCP/IP, SNA LU 6.2, IPX, DECnet, OSI, NetBIOS ...
In general
an SNA Network
Really Does Not
Want to Support
Much More than
SNA/APPN
and Some
TCP/IP Traffic.

FIGURE 1.5 The crux of the problem is vis-à-vis non-SNA traffic and SNA backbones.

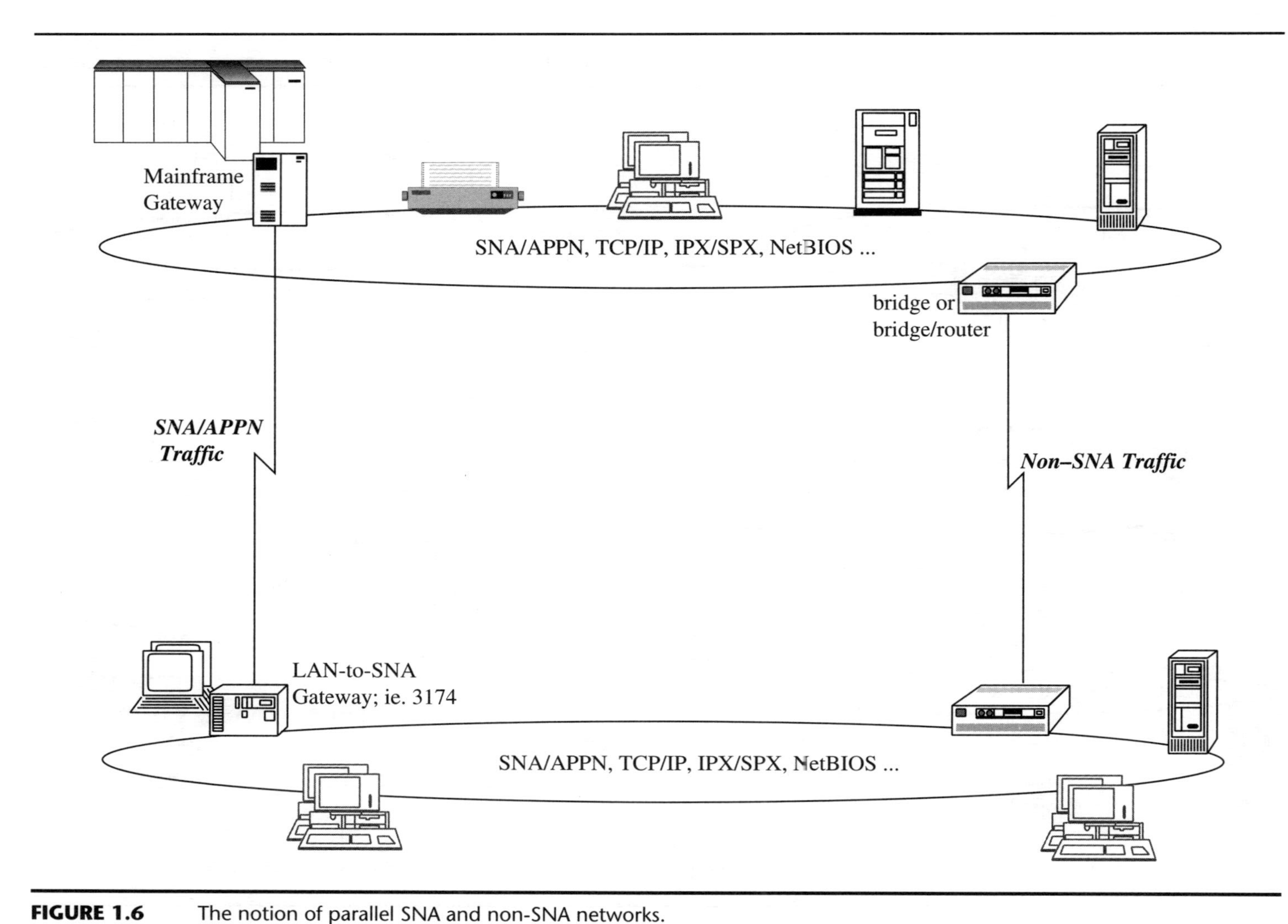

FIGURE 1.6 The notion of parallel SNA and non-SNA networks.

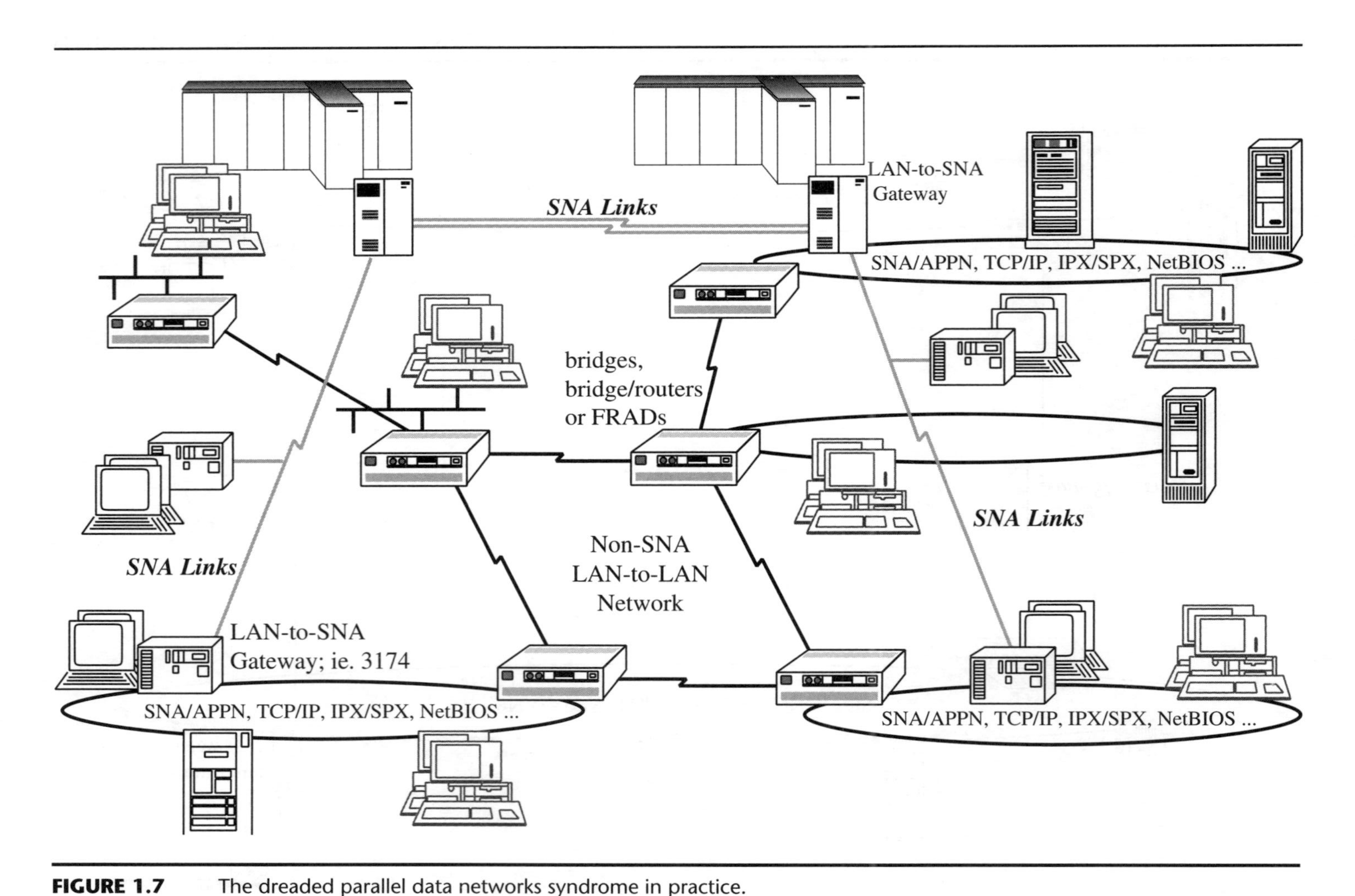

FIGURE 1.7 The dreaded parallel data networks syndrome in practice.

parallel networks. In 1995, such parallel networks are widespread and represent the most common approach for realizing multiprotocol networks in what were hitherto SNA-only environments.

Parallel networks are obviously far from optimal, both in terms of costs and operational effort. Instead of being able to concentrate on one production network, two networks now have to be administered, monitored, sustained, and managed. In addition, each network could end up having its own dedicated pool of bandwidth (e.g., leased lines). The drawback of this arrangement is that there is no possibility of one network being able to make use of unused bandwidth allocated for the other. It is thus possible that each network has been provisioned with excess bandwidth to take care of peak traffic loads. If the two networks could be consolidated, some savings in bandwidth costs could be realized by eliminating the need for such overbooking in each network.

One way to at least rationalize some of the bandwidth costs of running two parallel networks is to overlay the two parallel networks on top of a TDM multiplexer, frame relay, or (in the future) ATM network, rather than using a separate set of leased lines for each network. While this approach might defray some of the bandwidth costs, it does not ease the operational and management aspects. If anything it might exacerbate them, because now three networks are involved all together. However, *in some cases maintaining logically parallel networks (i.e., the SNA and non-SNA), across a common frame relay or ATM backbone, might prove to be the safest, most expedient, and the most effective option. This would especially be true if the cost of frame relay and, in time, ATM bandwidth continues to drop.*

If no steps are taken to consolidate these parallel networks in the short- or mid-term (say 5 years), two other factors will in time offset the increased cost of maintaining the two separate networks. One factor will be the reduction in bandwidth costs as a result of improved tariffs. The other is the potential reduction in lost opportunity (i.e., revenue lost during network outages) costs by eliminating any unstability that might have resulted in trying to combine the SNA and non-SNA traffic.

1.3.1 Consolidating Parallel Data Networks

Fortunately, parallel data networks are not the only way to effectively accommodate SNA/APPN and non-SNA traffic. If anything, there is a surfeit of technologies that can be used to collapse parallel networks into a single integrated multiprotocol network that is capable of supporting both SNA/APPN and non-SNA traffic. The primary technologies that can be used to integrate SNA/APPN traffic on LANs (i.e., from LAN-attached SNA devices) with non-SNA traffic include the following:

1. Most forms of LAN bridging, and in particular Source-Route Bridging (SRB) in the case of token-ring LANs
2. The encapsulation of SNA/APPN traffic within TCP/IP message units and the subsequent routing of these TCP/IP message units across a multiprotocol backbone. IBM's DLSw and Cisco's Remote Source Route Bridging (RSRB) are examples of IP-encapsulation schemes for SNA/APPN traffic. Cisco's DLSw+ facility, in essence, permits the coexistence of RSRB and DLSw within the same Cisco bridge/router based multiprotocol LAN/WAN internetwork.
3. Native encapsulation of SNA/APPN and non-SNA traffic within frame relay frames, per the RFC 1490 encapsulation standard. IBM's Boundary Access Node (BAN) feature on IBM 6611 and 2210 multiprotocol bridge/routers is a nonstandard variant of RFC 1490 encapsulation for SNA/APPN traffic.
4. APPN Network Node, and in the future HPR, routing on multiprotocol bridge/routers, that in conjunction with so-called Dependent LU (DLU) support for handling non-LU 6.2 traffic, permit SNA/APPN traffic to be natively routed alongside non-SNA traffic.
5. IBM AnyNet Gateways, à la the IBM 2217, that either convert IP traffic to SNA/HPR, or encapsulate other non-SNA traffic within SNA LU 6.2 message units. This leads to the creation of an APPN/HPR-based single protocol backbone that can support a limited set (e.g., IP, IPX/SPX, and NetBIOS) of non-SNA protocols alongside SNA/APPN traffic.
6. LAN-over-SNA, à la the now defunct IBM LAN-to-LAN Wide Area Network Program (LTLW) is essentially a variation of the AnyNet Gateway technique described above. LAN-over-SNA encapsulates all non-SNA traffic with LU 6.2, whereas an AnyNet Gateway when dealing with IP converts it to LU 6.2. LTLW is now bundled in the IBM AnyNet Gateway family. Other LAN-over-SNA solutions, such as ANR's L2-SR or CCI's Eclipse range, never captured adequate market following. This technique can now be seen as superseded by the 2217 like AnyNet gateway approach.
7. SNA Session Switching across multiprotocol networks as promoted by CNT/Brixton. This in effect is direct mainframe-to-mainframe switching of SNA traffic in conjunction with IP-encapsulation, as in item 2 above—and is targeted at networks with multiple mainframes.
8. CrossComm's Protocol Independent Routing (PIR), which during 1992–1994, was one of the premier techniques for integrating SNA/APPN traffic with non-SNA traffic. PIR, which could be

thought of as a highly refined form of SRB, with many value-added facilities and none of the limitations of SRB, is no longer being aggressively promoted by CrossComm.

Table 1.1 compares the pros and cons of these eight technologies, as well as indicating the primary sources for each. These technologies are described in detail in Chapters 5 to 8.

Any one of the above eight technologies can be used to construct a multiprotocol network that consists of a WAN backbone that interconnects multiple dispersed LANs. Figure 1.8 shows such a LAN/WAN network and also highlights the fact that these technologies are LAN-specific and as such do not handle link-attached SNA devices.

If the technology chosen to build the LAN/WAN network is either bridging, TCP/IP encapsulation, RFC 1490, APPN, HPR, Session Switching, or PIR, the backbone WAN will be truly multiprotocol. Assorted protocols (e.g., IP, IPX, SNA/APPN) and different traffic types will continually flow across the backbone on a serial basis. With this type of multiprotocol backbone, all the so called routable protocols (e.g., IP, IPX, Vines, XNS, OSI, AppleTalk, APPN and HPR) will be, in general, individually routed across the backbone. The so-called non-routable protocols (e.g., SNA and NetBIOS) are likely to be transported using either bridging, TCP/IP-encapsulation, or RFC 1490, if APPN or HPR routing is not being used.

The AnyNet Gateway and LAN-over-SNA approaches, on the other hand, opt for the notion of routing multiple protocols (e.g., TCP/IP, IPX/SPX, SNA, and NetBIOS) across a *single* protocol backbone, in this instance either APPN or HPR. With the AnyNet gateway technology it would, however, also be possible to have IP as this single protocol—even though IBM, understandably, does not aggressively advocate this option. This single-protocol backbone approach for supporting multiprotocol traffic is discussed in detail in Chapter 8.

Unfortunately, in most cases consolidating the multiprotocol LAN traffic, using one of the above technologies is only half the battle. In some enterprises more than 50% of the SNA devices may still be link-attached, as opposed to LAN-attached. While many of these links are likely to be SDLC, or X.25 with QLLC in Europe and Asia, there is still a fair amount of 3270 BSC devices in active use. BSC is particularly prevalent in banking environments where many automated teller machines still use 3270 BSC. Fortunately there are three viable technologies that can be used, in conjunction with any of the LAN-consolidation technologies discussed above, to integrate SDLC, X.25/QLLC, and 3270 BSC traffic into a multiprotocol LAN/WAN network. These three technologies are:

TABLE 1.1 Comparison of the Techniques for Integrating SNA/APPN LAN Traffic into a Multiprotocol LAN/WAN Network

	Bridging	*TCP/IP Encapsulation á la DLSw*	*Frame Relay Encapsulation per RFC 1490*	*APPN/HPR Network Node Routing*
Basic concept	LAN packets forwarded to destination LAN via previously discovered route	SNA/APPN message units first encapsulated within TCP/IP packets. The TCP/IP packets then routed to a bridge/router on the destination LAN using IP Address.	SNA/APPPN message units encapsulated within an FR frame with an RFC 1490 header. Frames forwarded to FR Router serving desination via preconfigured Virtual Circuits	SSNA/APPN message units routed to dynamically located destination using a route calculated to satisfy a COS that can consist of up to nine criteria
Routing criteria for SNA/APPN traffic	LAN MAC address	LAN MAC address	LAN MAC address	SNA/APPN application (LU) name
Locating of SNA/APPN destination	Dynamically located	Destination MAC addresses dynamically located, but IP addresses of the routers involved (may) need to be predefined.	Predefined mapping between destination and VCs	Dynamically located
Caching of previously discovered locations	No for token-ring SRB	Yes	N/A	Yes
SNA subarea routing support	No	No	No	No
SNA routing emulation	No	No	No	Yes
Dynamic alternate routing	No	Yes—but with some caveats	Dependent of frame relay network or router specific extensions	No for APPN—but some bridge/router implementations have proprietary work-arounds. Yes for HPR.
SNA-like parallel link transmission groups	No	Yes with certain IP routing techniques such as OSPF	No	No for APPN. Implementation dependent with HPR.

TABLE 1.1 *(cont.)*

	Bridging	*TCP/IP Encapsulation á la DLSw*	*Frame Relay Encapsulation per RFC 1490*	*APPN/HPR Network Node Routing*
Setup requirements	Minimal—can be close to plug-and-play	Moderate to high depending on the number of IP addresses that have to be defined and the means that they have to be defined.	Some to moderate depending on the number of destinations involved. IBM's BAN variant similar to SRB bridging.	Some but bordering on plug-and-play
Degree of openness to facilitate multivendor interoperability	De facto standard	DLSw is a de facto standard Other implementation proprietary	Official IETF standard	De facto standard but IBM license required in order to implement.
Key strengths	• Simple • Well-proven	• Can automatically locate destinations • Available from most, but not all, bridge/router vendors • Works across any WAN	• Very low encapsulation overhead. • Applies to all protocols. • Can be customized to avoid the discarding of packets in the event of congestion.	• SNA application name based routing • Sophisticated multicriteria path selection scheme • Automatic registration of end node LUs
Main weaknesses	• No dynamic alternate routing • No support for parallel links • Absence of address caching SRB results in repetitive searches for the same destination. • Best suited for networks with less than 100 remote sites.	• High encapsulation overhead • Dynamic alternate routing cannot deal with failed central-site bridge/routers. • Scalability issues may limit number of remote sites that may be supported. • May involve two-hop routing in networks with multiple mainframes.	• Currently restricted to frame realy networks. • Best suited for networks with few destinations.	• APPN does not support dynamic alternate routing. • Need extensions (known as DLUS/DLUR) in order to support 3270 SNA traffic without configurational restrictions. • Scalability can become an issue unless Border Notes are used.
Key vendors	Ubiquitous	Cisco, Bay, IBM, Proteon, 3Com, ACC, and others	Hypercom, Motorola, Sync, Research, NetLink, Cisco	IBM, Cisco, Bay, 3Com

TABLE 1.1 *(cont.)*

	IBM AnyNet Gateway à la IBM 2217	*LAN-over-SNA*	*SNA Session Switching*	*CrossComm's Protocol Independent Routing*
Basic concept	SNA, APPN, or HPR based WAN backbone. IP traffic converted to SNA LU 6.2 message units for transportation across WAN. Converted back to IP at other end. IPX and Net BIOS traffic encapsulated within SNA LU 6.2.	Similar to AnyNet Gateway approach but IP traffic rather than being converted is also encapsulated within SNA LU 6.2 along with IPX and NetBIOS traffic	Captures and terminates SNA sessions at remote sites using a minihost emulation. Encapsulates SNA traffic within IPO similar to DLSw. Directly routes the traffic to appropriate mainframe.	Highly refined bridging scheme with dynamic alternate routing, load-balancing, and address caching
Routing criteria for SNA/APPN traffic	SNA/APPN application (LU) name	SNA/APPN application (LU) name	SNA/APPN application (LU) name	LAN MAC address
Locating of SNA/APPN destination	Dynamically located, in the case of SNA using T2.1 protocol.	Dynamically located, in the case of SNA using T2.1 protocol.	Predefined mappings	Dynamically located
Caching of previously discovered locations	Yes	Yes	Yes	Yes
SNA subarea routing support	Yes, if WAN is SNA	Yes, if WAN is SNA	No	No
SNA routing emulation	Yes	Yes	Yes	No
Dynamic alternate routing	No if SNA or APPN WAN. (See APPN column) Yes for HPR.	No if SNA or APPN WAN. (See APPN column). Yes for HPR.	Yes—but with some caveats	Yes
SNA-like parallel link transmission groups	Yes with SNA or HPR WAN. No with APPN.	Yes with SNA or HPR WAN. No with APPN.	Yes with certain IP routing techniques such as OSPF	Yes

TABLE 1.1 *(cont.)*

	IBM AnyNet Gateway à la IBM 2217	*LAN-over-SNA*	*SNA Session Switching*	*CrossComm's Protocol Independent Routing*
Setup requirements	Moderate	Moderate	Moderate to High	Minimal—can be close to plug-and-play. Similar to bridging.
Degree of openness to facilitate multivendor interoperability	Implementation dependent encapsulation scheme	Vendor-specific encapsulation scheme	Proprietary	Proprietary
Key strengths	• Extends SNA/APPN/HPR WANs to provide multiprotocol support. • Leverages investment in SNA • SNA/APPN/HPR features now available to other protocols.	• Extends SNA/APPN/HPR WANs to provide multiprotocol support. • Leverages investment in SNA. • SNA/APPN/HPR features now available to other protocols.	• SNA application name–based routing. • Efficient encapsulation scheme. • Ideal for enterprises with multiple mainframes that wish to standardize on an IP-centric backbone.	• Close to plug-and-play • Dynamic alternate routing & load balancing. • Address caching.
Main weaknesses	• Limited protocol support. • APPN does not support dynamic alternate routing. • SNA/APPN links should be at least 56Kbps for adequate LAN-to-LAN performance. • Many of the advantages of using a APPN/HPR WAN backbone nullified if using a public frame relay network as core WAN.	• Limited protocol support. • APPN does not support dynamic alternate routing. • SNA/APPN links should be at least at 56Kbps for adequate LAN-to-LAN performance. • Many of the advantages of using a APPN/HPR WAN backbone nullified if using a public frame relay network as core WAN.	• Requires complex non-IBM software to remotely terminate the SNA sessions. Always a potential for compatibility issues down the road.	• CrossComm-specific. • Best suited for networks with less than 300 remote sites. • No longer actively promoted by CrossComm.
Key vendors	IBM	IBM, Cabletron	CNT/Brixton	CrossComm

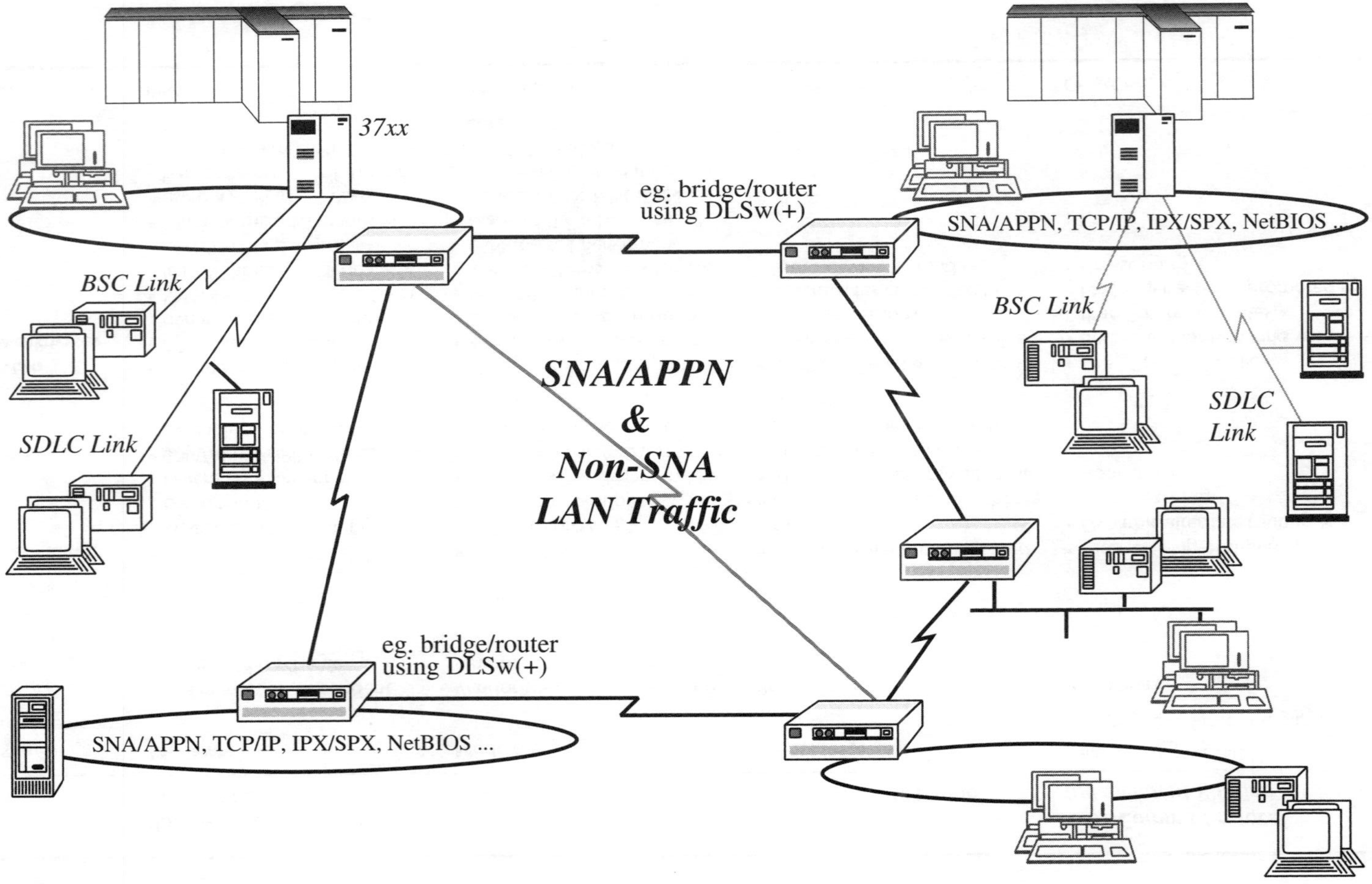

FIGURE 1.8 Semiconsolidated data network with the SNA/APPN-capable LAN-to-LAN subnetwork.

1. Synchronous Passthrough for SDLC traffic
2. Remote Link Polling for SDLC, X.25/QLLC, and 3270 BSC traffic
3. SDLC-to-LLC:2 and X.25/QLLC-to-LLC:2 conversion for SDLC and X.25/QLLC traffic, whereby link-attached devices are made to appear as if they are LAN-attached and using Layer 2 LAN-oriented LLC:2 protocol

Table 1.2 compares the pros and cons of these three technologies, as well as indicating the primary sources for each. These technologies are described in detail in Chapter 5.

Combining one of these three link-integration techniques with any one of the eight LAN-integration technologies now makes it possible to build a totally consolidated multiprotocol LAN/WAN network that includes comprehensive support for link-attached SNA and 3270 BSC devices. In many cases the repertoire of link-attached devices in use within an enterprise will not be limited just to SNA and BSC devices—particularly so in the financial and retail sectors. There is still a fair amount of async (i.e., asynchronous) traffic, much of it today coming from security alarm systems. In addition, there are still many devices, in particular point-of-sale terminals, that use protocols such as Burroughs Poll/Select or NCR BSC. The RFC 1490 FRADs typically excel at supporting such protocols as well as async. If it is important that the consolidated data network is also capable of handling async and non-SNA link traffic, RFC 1490 might prove to be the optimum solution.

Figure 1.9 shows such a consolidated LAN/Link/WAN network. *In many cases, successfully implementing such a consolidated multiprotocol network will be the first major goal of an IBM network reengineering initiative.*

1.3.2 Making Success the Mission of Consolidation

Many of the LAN- and link-integration techniques discussed above have been available since at least 1990—with bridging making its debut in the early 1980s on Ethernet LANs. Nonetheless, parallel networks were still the norm in mid-1990. There are obviously some sound reasons that there has not been more of a concerted effort, prior to now, to consolidate these parallel networks. Running two parallel networks, each with its own pool of dedicated bandwidth, is a costly proposition. Though bandwidth is getting more affordable, it is still far from being inexpensive.

A major disincentive was the perceived reliability, resilience, and robustness of some of the integration technology—particularly in their

TABLE 1.2 Comparison of the Techniques for Integrating SNA/APPN Link Traffic into a Multiple LAN/WAN Network

	Synchornous Passthrough	*Remote Link Polling*	*SDLC-to-LLC:2 or X.25/QLLC-to-LLC:2*
Basic concept	No Layer 2 (SDLC) intervention. Frames, in their entirety, forwarded end-to-end encapsulated within either HDLC or IP.	Eliminates polls having to be continually transported across WAN. Polling performed by remote nodes. Node adjacent to mainframe responds to polls from mainframe.	Variation of remote polling. Polling performed by remote nodes to capture and terminate Layer 2 link (SDLC) connections. Captured traffic made to appear as if originating from a LAN-attached device.
Layer 2 intervention	No	Yes	Yes
SNA level intervention	No	No	No
Transparent to mainframe/37xx communications controller software	Yes	Yes	Yes. But network management alerts sent by link-attached devices will have link information different from that defined for those devices at the mainframe. Does not, however, result in any problems.
Requires serial ports at the mainframe end	Yes	Yes	No
Suitable for 3270 BSC traffic	No	Yes	Not unless BSC device is made to look like an SDLC device.
Ability to consolidate multiple downstream devices and links to fewer number of links at mainframe side	No	Yes	Yes—in that all devices mapped to appear as being on one LAN.
Degree of openness to facilitate multivendor interoperability	Vendor-specific encapsulation schemes; thus remote and mainframe nodes have to be from the same vendor.	Vendor-specific; thus remote and mainframe nodes have to be from the same vendor.	Once converted to LLC:2, traffic can be treated like that from any other LAN-attached device. Thus, interoperability is possible.

TABLE 1.2 (*Cont.*)

	Synchornous Passthrough	*Remote Link Polling*	*SDLC-to-LLC:2 or X.25/QLLC-to-LLC:2*
Key strengths	• Relatively risk-free; immune to any changes that IBM may make to SNA or host/3745 software. • NetView has end-to-end visibility and SNA Control/Access. • Should support SNA Type 2.1 and APPN nodes.	• Eliminates polling sequences and retransmissions having to be done end-to-end across the WAN; thus considerable reduction in overall traffic transported. • NetView retains visibility and SNA access to the remote link-attached devices. • Potential for reducing serial ports used at the host/37xx end by mapping the actual physical link configurations into a multidropped virtual configuration. • Can be effectively used to support 3270 BSC.	• Eliminates the serial ports required at the mainframe/37xx and the router nodes at the mainframe end. • Eliminates polling sequences, and retransmissions, having to be done end-to-end across the WAN, thus considerable reduction in overall traffic transported. • NetView retains visibility and SNA access to the remote link-attached devices. • Can be used to support 3270 BSC. • Now the most popular technique for supporting SDLC links.
Main weaknesses	• All polling and responses continually transported end-to-end across the WAN. • All retransmissions involve end-to-end re-send across WAN. • WAN links need to be faster than the SDLC links being supported to ensure adequate performance.	• May not support for SNA Type 2.1 or APPN Nodes. • Requires serial ports at the host/37xx end. • NetView does not see any problems occurring on the actual remote link, nor any problems on the WAN.	• May not support for SNA Type 2.1 or APPN Nodes. • NetView does not see any problems occurring on the actual remote link, nor any problems on the inter-bridge/router trunks.
Key vendors	Cisco, Bay, CrossComm, and others	Cisco, CrossComm for 3270 BSC	Near ubiquitous from all bridge/router and FR RFC 1490 router vendors.

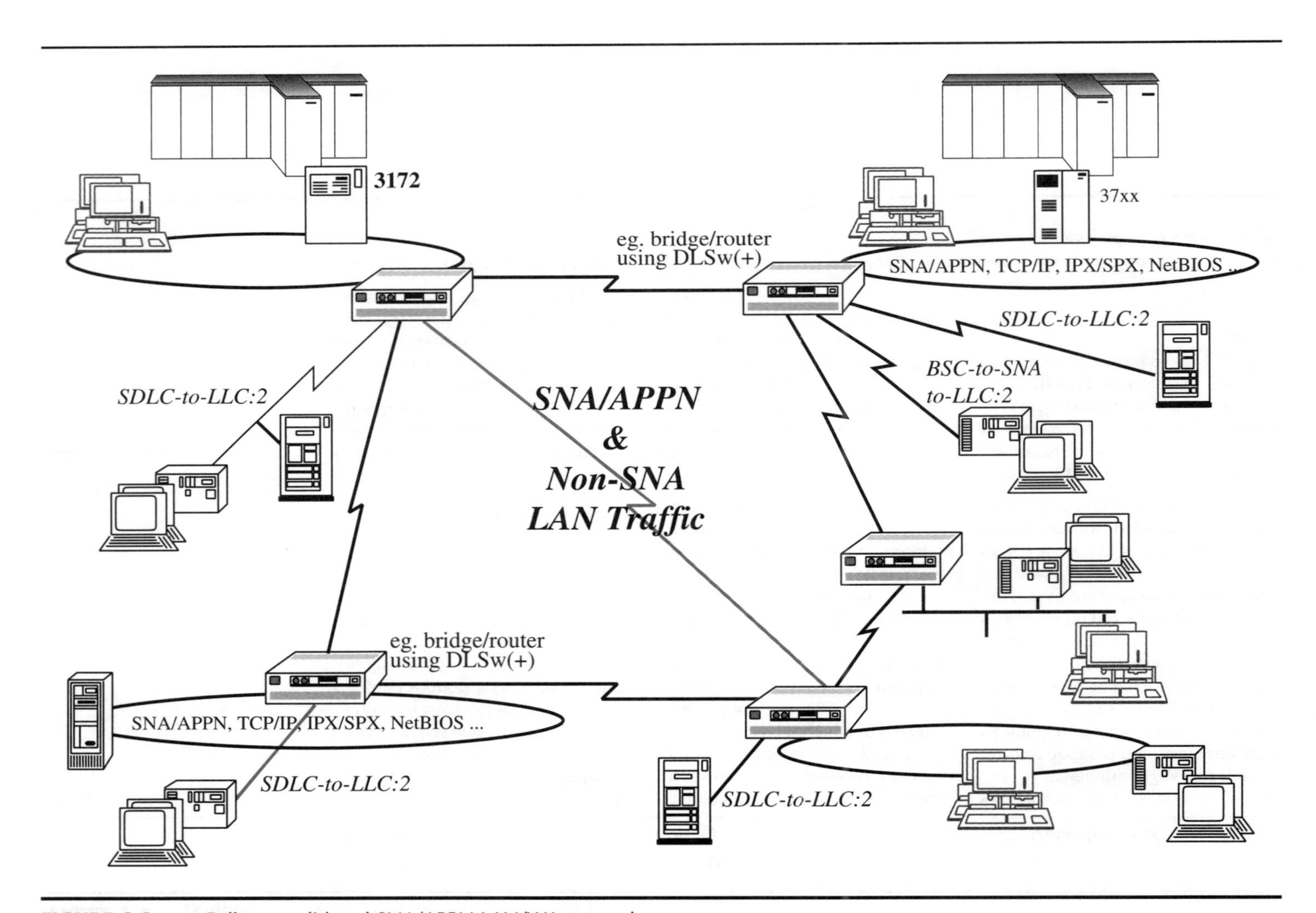

FIGURE 1.9 Fully consolidated SNA/APPN LAN/WA network.

initial versions in the early 1990s. Response times and the reliability of mission-critical SNA applications are sacrosanct. Continually compromising access to revenue-generating and business-critical SNA applications with a new network prone to errors or with erratic performance characteristics could prove to be as career-threatening as replacing functioning mission-critical applications with new and unproven ones. Consequently, one of the overriding concerns of moving SNA/APPN traffic from an SNA-only network onto a new multiprotocol LAN/WAN internetwork has been to ensure that the reliability and response attributes associated with SNA/APPN are in no way jeopardized.

Irrespective of their size or complexity, production-use SNA networks built around IBM 37xx Communications Controllers can easily boast of an annual uptime in excess of 97% for the core WAN backbone, high availability being the forte and the trademark of such SNA-only networks. In the early 1990s most SNA/APPN-capable multiprotocol solutions could not match the uptime characteristics of SNA networks—mainly because the technology was new and rapidly evolving.

Uptimes in the range of 92% for the central WAN tended to be the norm, with some of the downtime being due to the time taken to detect and fix a problem. At face value this looks like just a mere 5% difference in network availability. In reality this 5% difference in availability translates to 438 additional hours of network downtime, or 18.25 days, over one year. There is also no guarantee that this additional downtime would be even spread across the entire year. Instead, the infamous Murphy's Law will dictate that all this additional 18.25 days of network outage is liable to happen between Thanksgiving and Christmas, in part precipitated by the additional traffic during this period. For a mail-order, credit-card, mail/package delivery, financial services, or travel services company, 18.25 days of additional network downtime could have a very negative impact on the bottom line. The effects would be exacerbated if the downtime occurred during peak trading periods such as Christmas.

Most of the LAN- and link-integration technologies, as is to be expected, are significantly more robust and resilient than they were a few years ago. In general, the difference in availability between an SNA-only and an SNA/APPN-capable multiprotocol network is no longer 5%. While it is fair to say that most multiprotocol solutions still cannot match the uptime of an SNA network, the difference in uptime is now probably closer to 1 to 2%. This gap is getting narrower and narrower as the multiprotocol technologies mature. The mere existence of this gap, however small, is the key reason that some enterprises will continue to maintain parallel networks—ideally overlaid across a common frame relay or ATM backbone.

The AnyNet gateway and LAN-over-SNA approaches attempt to overcome this uptime issue by keeping the SNA backbone infrastructure in place and extending it to accommodate non-SNA protocols. At first glance this would appear to be the ideal way to ensure that the new multiprotocol network retains the characteristics of the old SNA network. In practice, however, this might not prove to be the case for two reasons. To start with, new and unproven devices (or software) will be required to add the necessary multiprotocol support. There is no guarantee that these new devices will, at least to begin with, have the same high-availability traits as other mature SNA products (e.g., 3745). Moreover, the responsibility for sustaining the core backbone will now fall upon these new and unproven multiprotocol devices.

Then there is the issue that traditional, hierarchical SNA networks, which invariably have a hub-and-spoke topology with mainframes acting as the hubs, do not usually have the direct, remote-site-to-remote-site links ideally required to facilitate LAN-to-LAN interactions. The optimum way to realize the necessary mesh network, with direct links between remote sites, is to move to peer-to-peer networking à la APPN or HPR. Thus, while both the AnyNet Gateway and LAN-over-SNA approaches will work over traditional SNA networks, the recommended networking fabric for both is either APPN or HPR. Though APPN has been around since 1986, it is still relatively new and unproven in large, mainframe centric networks. Till recently, AS/400 networks were its specialty. HPR right now is still in its infancy and has no track record at all. Therefore, it is important to note that the uptime characteristics of multiprotocol solutions built around AnyNet Gateways or LAN-over-SNA are not automatically and magically going to be blessed with the same high-availability traits of an SNA-only network. It is likely that their uptime will, in general, be similar to that of the other technologies. In addition, as with the other technologies, they will improve with maturity over the next half-decade to match the 97% of SNA.

1.3.3 Maintaining SNA Appearances

Maintaining high availability, though paramount, is not the only issue of import when evaluating technologies for consolidating parallel networks. There are other significant factors such as the possible need for SNA Subarea-to-Subarea routing, the desirability of being able to group multiple parallel links to create a single data transmission pipe à la SNA Transmission Groups, and minimizing the possibility of data packets being arbitrarily discarded because of network congestion.

Most of the integration techniques discussed above do not support conventional SNA routing, which since 1979 has been widely and routinely used by enterprises with multiple mainframes and one or more remote (i.e., link as opposed to mainframe channel-attached) 37xx Remote Communications Processors (RCPs). The absence of support for SNA routing in a new consolidated multiprotocol LAN/WAN will not, fortunately, preclude SNA users from being able to access applications on multiple, dispersed mainframes. What it may mean, however, is that some of the routes used to access the mainframes may prove to be nonintuitive and somewhat circuitous, and to involve two-hop, or two-stage, routing.

The APPN, HPR, and SNA Session Switching approaches, though not providing explicit SNA subarea routing functionality, or even trying to closely emulate such routing, will at least eliminate the need for any two-hop routing and instead route SNA/APPN traffic directly to the appropriate mainframe. Table 1.1 indicates which of the integration technique can support genuine SNA subarea node routing. The dynamics of SNA routing, and the potential work-arounds if support for it are not available, are discussed in detail in Chapter 2.

All of the network consolidation techniques discussed above permit multiple parallel links to be deployed either between the network integration devices (e.g., bridge/routers), or between these devices and a public network (e.g., frame relay). Many of these techniques, including APPN, do not in general offer the load balancing or link failure protection characteristics that SNA users have instinctively associated with multiple, parallel links forming an SNA Transmission Group (TG) since 1979.

A parallel-link SNA TG provides for continuous load-balancing across all of the links and bandwidth consolidation across those links, as well as protection against the failure of individual links. These SNA TG properties enhance network availability and throughput. SNA TG-like link-consolidation is not supported by all the integration techniques. HPR only supports TGs as an implementation-specific option. This means that some HPR implementations might not cater to TGs. IP can only offer TG-like facilities when being used with Open Shortest Path First (OSPF) routing, as opposed to Routing Information Protocol (RIP). Table 1.1 indicates which of the integration techniques can support SNA TG-type parallel links. This issue will also be discussed in Chapters 5 to 8 when each of the integration techniques are examined in detail.

1.3.4 Keeping Control of Congestion

Congestion control is another area that requires attention when consolidating data networks. Poor congestion control that could result in SNA/APPN message units being discarded within the backbone will invariably result in noticeable and unacceptable fluctuations in response times. In some instances it could possibly cause SNA/APPN session outages. Fluctuating response times and intermittent and unpredictable session outages are obviously unacceptable in the case of mission-critical applications—given that response times have a direct bearing on the amount of transactions that can be processed by such applications. Slower response times result in fewer transactions being processed since there will be longer delays between the start of each transaction.

Bridge/routers and frame relay networks, when experiencing congestion caused by unexpected traffic loads, typically try to ease the congestion by arbitrarily discarding traffic packets. Such congestion control based on the discarding of data-bearing packets is an alien and anathemic notion to the SNA/APPN community. In marked contrast, SNA, APPN and HPR take great care to prevent buffer-depleting congestion by throttling back the amount of new data entering the network. They manage this by using protocols such as dynamic Virtual Route pacing, LU 6.2 adaptive session pacing, subarea slowdown, and in the case of HPR a rather resourceful, closed-loop Adaptive Rate-Based (ARB) congestion control protocol.

If the backbone of a consolidated network relies on arbitrarily discarding data packets as its primary means of dealing with unexpected congestion, there is always a chance that some of the packets that get discarded will indeed be SNA/APPN message units. Fortunately, discarded SNA/APPN frames will not typically result in data loss or transaction errors. Higher-level SNA sequence numbers at Layer 3/4, as well as application-level transaction numbers, ensure that any SNA/APPN message units discarded by the network are retransmitted. However, even a single discarded message unit always results in the disruption of end-to-end data flow within the affected SNA/APPN session. This disruption is caused by the retransmission processes invoked to recover the lost message unit and resequence the data flow.

SNA, APPN, and TCP/IP do not support the notion of what is referred to as selective rejects, that is, the ability to request the retransmission of a single discarded or damaged packet. Instead, they have only a mechanism to acknowledge the packets up to the first packet that was discarded or damaged. Consequently all the packets starting from the discarded or damaged packet all the way up to the end of the acknowledgment window cycle (e.g., seven packets) have to be retransmitted. This occurs even if all the other packets following the one that was discarded (or damaged) had been successfully received at the other end in the first place. Selective reject is supported by HPR.

The retransmittal of previously delivered message units, which obviously increases the traffic volume, could exacerbate congestion, leading to even more packets being discarded. The amount of retransmissions required will depend on two factors: the size of the original acknowledgment window (e.g., 8, 14, or even 128), and the position within this window of the first discarded packet. If the window size is large (e.g., 128 vs. 8) and first discarded packet appeared early in the window (e.g., first packet in a window of 8 as opposed to the seventh packet in a window of 8), then a large number of packets will have to be retransmitted.

Such additional retransmissions could cause other message units within the block that is being retransmitted to be discarded. This, obviously, would result in even more retransmissions, which could worsen the congestion! Moreover, all these retransmissions will adversely impact response times, overall network performance, and efficiency, while all the time chewing up expensive bandwidth that could have been used for productive data transfer.

Delays in the in-sequence delivery of SNA/APPN message units as a result of retransmissions invariably lead to erratic response times. When there is a retransmission that affects the data flow of a terminal interaction, the response time, which might normally be .5 second for a given SNA mission critical application, could suddenly jump to 3.5 seconds or more. The amount of retransmissions taking place will, of course, depends on the amount of network traffic and the resultant levels of congestion at the various points within the network. The heavier the traffic load, the greater the chances that one or more nodes within the network may start to experience buffer-depleting congestion that could result in the arbitrary discarding of packets. Consequently, the need for retransmissions may occur on a regular and cyclic basis. Thus, each terminal operator may experience a sudden and dramatic increase in response times every few minutes. Such erratic response times frustrate terminal operators, disrupt their work patterns, and directly impact productivity.

Fortunately, overengineering the consolidated network with excess bandwidth to minimize the potential for congestion is not the only, or even the optimal, way to avoid packets being arbitrarily discarded. Recent software extensions on conventional bridge/routers as well as frame relay RFC 490 routers can virtually eliminate the danger of SNA/APPN packets being discarded, especially if the core backbone is frame relay-based. To do so, they first determine the various congestion thresholds of the WAN either by a dynamic network management query process, or via certain customization parameters defined when installing the network. Once these thresholds are known (e.g. the Committed Information Rate [CIR] for a given virtual circuit in FR), the bridge/routers or frame relay routers diligently meter the traffic being transmitted into the WAN backbone, particularly the SNA/APPN traffic. They do this metering to make sure that none of the congestion thresholds are exceeded.

Governing (or metering) the traffic being transmitted to the backbone network so that it does not exceed the bandwidth limits is also not necessarily contingent on the routers having large amounts of RAM for data buffering. Instead, the routers could use Layer 2 flow control techniques to govern the rate at which it accepts *input* data from its LAN and serial (e.g., SDLC or async) ports. By using such flow control, routers can balance the data input rate against the below-congestion output rate to the backbone network, thus minimizing the amount of data to buffer awaiting bandwidth to become available on the output side. The details of this technique are discussed further in Chapter 6.

This recent software extension to minimize the potential of SNA/APPN packets being discarded is a good example of the solid strides now being made to ensure that consolidated data networks can indeed meet the exacting demands of SNA/APPN traffic and the mission-critical applications reliant on that traffic.

1.3.5 Multivendor Networking Without Protocol

Multivendor networking per se is not a new notion to the IBM world. If anything, single-vendor networks have always been somewhat of a rarity, even if one excludes Layer 1 (e.g., modems) devices when counting the vendors involved. Plug-compatibles and emulators have always played a major role in IBM networking—particularly in the case of BSC and SNA. Since the late 1970s, it was more the rule than the exception that most large SNA networks would have non-IBM SNA peripheral devices (e.g., 3174 look-alikes), possibly from multiple vendors.

In the heyday of non-IBM minicomputers (e.g., DEC, Wang, Data General, Prime) in the 1980s, much of the traffic flowing across the

SNA networks of large enterprises originated from or was destined to these minicomputers. They masqueraded as 3274 control units or 3770 type Remote Job Entry (RJE) terminals. It would even be fair to say that it was the SNA prowess of these then ubiquitous minicomputers, and the resulting role of SNA as the de facto interoperability standard between them and mainframes, that served as the real catalyst in entrenching SNA as the means of enterprise networking. If these minicomputers had not supported SNA so well, there would have been a pressing need for another interoperability standard (e.g., OSI) to enable data interchange between these minicomputers and mainframes. And if that had come to pass, SNA would never have been as dominant as it is today.

There used to be large (i.e., 20,000 users or more) SNA networks that consisted entirely of non-IBM "iron"—plug-compatible mainframes (e.g., Amdahl) running IBM software, NCR Comten, or Fujitsu 37xx look-alikes as communications controllers and non-IBM 3270 and RJE terminals. Multivendor networking à la SNA, however, is very different from that of multivendor networking in future multiprotocol, multimedia networks. SNA provided an all-embracing, highly detailed bits-and-bytes architecture. This architecture covered all aspects of the entire network from the options for interconnecting various classes of devices to each other (e.g., peripheral node to subarea nodes or subarea nodes to subarea nodes) all the way to the highly sophisticated LU 6.2 application programming interface (API). It did not even stop there. SNA included a highly detailed network management scheme, the so-called SNA Management Services (SNA/MS), and even went on to specify architectures for application level services such as file transfer and data distribution (e.g., SNA/DS). SNA thus provided a flexible, extensive, and extensible framework, as well as all the necessary protocols to dictate and facilitate interoperability—whether between SNA devices and services from the same vendor or multiple vendors.

Unfortunately, there is no SNA-like overall architecture that covers all facets of the new generation of multiprotocol, multimedia networks. The indications are that there is unlikely to be one anytime soon. Given the diversity, complexity, and rapid evolution of these networks, it is highly unlikely that it would even be possible to synthesize an all-embracing and meaningful architecture that would be able to keep up with either the user demands or the advances in technology. The best that can be done is to have individual architectures, specifications, or standards, and in some cases a suite of related standards that address various parts and aspects of a contemporary network.

The IETF has become the focal point for many of the overarching specifications and standards related to network infrastructure. Other organizations such as the Frame Relay Forum and ATM Forum concentrate on specific technology-related aspects. Within this structure, the validity of certain so-called standards can end up becoming somewhat hazy. The so-called APPN, HPR, and DLSw standards are all administered by the IBM-sponsored APPN Implementors Workshop (AIW), which is not a recognized or bona fide standards ratifying organization. Nevertheless, nearly all, if not all, the vendors involved in IBM networking belong to the AIW and attend all the meetings. Hence, even if DLSw and HPR are not true standards in the conventional sense, they were at least developed in an open forum with active participation from all the key players.

APPN, DLSw, and HPR not being true standards should not in practice prove to be a major stumbling block to multivendor interoperability. SNA never was and will now never be a true standard. It was, however, for the longest time the predominant basis for multivendor interoperability within the commercial sector. In the case of APPN and HPR in particular, multivendor interoperability in practice will even be less of a problem than interoperability was with SNA. At present all the APPN network node implementations are based on code obtained from one of two sources—IBM or Data Connection Ltd. (DCL). This is also likely to be the case with HPR. Because IBM and DCL currently work hand in glove to ensure compatibility between their code bases, interoperability is not really an issue. This will continue to be the case until there are other source code vendors or there is a major rift between IBM and DCL. In addition, there have been regular "connect-a-thon" sessions where multiple vendors have been able to test the interoperability of their APPN implementations against those of others. In a similar vein, the consistency and compatibility of the various DLSw implementations have also been put to the test at various venues.

In the case of APPN, DLSw, and HPR, interoperability is further facilitated by the fact that all these protocols now include a capabilities-exchange handshake sequence. This capabilities exchange is invoked as soon as connection is established between any two nodes. With this capabilities exchange, two separate implementations of the same facility, say DLSw, can quickly and easily determine which features are supported by either side and establish a common denominator of functions available on both sides. Though not a guarantee of successful interoperbility or compatibility, such capabilities functions are a great boon in today's networking climate. (Until APPN, SNA never had a capability whereby different SNA implementations could dynamically and automatically hammer out a common ground for interoperability without recourse to human intervention and customization.)

The bottom line here is that multivendor interoperability will be an inescapable and quintessential facet of current and future networking. Unfortunately there will be no detailed, master blueprint for such networks that could serve as a basis for overall interoperability. Instead, there will be a myriad of function-specific specifications and standards relating to different parts, levels, and features of the network. In some instances, say with DLSw, the so-called standard will at best be de facto. Situations might also arise where, despite its pervasiveness, a technology might not be based on any recognized or readily available architecture or standard. SNA gateways, dealt with in the next chapter, fall into this category.

Despite these impediments, today it is indeed possible to design and build complex and wide-scale multiprotocol and to an extent even multimedia networks. This can be achieved using technologies, products, and services from multiple vendors. Building these networks requires more care, time, evaluation, validation, and rigorous testing than was ever the case in the past with SNA-only networks. The need for this care, patience, and iron will in order to realize these multivendor networks is now a fact of life. In the remainder of this book, but especially in Chapters 5 to 8, particular care will be taken to highlight issues that have a direct bearing on multivendor interoperability.

1.4 REFLECTIONS

Enterprises that have hitherto relied on IBM networking technology, in particular SNA, are now at a crucial crossroads. SNA-only networks, though still the lifeblood of most mission-critical applications, are now rapidly becoming anachronisms. The growing need today is for multiprotocol, medium to high bandwidth, multimedia networks. ATM-based networks are seen by most as their end-goal, at least for the next decade and a half. It is, however, unlikely whether most will move directly from the networks they have today to ATM. Consolidation of today's prevailing parallel data networks is likely to be a relatively common interim step. There is a plethora of wildly dissimilar technologies ranging from IP-encapsulation at one end to AnyNet gateways at the other that can theoretically be used to realize such consolidated, SNA/APPN capable, multiprotocol, multivendor networks. Despite the surfeit and the diversity of the available technology, successfully realizing such a new-generation, high-availability, mission-critical network is rarely going to be trivial and headache-free. There is a multitude of issues ranging from congestion control mechanisms to system management methodologies that will require careful attention and consideration. However, the good news is that rapidly maturing technologies are now at hand to enable a systematic and concerted network reengineering initiative to begin with a very real hope of success.

CHAPTER 2

State-of-the-Art SNA Networking

A QUICK GUIDE TO CHAPTER 2

The severely endangered status of SNA-only networks does not in any way portend the imminent demise of SNA. SNA is destined to be around, in large volumes, well into the start of the twenty-first century as a vital end-to-end protocol between mainframe- or AS/400-resident, business-critical applications and LAN-attached PCs/workstations. The difference now is that SNA, rather than enjoying the privileges of having its own networks, will end up sharing the resources and bandwidth of multiprotocol, and in time multimedia, LAN/WAN networks.

This chapter is devoted to examining the implications of supporting SNA/APPN traffic across multiprotocol networks. It pays particular attention to four key issues. These are: SNA LAN gateways; SNA Subarea-to-Subarea Routing; giving SNA traffic priority over non-SNA traffic; and the possibility of deploying 37xx Communications Controllers around the periphery of bridge/router networks. This chapter, in Section 2.7, also has a preliminary discussion on the thorny but topical issue of replacing channel-attached and costly 37xx gateways with cost-compelling alternatives. This very important topic is further discussed in Chapter 11.

SNA LAN gateways are what permit LAN-attached PCs/workstations to access mainframe- or AS/400-resident SNA/APPN applications. They come in three very distinct forms: PU Controller, PU Passthrough, and PU Concentrator. Section 2.3 provides an exclusive and exhaustive review of the technologies behind all three gateway types, their respective strengths and weaknesses, and the criteria for choosing between them.

SNA routing is a much maligned and misunderstood notion. Some, even today, claim that SNA is not a routable protocol. This is patently false. SNA always has been an eminently routable protocol. Just as with any other protocol, SNA routing is contingent on a device containing the appropriate SNA-smart software that understands the dynamics of SNA. Mainframes and 37xxs have acted as SNA routers since the late 1970s. Multiprotocol networks, however, will not support SNA routing. In networks with multiple mainframes this could lead to nonoptimumal routing taking place. Moreover, APPN NN routing, now available on most bridge/routers, is not the same as SNA routing. All of these issues are hammered out in Section 2.4.

Fast, predictable, and consistent response times are a sacrosanct feature of SNA networking. In multiprotocol environments non-SNA traffic could swamp SNA traffic. This could have a disastrous impact on SNA response times. A way to overcome this problem is to employ a sophisticated traffic prioritization mechanism that permits SNA traffic to be given some level of priority over other types of traffic. Section 2.5 includes an in-depth study of all the various traffic prioritization schemes that may be used in multiprotocol environments. Sections 2.6 and 2.7 deal with the role of 37xx's vis-à-vis multiprotocol networks.

Technical innovation, albeit at what at many times appeared to be a snail's pace, has always been an abiding trait of SNA. Its other hallmark has been its chameleon-like ability to adapt to most prevailing networking trends—though yet again at an infuriatingly sedate pace. This is not to say that SNA always got it right, or that it eventually, but unfailingly, met all market needs. Some of SNA's and APPN's prevailing foibles, in particular their inability to provide dynamic alternate routing in the event of a path failure, are legendary. Despite such quirks, no one could ever deny that SNA has ever suffered from complacency or stagnation; slow evolution has always been afoot. This is obviously to be expected. If not for this ability to keep up with technology and demands, SNA was unlikely to have retained its iron grip on commercial networking for so long, irrespective of its close ties to mission-critical applications.

The original SNA architecture, which first saw the light of day in the summer of 1974, could only support a network consisting of a solitary mainframe, a single 370x communications controller, and leased SDLC lines. Today, in marked contrast, particularly with APPN, it is possible to have peer-to-peer networks bristling with mainframes, minicomputers, and PCs. The backbone WAN interconnecting these various nodes could be in the form of an interconnected mesh network with lots of intermediate nodes. This backbone WAN could even be made up of X.25 or frame relay.

SNA's remarkable metamorphosis over the last 25 years is highlighted in Table 2.1 in the form of a time-line showing when the major changes took place. Figure 2.1 provides an initial and high-level view

TABLE 2.1 The SNA Time Line

1995	HPR, 2217 AnyNet Gateway, Channel-Attached bridge/routers
1994	DLSw, 3746
1993	SNA Over Frame Relay, DLUS/DLUR, MMMTG
1992	Subarea APPN, Blueprint, 6611
1991	Frame Relay (DTE)
1990	APPC/MVS
1989	SNA/FS, SNA/MS II, SNA/DS II, Casual Connections
1988	31-bit Subarea Addressing, AS/400 APPN
1987	Type 2.1 Integration, APPCCMD, XRF, SAA
1986	SNA/MS, Token-Ring, NetView, XI, S/36 APPN
1985	23-bit ENA
1984	SNI, Native VM Support
1983	Type 2.1 Nodes, SNADS, NSI
1982	LU6.2, DIA, DCA, CTCA, Switched Subarea (X.25)
1980	X.21
1979	Network Management, Parallel Sessions, NTO
1978	Parallel Link, Mesh Networks
1977	Multihost, X.25
1975	RCPs, Channel-Attached Controllers
1974	SNA 0, SDLC

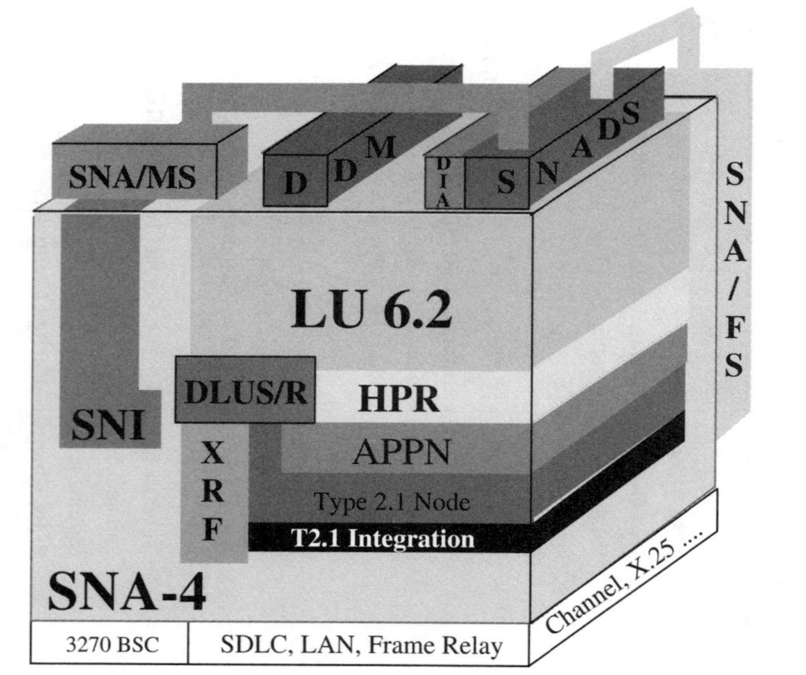

Key:
APPN – Advanced Peer-to-Peer Networking
DIA – Document Interchange Architecture
DDM – Distributed Data Management
DLUS/R – Dependent LU Server/Requester
HPR – High Performance Routing
SNA–4 – 5th SNA Spec. published in 1978
SNADS – SNA Distribution Services
SNA/FS – SNA File Services
SNA/MS – SNA Management Services
SNI – SNA Network Interconnection
XRF – Extended Recovery Facility

See Glossary for definitions

FIGURE 2.1 The SNA clan of architectures.

of what is now really the SNA clan of architectures—given that SNA is no longer a single monolithic architecture, but an interrelated clan of separate architectures. Figure 2.1 illustrates how these various architectures relate to each other. Despite its impressive resume of technical accomplishments and the growth in its clan of architectures, there is something very incongruous about the way that SNA has changed—or more to the point has not changed.

Plus ça change, plus c'est la même chose (the more things change, the more they remain the same) is a French adage whose applicability to SNA is uncanny. The topology of SNA networks have grown and changed. The types of peripheral device attached to these networks have also changed with time with PCs currently in vogue. *Nevertheless, the overall workings of these networks, or the nature of the predominant traffic flowing through these networks, have hardly changed since the 1970s!*

The bulk (i.e., at least 90%) of the traffic flowing across today's SNA networks is still 3270 datastream-based and continues to flow between mainframe applications and terminals, using protocols that have changed little, if any, since 1979. Although LU 6.2-based program-to-program traffic has grown, especially over the last couple of years, it is still comparatively meager. Till now the popularity of peer-to-peer APPN has mainly been restricted to the AS/400 community. This has been mainly due to the fact that the continuing 3270-centricity of mainframe-related SNA interactions minimizes the potential impact of the new networking possibilities offered by APPN.

The displacement of fixed-function 3270 terminals clustered around SDLC, or X.25-attached 3x74 control units by LAN-attached PCs/workstations has been one of the most profound and far-reaching changes to affect the nature of SNA networking since 1979.

2.1 DEJA VU: APPLICATION SWITCHING ALL OVER AGAIN

There is no doubt that today's huge installed base of PCs/workstations is definitely furthering the cause of LU 6.2-based SNA interactions. There is now a steady growth in LU 6.2-based file-transfer (including e-mail), software distribution (sometimes referred to as *change management*), database management, transaction processing, and even network management applications built around PCs/workstations. With a typical 3270 terminal, by definition being nonprogrammable, it was obviously not possible to build true program-to-program applications since the 3270 terminals per se (as opposed to, say, a controller in front of them) could not directly support the LU 6.2 protocol. Therefore, the current ubiquity of PCs/workstations definitely creates the right climate to finally usher in a new wave of non-3270 SNA applications.

Nonetheless, 3270 traffic and 3270-oriented mission-critical applications still prevail, and most PCs/workstations from an SNA context still end up acting as dumb 3270 terminals. The root cause of this, of course, is the sheer buisness-sensitive nature of typical SNA mission-critcal applications. Tinkering with such applications is not a trivial task undertaken lightly, prudent conservatism being by far the prevailing byword in the mainframe centric SNA community.

Despite their knack for 3270 mimicry and the ability to easily accomodate LU 6.2, LAN-attached PCs/workstations have forever changed the fundamental paradigm of SNA networking in a subtle but very far-reaching manner. *SNA now no longer has a dedicated population of SNA-only terminals as was the case with 3270s and SNA RJE terminals in the 1970s and 1980s.* Terminals attached to minicomputers precipitated this trend, but such terminals, rather than displacing 3270s in large quantities, tended to augment them. Plus, starting in the mid-1980s these terminals were also rapidly getting replaced by PCs.

From a PC/workstation perspective, SNA, rather than being the only, is just another one of the many applications supported. And there is the rub. PCs/workstations are not developed as SNA devices. They require auxiliary products, typically in the form of software, known generically as SNA LAN Gateways, to imbue them with SNA and 3270 terminal emulation capability. SNA LAN gateways are described in detail later in this chapter.

SNA becoming just another application accessible from PC/workstation is an ironic twist that harks back to the very origins of SNA. One of the primary rationales of SNA was to provide a consistent and comprehensive mechanism to permit a terminal operator, sitting at a single terminal, to use software to switch easily between multiple applications—on the same mainframe or on different mainframes. Prior to SNA, mainframe applications insisted on owning their own dedicated set of terminals and communications links. Unless one resorted to innovative non-IBM solutions such as NCR Comten's BSC-based Multiple Access Facility (MAF), access to multiple applications required duplicate hardware. At worst this took the form of multiple terminals. But in many cases it even involved deploying multiple 3270 control units with coax switches that enabled a single terminal to be switched between separate control units owned by different applications.

2.1.1 Dismantling the Old Faithful

In one swell swoop SNA eliminated the need for application-specific hardware. Instead, all the hardware—37xx communications controllers, communications links (e.g., SDLC links), control units, and terminals—now belonged to SNA (i.e., Advanced Communications Functions for the Virtual Telecommunications Access Method [ACF/VTAM] and Advanced Communication Functions for the Network Control Program [ACF/NCP] in cahoots). Now terminal users could at will switch between applications from the same terminal, typically using the SNA *Logon <<application name>>* convention. This was of course only true if all the applications were SNA-based. *PCs/workstations elevate application switching to the next logical plane—the ability to theoretically switch between, as well as simultaneously run, multiple disparate applications with SNA applications being but a subset.*

This universal application switching possibility is, however, not contingent or based upon SNA. Depending on one's orientation and background this could be viewed either as refreshingly fortunate or frustratingly unfortunate. Either way, this is now an inescapable fact of life. From a networking standpoint SNA will never again be as dominant as it used to be.

However, unlike with SNA, the one key problem with PC/workstation-based universal application access is the absence of a unifying architecture to provide a framework and bounds for this type of at-will switching between heterogenous applications. Each class of applications that use a given protocol (e.g., TCP/IP applications, IPX/SPX applications, NetBIOS applications) will have its own set of networking requirements and behavior. This is the crux of the challenge of reenginnering IBM networks, at least from a data-only perspective. The goal is to build a mission-critcal, PC/workstation-centric networking fabric that permits unrestricted interactions based on assorted protocols, with SNA being just one, albeit an important one, of the protocols involved. These multiprotocol, multivendor networks, as was discussed in the previous chapter, will over the next few years take the place of today's SNA-only WAN backbone networks.

The services and attributes taken for granted with an SNA-only network (e.g., SNA routing, SNA parallel link transmission groups or congestion control without the discarding of packets) now have to be provided, with as much veracity as possible, across a multiprotocol network. Moreover, in most cases the multiprotocol LAN/WAN network will not be built around an SNA/APPN infrastructure, the only exception to this being the SNA/APPN-centric, multiprotocol networks possible with AnyNet Gateway, à la IBM 2217 Nways MultiProtocol Concentrator, or LAN-over-SNA technology.

The replacement of SNA-only networks, with their SDLC- or X.25-based core WAN, with multiprotocol networks will have repercussions across the board. Some of the key issues have to be considered relative to this switchover, in addition to the perrenial quest for reliability, resilience, and robustness to ensure near 97% annual uptime, are as follows:

1. Selecting the optimum SNA LAN gateway technology to enable PC/workstations to gain access to mainframe- or AS/400-resident SNA applications—choosing between PU Passthrough Gateways, PU Controller Gateways, and PU Concentrator Gateways, which requires the balancing of cost and administrative efforts against functionality and extensibility.
2. Leveraging the move away from an SDLC- or X.25-based WAN, to replace high-maintenance 37xx Communications Controllers (e.g., the IBM 3745), whose forte was supporting large numbers of low- to medium-speed serial ports, with cost-effective LAN-centric, LAN-to-mainframe gateways such as IBM's 3172 Interconnect Controller or even multiprotocol bridge/routers with built-in, 3172-based, mainframe channel interfaces (e.g., Cisco).
3. Ensuring that the new multiprotocol backbone has adequate, flexible, and proven traffic prioritization and bandwidth allocation mechanisms that can be used to make sure that the response times of revenue-producing, real-time SNA mission-critical applications can be maintained irrespective of the other non-SNA traffic now vying for the services and bandwidth of the consolidated network.
4. Providing SNA Routing or some acceptable means of emulating such routing (e.g., APPN Network Node routing or SNA Session Switching), if there are multiple mainframes connected to the network and there is a need for users at remote sites to log onto different SNA applications resident at two or more of these mainframes. In the case of networks with multiple AS/400s there is likely to be a similar need for APPN NN-based routing.

All these issues are discussed in the remainder of this chapter.

2.2 THERE IS NO SUCH THING AS APPN TRAFFIC!

At this juncture, prior to delving futher into SNA/APPN related issues vis-à-vis multiprotocol networking, it is important to make a clear distinction as to the exact nature of the so-called APPN traffic. APPN, as described in detail in Appendix B, is a Layer 3 and 4, peer-to-peer networking scheme. It provides end-to-end routing, directory services, configuration services, network topology update schemes, and automatic resource (i.e., SNA Logical Units) registration facilities. APPN can therefore be viewed as a collection of services.

APPN per se, however, does not define any APPN-specific traffic types or message units. Instead, *SNA LU 6.2 program-to-program traffic is the native traffic scheme supported by APPN.* Other non-LU 6.2 SNA traffic, including the prevalent 3270 traffic (i.e., traffic-associated SNA LU-LU Session Types 2 and 3), is being supported via a technology known as Dependent LU Server/Dependent LU Requester (DLUS/DLUR), which is also described in detail Appendix B. In addition, the actual protocols used by APPN (e.g., to conduct a broadcast search or to convey changes in network topology) are also LU 6.2-based. There is, thus, in effect a recursive and symbiotic relationship between LU 6.2 and APPN. *APPN is an entirely LU 6.2-based networking scheme that in true native mode can only support LU 6.2 traffic. Consequently, there is no such thing as APPN traffic.* Many times, the term APPN traffic just refers to SNA LU 6.2 traffic. In a minority of cases it may be used to refer to all SNA traffic being supported across an APPN network, albeit in conjunction with DLUS/DLUR technology.

LU 6.2 is also the basis of HPR. HPR, like APPN, is also strictly a networking scheme, albeit only a Layer 2 scheme. Just as with APPN, LU 6.2 is HPR's native traffic scheme. HPR is discussed in detail in Chapter 7. Because both APPN and HPR are sets of services as opposed to explicit and unique traffic types, there are some intriguing relationships between APPN- and HPR-based routing and the other SNA LAN traffic integration technologies. These other integration technologies include bridging, TCP/IP encapsulation of SNA/APPN traffic à la DLSw, and frame relay RFC 1490-based encapsulation. Bridging, DLSw(+), and RFC 1490 all freely support LU 6.2, as well as all other SNA traffic types. Thus, APPN or HPR—LU 6.2 traffic flows—can be supported across a bridging, DLSw, or RFC 1490-based consolidated network, without there being an explicit need for APPN- or HPR-based NN. The one major exception to this would be networks with multiple mainframes or AS/400s. In such networks, APPN or HPR routing will permit remote users to directly switch between separate SNA applications running on different mainframes or AS/400s, precluding the possibility of any circuitous, two-hop routing.

2.2.1 Supporting APPN Without Using APPN Network Node Routing

Using bridging, DLSw, or RFC 1490 to integrate SNA traffic, as opposed to using APPN or HPR NN routing, obviates the need for DLUS/DLUR capabilities. DLUS/DLUR is only an issue when trying to support non-LU 6.2 traffic within the context of APPN or HPR. When APPN or HPR NN routing is not being used as the means for integrating SNA traffic into a multiprotocol network, any non-LU 6.2 traffic can be supported without any problems with either bridging, DLSw, or RFC 1490 without recourse to DLUS/DLUR.

One can easily implement a bridging, DLSw, or RFC 1490 consolidated multiprotocol network that supports all the end-to-end APPN interactions between any set APPN End Nodes (ENs) and NNs—such as a collection of AS/400s. In such a network, the backbone will not provide any APPN services (e.g., directory or LU registration) or support NN-based routing across the backbone. Instead, all the routes between the APPN nodes will be in the form of end-to-end point-to-point connections, across the backbone, that would have been established using a Destination MAC address-based bridging scheme.

In this type of non-APPN-based multiprotocol network, the backbone is in essence just providing a set of logical point-to-point links between the APPN nodes. There are existing bridge/router solutions that support APPN nodes in this manner—typically using Source Route Bridging (SRB) or DLSw. This ability to freely support APPN or HPR across a bridged, DLSw, or RFC 1490 consolidated backbone also leads to the possibility of deploying hybrid bridge/router networks. In such hybrid networks all bridge/routers or RFC 1490 routers do not have to perform APPN NN routing. The routers without APPN NNs will use bridging, DLSw, or RFC 1490 to transport the APPN traffic to/from the routers acting as APPN NNs. Figure 2.2 depicts such a hybrid network where APPN NN routing is only being performed by some of the routers even though SNA/APPN traffic is supported right across the network.

There can be some limitations in trying to support a series of APPN nodes across a multiprotocol network where the backbone—or at least the routers at the periphery of that backbone—are not providing APPN NN services. For a start, there will be no dynamic, Class-of-Service (COS)-based APPN routing across the network. In addition, if all the NNs are upstream of the backbone, and all the downstream nodes are ENs, all the NN services required by the ENs—such as LU registration or remote LU location—will involve interactions across the backbone. If the routers at the periphery of the backbone were NNs, all such services could be done locally, obviating the need for interactions across the backbone. If this type of EN-to-NN interaction is likely to be common, as might be the case with a large AS/400-centric APPN network, it would be advantageous to consider a APPN NN-based bridge/router solution. If the SNA traffic is very high even an AnyNet Gateway technology-based solution such as the IBM 2217 might be necessary. These notions will be futher elaborated in Chapters 5, 7, and 8.

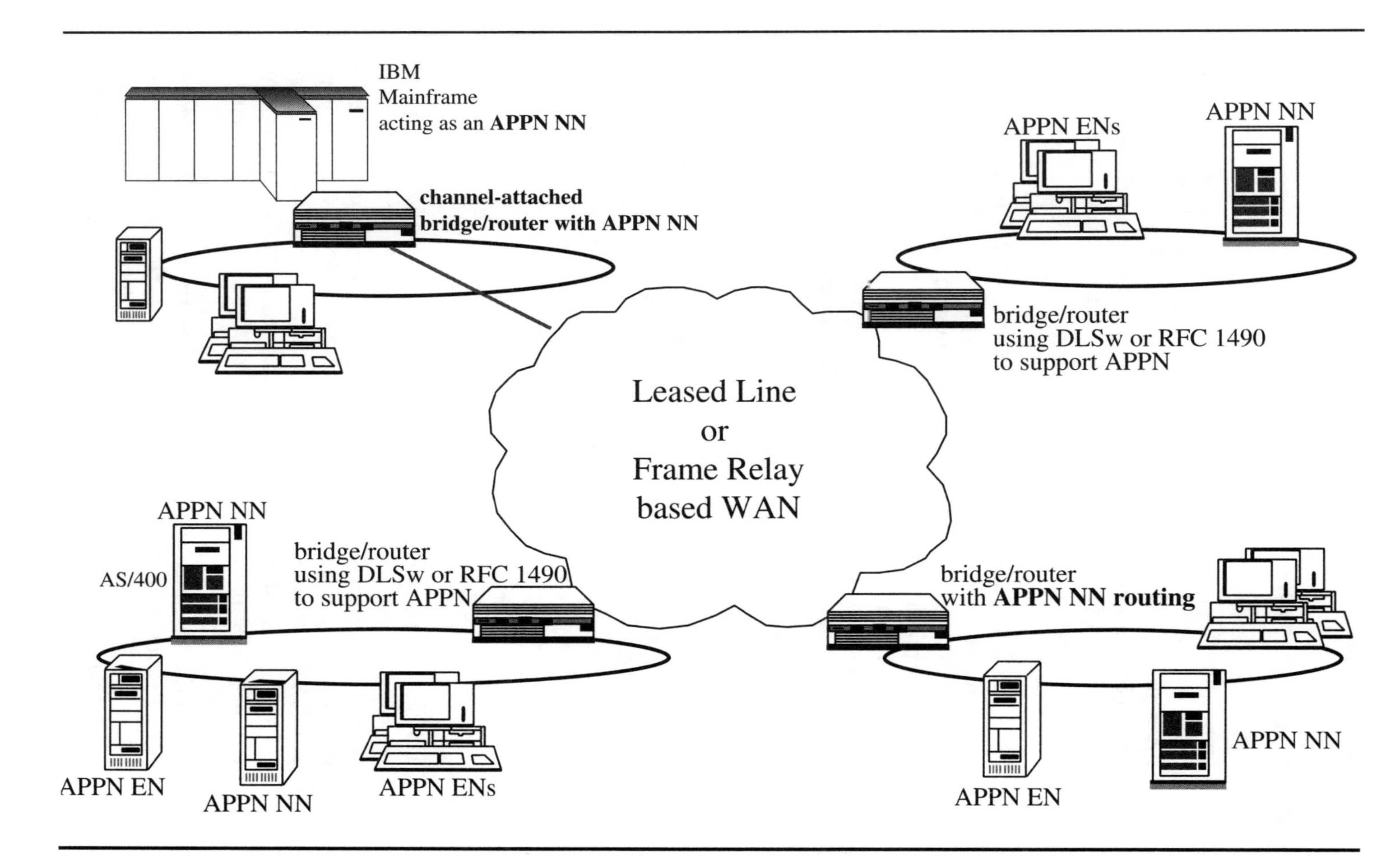

FIGURE 2.2 A hybrid network where both APPN NN routing, and DLSw or RFC 1490, is being used to freely support APPN traffic.

2.3 COMING TO TERMS WITH SNA LAN GATEWAY

PCs and workstations do not possess built-in SNA or 3270 terminal emulation (i.e., painting and reading data from screens with 3270 fields with attributes such as modifiable, protected, nondisplay, high-intensity, etc.) functions. In order for PCs/workstations to be able to access SNA applications, whether on a mainframe, AS/400, or another peer PC/workstation, they require at a minimum three SNA-related entities; these are as follows:

1. Basic SNA functionality, referred to as an SNA node, and typically provided in the form of a SNA Type 2 or 2.1 Node, which will include the SNA PU, LU, and Path Control Network functionality (see Appendix A).
2. One or more SNA client applications (or emulations) that work in conjunction with the SNA node functionality. The client application may be in the form of either a 3270 emulation for mainframe application access, a 5250 emulation for AS/400 application access, a 3270 datastream (e.g., IBM PC 3270 IND$FILE) or LU 6.2-based file-transfer support, a RJE terminal emulation, an LU 6.2-oriented API for developing program-to-program applications, a customized program-to-program application (e.g., a distributed database management system), or a combination of one or more of the above.
3. A LAN-to-SNA Network Gateway that will provide the physical interface between a LAN and either a SNA-only SDLC or X.25 WAN, a SNA-capable multiprotocol WAN that is using an SNA LAN traffic integration technique (e.g., bridging, DLSw, APPN NN) to transport the SNA traffic across the consolidated WAN, or Bus-and-Tag or ESCON II I/O channel on an IBM mainframe.

This SNA Node and SNA client application/emulation functionality is typically provided today via software products that use a PC/workstation's LAN interface (e.g., token-ring or ethernet adapter) to realize the necessary connections with the LAN-to-SNA Network Gateway. The LAN-to-SNA Network Gateway may be a dedicated or nondedicated PC, a multiprotocol bridge/router with a built-in gateway function (e.g., Cisco), or a dedicated communications processor such as an IBM 37xx or 3172.

In the the early 1980s, however, when the need to interface PCs to SNA networks first surfaced, and prior to the advent of token-ring LANs in 1986, the original and most widely adopted solution was totally hardware-based. It relied on the use of a dedicated PC adapter that enabled a PC to be connected to a standard 3270 control unit (e.g., IBM 3274 or 3174), via a standard coax-cable, and act as if it was a bona fide 3270 terminal. The Document Content Architecture (DCA) IRMA cards were the epitome of this approach. This hardware-based, 3270 coax-oriented, terminal emulation approach started to lose favor in the late 1980s as LAN-attached PCs began to proliferate and the appeal of coax-attached 3270 terminals began its downward spiral.

The so-called SNA LAN gateways provide the SNA node and SNA client application functionality in a variety of different permutations. The key variance between the various permutations is the relative location of the SNA node and the relevant client application/emulation. In some approaches the SNA node and the client application functionality run together on each and every LAN-attached PC/workstation. In such an approach, the SNA LAN gateway per se that will interface the LAN-attached PC/workstations to the network or a mainframe performs no SNA-related Layer 3 or above functions. Instead, it will devote itself to performing Layer 2 protocol conversion—for example, token-ring LLC:2 to SDLC or token-ring LLC:2 to mainframe channel.

A gamut of diverse SNA LAN gateways enables LAN-attached PC/workstations to access mainframe-resident SNA applications. Included in this array of diverse solutions are the IBM 37xx communications controllers, the IBM 3172, the IBM 3174 Establishment Controller, and the Bus-Tech 3172-BT1. Enterprises should also consider Cisco channel-attached bridge/routers and IBM and non-IBM PC gateways such as the IBM 3270 Emulation Program Version 3.0 and the market-leading Novell NetWare for Systems Applications Architecture (SAA). Another possibility is that of using minicomputers such as the IBM AS/400 as an SNA LAN gateway.

2.3.1 Different Flavors of SNA LAN Gateways

SNA LAN gateway solutions are divided into three distinct categories: PU Passthrough Gateways, PU Controller Gateways, and PU Concentrator Gateways. The differentiating factor between the three gateway types is the location of the SNA node software that provides the actual SNA functionality. The SNA node software could reside in the downstream LAN-attached device (e.g., PCs/workstations), in the LAN-to-SNA network gateway itself, or in both the LAN-attached devices, as well as the network gateway. The location of the SNA node software with these three types of SNA LAN gateway can be summarized as in Table 2.2.

TABLE 2.2 The Location of the SNA Node Software When Using the Different SNA LAN Gateway Types

	Individual LAN-Attached Devices (PCs or workstations)	*LAN-to-SNA Network Gateway (provides the physical and Layer 2 interface between the LAN and either a WAN or a mainframe)*
PU Passthrough Gateway	SNA Node	No SNA Node
PU Controller Gateway	No SNA Node	SNA Node
PU Concentrator Gateway	SNA Node	SNA Node

All three of these SNA LAN gateway solution types can be found in two very distinct forms: either in the form of local, mainframe channel-attached gateways (e.g., 3172), or as remote, link-attached gateways located at the periphery of a WAN. Figures 2.3 to 2.6 show both local and remote gateway configurations.

Local gateways act as the interface between a mainframe and LANs that are located in physical proximity to that mainframe. Remote gateways act as the interface between a mainframe and LANs located a considerable distance away from the mainframe. In the case of remote gateways, a WAN is used between the gateway and the mainframe. The WAN does not have to be SNA-based, nor is it always the case that the protocol used across the WAN is SNA. Both of these factors are of particular importance now as SNA-only networks start to give way to SNA-capable multiprotocol networks. Novell, for one, has been actively promoting the concept of using a non-SNA protocol, across the WAN, between the remote LANs and the mainframe(s).

In the case of Novell, the protocol original used across the WAN was Novell SPX/IPX. A channel-attached gateway running NetWare for SAA typically from Bus-Tech or Memorex-Telex then converts the SPX/IPX traffic to SNA. The rationale of the Novell approach is to eliminate the need for supporting any SNA traffic across the WAN. By eliminating the need to transport SNA across the WAN, one could adopt a bridge/router-based solution for realizing a consolidated network. Although it would support SNA applications via the channel-attached gateway, this type of solution does not have to deal with the issues of having to integrate SNA traffic with non-SNA traffic within the WAN. Given that supporting SNA traffic across bridge/router networks used to be far from straightforward, and painless, there is some merit in the Novell approach. In a sense it is the opposite of the

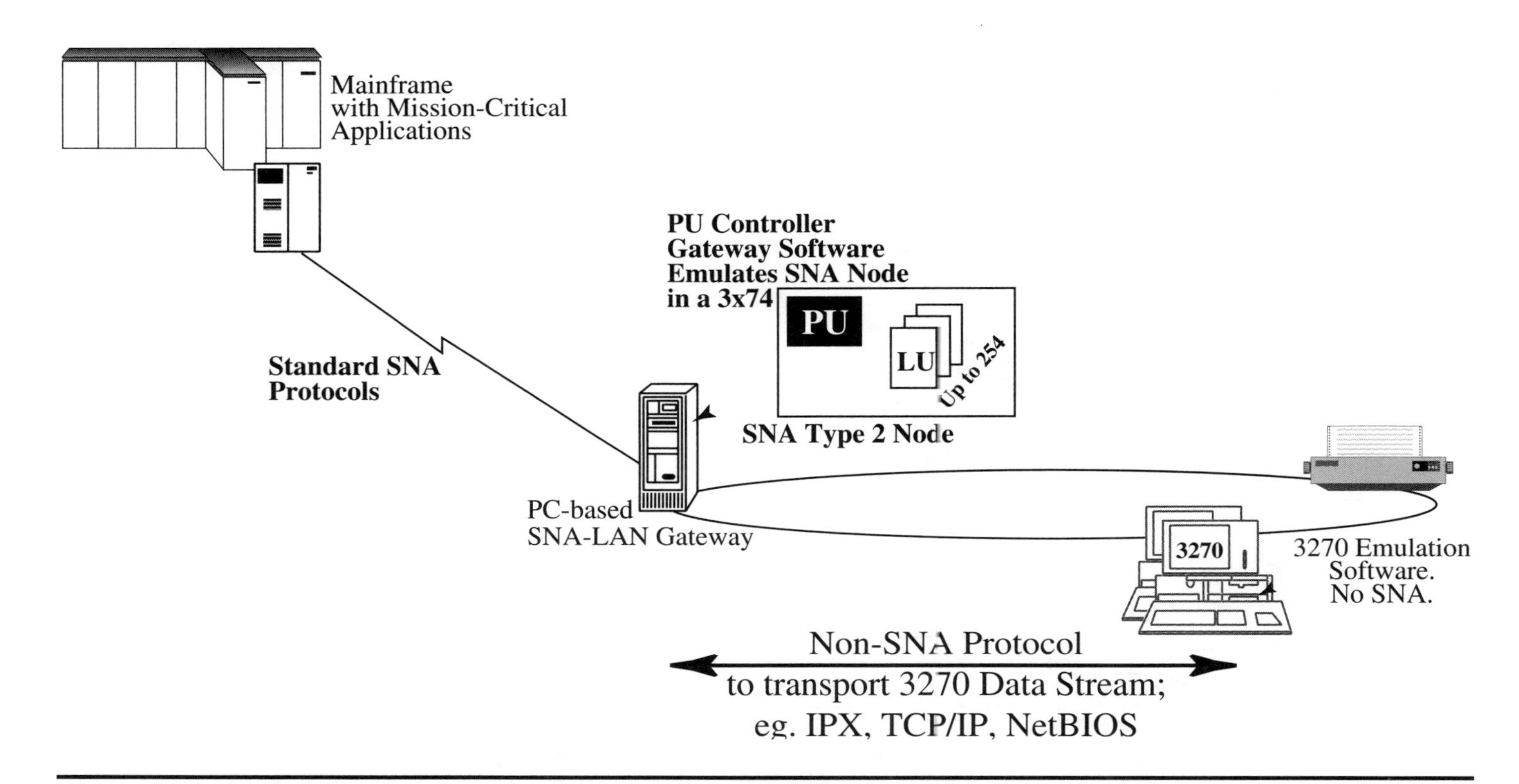

FIGURE 2.3 The workings of a PU controller-based SNA LAN/WAN network.

PU Controller
Gateway Software
Emulates SNA Node
in a 3x74
PU
LU
Up to 254
SNA Type 2 Node
PC-base
SNA-LAN
Gateway
G/W
3270
3270 Emulation
Software.
No SNA.
37xx acting
as a 'Passthrough'
Gateway
SNA
Non-SNA
SNA
3174 using SNA
through the 37xx
Bridging,
DLSw(+) or
RFC 1490
to transport SNA
over the WAN
PU Controller
Gateway Software
Emulates SNA Node
in a 3x74
PU
LU
Up to 254
SNA Type 2 Node
SNA over LAN
[ie. LLC:2(SNA)]
3270
3270 Emulation
Software.
No SNA.
Non-SNA Protocol
to transport 3270 Data Stream;
eg. IPX, TCP/IP, NetBIOS

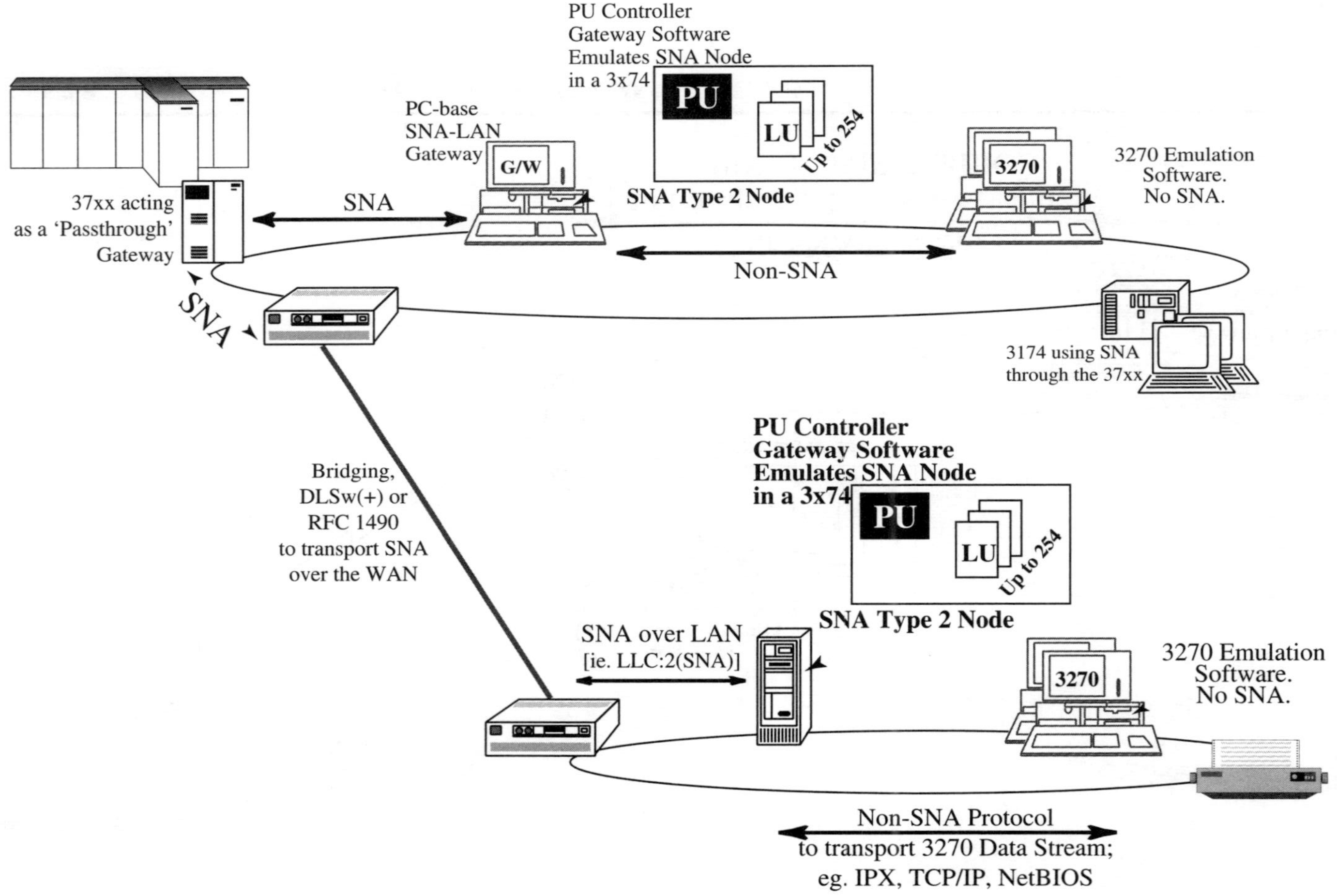

FIGURE 2.4 LAN-only PU controller gateways.

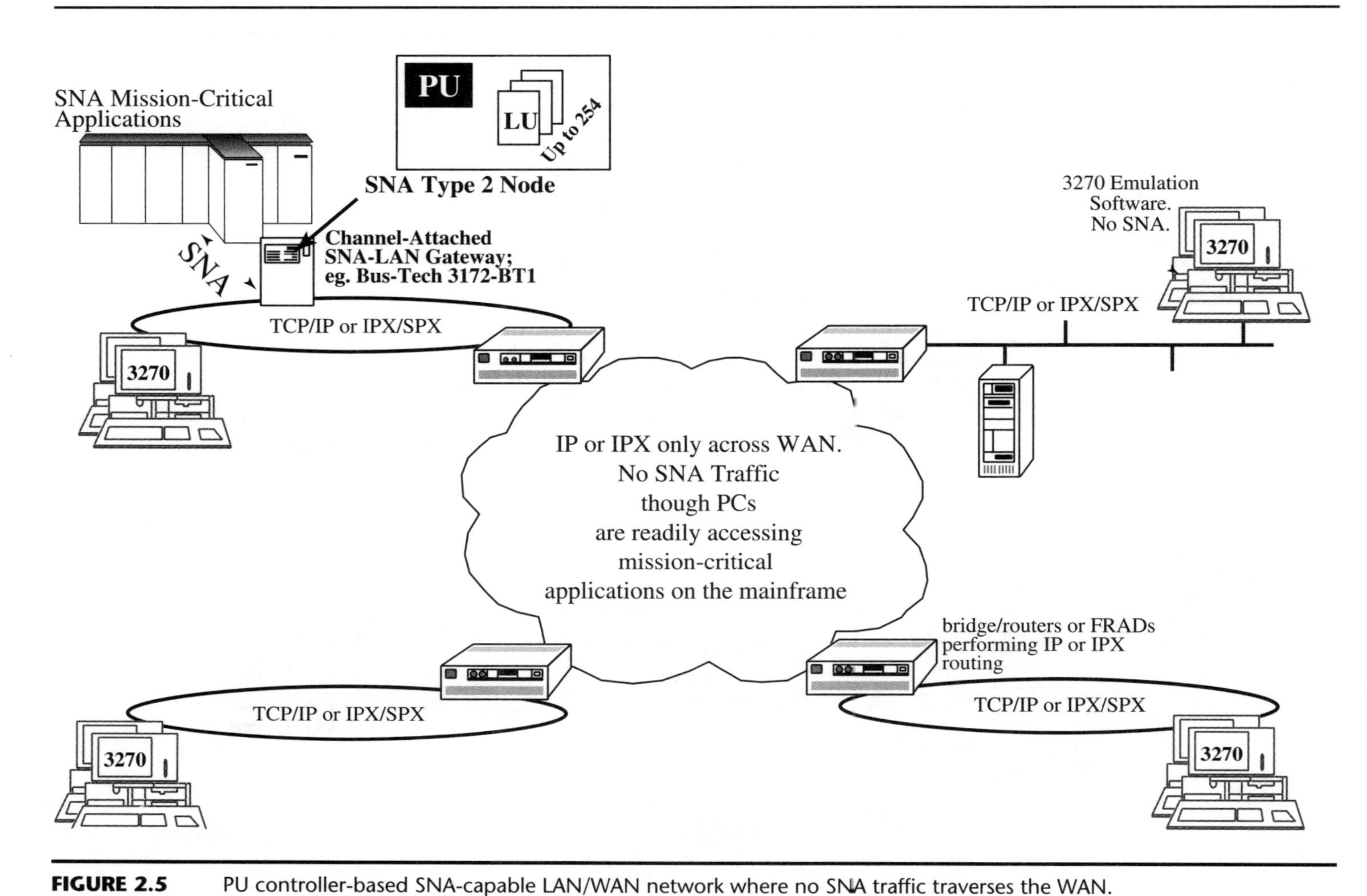

FIGURE 2.5 PU controller-based SNA-capable LAN/WAN network where no SNA traffic traverses the WAN.

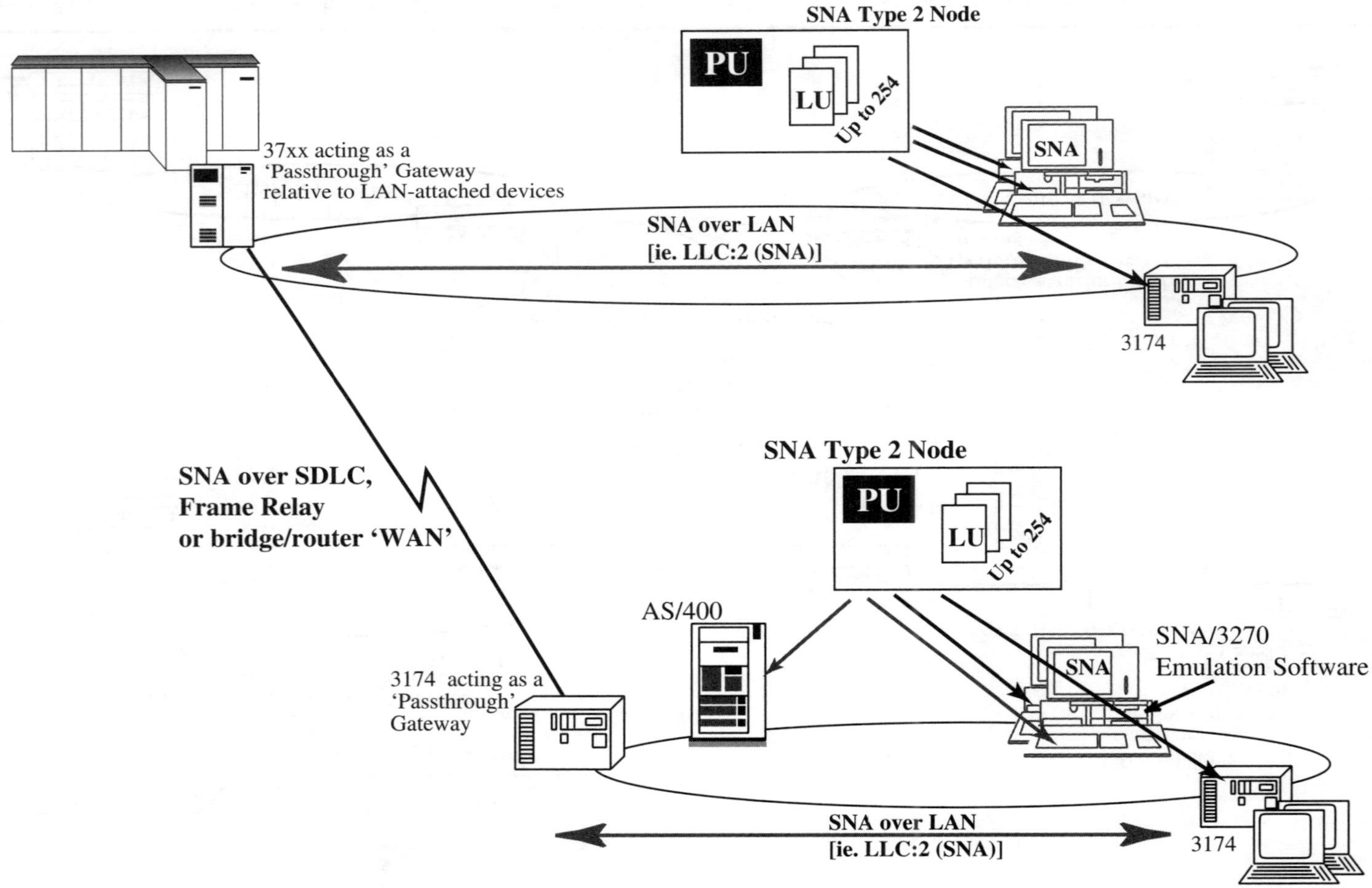

FIGURE 2.6 PU passthrough gateways in action.

APPN/HPR backbone-centric solution proposed by AnyNet Gateways and LAN-over-SNA solutions. The possibilities of being able to freely conduct SNA transactions across a network that does not support any SNA traffic is further described in Chapter 8.

There are, obviously, some drawbacks to the No-SNA-across-the-WAN approach. Otherwise, in today's multiprotocol-oriented climate this approach is likely to prevail and there would be little call for any other solutions. Eliminating the use of SNA across the WAN excludes its use end-to-end between the SNA applications on the mainframe and the LAN-attached PC/workstations. The LAN users are thus deprived of many of SNA's powerful features such as end-to-end congestion control and error-recovery. This loss of SNA functionality fortunately does not have to extend to network management. An Alert Relay mechanism available with PU Concentrator Gateways can ensure that NetView/390 based network management can still have visibility of the end-PCs even with the Novell SPX/IPX approach. NetView/390, however, will not see or understand the working of the multiprotocol-based WAN across which the SNA transactions are being conducted.

Novell's original implementations of the SNA LAN gateway were also contingent on the use of the SPX protocol. SPX is used on top of IPX, in the same way that TCP is used on top of IP, to achieve a connection-oriented guaranteed delivery mechanism. Such guaranteed delivery mechanisms are obviously ideal for supporting SNA interactions that rely on packets arriving at their destination in the order that they were sent with no packets lost or discarded within the network. SPX, like SNA and TCP, guarantees delivery by relying on an explicit end-to-end acknowledgment mechanism. Unfortunately, the SPX used to insist on one-for-one acknowledgments—that is, every SPX packet in each direct had to be explicitly acknowledged before another packet was sent. The need for such one-for-one SPX acknowledgments, across the WAN, used to in some instances degrade the overall performance of a SPX/IPX WAN-based SNA gateway configuration. Selecting the most appropriate SNA LAN gateway solution for a given network is never trivial and can involve issues that are not immediately obvious.

Having both local, channel-attached mainframe gateways and WAN-attached remote gateways means that it is possible to have *cascaded gateway* configurations. In such a configuration one has a hierarchy of gateways—typically a channel-attached local gateway (e.g., 3745) supporting multiple link-attached remote gateways (e.g., 3174s or PC Gateways). Unfortunately, despite its pivotal role in integrating LAN-attached PCs/workstations into the SNA world, SNA LAN gateway methodology has evolved without a uniform underlying discipline, let alone a strategic, comprehensive architecture. Thus, actual product implementations, with some key ones such as Novell's NetWare for SAA from outside IBM, have dictated, and molded, mainframe gateway concepts, technology, and trends. This product-driven approach has worked to date, and there are gateways in all shapes and sizes that adequately address every possible LAN-integration scenario. Nonetheless, SNA LAN gateway technology tends to be somewhat arcane and convoluted, and thus often requires considerable evaluation in order to select the optimal gateway solution for a particular network. When it comes to selecting such gateways there are few, if any, rules of thumb, and the requirements of each network should be individually evaluated against the pros and cons of each gateway type. Table 2.3 summarizes the pros and cons of these three SNA LAN gateway types.

2.3.2 PU Controller Gateways

PU Controller Gateways were the first generation of SNA LAN gateways for LAN-attached PCs that did not rely on using a coax adapter to connect PCs to a 3270 control unit. Instead, these gateways set out to provide the SNA node functionality previously provided by a 3270 control unit, hence the name *controller*. Rather than using coax attachments, these PU Controller Gateways, in tune with the move toward LANs, used LAN connections to realize the necessary interactions between the various PCs that required SNA capability and the gateway. The gateway thus had two interfaces: a LAN interface with the PCs and a WAN or mainframe channel interface with the SNA network. Figure 2.3 shows the overall structure of a typical PU Controller Gateway-based LAN/WAN network.

In the case of a multiprotocol network, remote gateways of this type may end up only using their LAN interfaces. The gateway will send and receive SNA traffic over the LAN, rather than over a SNA/SDLC link. It will then rely on a bridge/router or a RFC 1490 router, attached to that same LAN, to transport the SNA traffic to and fro across the multiprotocol network. In this type of LAN-only configuration, the PU Controller Gateway will be acting purely as a SNA Protocol Converter for the LAN-attached devices. The interface between the LAN and the multiprotocol WAN is handled by a bridge/router or a RFC 1490 router. This type of LAN-only PU Controller Gateway configuration is shown in Figure 2.4, with both remote and local LANs.

As was discussed above in the case of Novell's No-SNA-across-the-WAN approach, it is indeed also feasible not to have any remote SNA

TABLE 2.3 Examples and Pros and Cons of the Three SNA LAN Gateway Types

PU Controller Gateway	*PU Passthrough Gateway*	*PU Concentrator Gateway*
SNA functionality provided by the gateway, typically in the form of a 3x74 Control Unit	SNA functionality resides only in the LAN-attached end devices.	SNA functionality in BOTH gateway and the individual, LAN-attached end devices
Examples: IBM 3270 Emulation Program Ver. 3 Eicon SNA Gateways AS/400s Novell NetWare for SAA Microsoft SNA Server Most third-party SNA Gateways	**Examples:** IBM 3745, 3172, and 3174 Cisco channel-attached bridge/router	**Examples:** IBM OS/2 Comms. Manager IBM PC/3270 Novell NetWare for SAA Microsoft SNA Server Cisco bridge/routers
Pros: 1. Only the Gateway, rather than the individual LAN-attached end devices, needs to be defined to the host. 2. LAN-attached end devices do not have to run a full SNA stack. Instead, they run a "small" 3270 emulation program.	**Pros:** 1. End-to-end SNA control, with SNA-based error recovery, congestion control, etc. 2. Host has full visibility of the inidividual LAN-attached devices. NetView management extends all the way to the end devices	**Pros:** 1. Combines the best features of PU Controller and PU Passthrough Gateways. Gateway acts as a "mini" host. Network management data is conveyed to/from the actual host and the end devices via a relay mechanism provided by the gateway. End-to-end control and visibiliy is based on SNA.
Cons: 1. SNA does not go end-to-end from the host TP application to the LAN-attached device; e.g., PC. Instead, a LAN-oriented protocol such as NetBIOS or Novell IPX is used between the gateway and the LAN-attached devices. SNA functions such as Error-recovery, congestion control (ie., pacing), and data flow control do not extend all the way to the LAN-attached devices. 2. Host does not have visibility of the LAN-attached devices. Thus, NetView's management domain ends at the Gateway 3. Cannot be used to support any bona fide SNA devices (e.g., 3174) that may also be on the LAN alongside the PCs. 4. The LAN protocol used between the gateway and the LAN-attached devices might not be as efficient as SNA—particularly true with SPX.	**Cons:** 1. Each end device needs to run a full SNA/3270 emulation 2. Each end device has to be individually defined to the host	**Cons:** 1. Each end device needs to run a full SNA/3270 emulation.

LAN gateways in multiprotocol scenarios. Only a local (i.e., either channel-attached, or attached to a LAN connected to the mainframe) gateway may be necessary. Non-SNA protocols, most likely TCP/IP, will now used between the client PC/workstations and the local gateway, which will then perform the conversion to SNA; see Figure 2.5.

It is also worth noting that such gateways may, if necessary, also be connected to bridge/routers or RFC 1490 routers forming a multiprotocol WAN, using any of the link-integration techniques such as SDLC-to-LLC:2 conversion. In this type of setup, the gateway, as in the case of a SNA-only network, will continue to use its serial port, as opposed to the LAN, for all of its SNA interactions. This serial port, however, will now be connected to a router that again will be responsible for transporting the SNA traffic.

With PU Controller Gateways the SNA Node (which is typically a Type 2 node) is implemented within the gateway. No SNA node functionality is installed or required in the individual downstream devices. Instead, the downstream devices will execute an SNA client application—that is, 3270 emulation, RJE emulation, file transfer, and so on. In the absence of SNA node functionality, the downstream devices will not have any SNA-specific capability or cognizance. They will communicate with the PC Controller Gateway using a LAN protocol such as NetBIOS, Novell's SPX/IPX, or TCP/IP, rather than SNA.

SNA traffic per se only flows between the gateway and the WAN, or, in the case of a channel-attached, local gateway, between the gateway and the mainframe. SNA does not flow across the LAN, between the gateway and the downstream LAN-attached devices that are actually interacting with the SNA applications. IBM's PC-based 3270 Emulation Program Version 3, the AS/400 when used as a SNA LAN gateway, and many third-party SNA LAN gateways such as those offered by Eicon Technology are examples of PU Controller Gateways. So are tn3270-based solutions that use an external tn3270 server, as opposed to TPC/IP-based tn3270 software resident on a mainframe.

Tn3270 is a TCP/IP, telnet virtual terminal emulation protocol scheme to enable PC/workstations using TCP/IP to gain access to SNA applications. The original implementations of tn3270 used mainframe-resident TCP/IP-based software to do the protocol conversion between telnet and the 3270-based mainframe applications. In such implementations, relative to tn3270, the only traffic that went to and fro between the mainframe was TCP/IP traffic. It is now possible to get external, tn3270 servers or gateways, in both channel-attached and link-attached versions, from a variety of vendors that include CNT/Brixton, Apertus, and OpenConnect Systems.

With such external tn3270 servers, neither TCP/IP software nor tn3270 software is required on the mainframe. Instead, standard SNA protocols are used between the mainframe and the external server. The external server acts as an SNA-to-telnet protocol converter, and converts the telnet interaction to SNA and vice versa. Such external tn3270 servers, from the perspective of LAN-attached devices using TCP/IP telnet, are acting as SNA LAN PU Controller Gateways. The need for and the desirability of having mainframe resident TCP/IP software to facilitate TCP/IP-based terminal access (i.e., telnet) and file transfer are discussed in detail in Chapter 8.

A key advantage of the PU Controller Gateway approach is that it eliminates the need to define each LAN-attached device to the mainframe. This could be very important in a network with a large number of PCs/workstations, especially if not all them are going to be accessing mainframe SNA applications at the same time. Nondedicated access to SNA is increasingly becoming the case as the population of non-SNA applications being used by enterprises continue to grow. With this type of gateway, only the gateway with 254, or in some cases even more, SNA LUs needs to be defined to the mainframe. (SNA's addressing mechanism for Type 2 nodes has a ceiling of 254 LUs per node. To get around this 254 LU limitation, this type of gateway is capable of emulating multiple separate Type 2 nodes. This type of multinode capability is known as *Multiple PU* support given that the mainframe requires a PU, with a separate link [e.g., SDLC] address, to be defined for each group of 254 consecutive LU definitions.) The gateway provides a LU-allocation mechanism whereby a large population of PC/workstation users can contend (i.e., bid) for one or more of these LUs, if and when they are required to interact with a mainframe application.

In addition to minimizing the number of mainframe definitions required, PU Controller Gateways restrict all the SNA node functionality to the gateway. Thus they require only a relatively small client application/emulation package to run on all the LAN-attached downstream PCs/workstations. Because client emulation software is less complex than SNA node software it is typically less expensive. Therefore, this approach can prove to be very cost-effective. By not requiring SNA node software to run on each device, it also ensures that maximum memory and processing power is available for the SNA client application/emulation (e.g., distributed database manager). This parsimonious use of memory was particularly advantageous in the days when PCs had the DOS-imposed limitation of only being able to use 640K of active memory.

The attractiveness of this type of gateway is, however, somewhat diminished by the fact that it unfortunately suffers from two significant weaknesses. The main, and most obvious, is that it does not support SNA end-to-end all the way into the LAN-attached devices. Thus, all SNA protocols—for example, congestion control and error recovery—end at the gateway. Moreover, this SNA-stops-at-the-gateway characteristic also means that mainframe-based NetView/390 does not have any visibility of the LAN-attached end-devices. This can be galling for enterprises that rely on NetView/390 for a complete and incisive view of the operational status of their entire SNA network. This is particularly true if much of the remote population of SNA-devices consists of LAN-attached devices. The lack of visibility of these devices could mean that NetView/390 is now only seeing a very small part of the actual SNA network.

The other problem with this type of gateway is that it cannot support LAN-attached devices that can only talk SNA or that want to talk SNA. The basic premise of this type of gateway is that it does not use SNA, across the LAN, between itself and downstream devices. Thus, if a downstream device insists on using SNA, as opposed to another LAN protocol, a PU Controller Gateway is flummoxed. This is now likely to happen with increasing frequency as more and more native SNA devices (e.g., AS/400s) are attached to LANs. One way to accommodate this type of scenario involving SNA-speaking LAN-attached devices is to use a different type of gateway, such as PU Passthrough. Another is to provide each of these SNA-speaking devices with their own direct connection to the SNA or multiprotocol WAN network. Obviously, providing each of these SNA-speaking devices with its own connection to SNA or multiprotocol WAN could become cost-prohibitive and difficult to administer.

2.3.3 PU Passthrough Gateways

PU Passthrough Gateways were developed to eliminate the two major weaknesses of PU Controller Gateways. With a PU Passthrough Gateway, the SNA node functionality, as well as the client application (e.g., 3270 emulation), must reside within each and every LAN-attached device that requires to access to SNA applications. Running the SNA node software and the client application on top of each other within the same PC/workstation in this manner is often referred to as a *full-stack* implementation. This alludes to the fact that each device now contains a complete SNA protocol stack consisting of Layers 3 to 7.

As the name suggests, the PU Passthrough Gateway per se just passes the SNA message units between the LAN and a WAN or a mainframe channel. It transforms only the Layer 2 data link protocol within which the SNA message is encapsulated. Thus, a PU Passthrough

Gateway will perform LAN LLC-to-SDLC, LAN LLC-to-X.25, or LAN LLC-to-mainframe-channel-protocol conversion. This protocol conversion will always involve an explicit swap of the Layer 2 addresses, for example, from token-ring MAC address to SDLC, and so on.

The PU Passthrough Gateway approach, with its use of full-stack SNA implementations in every LAN-attached PC/workstation, obviously ensures that SNA is now used end-to-end between the mainframe applications and the LAN-attached device. This also guarantees that NetView/390 now has complete visibility of the LAN-attached devices—albeit at least when they are using SNA. In addition, with this type of gateway, any LAN-attached SNA device, such as AS/400s, can now be supported through the gateway.

If PC/workstations with a full stack are to be used to realize end-to-end SNA, there is no option but to use either this type or a PU Concentrator Gateway in an SNA-only network to interface the LANs to the SNA network. This type of PU Passthrough Gateway-based network, with both local and remote gateways, is shown in Figure 2.6.

It is important to note, however, that if a multiprotocol WAN is being used, full-stack PC/workstation implementations, on remote LANs, will not require a PU Passthrough Gateway—or for that matter any other type of SNA LAN gateway—in order to send or receive SNA traffic across the WAN. Instead, any of the SNA LAN traffic integration techniques—for example, bridging, DLSw, RFC 1490, or even an AnyNet Gateway—could be used to transport the SNA traffic between the LANs and the mainframe. The router or AnyNet Gateway providing the necessary SNA LAN integration technique, in this instance, ends up doing the necessary LAN-to-WAN conversion support.

IBM 3745, 3174, and 3172 are all examples of PU Passthrough Gateways when it comes to LAN-attached SNA devices. This indicates IBM's overall preference for this approach, which does obviously have the indubitable advantage of ensuring that SNA flows end-to-end as was always envisaged by the original SNA architecture. The SNA node and the client emulation, in the case of a PC, will be implemented using a 3270 emulation package such as IBM's 3270 Workstation Program Version 1.1.

Group Poll, a technique pioneered by the 3174, is a very useful and powerful feature—albeit only in SDLC-based, SNA-only networks. Nonetheless, it is now a feature often provided with remote PU Passthrough Gateways. In a remote, PU Passthrough Gateway configuration, in an SNA-only network, the gateway (e.g., 3174) is attached to the mainframe, typically through a 37xx, via an SDLC link. The LAN-attached full-stack devices supported by the remote gateway are defined to the host not as LAN-attached devices, but as SDLC devices multidropped on the SDLC link alongside the gateway. Thus, the mainframe side (i.e., ACF/NCP in the 37xx) assumes that it has to poll each of these devices individually.

Individually polling each device would be very inefficient in terms of optimum link utilization and could severely degrade performance especially if there are LAN-attached devices involved. To avoid the need for individual polls, the gateway can support a Group Poll Address. When polled with the Group Poll Address, the gateway can respond with data from any of the LAN-attached devices. Thus, rather than having to iteratively poll individual devices the mainframe can just issue Group Polls on a regular basis.

Whereas PU Controller Gateways have two major deficiencies, Passthrough Gateways suffer from only one major weakness, this being the need for each LAN-attached device to be individually defined to the mainframe, that is, to ACF/VTAM and ACF/NCP. Even with some of the new dynamic definition techniques now available with ACF/VTAM, since Version 3 Release 2, this need for individual definitions could prove to be daunting in a network with thousands of LAN-attached devices. PU Concentrator Gateways were developed to alleviate this need to individually define each LAN-attached device.

The cost of implementing a full stack of SNA on each PC/workstation is also, obviously, an issue even though some may argue that the advantages of this approach outweigh the cost penalty. This is indeed a situation of having to pay for what you want and get. If end-to-end SNA is required, there is really no other way to get around having to have an SNA node within each downstream device. PU Concentrator Gateways also require full-stack implementations. The good news, however, is that the cost of full-stack SNA implementations is coming down to near commodity-item pricing (i.e., under $200)—now that the development cost for much of this software has been recouped. Therefore, while there is no doubt that any full stack is inherently costly, the actual cost is coming down to take away some of the sting.

2.3.4 PU Concentrator Gateways

PU Concentrator Gateways can be thought of as consisting of the best features of both PU Passthrough and PU Controller Gateways. With this type of gateway, SNA nodes are implemented both in the Gateway as well as in each downstream LAN-attached device, as shown in Figure 2.7.

A PU Concentrator gateway appears to the downstream LAN-attached devices as being a mini-mainframe. It includes a subset of SNA SSCP functionality, and sets out to capture and control the SNA

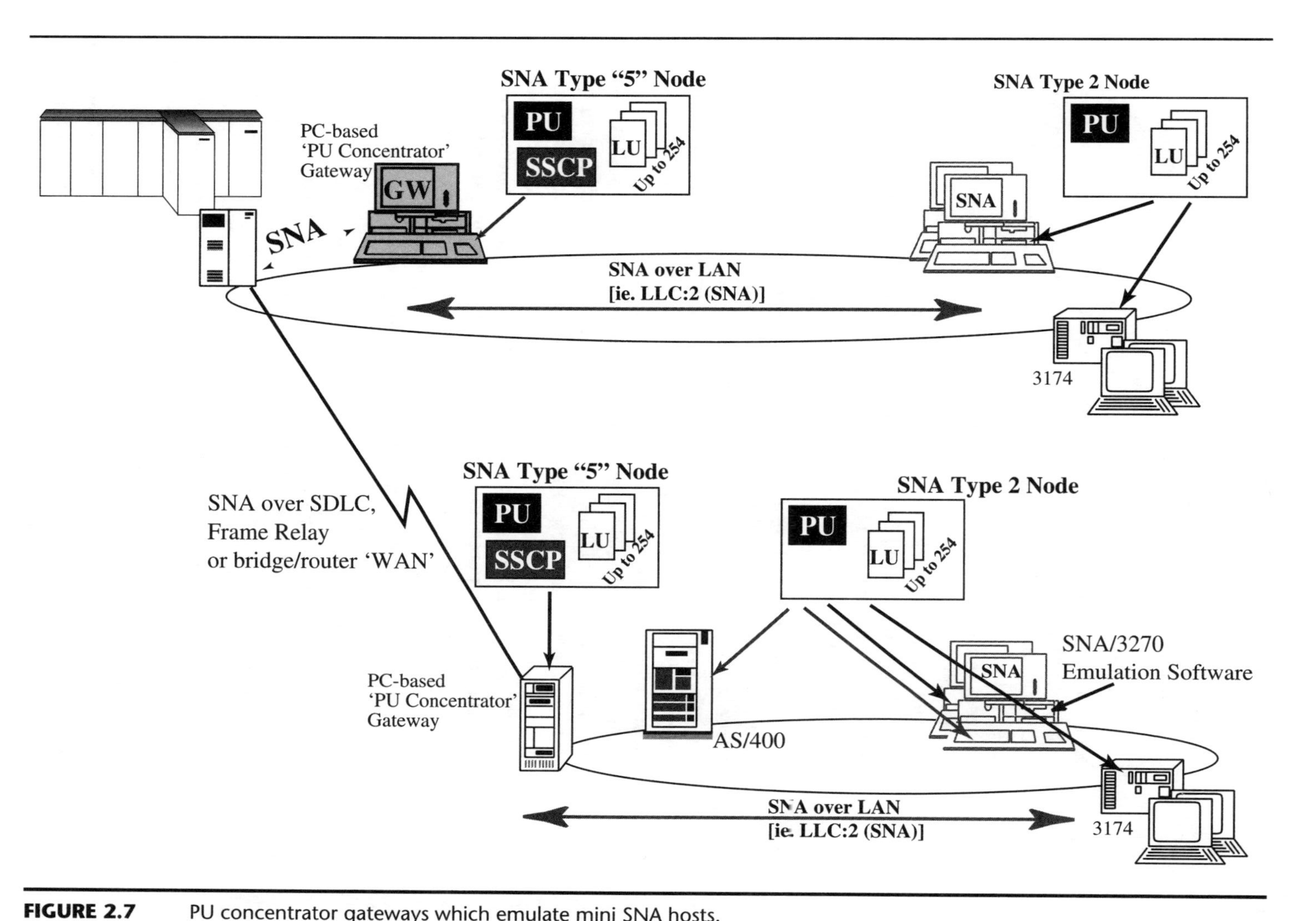

FIGURE 2.7 PU concentrator gateways which emulate mini SNA hosts.

SSCP-related Control Sessions (i.e., SSCP-PU and SSCP-LU) that have to be established and maintained by every downstream SNA device—for example, PCs/workstations with an SNA full stack.

Given that the SSCP function in a PU Concentrator Gateway sits between the mainframe(s) and the LAN-attached devices, the real mainframes only end up seeing the PU Concentrator Gateways (each with up to 254 SNA LUs per the number of SNA nodes it is emulating). The real mainframes do not end up seeing the individual downstream devices. In this respect a PU Concentrator Gateway acts just like a PU Controller Gateway.

However, in marked contrast to Concentrator Gateways, SNA is the only protocol used between the downstream devices and the Concentrator Gateway. It is also obviously the protocol used between the gateway and the mainframes. Consequently, Concentrator Gateways, like Passthrough Gateways, end up supporting end-to-end SNA. *To overcome the presence of the Concentrator Gateway between the mainframes and the LANs, an Alert Relay technique is used to ensure that all network management information from the LAN-attached devices are conveyed end-to-end.* Textual information is included with the Alert to identify the actual downstream device that issued the alert by means of a symbolic name.

In terms of using SNA end-to-end, and ensuring that NetView/390 has full visibility of the LAN-attached devices, a PU Concentrator Gateway looks, and acts, like a PU Passthrough gateway. In terms of the need for mainframe definitions, it looks like a PU Controller Gateway. Thus, it can be thought of as providing the best of both worlds. In reality, the true potential and power of PU Concentrator Gateways goes even further. Just as with PU Controller Gateways, Concentrator Gateways, in addition to their alert relay mechanism, have a contention-based scheme where downstream devices can contend for use of the pool of LUs provided by the gateway.

These pooled LUs in the gateway, in conjunction with the alert relay facility, can be used to enable a large number of actual full-stack SNA devices to gain access to a mainframe. This can be accomplished even though only a relatively few nodes (i.e., the PU Concentrator Gateways themselves) have been defined to the mainframe. *In essence, this type of gateway, true to its name, permits a large number of SNA nodes (i.e., downstream PUs) to be concentrated on—or mapped onto—a smaller number of SNA node definitions.*

This concentration is based on the fact that each SNA Type 2 Node can support 254 LUs. In reality, however, a downstream SNA device is unlikely to use much more than 10 LUs, particularly if it is a single-user PC or workstation. Even though an SNA device may only need to use 1 to 10 LUs, given that each device is a bona fide SNA node, each such device from a mainframe perspective is no different to another node that does intend to use all 254 LUs. This could become an issue if there are situations where the mainframe is willing to support only a limited number of SNA nodes. This situation used to arise when trying to use a 3172 as a LAN-to-mainframe gateway.

Due to initial memory limitations, IBM restricted the 3172 to being able to support only 1,020 downstream SNA nodes. In reality, the limitation was on a per LAN adapter basis. A single 3172 LAN adapter could only support a maximum of 255 SNA nodes. Thus, in order to accommodate 1,020 downstream nodes, a 3172 would require its full complement of four LAN adapters. Support for such a relatively small number of SNA nodes could be extremely restrictive especially with PCs/workstations running full-stack SNA. Even though each PC/workstation may be using only a few LUs, each PC/workstation counts as one node. PU Concentrator Gateways can be used to overcome such limitations. With PU Concentration, only the gateway node, with a full complement of 254 LUs, would count as a single node, rather than each PC/workstation. If each PC/workstation was using only one LU, which might be the case if only 3270 emulation is being used, there is an immediate 1-to-254 multiplier factor. *Now, a 3172, rather than being able to support only 1,020 PCs/workstations, could in theory support the work flow of 259,000 PC/workstations!*

With cost-effective LAN-to-mainframe gateways such as the 3172 gaining increasing attention, the importance of the PU Concentration capabilities of this type of gateway will continue to grow. Recognizing this, and in order to also make their built-in channel-attached mainframe interface even more compelling, Cisco now provides this type of PU Concentration capability within its bridge/routers! Having this type of PU Concentration capability within a bridge/router has other vicarious advantages. An automatic and obvious byproduct of node concentration is the reduction of the number of SSCP-related control sessions that have to be supported across the network. Rather than having one SSCP-PU session for each of 254 PCs/workstations, it would now be possible to have one such SSCP-PU session serving that same population. The availability of alert relay ensures that this reduction in SSCP-PU sessions, does not, however, compromise NetView/390 visibility. The reduction of such control sessions could noticeably speed up network activation in large networks.

PU Concentrator Gateway implementations that are now available include: IBM OS/2 Communications Manager/2, IBM's PC/3270, Novell's NetWare for SAA, Microsoft's SNA Server, and Cisco bridge/routers.

2.4 THE MYSTIQUE OF SNA ROUTING

Contrary to any and all widespread assertions, SNA is eminently routable. Just as with any other protocol (e.g., IP or IPX) the appropriate routing software is required in order to be able to route SNA. SNA's equivalent of routers—the 37xx communications controllers with IBM's ACF/NCP software—have been routing SNA traffic between 3270 terminals and applications resident in geographically dispersed IBM mainframes since 1979.

SNA routing, also referred to as SNA Subarea routing, is a fixed path (i.e., static) routing scheme-based on manually predefined routes specified via ACF/VTAM and ACF/NCP PATH tables. The fixed route to be used by an SNA session is established when the session is being initiated-based on the COS associated with that session.

Neither multiprotocol bridge/routers nor RFC 1490 routers, irrespective of vendor and despite any claims stating otherwise, can perform SNA Subarea routing. This is an unequivocal fact: Today's routers just do not have the software to enable them to support SNA routing. The best that today's routers can do is to route SNA traffic or perform APPN NN routing. Routing SNA traffic is not the same as SNA routing. Routing SNA means that the router transports SNA traffic between an SNA device and an SNA LAN gateway (e.g., a 37xx) using a Layer 2 routing mechanism such as bridging, DLSw, or RFC 1490 encapsulation. SNA per se works above Layer 2, that is, SNA is independent of, and operates above, the Data Link Control layer (e.g., SDLC, LLC:2, mainframe channel).

SNA routing per se is best thought of as occurring at Layers 4 and 5. The Layer 2 transport mechanisms used by routers to route SNA are thus outside the scope of SNA, and are essentially transparent to SNA. These transport mechanisms invariably use the MAC address of the destination gateway as their routing metric, as was shown in Table 1.1. MAC addresses, being Layer 2 entities, are well outside the domain of SNA and SNA-based routing. Hence, the end-to-end routes used by routers to transport SNA traffic in networks with multiple mainframes may differ, sometimes very significantly, from those that would have been used if SNA routing was used. The end-to-end routes used by the routers as previously discussed may involve two-hop routing. SNA traffic, however, will in general still reach its intended destination—but in some instances via an unanticipatedly circuitous route.

SNA routing only occurs between subareas. An SNA subarea is formed around a mainframe with an SNA Type 5 node or a 37xx with a Type 4 node. ACF/VTAM in a mainframe implements a Type 5 node while ACF/NCP in a 37xx implements a Type 4 node. Thus, SNA routing takes place between: 37xx's, 37xx's and mainframes, or mainframes. No SNA routing takes place between peripheral devices (e.g., a 3174 control unit) and a 37xx—given that a single link, whether point-to-point, multipoint, or through an X.25/frame relay packet switching network, is the only path between these two types of node.

SNA routing is based around SNA subareas and SNA Virtual Routes (VRs) between subareas. An SNA subarea can only be formed around an SNA Type 5 or Type 4 Subarea Node. For a router to be able to perform SNA routing it would have to contain the software that would implement a Type 4 node. To its credit, Cisco, recognizing the role of SNA routing, explored the possibility of implementing a Type 4 node within its bridge/routers back in 1991 as a part of its original five-phase SNA internetworking strategy. There is a significant amount of real-time processing, buffer space, and memory for software code required to realize a type node. In addition, some of the newer protocols used by subarea nodes are not in the public domain. Thus, implementing a Type 4 node on a Cisco bridge/router, particular during 1991–1992 when Cisco was still saddled with a single-processor scheme, was far from easy. Recognizing the difficulties, Cisco abandoned this initiative in early 1992. Since then no other bridge/router vendor has embarked along this route of trying to implement a Type 4 node within a bridge/router.

2.4.1 Going Across Europe by Crossing the Atlantic Twice

Figure 2.8 shows a quintessential example of SNA routing through a remote 37xx. The remote 37xx in Geneva can act as a two-leg Y-switch and route SNA traffic either to New York or London. This routing will be done on an SNA-session-by-session basis. The path to be used by each session will be determined by ACF/VTAM in a mainframe when the session is first being established. The mainframe in which the destination application (e.g., Customer Information Control System/Virtual Storage [CICS/VS] in NYC or Timesharing Option [TSO] in London) is resident will determine the end point of each session—and thus the end points of a SNA route required to support that session.

The route that can be taken from the PC in Geneva to one of the two mainframes consists of, and is specified in terms of, a series of hops between subarea nodes—for example, Geneva 37xx to London 37xx to the London mainframe. Once the subarea of the destination application has been located the route used to reach it would be selected from prespecified PATH tables located in the mainframes and all the 37xx's, based on a COS associated with that session. Figure 2.9 shows a representative view of all the various table lookups that are necessary to establish an SNA Route.

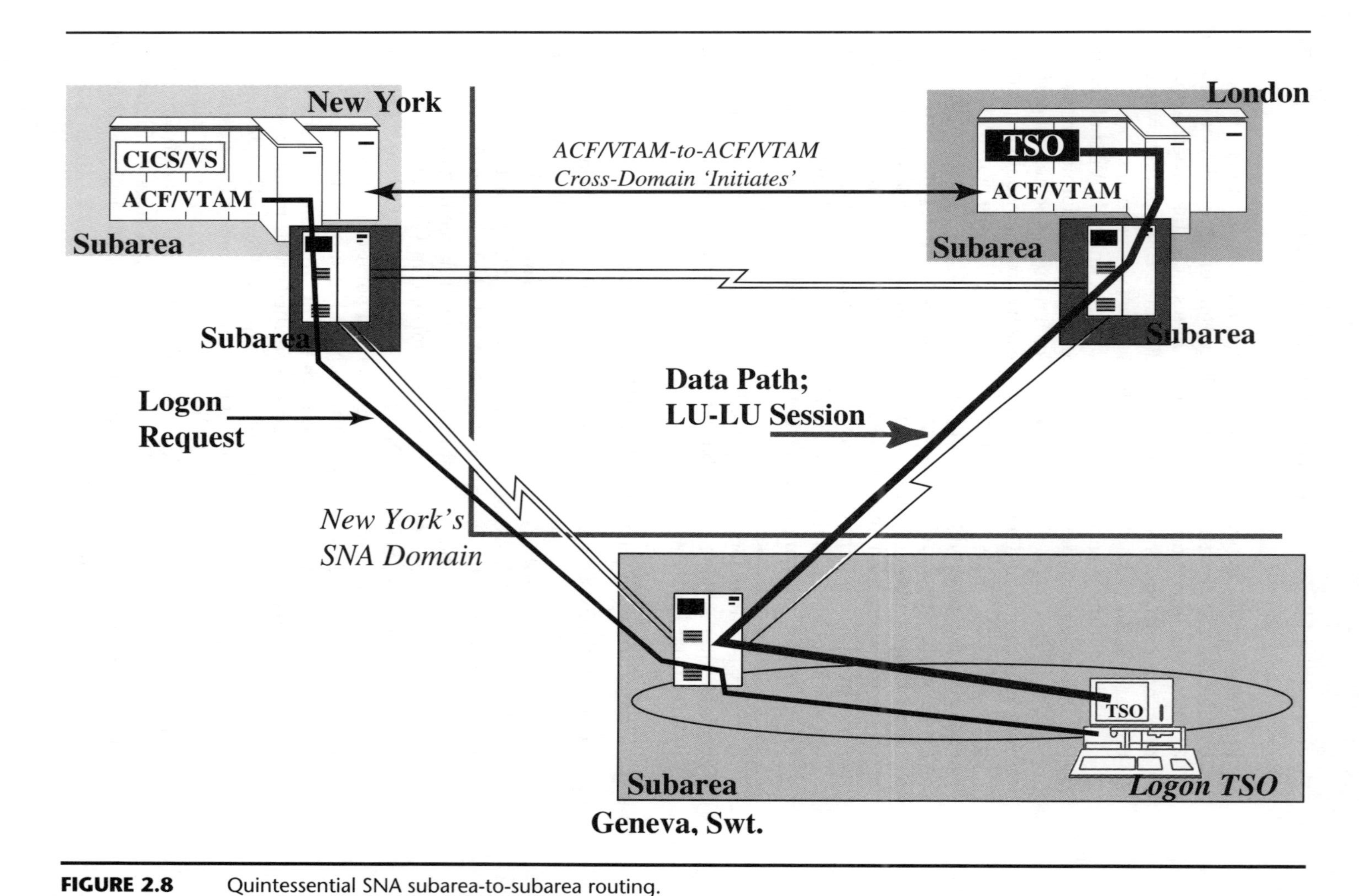

FIGURE 2.8 Quintessential SNA subarea-to-subarea routing.

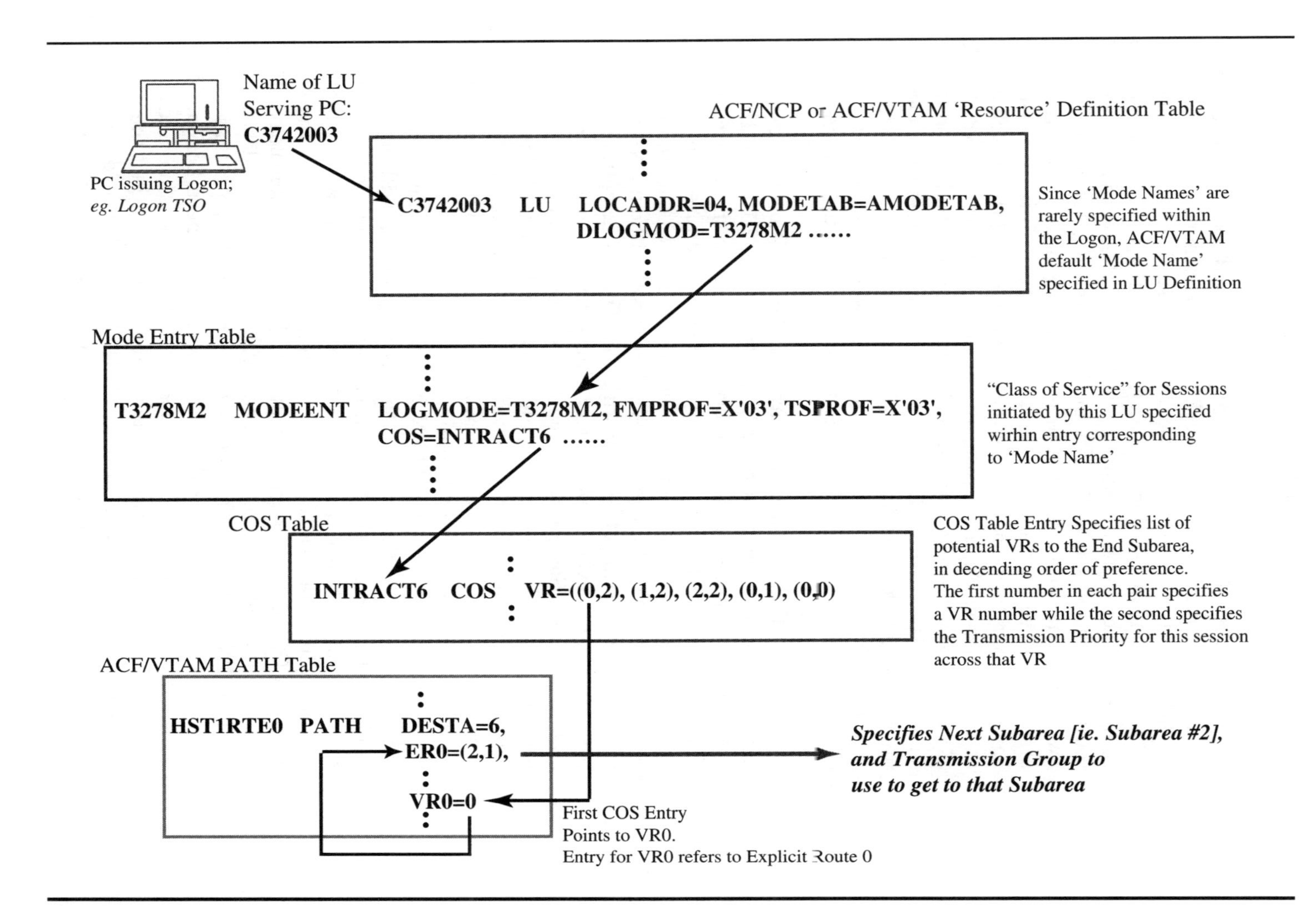

FIGURE 2.9 The table lookups involved in establishing an end-to-end path in SNA.

Figure 2.10 illustrates what happens if the 37xx in Geneva is discarded and a router-based network that uses either bridging, DLSw, or RFC 1490 to support the SNA traffic is installed between Geneva, NYC, and London. Sessions to London will still get there, but they do so by going across the Atlantic twice! It is a classic case of two-hop routing in the absence of SNA or APPN routing, or SNA Session Switching. The reason for this two-hop routing is the loss of the subarea in Geneva. Without this subarea, the SNA devices in Geneva have to belong to one of the subareas formed around the two channel-attached 37xx's. Figure 2.10 shows Geneva being in the New York 37xx's subarea. Geneva could have been put inside the London 37xx's subarea. In that case, a direct Geneva-to-London route that does not involve going across the Atlantic twice would have been possible when accessing TSO. This, however, is not the answer to the problem. In this configuration, accessing CISC/VS in New York would involve going through London.

Also note that in each configuration, there will be an interrouter path that is not being used by SNA. In Figure 2.10 it is the link between the Geneva and London bridge/routers. In the alternate configurations it would be the link between Geneva and New York.

The crux of the matter is that you need an SNA subarea node in bridge/routers in order to get true SNA routing. Without SNA routing, SNA traffic will still get to its destination. The traffic patterns across the network, however, could be very different from what was expected and planned for. Some routes might have more traffic than was budgeted for them while others may not have any SNA traffic at all. And this is a case where being forewarned allows one to be forearmed. When designing bridge/router networks always remember that bridge/routers cannot perform SNA routing.

When planning to use APPN NN routing in a multimainframe network it is imperative to note that APPN routing will not exactly replicate the traditional SNA subarea-to-subarea routing. Subarea-to-subarea routing takes place across fixed, predefined paths. APPN in contrast will dynamically select the optimal path based on the nine-criteria COS and the current state of the network topology. This means that a path selected between a given source and a destination using APPN routing might not be the same as the path that would have been used with SNA. For example, in the Geneva-London-New York network shown in Figures 2.9 and 2.10, there is no guarantee that APPN NN routing would have automatically selected the direct Geneva to London router to reach TSO. This path would have only being selected if it met the COS criteria associated with that TSO session.

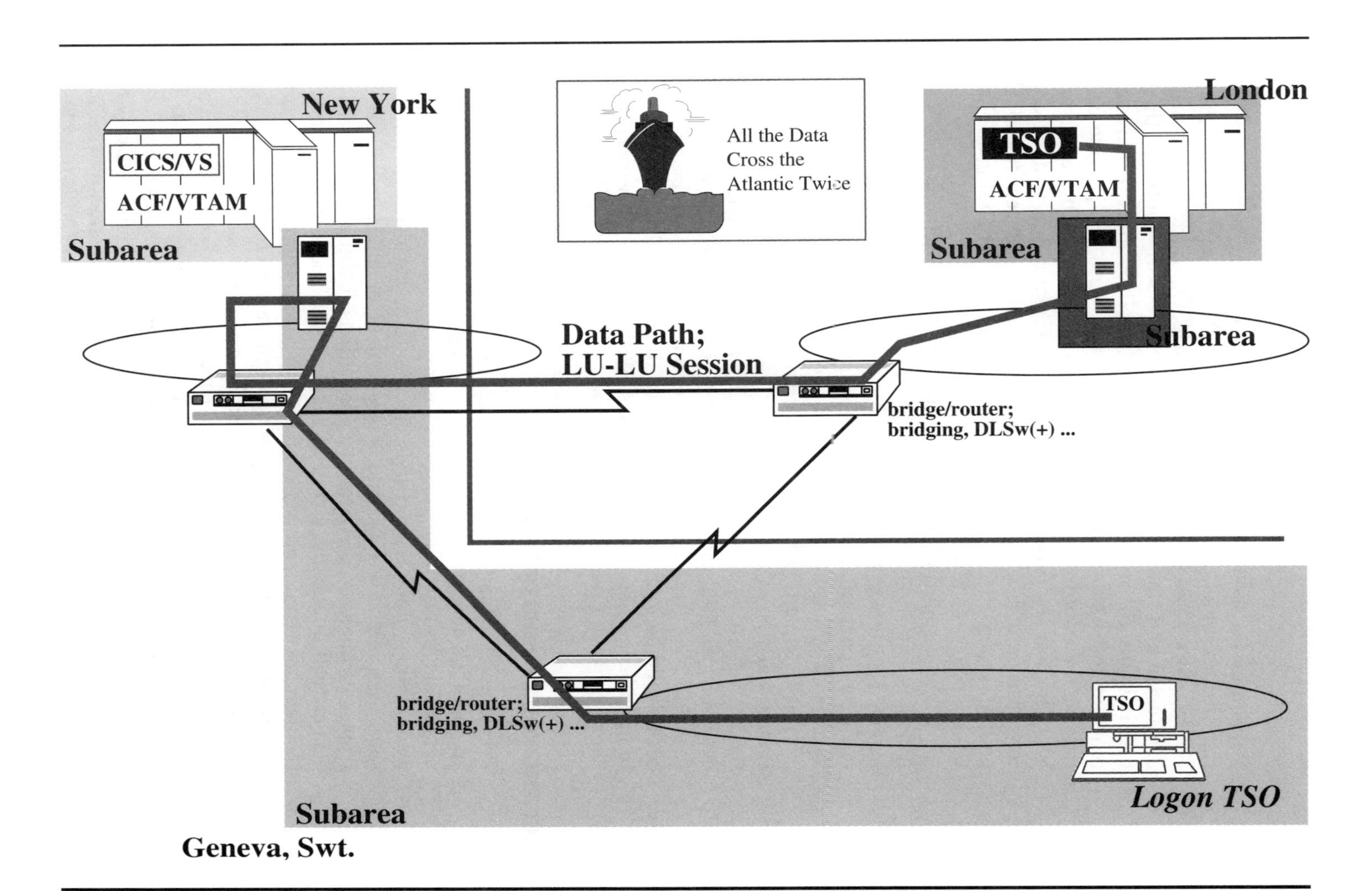

FIGURE 2.10 Two-hop routing in the absence of SNA routing.

Depending on the criteria specified, and the actual characteristics of the various links, it is possible that APPN may have selected the Geneva-New York-London route to reach TSO, for example, if cost was a criterion and this happened to be the least costly route. (In this context it is interesting to note that the APPN architecture, in marked contrast to LAN protocol routing schemes, does not provide criteria that can be used to specify the selection of a path with the least number of hops or the theoretically fastest path, bar congestion, based on a combination of link speed and number of hops.) Moreover the APPN path may vary each time a new session is set up depending on the network topology at that instance.

This incompatibility between the two routing schemes of SNA and APPN is not really a major issue provided that network designers involved in network reengineering are cognizant of it. However, they must pay particular attention when provisioning bandwidth for the various APPN routes across the network. If a new multiprotocol network is designed with the assumption that APPN will use the same routes that SNA would have used, the actual traffic patterns within the network may prove to be very different. Some routes will experience overload while others will hardly be used. The only way to ensure that APPN routing follows the same paths that were being used by SNA would be to manually customize all of the APPN NN COS tables to provide the necessary weightings. With a large network, this may not be a trivial task.

2.5 GETTING YOUR PRIORITIES RIGHT

Predictable, consistent, and in general fairly swift response times are an indelible feature of SNA networking. In general when an SNA network is first installed, and subsequently whenever there is a significant change made to that network, Systems Programmers and Network Administrators expend a great deal of effort to ensure that response times meet stringent expectations, because with most mission-critical applications time is money. Significant changes made to the network could include expanding its reach, adding a large number of new users, adding a major new application, and topology or link configuration. Slow response times can be a major cause of tangible lost-opportunity costs in that fewer transactions are processed over a given period of time, and in the case of reservation systems (e.g., car hire) customer dissatisfaction results in loss of repeat business.

Fine-tuning SNA networks, though an art, is comparatively easy. Well-established and proven tools are available to measure traffic volumes (e.g., IBM's NetView Performance Manager [NPM]) and simulate potential network performance (e.g., IBM's Teleprocessing Network Simulator [TPNS]). In addition, a plethora of parameters is provided by ACF/NCP and ACF/VTAM to fine-tune link polling (e.g., SDLC), SNA pacing (i.e., SNA's congestion control scheme), acknowledgment window size, segmentation (i.e., splitting a long message unit into smaller packets), data blocking (i.e., bundling together small message units into a single packet), and various buffer size values.

Provided adequate link bandwidth to handle the anticipated traffic volumes on a given link, these ACF/NCP and ACF/VTAM parameters can be iteratively modified until the desired response levels are achieved. To further facilitate response time management, most 3174s and lookalikes, as well as some of the SNA LAN Gateways, have a Response Time Monitor (RTM) feature. RTM can be used to obtain accurate histogram-based maps of the end-to-end response time—that is, terminal-to-application-back-to-terminal—experienced by that control unit or gateway over a given period of time.

Maintaining the required response time characteristics of SNA applications becomes more challenging when the SNA traffic is moved over from an SNA-only network to a multiprotocol network. The main problem is that SNA traffic no longer enjoys exclusive use of preallocated, dedicated, and finely tuned bandwidth, not just on the WAN but also on individual LANs. Instead, SNA traffic may now have to contend for bandwidth with high-intensity, high-volume LAN applications. This contention for bandwidth will begin to occur on the LAN, when an SNA device will have to wait its turn to receive the "permission-to-transmit" token on a token-ring LAN or wait for a chance to gain control of an ethernet LAN, so that it can send its message units to the LAN-to-WAN "gateway" unit (e.g., a bridge/router). In general this has not been too severe a problem, as yet, given that the token-ring LANs used by most SNA-oriented enterprises are rarely saturated. Moreover, the token passing polling scheme used by token-ring LANs typically ensures smooth and acceptable performance levels until LAN usage reaches 85% or so of capacity. (With Ethernet LANs, which use the Carrier Sense Multiple Access with Collision Detection [CSMA/CD] technique for controlling LAN access, performance degradation may begin to occur even before the usage reaches 50% of capacity.)

The nascent LAN switching technology, for example, token-ring switching, as described in Chapter 4, will provide a cost-compelling and nondisruptive means of improving LAN performance, if and when this becomes a major issue. The other option, which will require much more reengineering and is likely at least initially to be more costly, is to consider using broadband ATM, say 155Mbps, at the campus level instead of conventional LANs.

2.5.1 Avoiding Bandwidth Starvation Pains

The contention for WAN bandwidth can be considerably more severe than that for LAN bandwidth. Today's multiprotocol WANs typically have at least one order of magnitude less bandwidth than LANs (e.g., 1.544Mbps with T1 compared to 16Mbps token-ring LAN). While widespread and cost-effective broadband ATM will eventually cure this problem, this is, unfortunately, unlikely to happen for a few years yet. Multiprotocol WANs can thus prove to be the graveyard for SNA response times, with SNA traffic getting buried behind large volumes of non-SNA traffic waiting to cross the WAN.

One obvious way to mitigate this problem would be to filter out the SNA message units when they reach the LAN-to-WAN gateway, and then queue them ahead of the other traffic for transmission across the WAN. If SNA traffic was given such preferential queuing at each hop along the way across the WAN it would reach its destination faster than if it had to continually wait behind other traffic before it was transmitted. *Bridge/routers and RFC 1490 routers now provide relatively powerful and flexible traffic prioritization schemes that can be customized to provide exactly this type of preferential, priority queuing for SNA traffic.*

These traffic prioritization schemes, however, are not geared or designed exclusively for SNA—though SNA was the primary target and the main benefactor. They are generic schemes that can be used to give any protocol or a class of message units (e.g., TCP/IP telnet traffic) preferential priority over another protocol or another class of message units (e.g., TCP/IP File Transfer Protocol [FTP] traffic). In this context, SNA or certain types of SNA message units (e.g., those flowing between 37xx's with an SNA Format Identifier [FID-4] Type Transmission Header), are just instances of another protocol and a class of message units that can be prioritized relative to other protocols and message units.

Today there are two distinct schemes available for multiprotocol traffic prioritization:

1. Assigning High, Medium, or Low priority to various types of traffic and having three queues corresponding to the three priorities for the traffic waiting to cross WAN links. With this scheme, all the traffic queued against the high-priority queue will be transmitted first before the medium-priority queue is serviced. With this type of scheme there is a danger that in the event of high volumes of high-priority traffic, the other traffic will not get serviced and languish for excessive periods of time on the medium and low queues.

In essence, the medium- and low-priority queues end up getting starved of WAN bandwidth while the high queue uses up all the available bandwidth. This type of scheme is referred to as providing *unequitable bandwidth allocation*. The way to overcome this problem is to have some means of guaranteed bandwitdh allocation.

2. Bandwidth allocation schemes, typically with 10 queues or more, where each queue is guaranteed a minimum amount of bandwidth over a given period of time (e.g., 15% of bandwidth every 2 seconds). Bandwidth unused by one queue can be used by another to preclude any bandwidth from going unused. With this type of scheme none of the queues end up being starved on bandwidth. Bandwidth allocation is obviously a better and more powerful scheme than the high-, medium- and low-priority scheme described above. Bandwidth allocation schemes are now slowly but surely becoming the standard means for multiprotocol traffic prioritization. Most of the leading bridge/router and RFC 1490 routers now have bandwidth allocation schemes. Some offer it alongside the older three queue schemes to cater for backward compatibility and phased migration.

2.5.2 Preferential Treatment for SNA Traffic

It is theoretically possible to have two very different types of SNA/APPN traffic within a multiprotocol LAN/WAN network, these two types of traffic being:

1. Traffic with SNA FID-2 Type Transmission Headers (THs) flowing between either SNA Type 2/2.1 Peripheral Nodes and an SNA Subarea Node Gateway (e.g., 3745), or peer-to-peer APPN/Type 2.1.
2. Traffic with SNA FID-4 Type THs flowing between remote 37xxs and channel-attached 37xxs (or a mainframe).

Figure 2.11 depicts these two potential scenarios for SNA traffic prioritization within multiprotocol networks.

Traffic flowing between Type 2/2.1 nodes (e.g., 3174s or PC/workstations) and mainframe gateways (i.e., 37xx) is by far still the most common type of SNA traffic that will be encountered in multiprotocol networks that have mainframes attached to them. Unfortunately, however, SNA per se does not assign any kind of transmission priority to this type of traffic. The FID-2 THs that prefix such traffic do not contain a Transmission Priority Field (TPF). (Neither does the new HPR FID-5 TH.) The SNA TH types are described in Appendix A.

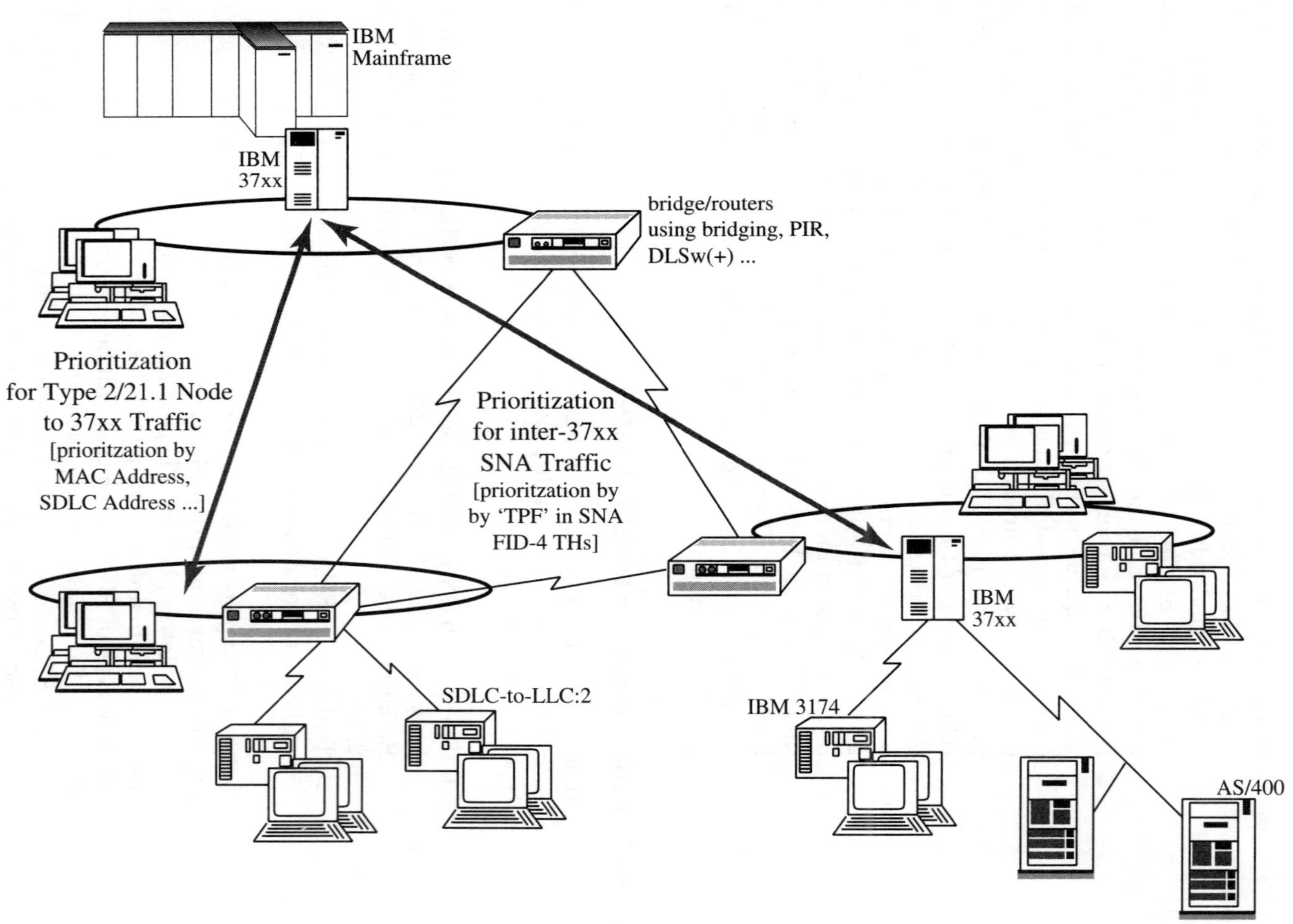

FIGURE 2.11 Two distinct scenarios for SNA traffic prioritization.

FID-4 THs used to prefix all traffic between subarea nodes, however, do contain a two-bit TPF (Byte 3; bits 6 & 7) that is set to indicate high, medium, or low priority for the SNA message unit they prefix. The TPF settings in the FID-4 TH are derived from the transmission priority assigned to each session—relative to the VRs that they happen to be using. This transmission priority is assigned via the COS associated with every session through the Mode Name specified for that session (usually by default) when a session is being initiated. Figure 2.9 shows how the Mode Name associated with a given session serves as a pointer to a series of table lookups that eventually provide the TPF for that session. Cisco and CrossComm among others can actually read and react to this FID-4 TPF when prioritizing traffic flowing between 37xx's.

Given the absence of an SNA-prescribed means for determining the priority of traffic flowing between Type 2/2.1 nodes and subarea nodes, auxiliary criteria such as Media Access Control (MAC) Addresses have to be used if there is a need to prioritize creation types of SNA traffic relative to other SNA traffic, rather than just against other non-SNA traffic.

All of today's traffic prioritization schemes use protocol type as the first, or master, "sieve" that determines the relative precedence of traffic queuing. A priority (high, medium, or low), or, in the case of bandwidth allocation, a minimum guaranteed percentage of output WAN capacity, is assigned for all the traffic corresponding to given protocol type. Thus, for example, SNA can be specified as having high priority (or a minimum bandwidth allocation of 40%), TCP/IP and IPX as having medium priority, and NetBIOS as having low. A default, or "catch-the-rest," definition is used to cater for protocols that may not have been assigned a priority by the Network Administrator. This simplifies the amount of manual definition that needs to be done.

This first-cut by protocol type prioritization is not optimal. It, however, has the virtue of being easy to define and implement, and somewhat intuitive to understand. Ideally, transmission priorities for a given LAN should be specifiable first on a per user (e.g., MAC Address) basis, and then on a per-protocol basis relative to each user. Such a scheme, in addition to being more cumbersome to define, requires more "on-the-fly" processing for each input frame before it can be queued for transmission across the WAN.

Most of today's prioritization schemes permit traffic using the same protocol be further prioritized according to other criteria. Such schemes, in theory, can be used to provide preferential treatment for set select users or applications. These schemes are not intended as a mechanism whereby each user is given a specific priority for each type of protocol. Trying to do so for multiple protocols, even on a LAN with only 30 PCs, could be cumbersome, frustrating, and in the end most likely counterproductive.

When considering traffic prioritization schemes it is always important to remember that the processing power and time taken to deal with prioritization always increase the time taken to forward packets from the LAN onto the WAN. This will be particularly true if elaborate hierarchical criteria are being used. In some cases, especially when T1 or greater WAN links are being used, and the traffic volumes are not that great, it might be prudent not to activate prioritization at all! Thus, when considering means of doing traffic prioritization always note that a complex prioritzation scheme may end up slowing all the traffic down.

Certain criteria relative to a specific protocol—such as a MAC Address, SNA Local LU Address, or TCP/IP application (i.e., port)—can typically be used to provide prioritization within a given protocol. If no prioritization criteria are defined for a specific protocol, which will always be a prerogative of the network administrator, all traffic associated with that protocol will be queued First-In, First-Out (FIFO)—on the queue associated with that protocol.

Trying to provide prioritization within a protocol can be limiting when used with bandwidth allocation and should not be contemplated. If such a technique is to be used, each preferred user needs to be assigned a minimum amount of guaranteed bandwidth—that is, be assigned to a separate queue. Trying to generalize this scheme could result in having to support a huge number of queues. Most schemes today do not provide more than 30 queues. Even 30 queues does not really provide enough scope for having preferred users per protocol. A better approach is to allocate bandwidth just for each protocol—that is, just have a queue per protocol, with one or two additional queues to handle ultra-high priority network management-related traffic (e.g., SNMP).

2.5.3 Jostling for Priority Within the Same Protocol

If it is deemed that prioritization within a given protocol is imperative, there is a variety of techniques that can be used to realize this. Yet, again it should be noted that complex prioritization may end up doing more harm than good. Thus, prioritization within a protocol should be used judiciously and with prudence—and preferably a willingness to abandon it all together if it does not deliver the desired results.

Prioritization by MAC address can be used, effectively and meaningfully, with and across all protocols. Another universal, potentially powerful, and protocol-independent scheme is prioritization by *Pattern Matching*. With this scheme a bit mask up to 60 to 80 bytes long is applied to the front of every frame to select the relevant frames. In general, however, network administrators should not be expected to use this scheme without support from the router vendor or "cookbook" of appropriate masks per protocol.

With bandwidth allocation, Cisco for one permits TCP/IP traffic to be prioritized by port number—that is, TCP/IP application type. Thus, for example, SNMP traffic can now be prioritized relative to FTP traffic. This is a valuable and powerful feature, which will no doubt have major appeal to most TCP/IP customers. Other types of criteria could also be meaningfully used with TCP/IP, such as destination IP-address.

The typical prioritization criteria now available for SNA traffic include the following:

1. By MAC Address.
2. By Local, Type 2 Node LU Address (i.e., Origin Address Field [OAF] or Destination Address Field [DAF] field, depending on direction of flow, in the FID-2 TH) for a given MAC address—with this scheme it would be possible to assign specific priorities to individual LUs on a 3174, AS/400, SNA LAN gateway, or even a single OS/2 PC using Communications Manager/2. However, this scheme will work successfully only if the router doing the prioritization is willing and able to resequence all of the MAC frames being prioritized. To resequnece the frames the router assigns them new LLC:2 sequence numbers. Many routers do not support this type of resequencing. Without resequencing, this type of prioritization can result in chaos and session outages because there is a constant possibility that message units will be delivered to the destination with their LLC:2 sequence numbers out of order!
3. By SDLC Address, and by LU Addresses relative to a given SDLC Address.
4. Through pattern matching, which could be used to select certain sets of SNA DAF/OAFs Local Addresses using the appropriate masks.
5. By TPF settings in FID-4 THs in the case of inter-37xx traffic.

Prioritization by Destination Application Name—prioritization between CICS, TSO, and Database/2 (DB2) traffic—would be a very pertinent and powerful scheme that many enterprises would wish to use. Unfortunately, there is no explicit support, as yet, for being able to specify application names as criteria for prioritization. The best that can be done at present is to use the DAF/OAF pattern matching scheme mentioned in item 4 above.

2.5.4 Resequence or Perish

When prioritizing traffic within SNA as discussed above, there are two scenarios during which a stream of SNA message units issued by a device with single MAC Address (e.g., 3174) may have to be prioritized prior to transmission over the WAN. These two scenarios are as follows:

1. Inter-37xx traffic being prioritized by FID-4 TH TPF settings.
2. When individual LUs within the same device (e.g., 3174), all of which have separate sessions to the same destination (e.g., mainframe), are assigned different priorities.

The only way to deal with either of the above scenarios is to resequence the MAC/LLC frames—that is, repackage the SNA message unit inside a new MAC/LLC frame. Otherwise, there could be situations where MAC/LLC frames, each with their own LLC:2 sequence numbers, end up being delivered to the destination out of order, because Layer 2 reordering was done without regard to the LLC:2 sequence numbers.

This type of out-of-sequence delivery is, obviously, not desirable. The destination will discard the out-of-order frames and demand a retransmission. The retransmitted frames, because of prioritization, could also end up more than likely being out of order! This could result in a stalemate with no traffic getting through. Unfortunately, this type of gridlock has already happened, albeit at just a very few sites, when inter-37xx TPF prioritization schemes were first tried out. (Contrary to expectations, Local LLC:2 Acknowledgment, a technique supported by nearly all routers, by itself does not solve this need for resequencing.) If resequencing is not supported, do not attempt to implement this type of SNA prioritization scheme.

Lack of resequencing is not the only type of problem that may be encountered with traffic prioritization schemes. Anomalies could also occur if a WAN consists of multiple intermediate nodes and each node does its own traffic prioritization. With some bandwidth allocation schemes, each node along the end-to-end path may end up reprioritizing the traffic according to the way the queues have been defined at that node—rather than at the source node. Thus, it is possible to have situations where traffic prioritized according to one set of rules at one node (e.g., with SNA getting the most bandwidth) could get reordered according to a vastly different set of rules at an intermediate node (e.g., with NetBIOS getting precedence over SNA).

The only way to ensure consistent prioritization across the network is to make sure that all the bandwidth allocation queues are defined identically at each node. In a medium to large network, maintaining this integrity across all the nodes may be somewhat difficult.

2.6 SUPPORTING SNA TGS ACROSS A MULTIPROTOCOL NETWORK

Deploying a multiprotocol LAN/WAN network, essentially inside an SNA subarea network made up of multiple 37xx Communications Controllers, appears at first sight to be incongruous, and possibly even redundant. Figure 2.12 illustrates the notion of having a bridge/router or RFC 1490 router-based multiprotocol network inside a 37xx network. It should also be noted that the right-hand side of Figure 2.11 also depicts this type of routers within an SNA network. There could be some scenarios where there may be a need for this type of hybrid network—albeit in many cases just as a short-term, tactical solution.

This type of hybrid network does have the indubitable distinction of providing multiprotocol capability while preserving certain key SNA traits—true subarea-to-subarea SNA routing being key among these. In addition, the 37xx's, especially link-attached RCPs, could be used as an effective, but relatively expensive, way to provide proven support for downstream SDLC, BSC, and even asynchronous links. It could even possibly provide a means for extending NetView/390 visibility and control further into the periphery of the network.

The FID-4 TH traffic prioritization schemes for inter-37xx traffic discussed above were obviously designed for this type of hybrid network where 37xx-to-37xx traffic would be flowing across a multiprotocol network. A complementary facility available from Cisco and CrossComm for this type of hybrid network is the ability to explicitly support SDLC-based SNA Transmission Groups across the multiprotocol network.

SNA TGs have been an intrinsic and powerful fault-masking, resilience-inducing feature of most large SNA subarea backbone WANs since 1979. TGs permit connections between subarea nodes, particularly those between 37xx communications controllers, to be implemented using multiple, parallel links. The multiple, parallel links will appear and act as a single, consolidated pipe that provides that intersubarea node connection with the combined aggregate bandwidth of all of the links that together form the TG. Thus, for example,

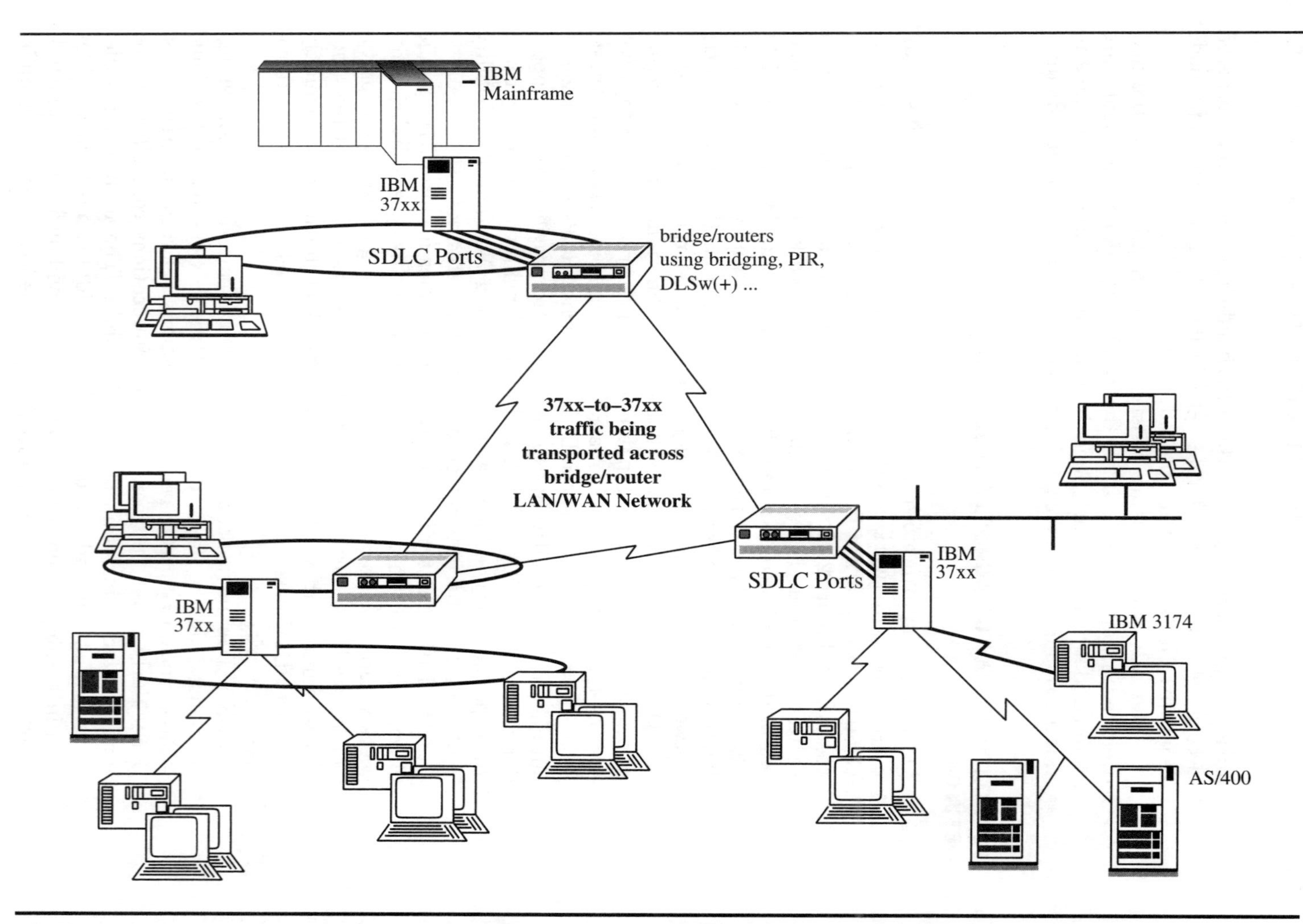

FIGURE 2.12 The notion of an SNA 37xx subarea network that is interconnected across a bridge/router network.

if four 19.2Kbps SDLC links were used between two 3745s to form a TG, that TG would essentially act as a single connection with 76.8Kbps of bandwidth (i.e., 4 × 19.2Kbps).

Moreover, a TG will continue to be fully active and productive provided that at least one of the links forming it is operational. Thus, with a multilink TG, single or even multiple link failures will only degrade the bandwidth available for the TG, rather than its operational status. Single or multilink failures on a TG thus do not disrupt the sessions flowing across that TG—apart from obviously slowing them down because of the reduced bandwidth. A session disrupting path failure will only occur if all the links forming a TG fail at the same time.

It is TGs that have permitted SNA to get away with its one conspicuous Achilles heel—its continued inability to support dynamic alternate routing in the event of path failures. By the judicious deployment of multilink TGs users can avoid catastrophic path failures and at the same time guard against failures to single links. TGs thus permit multiple links to be grouped together to form a single pipe that provides the combined bandwidth of the links being aggregated, resilience against single-link (or even multilink) failures, and continuous, dynamic load balancing across the multiple links.

An error occurring on one link will automatically cause the affected frame to be retransmitted on another link. However, SNA will continue, up to a predefined retry count, to attempt to transmit the subject frame across the original link. This multilevel retransmit scheme permits SNA to determine, for network management purposes, if the error was transient, permanent, or even dependent on the makeup of a single link (i.e., the so-called context-sensitive errors) without delaying the delivery of the subject frame. In the event of errors occurring on multiple links with the same frame it is possible to have scenarios where that frame is being repeatedly retransmitted on multiple links. Duplicate frames received due to this multilevel retransmit technique are discarded and there is a very sophisticated protocol known as a *TG Sweep* to protect against duplicate frames whenever the TG specific sequence numbers (which go from 0 to 4,095) wrap around back to 0.

TGs form the lowest layer of SNA's Layer 3 Path Control Network component within subarea nodes. TGs guarantee an always in-sequence, FIFO, packet-delivery mechanism to the sublayer above. The FIFO aspect is even maintained when the links forming a TG have different link speeds. Implementing TGs with links that have different speeds should be avoided as this could severely degrade the overall performance of the TG. The TG's FIFO sequencing mechanism will continually have to wait for packets sent on the slowest link even though it has already received subsequent packets sent on the faster links.

Up until ACF/NCP Version 6 Release 2, in mid-1993, TGs could consist of up to eight parallel, SDLC links between 37xx's. Up to 16 separate TGs, each consisting of multiple parallel links, could also be implemented between a given pair of 37xx's. With ACF/NCP Version 6 Release 2, it is possible to have Mixed-Media, Multilink TGs (MMMLTGs). With MMMLTGs it is possible to have a TG consisting of a combination of SDLC, token-ring LAN, and FR Virtual Circuits (VCs). In practice, MMMLTGs are likely to consist of either parallel token-ring LANs, or FR VCs. (Combining SDLC links, token-ring LANs and FR VCs could be considerably worse than mixing SDLC links of different speeds, as described above, given that a token-ring LAN could be running at 16Mbps alongside a SDLC link running at 56Kbps.) Since ACF/VTAM Version 3 Release 3, in mid-1989, it has also been possible to have up to 16 separate TGs, each consisting of a single mainframe channel connection, between a 3745 and ACF/VTAM running on the mainframe.

Multiple SDLC link-based TGs are often used between 37xx's. Even when deploying a multiprotocol LAN/WAN network between such 37xx's, network administrators may still wish to take advantage of the fault-masking, bandwidth-consolidating, and load-balancing characteristics of a multiple-link TG across the multiprotocol network. This may be particularly attractive if the multiprotocol network contains multiple separate routes between the 37xx's. Figure 2.12 illustrates a network where 37xx's are deployed around the periphery of a bridge/router-based multiprotocol network.

Cisco and CrossComm bridge/routers include a TG support feature to enable TGs to be explicitly supported across a multiprotocol network. With this feature, each SDLC link forming a multiple link TG can be specifically mapped to a separate route across the multiprotocol network in such a way as to emulate the multilink nature of the TG. CrossComm's TG capability is realized through their PIR technology that intrinsically supports facilities such as dynamic alternate routing. Today the TG feature is restricted to SDLC link-based TGs.

Though still on offer, the true applicability of this facility is limited and essentially artificial. For a start, most enterprises are slowly but surely getting rid of remote 37xx's with their high maintenance costs, as they move toward LAN-centric multiprotocol networking. In addition, a TG that is mapped across different routes within a multiprotocol network does not provide the same "not-even-a-hiccup" fault-masking characteristics as a TG using dedicated SDLC links. In the event of a route failure in a bridge/router-based multiprotocol network, the bridge/routers immediately set about reconverging the network to bypass the failed route and establish the new topology.

In a large network consisting of 50 or more bridge/routers this reconvergence process could end up taking a few minutes or more. Typically, most data transfers are suspended during reconvergence. This reconvergence delay minimizes the advantage of having a multipath TG across the network. In the event of any path failure, data transfer across all paths are likely to be interrupted while the network is reconverged. Consequently, in most cases it is just as effective to have a higher-bandwidth single-link TG across the network, and rely on the dynamic alternate routing capabilities of bridge/routers (e.g., by using DLSw) to guard against route failures.

2.7 THE QUEST FOR A COST-EFFECTIVE, MULTIPROTOCOL MAINFRAME GATEWAY

From day one, SNA and 37xx communications controllers have been inexorably intertwined. Invariably, 37xxs were a prerequisite for implementing a SNA WAN. It was only in the late 1980s, with the advent of source-route bridges and channel-attached 3174 LAN gateways that it was possible to contemplate having a SNA WAN that did not involve a 37xx or lookalike from NCR Comten or Fujitsu.

The need for 37xx's is, however, now in decline as SNA-only networks start to be reengineered to become multiprotocol networks. Supporting large numbers of low- to medium-speed serial ports, usually SDLC, was indubitably the forte of 37xx's. With the move toward multiprotocol networking, few, if any, SDLC or X.25 links are going to reach as far back as the data centers in which the mainframes reside. Instead, these serial links are going to be invariably converted at the remote sites, via techniques such as SDLC-to-LLC:2 or X.25/QLLC-to-LLC:2, to present a LAN appearance.

Rather than needing a mainframe gateway that can support large numbers of serial ports, the new requirement is for reliable, proven, and cost-effective LAN-to-mainframe gateways that can support both SNA/APPN and TCP/IP traffic to and from the mainframe. Typically, such a LAN-to-mainframe gateway will not require any SNA-specific serial ports since all SNA serial traffic would have been taken care of downstream. Instead, all it would need would be two or more LAN ports and a few high-speed serial ports for multiprotocol WAN links, which in most cases are likely to be frame relay-based. There is a variety of IBM and non-IBM products already on the market that meet these requirements—albeit in some cases without thoroughly proven reliability and veracity.

These potential LAN-to-mainframe gateways include the IBM 3745, IBM 3746-950, IBM 3172, IBM Open Systems Adapter (OSA), IBM 3174, Bus-Tech 3172-BT1, and Cisco channel-attached bridge/routers. In each case, these gateways, from the perspective of SNA traffic, act as PU Passthrough Gateways—even though the 3172-BT1 is capable of acting as PU Controller Gateway for NetWare for SAA or MicroSoft LAN server clients.

Of this cast of potential LAN-to-mainframe gateways the 37xx and the 3172 are the best known. However, both the Ciscos channel attachment, as well as the OSA, emulate the 3172. The 3745 and 3172 are both capable of acting as SNA, APPN, TCP/IP, or OSI gateways. This apparent similarity in their capabilities, when acting as LAN-to-mainframe gateways, has resulted in many enterprises seriously contemplating replacing their relatively expensive 3745s, with their hefty maintenance contracts, with the much more cost-attractive 3172s. Another option is to consider a channel-attached bridge/router as an alternative to a 37xx or 3172. Some are even looking at channel-attached 3174s even though the 3174s gateway capabilities are not as extensive as those of a 3172.

The current 3172 gateways, due to memory limitations, can only support a maximum of 1,020 downstream SNA nodes. Even this relatively small limit can, however, only be realized by configuring a 3172 with its full complement of four LAN adapters, since each 3172 LAN adapter can support only 255 Downstream Physical Units (DSPUs). (Using the optional 3172 Operator Facility across a 3172 token-ring adapter will reduce the number of DSPUs that can be supported by that adapter by 10 to 245.) A channel-attached 3174 can currently support around 250 DSPUs. Each DSPU supported via a 3174 does, unfortunately, require a dedicated subchannel address on the mainframe channel. With a mainframe channel typically being able to support only 256 subchannels, the 3174 DSPU limit cannot be readily extended.

The 3172, in contrast, requires only one subchannel address per LAN adapter, rather than one per DSPU. (An additional subchannel needs to be defined if NetView/390 is to have explicit visibility of the 3172 Gateway per se, rather than just the DSPUs defined through it.) A new ACF/VTAM resource definition structure, known as an *External Communications Adapter Major Node*, was introduced with ACF/VTAM Version 3 Release 4 to enable 3172-based DSPUs to be defined to the mainframe without the need for a subchannel address per DSPU. The actual DSPUs that may be supported via a 3172 or 3174, as well as a channel-attached bridge/router, may of course, be considerably increased by the use of PU Concentrator Gateways as was discussed earlier. A 3745 today can, at least in theory, support up to 9,999 LAN-attached, SNA DSPUs—and do so without requiring a subchannel address per DSPU.

The DSPU limitation, which can be mitigated by using PU Concentration, is, however, not the only potential drawback of using a non-3745 SNA gateway. IBM estimates that SNA DSPU sessions that do not go through a 3745 require 20% processing, per session, on the mainframe by ACF/VTAM, compared to those going through a 3745! Such additional networking-specific processing could severely disrupt a mainframe's application processing workload and degrade response times. The need for the additional processing on the mainframe is understandable. ACF/NCP software on 3745s performs all of the Layer 2, and much of the Layer 3, processing required by SNA sessions, in particular checking for and taking care of all SNA message unit segmentation. In contrast 3172s, and 3174s act as strict passthrough gateways that only deal with Layer 2 LAN traffic-processing issues. There is, as yet, no concept of an SNA Offload capability on 3172s that could be activated to reduce the amount of mainframe processing required.

In the case of TCP/IP traffic, the 3172 Model 3 does, however, offer a TCP/IP Offload facility, which, true to its name, permits some of the processing associated with supporting TCP/IP-based transactions to be performed on the 3172, rather than on the host. The current TCP/IP Offload facility, however, is not all that one would expect it to be. For a start, IBM does not claim that the offload facility improves overall performance or response times. That alone is counterintuitive. But despite the implied connotation that the 3172 offload feature will improve performance by offloading the processing from the mainframe, this is not one of its goals.

Instead, TCP/IP strives only to eliminate mainframe cycles being used for processing TCP/IP-related protocols. To add to this incongruity, the 3172 TCP/IP Offload facility is not provided within the framework of the standard 3172 control software, which is referred to as the 3172 Interconnect Controller Program (ICP). Instead, when running in offload mode a 3172 Model 3 runs a corrected version of the OS/2 TCP/IP Version 1.2.1 (or greater) software on top of OS/2 Version 1.3 (or greater). Running OS/2 software in place of the ICP precludes the 3172 from being a multiprotocol gateway when running in TCP/IP Offload mode—in that mode a 3172 Model 3 can only be a dedicated TCP/IP to mainframe gateway.

IBM's NPM is another facility that is currently is supported just by the 3745—and not even by the 3746-950. NPM's original charter was to provide SNA traffic volume measurements to facilitate capacity planning. It has, however, been widely used for the last decade or more by medium to large enterprises as a means of establishing accurate traffic usage statistics on each SNA user, à la itemized telephone bills, that could then be used as the basis of cost charge-back billing. Displacing a channel-attached 37xx with another LAN-to-mainframe gateway would mean that the mainframe-resident NPM would not be able to obtain the usage statistics required for charge-back billings. There is, however, a very high possibility that NPM support will be available around mid-1996 on either the 3746-950, the 3172, or both.

The NPM, DSPU, and Off-Load issues clearly demonstrate that selecting a LAN-to-mainframe gateway is far from straightforward, even when one restricts the choices to IBM-supplied, supposedly strategic gateways. Nonetheless, selecting a cost-effective and strategic LAN-to-mainframe gateway is pivotal to the network reengineering effort. The additional issues related to evaluating such gateways, and in particular, channel-attached bridge/routers will be discussed further in Chapter 11 once all the other relevant technology issues have been covered in detail.

2.8 REFLECTIONS

The overall workings of SNA networks have changed little over the last three decades despite SNA's impressive repertoire of technical accomplishments and the growth in the SNA clan of architectures. The bulk of traffic still flowing across SNA networks is 3270 datastream-based, even though PCs/workstations have now usurped 3270 terminals to become the predominant device on such networks. Ironically, PCs/workstations are, however, not inherently SNA devices or in anyway restricted to SNA-only applications. SNA LAN gateways, which come in at least three major flavors, are required to meld PCs/workstations into SNA environments. SNA LAN gateways, though integrating PCs/workstations with SNA, are not the only SNA-related technology or issue required to support SNA device and traffic within a multiprotocol network. Some of the other aspects that require evaluation and attention include: SNA routing, traffic prioritization, SNA TG support, and LAN-to-mainframe gateways. The good news, however, is that most of the methodologies and issues are at least now well known and even well understood. Though it would not be fair to say that it is all likely to be plain sailing, the technology required to accommodate SNA specific features within a multiprotocol network is now available. There is, thus, ample opportunity and motivation to take the necessary steps to move away from an SNA-only network toward a next-generation network that will eventually become an ATM-based multimedia, all-in-one network.

CHAPTER 3

The Blue Brick Road to ATM

A QUICK GUIDE TO CHAPTER 3

ATM has now become the "endgame" in contemporary networking. Networking vendors, with IBM in the fore, and telco carriers adamantly promote it as the future of networking. The appeal of ATM is obvious: It promises to deliver large amounts of cost-effective bandwidth and to permit effortless multimedia networking.

Migrating to ATM from today's IBM-centric networks is not, however, going to be a cakewalk. There are many issues that need to be resolved and challenges to be overcome. For example, bringing ATM to the desktop is likely to be much more costly than many have envisaged. The $1,000 cost for an ATM adapter will only be a small part of the total cost. The majority of PCs in use in enterprises today do not have a bus scheme that is fast enough to accommodate 155Mbps ATM.

Hence, to realize 155Mbps to the desktop, new PCs with high-speed buses will be required in addition to the ATM adapters. Fortunately, there are other options, such as 25Mbps ATM, 100Mbps LANs, and switched LANs. This chapter unearths all the issues related to making the switch to ATM and presents solutions, options, and work-arounds. The chapter begins with a brief introduction to ATM. It then highlights all the issues that should be considered vis-à-vis the migration to ATM.

In light of these issues, Section 3.1.1 lays out a realistic migration schedule starting with campus-level ATM in the 1996 time frame. Section 3.2 concentrates on ATM at the campus and desktop level and explores the pros and cons of desktop ATM. This section also covers IBM's 25Mbps ATM initiative and its viability. Once ATM is in use at the campus level, there will be a clamor for WAN ATM for internetworking between the campuses. Mid- to large-size enterprises have two distinct options for realizing WAN ATM. They could either subscribe to a public ATM service or build their own in-house ATM WAN network. Section 3.3 deals with how an enterprise should decide between a private network and a public service.

Most enterprises that have an IBM-oriented network are unlikely to get to ATM in one step. Invariably they will first have to consolidate their dreaded parallel data networks. Section 3.4 focuses on this challenge and previews the eight technologies that can be used to achieve this consolidation. It then goes on to discuss a unique scheme for choosing between these technologies. This scheme relies on the WAN traffic mix, in the 1997–1998 period, as the exclusive metric for selecting the optimal technology for implementing an SNA/APPN-capable multiprotocol LAN/WAN network, as the first step on the road to ATM.

Asynchronous Transfer Mode is currently perceived as the ultimate holy grail of contemporary networking. This perception that ATM is indeed the next big thing in networking is being propelled by the fervent and incessant endorsement of both the vendor and the telco carrier community at large, with IBM one of its most ardent champions. The lure of ATM is obvious. For a start, it promises cost-compelling, virtually error-free, scalable (i.e., wide spectrum from low to high), and manageable bandwidth, that extends well into the gigabit range. Quality of service characteristics can be associated with this bandwidth to realize networks with consistent and predictable behavior. Examples of the quality of service characteristics that may be sought include: the peak and average bandwidth, propagation (i.e., latency) delay, and acceptable transmission error rates. Moreover, ATM can adroitly support concurrent data, voice, and video traffic streams across a single composite network. ATM is thus the ultimate panacea for eliminating parallel networks.

SNA and non-SNA data networks are unfortunately not the only instances of parallel networks in enterprises today. Most enterprises that have multiple, geographically dispersed campuses, divisions, or even large branch offices are likely to also have an in-house voice network with tie-lines between the various locations. With the growing popularity of videoconferencing and in-house company news television broadcasts, many of the larger enterprises are also ending up with dedicated bandwidth for video traffic. Digital T1 trunks, with their 1.544Mbps bandwidth, have for the last decade or so been the staple means of obtaining bandwidth for intraenterprise, private networks. T1 multiplexors, which in general rely on Time Division Multiplexing (TDM), are used to share this T1 (and in some cases even 45Mbps T3) in-house bandwidth between the voice and data networks.

ATM in time is expected to make such T1-based networks obsolete. For a start, ATM will offer more bandwidth. This ATM bandwidth, from public ATM networks à la X.25 and frame relay networks, is likely, after a few preliminary "trial balloon" iterations, to be attractively tariffed. Its cost in terms of bits/second will be proportionally much lower than today's T1 and T3 rates (i.e., 155Mbps will not cost 100 times today's rates for a T1 connection). In addition, with ATM, in marked contrast to TDM, there is no need whatsoever to divide, reserve, and preallocate portions of the available bandwidth to each type of traffic stream—that is, data, voice, and video.

ATM shuns the notion of reserved, preallocated bandwidth. Reserving bandwidth for exclusive use by particular traffic types can prove to be very inefficient. With reserved bandwidth, bandwidth not being fully utilized at any given time by one traffic type is not dynamically available to another traffic type that could gainfully make use of that additional capacity. ATM on the other hand can dynamically share the total available bandwidth between all the various traffic streams, on an equitable, per-demand basis. This results in optimal bandwidth utilization with no possibility of any portion of the total bandwidth going unused.

Adequately accommodating voice and full-motion video traffic becomes a challenge when bandwidth is not being reserved for each traffic type. Voice and video traffic, sometimes referred to as *constant-bit-rate* (CBR) traffic, is heavily real-time dependent and cannot typically tolerate arbitrary delays in transmission caused by other traffic types hogging all the bandwidth. Delays in voice traffic transmission would result in disconcerting breaks in conversations. In the case of video, any delays would cause the picture to jerk or even to freeze. ATM circumvents the problem of CBR traffic experiencing unexpected delays due to the presence of other traffic by using a very fast, data-cell switching technique.

With ATM, data (i.e., digitized bit streams including digitized voice) from all the various traffic streams is mandatorily and automatically split into 53-byte cells, with each cell consisting of a 5-byte ATM header and a 48-byte databearing payload. These cells are then switched across the network as fast as possible by hardware incorporating specialized and dedicated ATM chips, over megabits per second broadband links with minimum intermediary processing (e.g., no checks for data errors or error recovery) at each switching node. Cells from different traffic streams are continuously and statistically (i.e., based on presence and demand) multiplexed across the various broadband links.

The rapid switching and quick transport of small cells on an end-to-end basis ensures that no one datastream hogs all the bandwidth. Moreover the cells associated with each datastream arrive at their intended destination on a consistent basis with only a very small (i.e., millisecond or lower range) and predictable delay between each cell. Since the intercell gap, or delay, is so small, CBR traffic is not subjected to unacceptable "hiccups" in data flow. The CBR traffic thus enjoys a relatively smooth and fairly constant flow of movement that does not disrupt voice or video transmissions in any tangible manner. This concept of providing datastream continuity and persistence by rapidly transmitting a series of cells is akin to the way motion is created in movies through the rapid projection of individual freeze-frame cells on to a screen at 16 frames a second.

Despite its formidable prowess in supporting broadband multimedia traffic, ATM is not a full-stack, top-to-bottom, full-function networking scheme, à la OSI, SNA, or TCP/IP, that spans all or at least most of the seven-layers of the OSI reference model. Instead, befitting its quintessential image of agility and speed, ATM is a lean-and-mean, low-level protocol that straddles just Layer 1 and a part of Layer 2 of the seven-layer model; see Figure 3.1. It is thus essentially a new-generation data link protocol for multimedia applications that supersedes frame relay, High-Level Data Link Control (HDLC), and SDLC. Higher-level services and protocols are interfaced to ATM via the Layer 2 ATM Adaptation Layer (AAL). AAL serves a function similar to that of an SNA LU. An SNA LU interfaces end users to SNA. Similarly, the adaptation layer interfaces ATM end users to an ATM network. The so-called slicing-and-dicing of data traffic frames into ATM cells is performed as part of the AAL function through a process referred to as Segmentation and Reassembly (SAR).

In the interest of speed and reflecting the fact that contemporary transmission media is relatively error-free, ATM does not at any stage check the data portion of cells (i.e., the payload) for any data corruption during the transmission process. There is no parity or frame check sequence (FCS) error-checking. Checking the integrity of the data

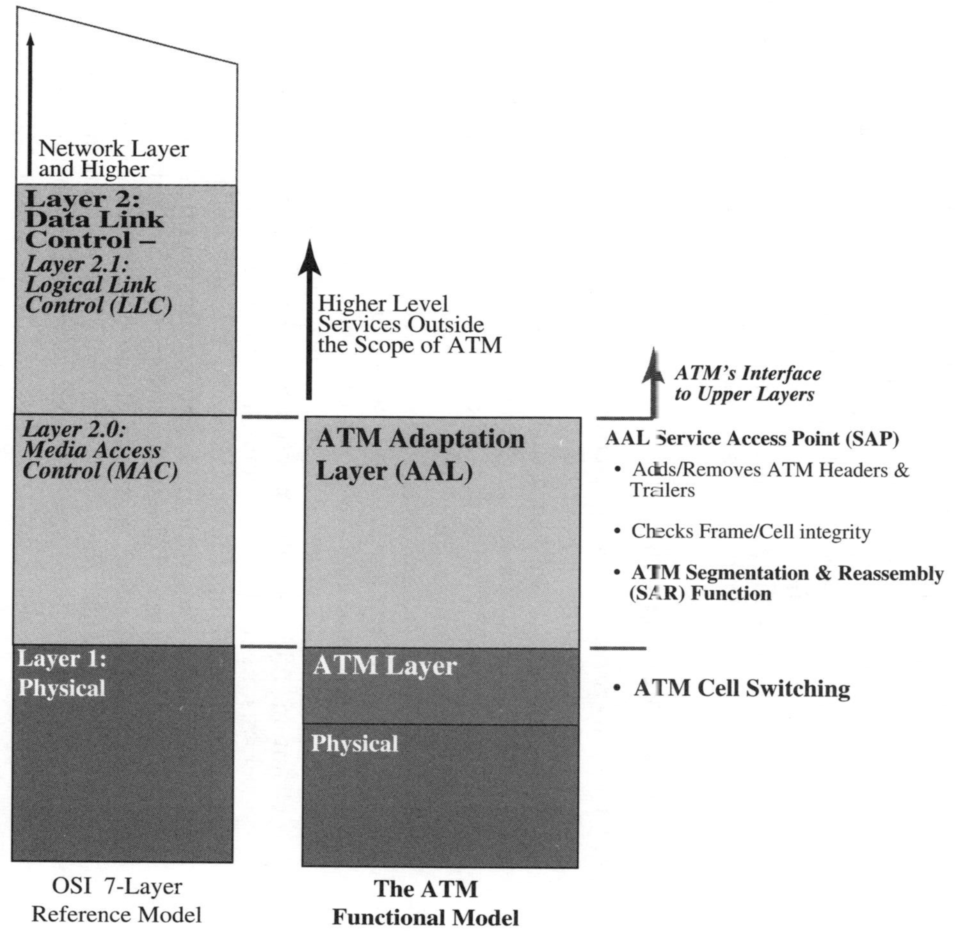

Network Layer and Higher

Layer 2: Data Link Control –
Layer 2.1: Logical Link Control (LLC)

Layer 2.0: Media Access Control (MAC)

Layer 1: Physical

OSI 7-Layer Reference Model

Higher Level Services Outside the Scope of ATM

ATM Adaptation Layer (AAL)

ATM Layer

Physical

The ATM Functional Model

ATM's Interface to Upper Layers

AAL Service Access Point (SAP)

- Adds/Removes ATM Headers & Trailers
- Checks Frame/Cell integrity
- **ATM Segmentation & Reassembly (SAR) Function**

- **ATM Cell Switching**

FIGURE 3.1 ATM's functional model vis-à-vis the seven-layer reference model.

received and requesting any retransmissions if the received data is found to be corrupted is left to the ATM end users. ATM at each cell-switching point does, however, check the 5-byte header at the front of each cell for any errors. To facilitate this a 1-byte Header Error Check field appears at the end of the header. Any cells with a corrupted header are simply discarded. ATM per se does not request the retransmission of discarded cells. This again is left to the ATM end users. ATM, like frame relay, will also arbitrarily discard cells if it experiences unexpected congestion at a switching node or a link between nodes. ATM end users are not explicitly notified if and when cells are discarded either due to congestion or header errors. Higher-level functions, at or above the adaptation function, are responsible for detecting and recovering from discarded or lost cells.

3.1 MIGRATION PLANS FOR ATM

Given its promise of cost-effective, broadband bandwidth and its support of multimedia, it is not hard to fathom why most networking professionals across the board and even some power users with their insatiable craving for raw performance are now so anxious to embrace and exploit ATM technology. For enterprises, especially in North America and Europe, the primary question when it comes to ATM is no longer "if" but "when." Independent of ATM's persuasive charms, the supply side of the market (i.e., the vendors and carriers in this case) has ensured—one could even say dictated—that ATM and ATM alone is the eventual end game in the *current cycle* of enterprise network evolution. ATM is being positioned to be the pervasive networking mechanism and the preeminent means of realizing mega- to gigabit bandwidth well into the first decade of the next millennium.

For enterprises with IBM-centric mission-critical networks, however, the move to ATM-based networking is not going to be exactly straightforward—both in terms of the best migration path and the optimum schedule. There are some crucial issues to be hammered out relative to ATM before any concrete steps can, or should, be taken to carefully map out a definite plan and itinerary for this migration. These crucial issues are:

- When is ATM technology, particularly as it applies to WANs, going to be proven, cost-justifiable and, in the case of public networks, aggressively and attractively tariffed? In general, the realistic answer to this is now at least 1997.
- Should the migration be phased such that ATM is first used just for data and then incrementally extended to embrace video and

then as the final step voice? Most enterprises are likely to opt for this cautious, least-risk approach.

- Should ATM be first used just at the campus level as a means for providing additional bandwidth for data-only LAN applications and intracampus LAN-to-LAN interactions? *If so, is LAN switching a potential alternative that is likely to be much less disruptive and more cost-justifiable? Moreover, ATM is not a particularly efficient transport mechanism for data-only applications given that it was designed to support concurrent, multimedia traffic.* This somewhat sensitive but nonetheless significant issue that ATM is an inefficient means for accommodating data-only traffic is discussed in detail in Section 3.2.
- Would a hybrid network that melded together LAN switching and ATM be the most optimum solution—at least for the next four to five years—for realizing additional bandwidth for campus applications?
- Is there a need for every PC and workstation in an enterprise to be ATM-capable (i.e., ATM to every desktop)? Could the interface to ATM be restricted to hubs, LAN switches, and bridge/routers? At this juncture it is also worth noting that the majority of the installed base of PCs today have the Industry Standard Architecture (ISA) bus. (The ISA bus is also known as the PC AT bus since this bus type was first used on the IBM PC AT that was introduced circa 1984.) The ISA bus cannot accommodate data transfer rates much in excess of 25Mbps. In order to use 100 to 155Mbps ATM speeds for data transfer, new bus types such as the very fast Peripheral Component Interconnect (PCI) bus are required. Given that new bus types cannot be readily or cost-justifiably retrofitted into existing PCs, the only realistic way to get PCs with buses that are fast enough to deal with ATM is to obtain new PCs. For enterprises with thousands of PCs this could be a costly upgrade exercise.
- Is a public network, provided by a major carrier such as AT&T or MCI, the only viable means of realizing ATM across a WAN? Does it make sense to implement in-house, private ATM WANs along the lines of the private X.25 networks implemented by some enterprises? Or for that matter the far-flung private SNA networks replete with dozens of 37xx communications controllers? IBM for one, with its 2220 Nways BroadBand Switches, is actively promoting the notion of private ATM, and for that matter frame relay, networks.
- What role will frame relay play in this migration scenario? At least for the next few years, public frame relay networks, particularly in North America, are likely to be the most cost-effective means of

obtaining bandwidth up to 2Mbps. Many data networks, especially SNA networks, are already in the process of cutting over from sub-19.2Kbps leased lines to 56Kbps (or above) frame relay virtual circuits. Some frame relay router vendors, such as Motorola and Hypercom, are already supporting voice over frame relay over 8Kbps or 16Kbps VCs. As with LAN switching and ATM, it is highly possible that the optimum solution for many enterprises is going to be a hybrid environment consisting of both frame relay and ATM. Standards such as the ATM Forum's Data Exchange Interface, the ATM Forum's Frame-based User-to-Network Interface, and the Frame Relay Forum's FRF.8 Frame Relay/ATM Service Internetworking are already in place to facilitate such frame relay to ATM, and frame relay over ATM, interworking.

- Should today's parallel data networks be consolidated into a single integrated data network prior to the move to ATM? Or does it make sense to keep the parallel networks in place and run them across a common ATM network? *While this may appear to be an option, the cost as well as the availability of the conversion boxes (i.e., the ATM equivalents of SNA/SDLC to FRADs) that will be required to attach SNA devices to ATM networks may dictate that this is not a realistic option.* The least costly and most expedient solution may be to connect the LAN- or link-attached SNA devices to an ATM-capable bridge/router or FRAD. Connecting SNA devices to a multiprotocol router or FRAD in order to provide them with access to ATM is still tantamount to consolidating the SNA and non-SNA networks.
- If the parallel data networks are to be consolidated prior to ATM, what would be the optimal way to realize this? Would one technique or technology be applicable and effective in every instance or are there different approaches depending on the eventual protocol mix or the size of the consolidated network?

Network professionals tasked with migrating existing networks to ATM need to resolve most if not all of the above issues in order to feel truly comfortable that they have not overlooked any important facets of this significant network overhaul. Unfortunately, there are no rote and but a few rule-of-thumb answers to most of these questions. Invariably, the appropriate resolution for all of the above issues is in the end going to be enterprise-specific, and possibly in some cases even specific to individual divisions or campuses. For example, a division or campus involved in high-end scientific imaging or complex CAD/CAM applications may already need ATM to satisfy the demand for megabit bandwidth between individual workstations and a CRAY or IBM SP/2 ilk supercomputer.

However, other divisions or campuses of that enterprise that are not using such bandwidth-intensive applications may find that LAN switching alone is sufficient to eradicate any LAN throughput bottlenecks in the short to mid-term. Over the next few years as the real cost of ATM technology starts to fall (or even plummet), as is inevitable, this enterprise will no doubt decide to standardize on ATM across the board. It will also start to use ATM as the WAN fabric for interconnecting all of its dispersed locations. Thus, it is clear that the master ATM plan for many a large enterprise, rather than being monolithic, will end up consisting of a series of time-staggered miniplans that will eventually converge four to five years down the road to produce an enterprisewide ATM infrastructure.

Given the futility of trying to provide generic answers to the above issues, the best alternative is to thoroughly explore all aspects of each issue and present all the various options with the pros and cons of each. This is the intent of the following sections in this chapter. Figure 3.2 depicts an overall road map for ATM migration highlighting the various landmarks along the way, while Figure 3.3 illustrates the multistep process for moving from SNA-only networks to multimedia ATM networks.

3.1.1 A Realistic Schedule for ATM Migration

Though general market awareness and interest in ATM only really began around early 1992, ATM's roots can be traced back to at least 1988 when the telephone companies started working on Broadband Integrated Services Digital Network (B-ISDN) technology. ATM products, albeit predominantly for campus scenarios, have been readily available from a variety of vendors since at least 1993, with companies such as Fore Systems, NetEdge Systems, and Newbridge leading the charge. Though not noted for being an early adopter of public-domain Layer 2 to 3 networking schemes, even IBM started shipping its first ATM products—the TURBOWAYS ATM adapters for PCs and RS/6000 workstations as well as the 2220 Nways BroadBand ATM-cum-frame-relay switches—in mid- to late 1994. (See Chapter 9 for details of all of IBM's ATM products.) The pioneering ATM networks, mainly in medical imaging (i.e., being able to read very high-resolution X-ray, CATscan, and MRI images), scientific/engineering imaging, and in-house broadcasting systems for interactive-video-based training arenas, also started to go on-line around this time.

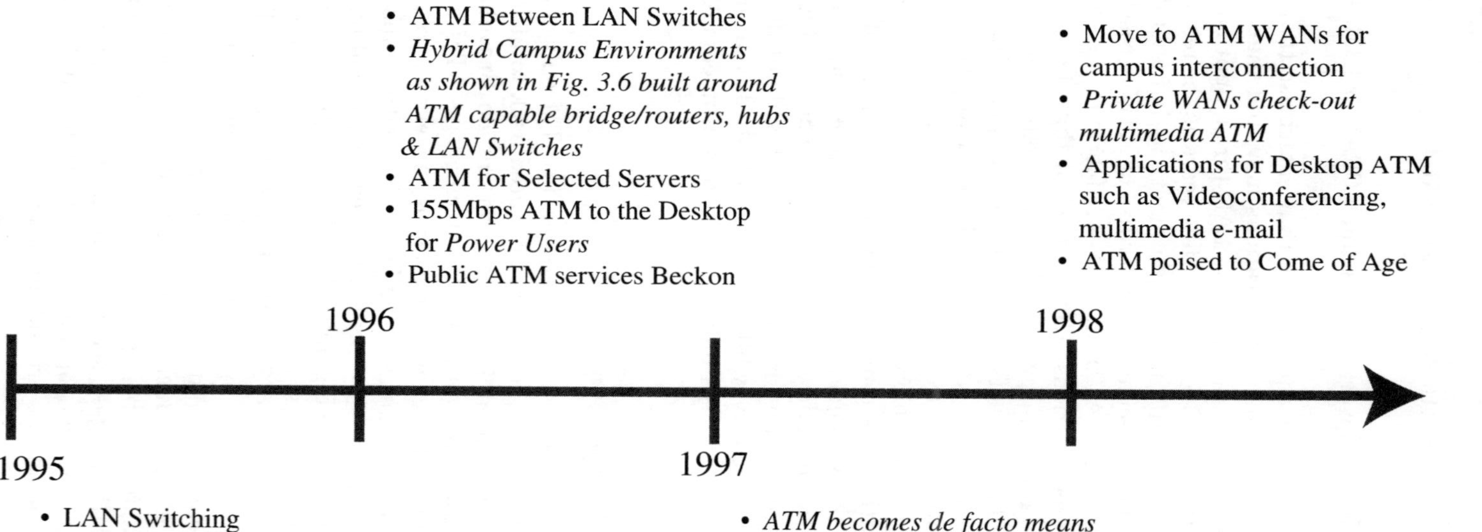
1995
LAN Switching
ATM Cabable Bridge/Routers
ATM Capable Hubs
ATM Adapters
LAN Emulation
ATM WAN Switches
Early Adopters
1996
ATM Between LAN Switches
Hybrid Campus Environments as shown in Fig. 3.6 built around ATM capable bridge/routers, hubs & LAN Switches
ATM for Selected Servers
155Mbps ATM to the Desktop for Power Users
Public ATM services Beckon
1997
ATM becomes de facto means for building campus backbones
Prices of ATM Adapters continue to fall making ATM to the Desktop tempting if not for the need to upgrade PCs
Large enterprises protype private ATM WANs -- Data Only to begin with
Public ATM services start to take steps to make broadband bandwidth cost compelling
1998
Move to ATM WANs for campus interconnection
Private WANs check-out multimedia ATM
Applications for Desktop ATM such as Videoconferencing, multimedia e-mail
ATM poised to Come of Age

FIGURE 3.2 A realistic road map for ATM migration.

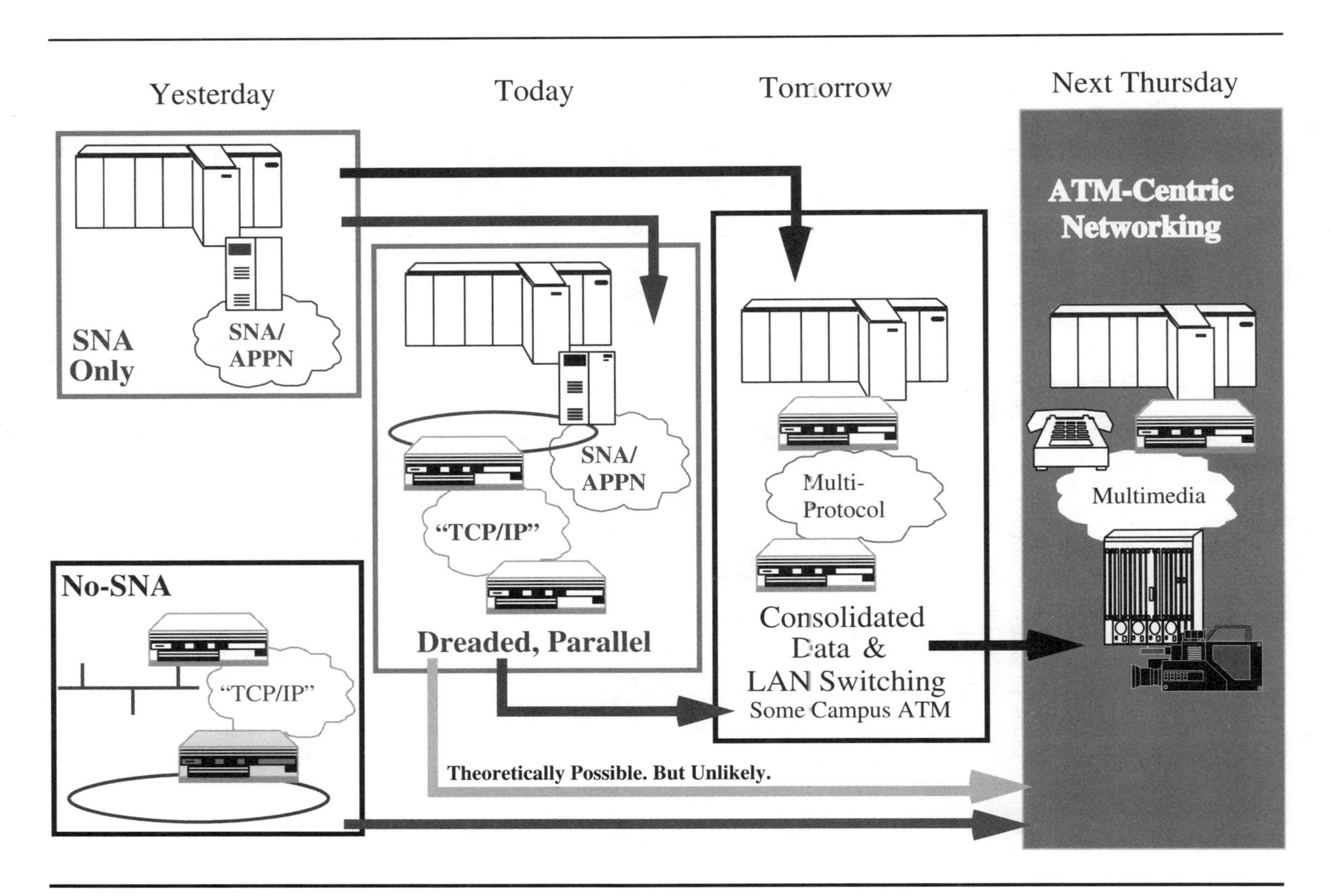

FIGURE 3.3 A graphical representation of the multistep process to get from SNA-only networks to ATM.

However, even in late 1995, ATM had yet to be widely adopted by the traditional IBM networking community—that is, enterprises with mid- to large-scale, mission-critical SNA/APPN networks. Nonetheless, there is significant consensus within this base that ATM is indeed the end game. The only marginal interest shown to date by this community toward actually implementing ATM, as opposed to planning for ATM, has been by and large restricted to limited campus applications. In many instances even these have been on a prototypical basis, primarily to evaluate the technology.

ATM's slow start within the IBM networking sector can be attributed to at least four reasons:

- Mission-critical networking and as yet not fully proven complex technology are mutually exclusive.
- Consolidating the dreaded parallel data networks—that is, the well-entrenched, mission-critical SNA network and the newer, more strategic multiprotocol network—has precedence over migrating to ATM.
- Campuses with token-ring LANs, especially 16Mbps LANs, are not yet as deprived of bandwidth as is the case with many 10Mbps Ethernet networks.
- SNA and APPN networks are inevitably WAN-centric and the tariffs for public ATM WAN networks have yet to fall below the threshold for making them compelling.

These four key reasons why ATM has not being wildly and widely embraced by the IBM networking community are somewhat self-explanatory. However, some further elaboration is appropriate to highlight some of the embedded issues.

Given the mission-criticality of most IBM-centric networks, there is a historical and marked reluctance to experiment with nascent technology until it has been thoroughly tested and tried in other networking environments. Such prudent conservatism vis-à-vis new technology is understandable since any unreliability in such a network can impact enterprise revenue and affect that enterprise's bottom line. It is this obvious reluctance to compromise in any way the high-availability characteristics of the mission-critical network that has resulted in most of these enterprises maintaining parallel SNA and non-SNA data networks, despite the technology being available to consolidate these two networks using bridge/routers or RFC 1490-based FRADs.

This cautious approach to new technology acceptance and assimilation, typically with a four- to five-year lag behind the rest of the market, can have a knock-on effect. This knock-on effect is another factor that is impacting the prevalence of ATM. The availability of ATM has occurred right on the heels of bridge/routers finally gaining acceptance within this community. With much of today's energies now being devoted to migrating SNA/APPN traffic to router-based networks there is no time or "bandwidth" left to explore and evaluate the potential and promise of ATM.

By mid-1995, a growing number of enterprises were just beginning to embark on their painstakingly planned data network consolidation programs—typically using bridge/routers or frame relay routers. The availability of ATM adds another convoluted dimension to this network consolidation effort. Ideally the data network consolidation plans should be reevaluated relative to ATM. Public ATM WAN networks, within the next two years, are poised to offer broadband bandwidth at what should be very compelling tariffs. Hence, there is a possibility that data network consolidation at this juncture in order to optimize bandwidth utilization may be futile.

With the availability of the more cost-effective ATM-based bandwidth, it may make more sense—at least in theory—just to keep the data networks parallel and thereby also ensure that the mission-critical SNA/APPN traffic is not impinged upon by non-SNA traffic. *There are, however, two major potential impediments that dictate that trying to run parallel SNA and non-SNA networks across ATM may not be as feasible as it first appears.* For a start, cost-compelling ATM WAN tariffs are still but a promise. It may be awhile before these tariffs drop low enough to justify running a parallel network across ATM. Tariffs and bandwidth cost may, however, not be the only stumbling block.

With parallel networks the SNA WAN component is made up of either SDLC links, X.25 connections (particularly in Europe), or frame relay connections. LAN-attached SNA devices (e.g., PCs) are connected to the SNA WAN via some type of LAN-to-SNA gateway (e.g., 3174 or a PC running a NetWare for SAA-type gateway program). Link-attached SNA devices (e.g., 3274 or AS/400) are connected to the WAN in one of two ways. One is to attach them natively using one of their SDLC, X.25, or frame relay ports. The other option is to connect them to X.25 or frame relay networks via SNA/SDLC-to-X.25 Packet Assember/Disassemblers (PADs) or SDLC-to-Frame Relay FRADs.

The key problem of trying to run this SNA network across ATM is that of interfacing the LAN-to-SNA gateways and the link-attached SNA devices directly to ATM. Today, many of the quintessential SNA devices such as 3174s and 3745s do not have ATM interfaces. IBM has already stated that the 3745 communications controller, the preeminent WAN-to-mainframe gateway in SNA networks, is unlikely to ever have a 155Mbps ATM interface. 155Mbps ATM instead will be offered only on 3746 Model 900 and 950 controllers. If and when ATM will be supported by the 3174 is also unclear. Even if IBM gets around to supporting ATM on 3174s, much of the installed base of 3174s falls into the category of being plug-compatible units supplied by other vendors. Some of these vendors may not have the resources to develop the necessary hardware and software to support ATM on their plug-compatible boxes.

The bottom line here is that trying to interface an existing SNA network to ATM is not going to be smooth sailing. *The optimal way to interface SNA devices to ATM, in many cases, may be to just connect the LAN- or link-attached SNA devices to an ATM-capable bridge/router or FRAD.* It is, of course, possible to contemplate scenarios where routers or FRADs are used just to interface SNA devices to ATM. In other words, these routers and FRADs become relatively expensive ATM-CSARs (viz. ATM Cell Segmentation and Reassembly units) for the SNA devices. The use of routers or FRADs just to interface SNA to ATM is improbable. What is, of course, more likely is that these routers and FRADs handle SNA traffic alongside non-SNA traffic vis-à-vis the ATM network. Having a router or FRAD handling SNA traffic alongside non-SNA traffic is in essence the consolidation of the two data networks.

Irrespective of the essentially open issue of trying to interface SNA devices directly to ATM, enterprises have also persevered with parallel data networks for long enough. They have conducted enough evaluations and "bake-offs" to test the mettle of contemporary routers. They have diligently worked on data network consolidation strategies and plans for the last few years. They do not want to postpone this initiative and wait for ATM. Most enterprises would rather tackle data network consolidation right now (i.e., in the 1995 to 1996 time frame), given that router technology at last appears to be robust and resilient enough for the task—and then think about ATM migration as the next step. To further encourage this school of thought, all router vendors offer ATM upgrade paths that in general ensure that any investment made in router technology today will not have to be written off in a few years in order to migrate to ATM. Thus, the bottom line here is that data network consolidation right now has priority over ATM migration. *Consequently, it is safe to assume that widespread migration to ATM, and in particular WAN ATM, by this hitherto IBM-biased community, is not going to begin in earnest until 1997.* Given that this community includes at least 30,000 enterprises worldwide, this migration to ATM will not all take place in 1997 per se. The migration will be staggered and spread over a few years with some enterprises only getting around to it toward the end of this millennium.

The third factor hampering the embracement of ATM by the IBM-centric networking community has to do with the type of ATM bandwidth currently on offer. ATM's current forte is that of providing 100 to 155Mbps bandwidth, at a campus level, across fiber links. Unfortunately, in the context of this community, this is generally not the area that is in need of any major assuagement when it comes to bandwidth. Most of this community, given its leanings toward SNA/APPN, tends to use token-ring LANs. Token-ring LANs on the whole, and in particular the 16Mbps ones, have not experienced the same levels of performance sapping congestion that has generally occurred, over the last couple of years, in comparable size Ethernet LANs. There are three tangible reasons why token-ring LANs have fared so much better compared to their Ethernet brethren. These three reasons are as follows:

1. Token capture-based media access scheme used by token-ring ensures controlled, deterministic, and equitable use of the LAN. This token-capture scheme, somewhat analogous to the polling mechanisms used by BSC and SDLC, ensures that there is no marked drop in performance until sustained traffic levels begin to exceed 70% of capacity. In contrast, the contention-based access scheme used by CSMA/CD-based Ethernets results in marked performance degradation even before traffic levels exceed 50% of capacity, as the rate of collisions caused by multiple users trying simultaneously to access the LAN increases. Each collision causes the data on the LAN to be invalidated. In addition, after each collision, the devices affected have to back off from trying to access the LAN for a certain period of time. The data lost due to collisions, as well as the mandatory back-off penalty, results in blackout periods when there is no productive data transfer across the Ethernet.

   It is the accumulation of these blackout periods as the rate of collisions increases, which in turn is proportional to the number of users on the LAN and the amount of traffic they are trying to exchange, that starts to impact Ethernet performance as traffic levels increase. The deterministic access of token-rings preclude any data loss due to collisions as well as the need for back-off penalties. Performance, relative to a given user, is only impacted if the time interval between the token becoming available lengthens. Such delays in token availability would be caused by an increase in either the number of users on the LAN or the amount of traffic.
2. The bulk of the traffic on token-ring LANs tends to be related to transaction processing (e.g., database access or update)—with most mainframe- or AS/400-resident, mission-critical SNA/APPN applications being transaction-oriented. In general the amount of

data associated with each transaction tends to be relatively short. It is not uncommon for the data inbound to an application not to exceed 150 bytes while the data outbound from the application to the PC/workstation may be less than 1,000 bytes. Thanks to the relative brevity of these transaction related message units (i.e., packets), a single token-ring user rarely monopolizes the LAN for any significant time. Consequently, multiple users can routinely gain equitable access to the LAN without undue delay. Traffic patterns of most Ethernets tend to be significantly different, with large data transfers, spanning multiple 1,500-byte Ethernet frames, being quite common. This more intense use of the LAN, coupled with data blackouts due to collisions and back-offs, decreases the amount of productive access that can be enjoyed by each Ethernet user, compared to that of a token-ring user.

3. A 16Mbps token-ring has at least 2Mbps more usable bandwidth than a 10Mbps Ethernet.

   If and when performance does become an issue in token-ring environments, the throughput bottlenecks invariably tend to relate to either interconnecting individual LAN segments to a LAN backbone or LAN server access. In both instances, token-ring LAN switching has proved to be a tactical and cost-effective solution that involves the minimum of disruption and new hardware. If full-duplex adapters at around $800 (and falling) each are also installed on the LAN servers, token-ring LAN switching can ensure that these servers can operate at sustained full-duplex data rates approaching 32Mbps. LAN switching, and token-ring LAN switching in particular, are discussed at length in Section 3.2.

The net result of all of these factors is that token-ring environments in general, as yet, are not in dire need of large amounts of additional bandwidth at the LAN or campus level. Any bandwidth shortfalls are likely to be first addressed with LAN switching. Consequently there is unlikely to be an all-out rush toward ATM in 1996 just to gain additional campus-level bandwidth. *ATM, in most cases, is likely to get introduced, in 1996, as a means for realizing 100Mbps plus bandwidth between LAN switches located within the same building or campus.* Even then, widescale adoption of ATM at the campus level is unlikely to occur before 1997.

Unfortunately, however, the sufficiency of bandwidth at the token-ring level does not also hold true at the WAN level—particularly when there is a need for real-time LAN-to-LAN interactions across the WAN. Moreover, with the IBM-networking community now trying to converge their parallel data networks into a consolidated network, there is a growing need for reliable and cost-effective WAN bandwidth to cope with the aggregated traffic volumes without any loss in performance.

ATM-based, public WAN networks should, in theory, be an ideal source for this bandwidth, especially given the possibility of also supporting video and voice traffic alongside the data traffic. Alas, in 1995 ATM-based public WAN offerings were in their infancy. As was the case with the original frame relay tariffs in the early 1990s, the current ATM WAN tariffs are not exactly aggressive. They are really targeted at the early adopters who are anxious to explore this promising technology irrespective of the cost. Just as with frame relay, these tariffs will be iteratively revised downwards over the next couple of years to attract a larger clientele.

At present, at least in North America, public frame relay networks are offering very attractive rates across the spectrum all the way up to T1 speeds. The guaranteed (i.e., CIR) bandwidth in the 16Kbps ($30/month) to 64Kbps ($100/month) being especially persuasive. For the next couple of years frame relay will continue to be the most cost-effective means of realizing sub-2Mbps bandwidth. However, the widespread expectations are that by the end of 1996, public ATM WAN services will start offering very competitive tariffs at 45Mbps and above. This late 1996 date again dove-tails very nicely with the proposition that ATM is only going to start making its mark felt in the IBM-centric networking community starting in 1997.

A realistic ATM migration scenario for much of the IBM-centric networking community is shown in the accompanying sidebar:

A REALISTIC TIMETABLE FOR MIGRATING TO ATM

1996:

- Parallel data network consolidation with ATM-ready bridge/routers or RFC 1490-based frame relay routers
- *Deployment of LAN switches, some possibly ATM-ready, to enhance backbone interconnect and LAN server throughput*
- *ATM-based, 100–160Mbps fiber fat pipes used to interconnect LAN switches colocated on the same campus*
- Some campus-level, ATM-based, new-generation switched broadband LANs, with certain LAN servers, Unix systems, RISC/Unix workstations, and PCs equipped with 100–155Mbps ATM adapters (It is unlikely that there will be a major demand for

ATM interfaces to mainframes, via either IBM 3172s, IBM 3746 Nways controllers, or channel-attached bridge/routers or hubs in 1996.)

These ATM-based LANs will exploit the LAN emulation capability now routinely included with ATM adapters. LAN emulation enables an ATM adapter to appear as a traditional token ring or Ethernet LAN adapter to the relevant LAN software resident in the LAN server, workstation/PC, or computer system (e.g., LAN Network Operating System [NOS] such as Novell NetWare or SNA LAN Gateway software such as IBM's 3270 PC Emulation program). Any interactions between devices with ATM adapters and devices on traditional LANs will be realized via ATM-to-LAN bridge function such as the one now available on IBM's 8260 Switching Hubs. Using LAN emulation and ATM-to-LAN bridging, PC/workstations can realize any of the following three types of ATM-related data transfer: ATM-to-ATM, LAN-to-ATM, and LAN-to-LAN across ATM; see Figure 3.4.

These switched ATM LANs will in general concentrate on data. A few will, in addition, venture to incorporate video for applications such as desktop videoconferencing.

- Compelling tariffs of public ATM WAN networks toward the latter part of the year
- At least in North America, public frame relay will provide much of the bandwidth up to 2Mbps to satisfy the initial demands of the consolidated data networks. Even some of the SNA-only networks will migrate to frame relay in order to capitalize on the very aggressively tariffed 56Kbps range bandwidth, while waiting for network consolidation or ATM.

  Enterprises with many dispersed remote locations (e.g., enterprises in the banking, retail, financial, travel, or insurance sectors) may start using voice over frame relay as supported by Hypercom, Motorola, and so on, in order to accommodate all branch-office traffic (i.e., SNA, LAN, BSC for automated teller machines, legacy point-of-sales protocols, async for security systems and voice) across one or more low-cost frame relay VCs.
- Enterprises already using frame relay will experiment with Frame Relay to ATM or frame relay over ATM interworking, using services provided by frame relay and ATM networks that conform to standards such as the Frame Relay Forum FRF.8 specification for Frame Relay/ATM Service Internetworking.

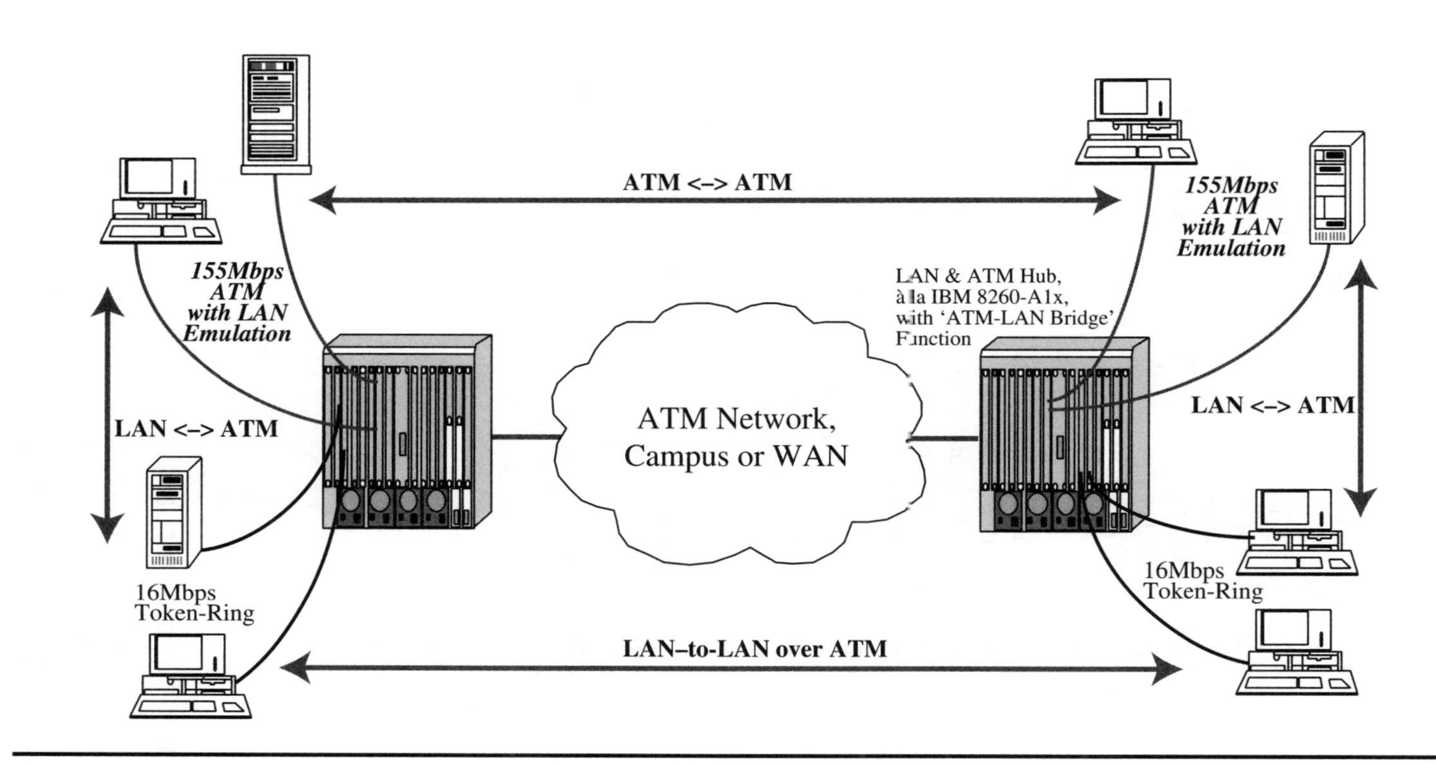

FIGURE 3.4 LAN and ATM interworking with LAN emulation and ATM-LAN bridging.

- Some large enterprises will start to evaluate private, in-house ATM WAN switches, for example, IBM 2220 Nways BroadBand Switches, Cisco LightStream 2020 Enterprise ATM Switch, possibly using them initially as frame relay switches. *Given that such switches typically also support mid-band links in the T1 to T3 range, there will be some low-speed ATM implementations, primarily for data, with ATM being used across 1.544, 2.048 (E2), 34.37 (E3), and 44.74 (T3) Mbps links.*

1997:

- Significant and sustained interest in ATM across much of the community
- Interest in campus level ATM will begin to intensify. Campus-level ATM switches will begin to win favor over LAN switches. In large enterprises or in enterprises involved in very high data traffic applications (e.g., scientific/engineering modeling), many LAN servers, PCs/workstations of power users, and computer systems will be equipped with 100–155Mbps ATM adapters. Desktop video applications will begin to capture the imagination, and possibly even some budgetary funds, of mid- to large-size enterprises.
- Where possible LAN switches and hubs will be upgraded to use 155(+)Mbps ATM fat pipes for intracampus interconnections.
- Gradual rise in the usage of public ATM WAN services
- More deployment and evaluation of in-house ATM WAN switches
- Some of the larger enterprises will pioneer mainframe ATM interfaces, most likely via channel-attached bridge/routers à la Cisco.

1998:

- ATM, within the IBM transaction processing biased networking community, finally comes of age.

3.2 CAMPUS AND DESKTOP ATM

ATM at the campus level and ATM to the desktop are currently being touted primarily as a means for turbocharging the performance of data-only LAN applications. This turbocharging is achieved by dramatically increasing the amount of bandwidth at the disposal of individual users, LAN servers, computer systems, as well as the bridges or bridge/routers being used to provide local, intracampus interconnection. This additional LAN bandwidth provided by ATM is derived through the synergistic synthesis of four separate techniques. These four techniques that together ensure that ATM can offer bandwidth galore at the campus level are as follows:

1. Use of dedicated media (i.e., Unshielded Twisted Pair [UTP], Shielded Twisted Pain [STP], or fiber) per user, as opposed to the shared media scheme favored by traditional LANs
2. Use of higher line speeds in the 25 to 622Mbps range rather than the 4 to 16Mbps line speeds used by traditional Ethernet and token-ring LANs
3. Providing each user with on-demand, maximum line-speed dedicated bandwidth to/from the ATM switch, now that the use of dedicated media obviates the need for the media access arbitration (e.g., waiting to capture a token or checking for collisions with CSMA/CD) that is mandatory with traditional LANs.
4. Support of full-duplex operation also made possible by the use of dedicated media, whereas conventional LAN adapters can only function in half-duplex mode

Figure 3.5 illustrates the notion of ATM to the desktop and that of a switched LAN-based on ATM.

This move away from shared media access is a fundamental precept and feature of campus-level switching technology, whether it be ATM or LAN switching. Maximum bandwidth gain is only possible with dedicated media. However, campus-level ATM or LAN switching does not have to be totally predicated on dedicated media to each and every user, particularly in the case of PC/workstation users. It is indeed possible to have hybrid scenarios. These hybrid scenarios could include: ATM on dedicated media along with shared media LANs, ATM on dedicated media interworking with LAN switching with the LAN devices also using dedicated media, as well as combinations of both of the previous configurations.

With a typical hybrid configuration, a campus-level ATM switch will support shared media LANs, either directly or via an ATM-capable bridge/router or LAN switch, on some ports while supporting native mode ATM connections, over dedicated media, on other ports. Figure 3.6 shows such an hybrid LAN-cum-ATM scenario. *In such hybrid scenarios ATM provides a broadband, campus backbone for intracampus LAN interconnection as well as high-speed, full-duplex server and computer access.*

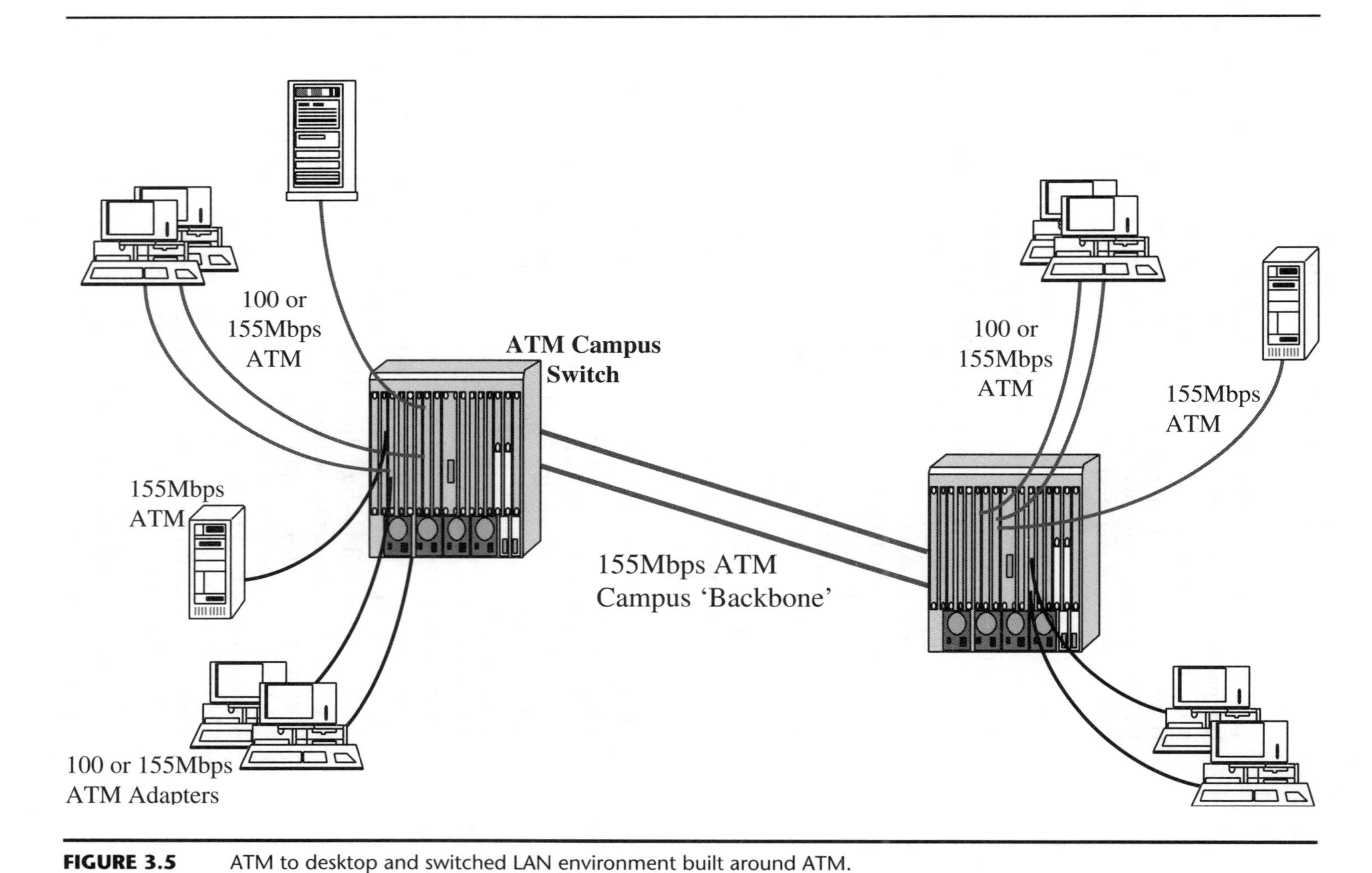

FIGURE 3.5 ATM to desktop and switched LAN environment built around ATM.

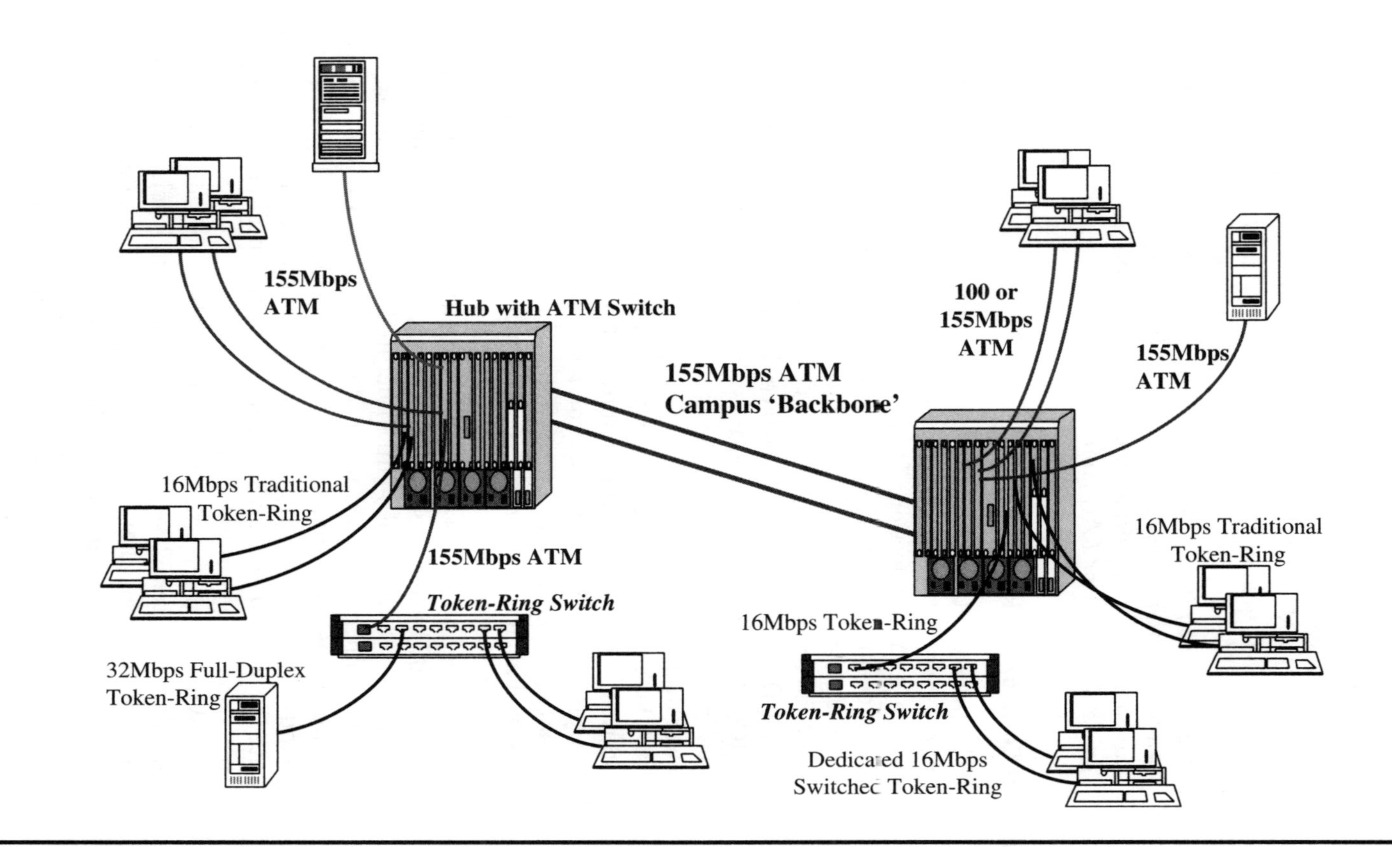

FIGURE 3.6 A hybrid campus environment consisting of desktop ATM, switched token-ring, and traditional token-ring.

Hybrid campus scenarios with ATM being initially used between hubs, bridge/routers, or LAN switches as a broadband pipe for intracampus, LAN-to-LAN transactions will in most cases provide IBM networking-oriented enterprises with their first taste of ATM.

Hybrid campus scenarios consisting of ATM backbones interconnection traditional LANs will in most cases provide DBM networking-oriented enterprises with their first taste of ATM. These ATM backbones will be made up of ATM "fat-pipe" connections between hub, bridge/routers, or LAN switches.

3.2.1 Pros and Cons of Desktop ATM

The pivotal factor when it comes to campus-level ATM is whether each and every device on the campus is going to have a native, dedicated media ATM connection. This is obviously the notion of *desktop ATM*. With desktop ATM, every PC and workstation, in addition to the LAN servers, computers, routers, and LAN switches, will be equipped with an ATM adapter. ATM adapters will displace Ethernet, token-ring and Fiber Distributed Data Interface (FDDI) LAN adapters—the so-called NICs (Network Interface Cards). PC and workstation users will only really be able to enjoy the true potential of ATM when it is available at each desktop.

Desktop ATM, in addition to making it possible for each PC/workstation to enjoy 100+Mbps bandwidth will open the door to the exhilarating possibilities of *interactive*, full-motion, CD-sound quality, desktop multimedia applications. Desktop videoconferencing, live news video feeds (for professionals such as stock brokers who can benefit from immediate access to fast-breaking news), and interactive, video-based wall-less classroom instructions are alluring examples of what could be achieved even with first-generation multimedia applications. (Voice and video-clip annotated documents, the so-called compound multimedia documents, do not require desktop ATM. They will, however, benefit from the added bandwidth of ATM. This additional bandwidth will certainly reduce the time taken to access and distribute these documents, which will typically contain many more bits than text- and image-only documents.)

Whether there will be any justifiable applications for voice, independent of interactive video, at the campus level, at least initially, is debatable given that any campus electing for desktop ATM is also likely to have a fairly good, PBX-based intracampus voice and voicemail system. There is, however, a real possibility that in time voice through desktop ATM could displace PBXs. This could lead to much tighter integration between voice and data applications. With this level of integration it would be possible to dynamically convert voice-mail to e-mail or vice versa to eliminate the need to process two distinct types of mail systems.

Desktop ATM is thus indubitably attractive and innately appealing. The rush to embrace desktop ATM, however, is currently being severely curtailed—particularly so within the transaction processing-biased IBM networking community—by five crucial factors. These are as follows:

1. Cost of equipping large numbers of PCs and workstations with individual ATM adapters: To achieve desktop ATM, the existing NICs of all the PCs and workstations have to be replaced with ATM adapters. Today, 25Mbps ATM adapters from IBM and others are available for just under $900 per PC/workstation ($395 for the 25Mbps ATM adapter per se and $499 for each port on the mandatory IBM 8282 Workgroup Concentrator. The concentrator is required since none of today's ATM switches directly supports 25Mbps ports. Hence, 25Mbps feeds have to be concentrated up to at least 100Mbps—the slowest ATM speed supported by most switches. Workgroup concentrators such as the 8282 perform this speed reconciliation function; see Chapter 9).

   Faster ATM adapters are obviously more expensive than 25Mbps adapters. The current prices for 100Mbps adapters are around $1,500, while 155Mbps adapters are close to $2,000. All these prices will drop—in some cases even plummet—over the next few years as demand and competition heats up, and when even faster, say 622Mbps, adapters will start appearing. By the end of 1996, it is likely that 25Mbps solutions will be available for around $200 per PC/workstation. Both 100 and 155Mbps adapters should be available for under $1000, with some 155Mbps adapters, benefiting from predatory pricing to secure market share, even approaching the $500 mark.

   Even at $500 per PC/workstation, the cost of upgrading all the desktops of an enterprise can be hefty, especially when thousands or tens of thousands of desktops are involved. Moreover, the upgrade cost is unlikely just to be restricted to securing ATM adapters. The majority of today's installed base of PCs and even some workstations do not have I/O buses fast enough to accommodate data transfer rates much in excess of 25Mbps. Hence, the upgrade to ATM may also entail a large-scale upgrade of the installed PC base.

2. Cost of replacing older PCs with ISA-type buses with new PCs with faster buses such as PCI: The majority of today's PCs do not have buses that are fast enough to handle 100+Mbps data transfer rates. Hence, just deciding to replace NICs with ATM adapters may not be sufficient when it comes to cost-justifying and allocating budgets for desktop ATM—unless of course the 25Mbps route is deemed to be adequate. New PCs, typically with PCI buses, will be required for use with 100 and 100Mbps ATM adapters. IBM, and a few others, currently offer both 100 and 155Mbps adapters with IBM Micro-Channel Architecture (MCA) buses. (See Chapter 8.) IBM's MCA-based PCs, the so-called PC/2s, were not that popular, and IBM discontinued making MCA-based PCs in early 1995.

   Having to replace PCs to realize desktop ATM will obviously increase the overall cost of this exercise dramatically. Even with significant volume purchase discounts, the minimum cost for a PCI bus PC with a 155Mbps ATM adapter, in 1996, is unlikely to be much below $2,000. In many cases the price may be in excess of $3,000. This $2,000 to $3,000 price per desktop significantly alters the cost-justification equation. Now it is no longer just a question of waiting for ATM adapters to fall below the $1,000 or $500 mark. The cost of upgrading large populations of PCs also has to be factored in and justified. It is unlikely that many, if any, enterprises in the IBM networking community have budgeted for this type of large-scale PC upgrade program in 1996 or even 1997. Consequently, ATM to the desktop will begin to occur within small "islands" with easily justifiable need for very high bandwidth. An example would be stock traders requiring multiple high-speed, up-to-the-second, market status data feeds; possibly a live CNN window on the PC; and the need to be able to fire off trading transactions with no hint of a delay.

   It is possible to conceive of multimedia 100–155Mbps ATM-cum-video-monitor ISA adapters that will process video and voice traffic directly within the adapter while also acting as the PC-to-monitor interface (i.e., the PC monitor will be plugged into this adapter as opposed to the existing VGA-type video card). With such an adapter, the ISA bus will only be used by data, rather than video and voice, applications. If data rates to the data applications can somehow be throttled to, say, 30Mbps or less, this type of approach may work. However, this type of adapter is unlikely to be inexpensive given that it is much more than just a straight ATM adapter. Hence, cost justification will continue to be an issue.

   LAN-switching, which can be used with existing NICs, and has a current per port cost in the $1,500 range—which will be down in the $500 to $750 range by the end of 1996—starts to look extremely attractive around this juncture as a bottleneck-eradicating and face-saving interim option to desktop ATM.

3. The absence of multimedia APIs and the reliance on LAN emulation: The standards for a native ATM data interface, as well as those for providing video and voice interfaces on ATM PC/workstation adapters, are currently being worked on. Consequently most of the ATM adapters available today, including all of IBM's current TURBOWAYS ATM adapter family, do not support a native ATM interface for data transfer or interfaces for voice and video. Instead, they offer LAN emulation. LAN emulation permits an ATM adapter to masquerade as a token-ring or Ethernet NIC.

LAN emulation has immediate and obvious advantages. With LAN emulation, existing LAN software can continue to be profitably used with ATM without any changes to the software or the LAN servers. The LAN software that can be used with ATM will include proven and prevalent LAN NOSs such as Novell's NetWare, IBM's OS/2 LAN Server, or even Unix. Thanks to LAN emulation, desktop ATM can be transparent and pervasive. An existing 10 to 16Mbps LAN environment, replete with file server, print server, and file transfer applications, can be upgraded to a 155Mbps ATM environment without any software or operational changes. This is obviously very appealing and possibly even tempting. The only negating factor preventing a stampede toward LAN emulation-based ATM is the cost of deploying ATM at every desktop as discussed directly above. (A native ATM data interface will permit data applications to directly interface with the ATM adapter and explicitly request ATM-specific options such as a particular Quality of Service [QOS]. It will be the ATM equivalent of today's NIC card interfaces such as Microsoft's Network Driver Interface Specification [NDIS] and Novell's Open Data-link Interface [ODI].)

LAN emulation also has a downside. By definition, as well as in practice, it is restricted to data-only, traditional LAN applications. Thus, ATM to the desktop based purely on LAN emulation becomes just a means of providing megabit, data-only bandwidth to PC/workstations. Some adapter vendors claim that it would be possible to upgrade their adapters to support a native ATM interface for data, as well as multimedia ATM once the relevant standards have been agreed upon. This at best is a vague promise and most of today's ATM adapters do not sport any physical interfaces to support video or voice feeds. In reality, the upgrade option may end up being a trade-in allowance against a new, multimedia-capable adapter. To exacerbate this already somewhat unsatisfactory situation, ATM is not a very efficient transport mechanism when dealing just with data.

4. ATM is not a particularly efficient when restricted just to data: ATM slices-and-dices all traffic streams into 53-byte, fixed-length cells in order to dexterously support concurrent instances of voice, video, and data traffic. In the absence of voice and video traffic, slicing-and-dicing data traffic into 53 bytes becomes an inefficient and even futile exercise. Trying to accommodate voice and video traffic is the only raison d'être for using fixed-length cells, as opposed to variable-length frames à la frame relay. Cell switching, whether ATM or otherwise, does not have a particularly efficient overhead-to-payload ratio—particularly so when dealing with data blocks in the 1,000-byte range, which are very common in SNA/APPN environments.

   The overhead-to-payload inefficiency of ATM is the result of four factors. The first and most obvious of these is the use of a 5-byte header at the start of a each 53-byte cell. A packet transfer mechanism such as frame relay uses at most a 12-byte header to prefix even a 8,192-byte data frame. The second factor that leads to the inefficiency is the need to include empty "pad" bytes to ensure that each cell is exactly 53 bytes long.

   The third and a relatively marginal cause of inefficiency is the inclusion of AAL trailers and in some instances even headers into the datastream, in addition to the 5-byte ATM headers per cell. For example, AAL Type 5 (AAL-5), also known as the Simple and Efficient Adaptation Layer, which is used with data traffic adds an 8-byte AAL-5 trailer to each data block. In addition, it pads the data block to ensure that the data block plus the 8-byte trailer is a multiple of 48 bytes (i.e., cell payload size). Figure 3.7 depicts the various overhead factors involved in ATM-based cell switching, while Table 3.1 compares the efficiency of ATM with that of frame relay for various data block sizes. The fourth factor that contributes to the overall inefficiency is the processing effort consumed to create the 53-byte cells with their 5-byte headers.

   Given the mega bandwidth involved and that the cell switching is done extremely rapidly via dedicated silicon chips, the inefficiency of ATM vis-à-vis data can be ignored or dismissed as being irrelevant. This is the stance taken by most ATM advocates any time this inefficiency issue is raised. There is some defensible merit to dismissing this inefficiency as the price one has to pay for the other advantageous of ATM. *However, an alternate, more enlightened approach would be to seek an automatic, instanta-*

neously interruptable cut-over from cell-switching to packet-switching when ever there is no voice or video traffic present. IBM refers to this mode of operation as Packet Transfer Mode (PTM), while others such as Northern Telecom refer to it as the Frame/Cell facility.

PTM or Frame/Cell is a feature available on certain ATM switches. When PTM mode is activated the switch continually monitors its ports for the presence of any CBR traffic. When CBR traffic is present the switch acts as a standard ATM switch and assiduously slices-and-dices all the traffic streams into 53-byte cells. When the switch detects a temporary absence of CBR traffic it automatically ceases to slice-and-dice the data-only traffic streams since the need to intersperse CBR cells in between the data cells to ensure smooth voice or video traffic flow is no longer there. When slicing-and-dicing is suspended, the switch transfers the data traffic, utilizing the full ATM bandwidth at its disposal, using variable-length packets à la frame relay that best match the size of the data blocks. When in PTM or Frame/Cell mode, data traffic does not incur the overhead inefficiencies of standard ATM.

While forwarding data traffic in packet format the switch continues to monitor all of its ports for CBR traffic. As soon as it detects any sign of CBR traffic, it automatically and instantaneously terminates the transmittal of the current packet, in mid-block if necessary, as soon as it has appended an FCS to error-protect the bytes transmitted up to that point. The switch then immediately reverts back into standard ATM mode and stays in this mode until there is another lull in CBR traffic.

PTM or Frame/Cell provides the best of both worlds in standard ATM multimedia support when there is CBR traffic and high-efficiency, full ATM bandwidth frame-mode data transfer when there is no CBR traffic. PTM or Frame/Cell is also always an option that can be activated or deactivated by the customer. Thus, the availability of the PTM or Frame/Cell facility does not have to in any way detract from ATM compliance or multivendor interoperability. If there is a need for multivendor interoperability or a sentiment that it is imperative at all times to maintain slavish compliance to ATM fixed-length cell switching, PTM or Frame/Cell can be deactivated.

Despite being ideally suited to dispel the dilemma of data-only ATM to the desktop, PTM or Frame/Cell is unfortunately not as yet offered on PC/workstation ATM adapters. Instead, this facility is currently only available on large, stand-alone switches such as IBM's 2220 range and Northern Telecom's Magellan family. It is

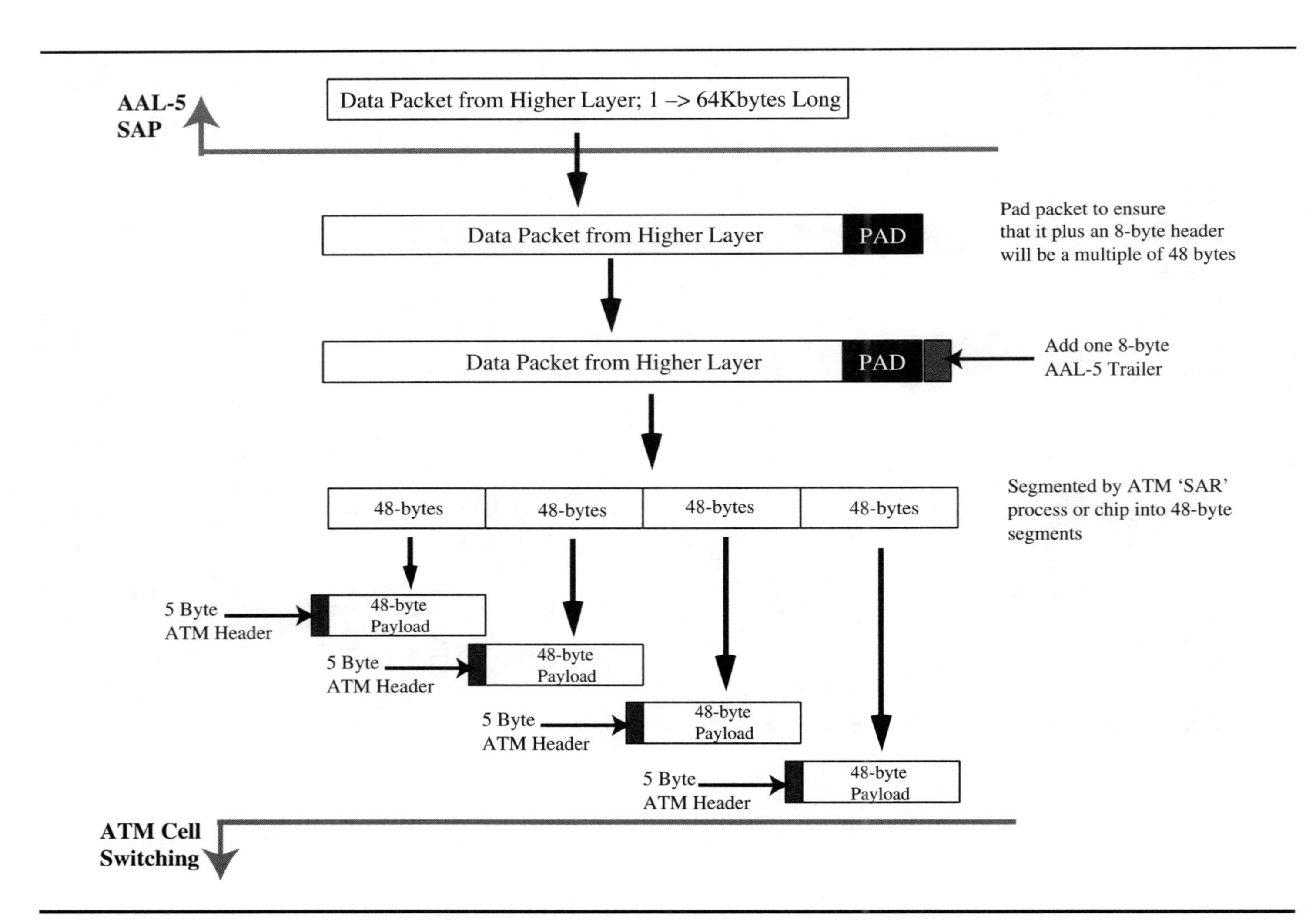

FIGURE 3.7 The process of converting a data packet into ATM cells and the various pads, headers, and trailers added.

TABLE 3.1 Comparison of the Header/Trailer Overhead of ATM vis-à-vis Frame Relay for Various Message Unit Sizes

Length of Message Unit (i.e., Payload) in Bytes	*Frame Relay Overhead with 10-byte RFT 1490 Headers*	*Ratio of Overhead to Payload in the Case of Frame Relay*	*Number of Cells Required in ATM When Using AAL-5 (see note 1)*	*Total Number of Bytes Transmitted; i.e., Number of Cells Times 53 bytes*	*Number of Bytes in Excess of Payload*	*Ratio of Overhead to Payload in the Case of ATM with AAL-5*
64	16	25%	2	106	42	65.63%
128	16	12.5%	3	159	31	24.21%
256	16	6.25%	6	318	62	24.21%
512	16	3.13%	11	583	71	13.87%
1024	16	1.56%	22	1166	142	13.87%
2048	16	.78%	43	2279	231	11.28%
4096	16	.39%	86	4558	462	11.28%

Note: AAL-5 first adds an 8-byte trailer to the payload. It then "pads" this payload + trailer so that its combined length is a multiple of 48 bytes. See Figure 3.7. This "padded" payload is then segmented into 48-byte cells—with each cell being prefixed with a 5-byte ATM header.

highly likely that IBM for one will in time get around to offering PTM on future TURBOWAYS ATM adapters. Adapters with PTM will, at least from a technical and aesthetic perspective, have an edge over those that can only deal with fixed-size cells.

Timing, however, could work against PTM. Even if PTM is available, the cost of ATM to the desktop is not going to be easy to justify in the short term. By the time ATM to the desktop becomes generally affordable, multimedia ATM to the desktop should also be a possibility if not a reality. With multimedia ATM the value of PTM becomes significantly depreciated compared to its lure in LAN emulation-based data-only applications.

The bottom line here is that at present ATM to the desktop is invariably restricted just to LAN emulation-based data-only applications. ATM is not a very efficient transport mechanism when restricted just to data. The PTM or Frame/Cell feature of ATM switches can be used to circumvent this inefficiency. Ironically, however, this feature is currently not available on ATM adapters. By the time it is likely to be available, ATM to the desktop may no longer be restricted just to data. In that case, PTM or Frame/Cell though still capable of increasing data transfer efficiency when there is no CBR traffic present, is not as key as it would be in data-only scenarios.

4. The long-term reliability under sustained loads of ATM adapters, and the across-the-board compatibility of the LAN Emulation code on these adapters with LAN software and LAN applications have yet to be conclusively demonstrated: This is but another facet of the overall mission-criticality issue that has been discussed at length in previous sections. With ATM to the desktop, the ATM adapter becomes the umbilical cord between a PC/workstation and the outside world. If there are any problems with an adapter, the affected PC/workstation immediately becomes isolated and unproductive.

   Backup configurations to cope with any ATM adapter problems are of course possible. One option would be to use multiple adapters. The other would be to use a traditional LAN or even a wireless LAN connection as the backup for ATM. Any such backup configuration will increase the overall cost of the solution and significantly increase its complexity in terms of administration, operation, and user friendliness.

   Given the justifiable conservatism inherent in mission-critical networking environments and the tangible lost-opportunity costs involved, network administrators responsible for running such networks understandably prefer to let others field-test new technology. In general, toward the end of 1995, ATM to the desktop is still a rarity. Consequently, customer-supplied as opposed to vendor-supplied data and testimonials on the robustness and resilience of ATM adapters and their LAN emulation code is still somewhat patchy. The mission-critical networking community is unlikely to embrace ATM to the desktop until there is widespread data from the trenches to vouch that ATM adapters are as reliable as their NIC counterparts. Concrete data on ATM adapter reliability should start to become available toward mid-1996.

5. The widescale disruption involved in moving from a traditional LAN setup to ATM at the desktop: Bringing ATM to the desktop is not a trivial exercise. It entails the installation of ATM switches as well as ATM adapters into PCs/workstations, LAN servers, and computers. In many cases existing PCs with ISA buses will have to be replaced with new PCs with faster bus types. In some instances, for example, if a shared cable Ethernet was being used, it would also require new dedicated UTP/STP or fiber wiring schemes to be implemented.

   Fortunately, thanks to LAN emulation, there will be no need for any software upgrades. Software upgrades will only become necessary when multimedia applications and the native ATM interface for data applications become readily available. Even without the headaches of installing new software, the scale and scope of the hardware upgrades involved is significant. In essence, the existing LAN plumbing on a workgroup basis as well as on a campus basis has to be dismantled and replaced with an ATM-based infrastructure. In theory it is indeed possible to systematically convert one PC/workstation at a time to ATM rather than an entire workgroup or campus. The ATM-to-LAN bridging function now available within some hubs (e.g., IBM 8260) could be gainfully used to provide transparent interoperability between the PCs/workstations that have been converted to ATM and those, including LAN servers, that have not. ATM-to-LAN bridging, with a current price range in excess of $16,000, is not exactly inexpensive. It does, however, provide a means of minimizing the overall disruption and permits the ATM conversion to be staggered.

Given the above issues pertaining to cost, disruption, and the initial absence of multimedia applications, ATM to the desktop is not likely to garner much enthusiasm from the IBM-centric networking community till well into 1997. In the interim, LAN switching and in particular token-ring switching will step into the breach to satisfy most of the demand for any additional LAN bandwidth. Compared to ATM to the desktop—particularly when restricted just to data only applications—token-ring switching provides a much less disruptive and cost-compelling solution, albeit also for considerably less gain in bandwidth. The technology, and issues, as well as the pros and cons of LAN switching, are described in detail in the next chapter.

Until ATM to the desktop becomes a reality, post-1996, ATM at the campus level will be restricted to backbone applications, as shown in Figure 3.6, where ATM will provide a broadband pipe between hubs, LAN switches, and bridge/routers servicing traditional token-ring or Ethernet LANs.

3.2.2 25Mbps ATM

The notion of 25Mbps ATM was formally advocated by IBM in June 1994 with the unveiling of the TURBOWAYS 25 family of ATM adapters for ISA and MCA bus-based PCs. The introduction of 25Mbps ATM caused much chagrin to ATM zealots who consider 25Mbps to be too slow to be deemed true ATM. The ATM Forum initially refused to accept IBM's proposed standard for 25Mbps ATM. It has since recanted and there is now an approved standard for 25.6Mbps (i.e., the precise line speed of the so-called 25Mbps) ATM. Moreover, there is now a consortium that includes luminaries such as: IBM, Chipcom (now a part of 3Com), Centillion (now a part of Bay Networks), Madge, H-P, Fujitsu, Olicom, Whitetree Network Technologies, and LSI Logic committed to the endorsement, promotion, and development of 25Mbps ATM technology.

The raison d'être for 25Mbps ATM is intuitive. It sets out to offer relatively low-cost, entry-level ATM to the desktop that will even work with PCs with ISA buses. Consequently, the success or failure of ATM was always contingent on the most basic and also the most potent of market forces—that of price versus performance. In mid-1995, IBM's TURBOWAYS 25 price of $900 per PC (i.e., $395 for the 25Mbps ATM adapter per se and $499 for each port on the mandatory IBM 8282 Workgroup Concentrator) appeared to be relatively attractive, and in line with the price/performance curve envisaged for ATM and high-speed LAN offerings. Figure 3.8 illustrates the relationship between TURBOWAYS 25Mbps PC adapters, the IBM 8282 Workgroup Concentrator, and an ATM network.

Despite its initially attractive price/performance there was no major demand for 25Mbps ATM in 1994 or 1995. This lack of demand reflected the fact that IBM-centric LAN environments with their 4 or

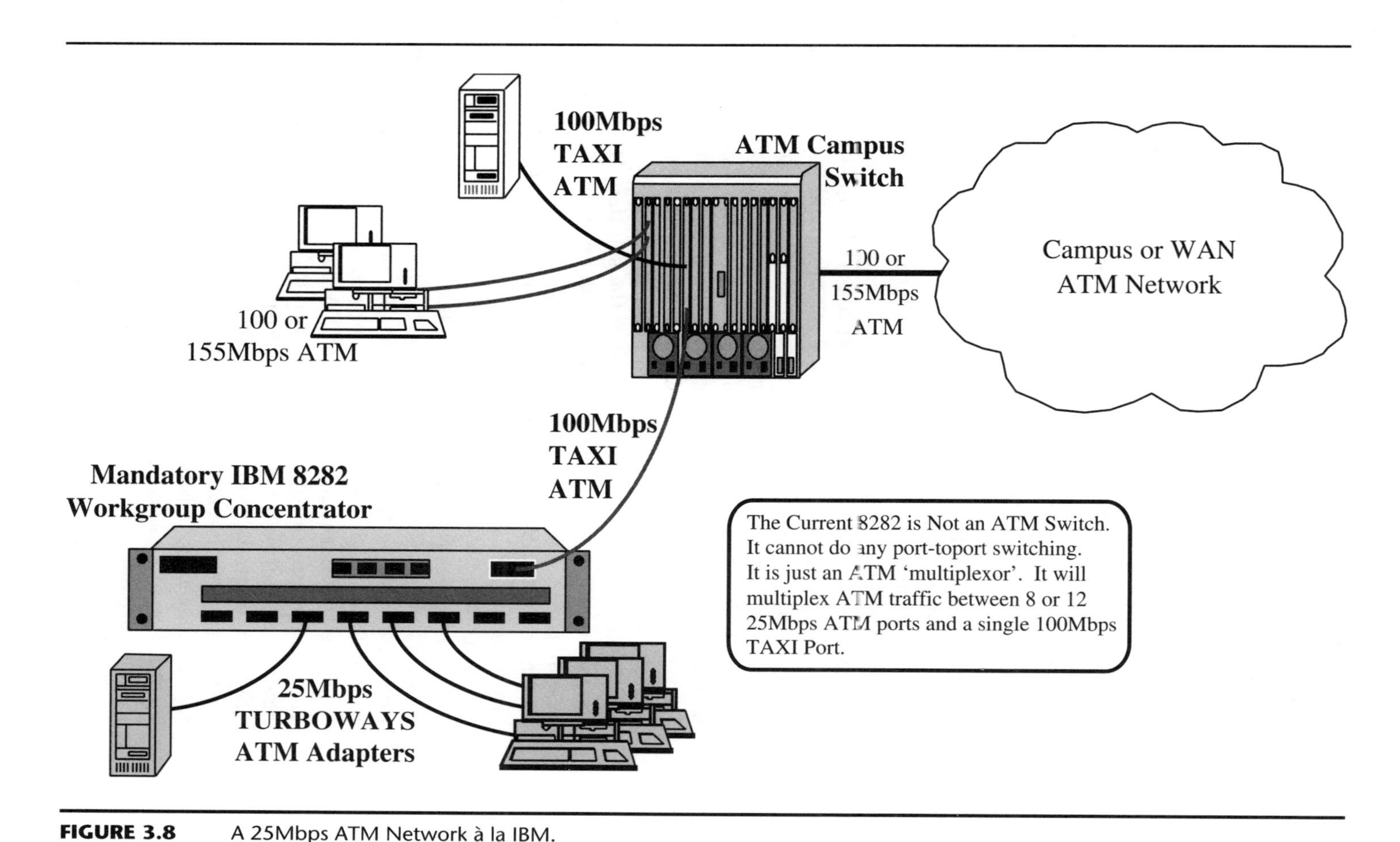
100Mbps
TAXI
ATM
ATM Campus
Switch
100 or
155Mbps ATM
100 or
155Mbps
ATM
Campus or WAN
ATM Network
100Mbps
TAXI
ATM
Mandatory IBM 8282
Workgroup Concentrator
The Current 8282 is Not an ATM Switch.
It cannot do any port-toport switching.
It is just an ATM 'multiplexor'. It will
multiplex ATM traffic between 8 or 12
25Mbps ATM ports and a single 100Mbps
TAXI Port.
25Mbps
TURBOWAYS
ATM Adapters

FIGURE 3.8 A 25Mbps ATM Network à la IBM.

16Mbps token-ring LANs did not have a dire need of any additional bandwidth at the desktop or workgroup level. If there was a need for more bandwidth it was for the LAN backbone, and 25Mbps ATM did not address this requirement.

In 1996, 25Mbps ATM is now under considerable price/performance pressure from a variety of alternate options for providing cost-effective, additional bandwidth to individual PCs/workstations as well as workgroups. These alternate solutions include: 100Mbps Fast Ethernet, 100Mbps 100Base-VG Ethernet and token-ring frame-compatible AnyLAN, LAN switching, and even aggressively priced 100 and 155Mbps ATM adapters.

25Mbps ATM, which is in reality 50Mbps given its full-duplex mode of operation, is still unfortunately limited to LAN emulation-based data-only applications. While a native ATM interface for data and multimedia interfaces has been promised, such capabilities are unlikely to be generally available at least till mid-1996. Even if one assumes that the prices for data-only 25Mbps will fall to around $400 in 1996, this is still relatively expensive compared to the $300 or less being asked for 100Mbps Ethernet or AnyLAN. In addition, there is already talk that 155Mbps ATM adapters will be available in 1996 for under $500. However, all of the 100Mbps or more solutions, whether Fast Ethernet, AnyLAN, or ATM, may require PCs with non-ISA buses. Having to replace a PC in order to gain the additional bandwidth does increase the overall cost of the solution. However, much of the additional cost may also be justified in terms of the additional processing power, faster I/O throughput, extra storage, and better multimedia (e.g., CD-ROM and speakers) capabilities that will no doubt grace the new PCs.

Thus, the bottom line when it comes to the future prospects for 25Mbps ATM is relatively simple. Its future is contingent on the same criteria that justified its creation in the first place—overall price/performance relative to other solutions. At present the relatively high price of 25Mbps vis-à-vis other solutions does not bode well for it. However, the one redeeming attribute of 25Mbps ATM is its ability to support the huge installed base of PCs with ISA buses. This might end up being a pivotal issue. Many enterprises may balk at the thought of having to upgrade thousands or even tens of thousands of PCs over a short period of time in order to gain additional bandwidth to the desktop. In that case, 25Mbps ATM could provide an interim solution, at least for certain select groups of power users, while new PCs or workstations are deployed gradually to replace aging PCs.

3.3 PRIVATE VERSUS PUBLIC ATM

The notion of, and the faith in, private data networks is an integral and ingrained facet of IBM-centric networking. Private ATM WAN networks in cahoots with some of the private frame relay networks that have already been built around ATM-capable IBM 2220 Nways BroadBand Switches set out to extend this proven and time-honored tradition. Though rarely thought of as such, the SNA-over-SDLC WAN networks that have been the cornerstone of commercial networking particularly in North America but also in certain European (e.g., UK) and Pacific rim countries (e.g., Australia) are the epitome of private data networks. With such SNA-over-SDLC networks public telco carriers just provide a given enterprise with leased (i.e., dedicated) 4.8Kbps, 9.6Kbps, 19.2Kbps, 56/64Kbps, and occasionally T1 links. In some cases these leased lines are augmented or backed up with dial-up (i.e., switched) connections made through the public switched telephone network.

The Layer 2 to 4 networking functions required to ensure controlled, reliable, and managed SNA/SDLC data interchange over these leased and switched links were provided by 37xx (or compatible) communications controllers running (ACF/)NCP software that front-ended the mainframes. In the larger networks, remote 37xx's, referred to as RCPs, would be deployed at strategic locations around the network to act as regional link-concentrators (i.e., consolidate the traffic of multiple slow speed links onto a high-speed trunk). They would also sever as branching-off points for implementing alternate data paths within the network. Figure 3.9 illustrates the composition of a typical mid- to large-scale private SNA/SDLC network. In SNA/SDLC networks, 37xx's, along with mainframe-resident ACF/VTAM, rather than a public carrier, are responsible for running and sustaining the entire network. All that the carrier provides is raw bandwidth—either across leased lines or switched connections.

Typically an enterprise with a private SNA/SDLC network would have multiple, geographically dispersed campuses, divisions, or even large branch offices that are served by this data network. Given this physical dispersement of personnel it is not uncommon, especially in North America, for such enterprises to also have an in-house, tie-line-based private voice network between the various locations. Rather than using two sets of leased lines for these two private networks, most enterprises obtain T1, T3, or Fractional T1 (e.g., 128Kbps) links from carriers and use private TDM-based T1 multiplexors to divvy up this leased line bandwidth between the SNA network and the in-house voice network.

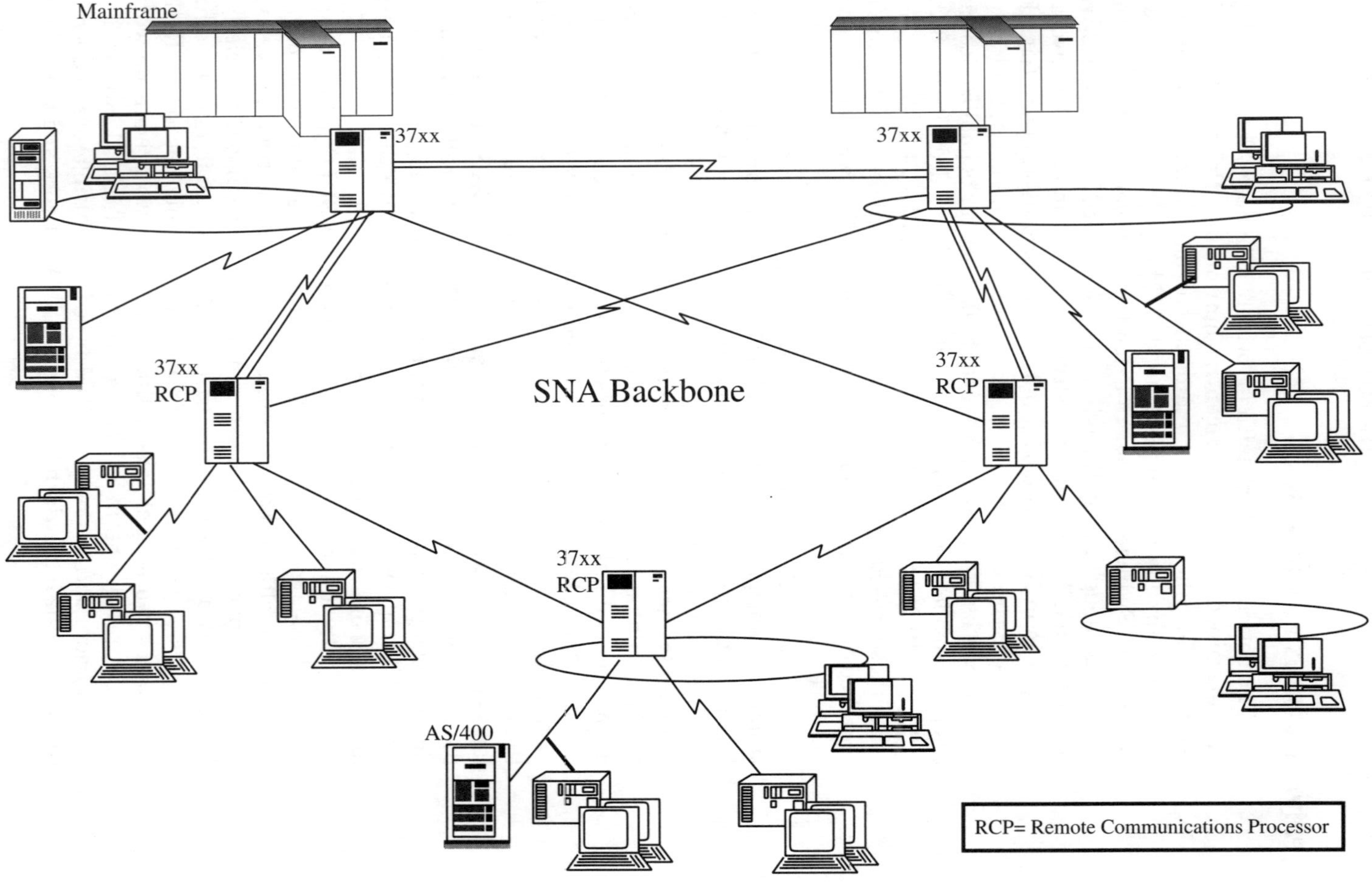

FIGURE 3.9 Private SNA network built around local and remote 37xx's.

In T1-based private networks the T1 multiplexors are responsible for running and managing the WAN network. An express goal of ATM is to displace such TDM-based, private voice-and-SNA networks. This alone explains the notion of private ATM WAN networks. Tens of thousands of large enterprises have been running private voice-and-data networks for at least two decades. Many relish and even rely on the near "total control," "being master of your own destiny," and the security of "no unauthorized users" aspects of private networks. Private ATM networks ensure that this in-house network heritage can be continued into the next millennium.

SNA/SDLC and T1 multiplexor networks are not the only types of private networks employed by large enterprises. Some enterprises, in North America, Europe, and Asia, rather than subscribing to private X.25 services, implemented their own in-house X.25 networks with private X.25 switches and carrier-supplied 56/64Kbps leased lines. A number of these private X.25 networks, especially those used for certain transport and credit-card authorization applications, are huge, straddle the globe, and service thousands of transactions per second. Some of these private X.25 networks support SNA traffic using either PADs or the integrated X.25 support available on certain SNA devices such as 37xx's, 3x74s and AS/400s. In Europe a few enterprises have also been known to use the optional X.25 SNA Interconnection adjunct to ACF/NCP to enable them to support bona fide X.25 interactions between native X.25 devices across a private 37xx-based SNA network!

3.3.1 Public Frame Relay Networks Pave the Way to Public ATM Services

Based on this relative popularity of private X.25 networks, IBM with the availability of ACF/NCP 6.2 in mid-1993 has been trying, albeit without much success, to convince SNA customers to implement private frame relay networks built around 3745 Communications Controllers. A 3745 would now act as full-fledged frame relay switches, as opposed to just being an SNA Controller. IBM, however, has been successful in convincing a few customers of late to implement private frame relay networks using the 2220 Nways BroadBand Switch as a frame relay switch. *Such 2220-based private frame relay networks could in time be upgraded to private ATM networks, or to hybrid private ATM-cum-frame relay networks where the frame relay portion may be used just to support certain data-only applications.*

Given the existence of so many private networks, especially within the IBM networking community, it is not surprising that IBM and of late even Cisco advocate the virtues of implementing private

ATM WAN networks, as opposed to relying on public ATM. The rationale for a private frame relay or ATM network is obvious. With a private network, the enterprise has total control of the network parameters and characteristics, and moreover can impose stringent security measures to control network access and safeguard the data flowing within the network. In addition, the enterprise can in theory rigorously control quality of service provisioning, packet or cell discard policies (i.e., the congestion control policies), and the bandwidth requirements for optimal performance. The enterprise will also have the option of being able to fine-tune the network, on an on-going basis if necessary, to obtain peak performance. With a public network an enterprise has little if any control over the performance or operation of the network—even though now with frame relay's CIRs and ATM's QOS an enterprise at least can negotiate for a minimum level of performance for a given price.

The downside of private networks, particularly in the case of frame relay and in future ATM, is cost. There are four major cost elements associated with running a private network. These are: the capital cost of acquiring the necessary networking products (e.g., frame relay or ATM switches à la the IBM 2220); the significant personnel costs to administer, operate, and manage the network; service, maintenance, and upgrade costs for the networking products; and last but not least the cost for acquiring bandwidth in the form of leased lines from one or more public carriers. Ideally, the control and security-related advantages of running a private network should outweigh the tangible cost elements of implementing and running that network.

With public networks, in particular frame relay services, now offering 9.6Kbps to T1 rate bandwidth at extremely attractive tariffs—in many cases at a fraction of the price of an equivalent capacity leased line—cost-justifying a private network is not as easy as it used to be in the past. Public ATM services, once they manage to get themselves established, which is likely to be around the end of 1996, are expected to continue this trend of offering cost-compelling bandwidth.

While the cost of bandwidth from public WAN services has been falling, the personnel costs for network administrators and operators have continued to rise. And in the end, these two cost elements—one going down while the other goes up—will become the crucial deciding factor when it comes to evaluating the viability of a private frame relay or ATM network. Out-sourcing as a means of containing or reducing corporate costs has now been in vogue for awhile and some swear by its effectiveness. *With a public frame relay or ATM WAN network, an enterprise can effectively out-source a significant amount of their network operation and management costs to the carrier.* This out-sourcing-related cost-saving is an added bonus.

Moreover, many carriers now provide a complete multiprotocol network out-sourcing option whereby they will supply, install, maintain, and remotely manage a complete bridge/router or FRAD-based network for a particular enterprise. The enterprise does not even have to acquire, install, and configure its own bridge/routers or FRADs. The service provider does all of this—albeit for a fee.

Today, by subscribing to a public frame relay service—either just for the bandwidth or for the entire infrastructure for multiprotocol networking—an enterprise can not only enjoy considerable savings in bandwidth costs but can also significantly reduce its network operations budget. Hence, it is not surprising many IBM networking enterprises have already switched to or are actively in the process of switching from private SNA networks to public frame relay services. They are either running SNA directly over frame relay (using SDLC-to-FR FRADs or the integrated SNA/FR support available on 3745s, 3174s, AS/400s, PCs, etc.), or running SNA traffic alongside other multiprotocol traffic using bridge/routers or RFC 1490 routers.

This migration of private SNA networks to public frame relay services is one of today's most significant and prevalent networking trends. North American frame relay carriers estimate that during 1995 and 1996 at least 50% of the enterprises migrating to their services will bring along SNA traffic that hitherto would have been running on a private network. This large-scale migration to public frame relay will indubitably have a knock-on effect on the popularity of public ATM service. Once an enterprise has moved its traffic to a public frame relay service it is unlikely to reconsider a private frame relay or ATM network—unless, of course, the service failed to deliver against its commitments. To date most SNA enterprises that have switched to public frame relay networks seem to be happy with the performance, reliability, and support provided by these networks. This is especially true once they have resolved how much CIR-based bandwidth they need to obtain for the various VCs to eliminate excessive amounts of frames being discarded because CIRs are continually exceeded. However, given that this migration really only started in earnest in 1995, it is still too early to tell whether this level of satisfaction will continue, particularly as the now relatively unsaturated and underutilized networks start to get more and more popular and start carrying heavier loads of traffic.

Public networks, however, are not for everybody. There will always be enterprises that will elect to maintain private networks despite their costs. They will do so due to either security concerns; the desire for total control; the need for lightning-fast, consistent, and predictable response times (as would be in the case of a travel reservation system where slow response translates to lost-opportunity costs); or the sheer volume of traffic involved. IBM, Cisco, and others are obviously cognizant of this and have already taken steps to address this demand for private frame relay and ATM networks.

The trend, though, without a shadow of a doubt is toward public networks. Frame relay is pandering to this trend and setting the pace that public ATM networks will obviously want to emulate. The bottom line here is that IBM, Cisco, and others will provide equipment and endorse private ATM networks. Some enterprises, particularly those with very large traffic volumes or very exacting networking requirements related to response times or security, will opt for private ATM networks. In some cases the private ATM networks will evolve from private frame relay networks. Private ATM or for that matter private frame relay networks will, however, not be as popular or pervasive as private SNA/SDLC or private T1 multiplexor networks have been in the past. With frame relay and ATM, public WAN networks have indeed become the way of the future.

3.4 CONSOLIDATING THE PARALLEL DATA NETWORKS ALONG THE WAY TO ATM

For the IBM-centric networking community, the consolidation of today's dreaded parallel data networks into a single, SNA/APPN-capable multiprotocol network will be a major milestone on their road toward ATM. In many cases, this consolidation effort and the inevitable "bedding-down" period that will follow this integration will in reality be the gating factor that dictates when an enterprise will have the time, the budget, and the true motivation to evaluate their migration plan to ATM.

At the start of 1995, there were still about 2,000 enterprises worldwide with SNA networks. Depending on one's networking bias, these enterprises were either very fortunate and astute, or very unfortunate and misguided in not having a parallel non-SNA network. Though some of these "dark blue" diehards will manage to persevere for another few years without embracing any other protocols, most during the next couple of years will be forced to adopt at least one other non-SNA protocol—most likely TCP/IP. Consequently, even the bulk of these hitherto SNA-only enterprises will have no choice but to contemplate multiprotocol networking while evaluating the potential of ATM.

As discussed in Chapter 1 there are eight proven technologies for integrating SNA/APPN LAN traffic into a multiprotocol network. These eight technologies are as follows:

1. Bridging in general, and token-ring Source-Route Bridging (SRB) in particular
2. Encapsulation of the SNA/APPN traffic within TCP/IP message units as epitomized by Data Link Switching (DLSw) and Cisco's Remote Source Route Bridging (RSRB)
3. Encapsulation of SNA/APPN traffic within Frame Relay per the RFC 1490 and FRF.3 standard
4. APPN Network Node or HPR Network Node implementations on bridge/routers that will permit SNA/APPN traffic to be routed alongside non-SNA traffic
5. IBM AnyNet Gateways, à la IBM's 2217 MultiProtocol Concentrator, that either convert IP traffic to, or encapsulate other non-SNA traffic within, SNA LU 6.2 message units to create an APPN- or HPR-based network capable of transporting non-SNA traffic
6. LAN-over-SNA, which has now really been superseded by AnyNet gateway solutions à la the IBM 2217
7. SNA Session Switching, which synthesizes TCP/IP encapsulation of SNA/APPN traffic with direct, mainframe-to-mainframe switching of SNA traffic realized by deploying downstream mini-SSCPs (i.e., SNA control-point logic as found in ACF/VTAM)
8. CrossComm's Protocol Independent Routing (PIR), which provides the value-added features of TCP/IP encapsulation such as dynamic alternate routing while maintaining plug-and-play ease of setup of SRB

In addition to one of the above technologies, SDLC-to-LLC:2, X.25/QLLC-to-LLC:2, remote polling, or sync passthrough will be used to integrate traffic from link-attached SNA devices with those from LAN-attached SNA devices and non-SNA devices. All of the above mentioned technologies are described in detail in Chapters 5 to 8.

Of the eight SNA LAN-traffic integration technologies listed above, LAN-over-SNA and PIR are now no longer actively promoted—though they are still viable techniques, with PIR in addition being well proven and once very popular.

The other six technologies (or for that matter the eight technologies in total) can be divided into two very distinct categories: non-SNA technologies that accommodate SNA/APPN traffic, and SNA/APPN/HPR-oriented technologies that provide an SNA-centric solution for multiprotocol networking. Table 3.2 shows the division of these eight technologies into these two categories:

TABLE 3.2 Demarcating the Eight Parallel Network Consolidation Technologies Based on their Perspective of SNA

Non-SNA Technologies That Accommodate SNA/APPN Traffic	*SNA/APPN/HPR-Oriented Technologies That Provide an SNA-Centric Solution for Multiprotocol Networking*
1. Bridging	1. AnyNet Gateways à la IBM's 2217
2. DLSw, RSRB, etc.	2. APPN or HPR NN routing
3. Frame relay RFC 1490 cum FRF.3	(3. LAN-over-SNA)
4. SNA session switching	
(5. PIR)	

This ability to divide the network consolidation technologies into these two very distinct categories immediately demonstrates that all these technologies are not similar and that they have different roots, heritage, and values. The SNA/APPN/HPR-oriented technologies, obviously, propagate SNA's penchant for value-added, highly secure, connection-oriented, and deterministic networking. The non-SNA technologies, such as bridging and RFC 1490, on the other hand, represent a more laid-back approach to networking.

The existence of these two categories when it comes to SNA/APPN integration technologies should dispel the notion that "one size fits all" when it comes to parallel data network consolidation: hence the irony of a sentiment that has prevailed within the IBM networking community that multiprotocol bridge/routers are the only strategic means for realizing a consolidated data network—irrespective of the protocol mix of the resultant combined network. An enterprise whose 1997 WAN traffic is going to consist of 80% mission-critical SNA traffic is likely to have a very different set of networking requirements to an enterprise whose 1997 WAN traffic is going to consist mainly of e-mail and file-transfer traffic. To the former, near zero-failure high-availability, predictable and consistent response times, and networking accounting (i.e., the ability to determine and charge-back network costs to actual network users) are sacrosanct. Network availability and performance is also obviously important to the latter, but not to the same extent that it is to the customer with 80% mission-critical SNA traffic.

E-mail, file-transfer, and software distribution are not real-time, store-and-forward operations. Network failures even as frequent as once a week, that typically require up to two hours to rectify, would not disrupt e-mail or file-transfer environments and would in many cases be deemed par for the course. This degree of downtime in a network supporting high volumes of SNA mission-critical traffic would be catastrophic and career-jeopardizing. Figure 3.10 depicts how networking requirements and values can differ depending on the mix of protocols supported by a converged network. Hence, when trying to select the best SNA/APPN integration technology to consolidate the parallel data networks of a given enterprise, it is best to use the envisaged 1997 or 1998 protocol mix of the converged network as the basis for making the decision. Figure 3.11 provides a preview of how to select an SNA/APPN integration technology-based on 1997–1998 protocol mix.

For enterprises whose SNA/APPN traffic content is already low or is rapidly falling, such that it will amount to less than 50% of the overall WAN traffic mix by 1997, the options for converging parallel data networks are straightforward, proven, and established. These options are bridging, TCP/IP encapsulation à la DLSw, and RFC 1490-based frame relay encapsulation. Of these, bridging should only be considered for relatively small networks consisting of no more than 100 remote sites or overall network nodes. If frame relay is to be the underlying mechanism for the WAN, RFC 1490, as discussed in Chapter 6, will invariably be more efficient than DLSw.

On the other hand, if an enterprise is expected to have a high volume of SNA/APPN traffic that approaches or exceeds 80% of the overall WAN traffic mix in the 1997–1998 period, it should consider the AnyNet Gateway or APPN/HPR NN routing options ahead of bridging, DLSw, or RFC 1490. For such SNA-heavy enterprises the AnyNet Gateway or the APPN/HPR NN routing solution will enable the implementation of an SNA-centric multiprotocol network. Such a network will pander to all the needs of the SNA/APPN traffic (e.g., supporting native SNA routing and COS-based path selection and traffic prioritization) while at the same time providing a highly reliable transport mechanism for the non-SNA traffic. The concepts and the pros and cons of implementing SNA-centric multiprotocol networks using AnyNet technology are described in length in Chapter 8.

3.5 REFLECTIONS

ATM is now without any doubt the endgame of contemporary enterprise networking. ATM's allure is very straightforward and addictively compelling. It promises to offer oodles of very cost-effective mega- to gigabit range bandwidth, and the possibility of being able to adroitly support concurrent data, voice, and video traffic streams across a single composite network. Despite its overtly seductive siren

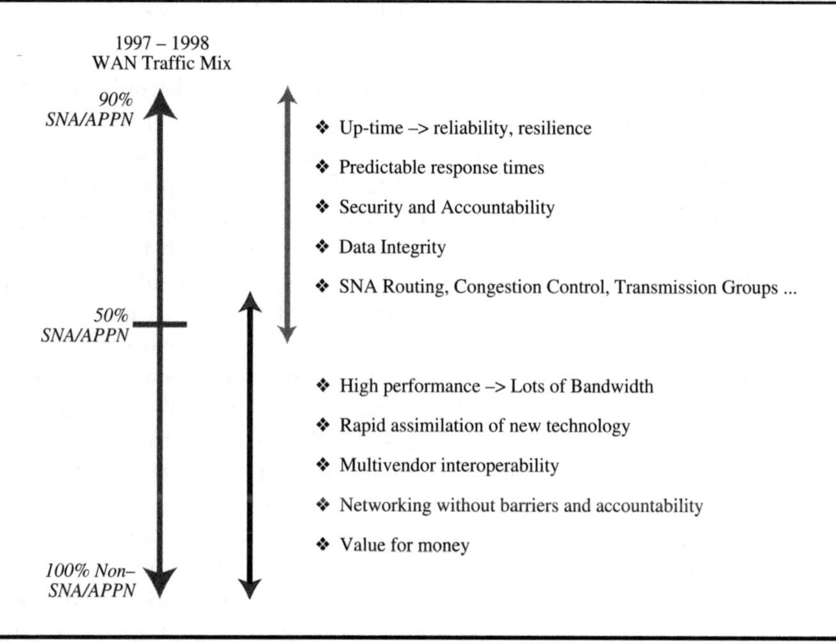

FIGURE 3.10 Profile of WAN traffic mix that can dictate networking values and requirements.

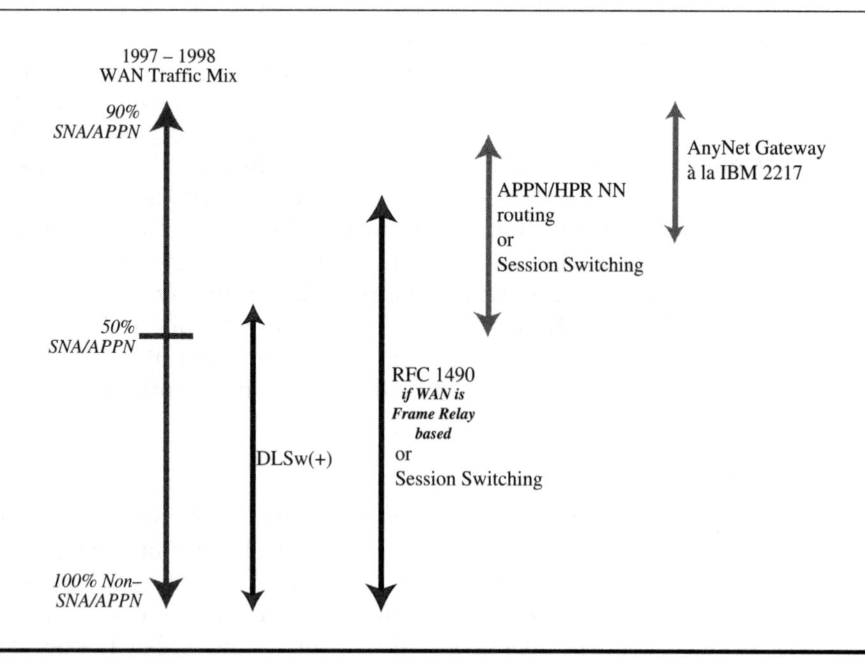

FIGURE 3.11 Selecting an SNA/APPN integration technology based on WAN traffic mix.

call, ATM is, however, not going to be widely embraced by the IBM-centric networking community till at least 1997. There is a variety of tangible reasons for this lag time. Some of these are: the innate and justifiable xenophobia of new technology among the mission-critical networking practitioners, the lack of a dire need for huge amounts of additional bandwidth in token-ring LAN-based campus environments, as well as the current relatively high costs of ATM to the desktop and WAN ATM. In addition, most enterprises are committed to consolidating their dreaded parallel data networks prior to thinking about ATM.

There are six to eight viable technologies that these enterprises can use to successfully realize this data network consolidation. These technologies can be split into two distinct categories: non-SNA technologies that accommodate SNA/APPN traffic, and SNA/APPN/HPR-oriented technologies that provide an SNA-centric solution for multiprotocol networking. Enterprises that expect their 1997–1998 WAN traffic mix to still consist predominantly (i.e., 80% or more) of SNA/APPN traffic should consider one of the SNA/APPN/HPR-oriented network consolidation technologies such as AnyNet Gateways or APPN/HPR NN routing. Enterprises whose 1997–1998 traffic mix is likely only to have 50% or less SNA/APPN traffic should consider non-SNA technologies, such as DLSw, RSRB, or RFC 1490, for converging their parallel networks. If frame relay is to be the WAN fabric for the consolidated data network, RFC 1490 in most cases is likely to be a more efficient solution than DLSw or RSRB.

Bridge/routers from nearly all of today's market leading vendors (Cisco, Bay Networks, 3Com) offer DLSw(+), RFC 1490/FRF.3-based encapsulation of SNA/APPN traffic, as well as integrated APPN NN routing. Hence, selecting one of these network consolidation technologies does not restrict one to a particular vendor or one type of bridge/router. Bridge/routers from any of these leading vendors can be configured to support either DLSw(+), RFC 1490, or APPN NN.

ATM will first appear in enterprises at the campus level. WAN ATM will follow one to two years later as the tariffs of public ATM networks start to drop. Some enterprises may even contemplate building their own ATM WAN networks using large ATM switches, such as IBM 2220 or Cisco LightStream 2020, that are interconnected via public carrier supplied T3 or OC3 links. Initially ATM will be used mainly for data applications with LAN emulation at the campus level, ensuring backward compatibility with today's LAN software. The cost and the disruption associated with ATM to the desktop will significantly diminish its appeal and viability in the short- to mid-term. The initial luster of 25Mbps data-only ATM to the desktop has already started to diminish given that its price/performance is now being challenged by

100/155Mbps ATM adapters, 100Mbps Fast Ethernet, 100Mbps Any-LAN, and LAN switching.

To begin with, ATM at the campus level will be restricted to backbone applications where ATM will provide a broadband pipe between hubs, LAN switches, and bridge/routers servicing traditional token-ring or Ethernet LANs. In many instances LAN switching, and in particular token-ring switching, will be used, prior to ATM, to provide additional bandwidth to individual PCs/workstations, LAN workgroups, and campus backbones. Hybrid campus scenarios with ATM being initially used between LAN switches, hubs, or bridge/routers as a broadband pipe for intracampus, LAN-to-LAN transactions will in general provide IBM networking-biased enterprises with their first true taste of the joys of ATM.

CHAPTER 4

Token-Ring Switching: The On-Ramp to ATM

A QUICK GUIDE TO CHAPTER 4

Token-ring switching is a compelling technology for economically enhancing the bandwidth of existing LANs with minimum disruption and very little risk. With Token-ring switching, additional bandwidth can be made available to LAN backbones, LAN workgroups, or even individual PCs. Moreover, the additional bandwidth to PCs is delivered without the need for any new LAN NICs or adapters. Token-ring switching will be the campus bandwidth-enhancing precursor to ATM. In many cases, ATM will get introduced into enterprises as a means of building a high-capacity backbone between token-ring switches. Token-ring switching will also complement ATM at the campus level for many years to come.

This chapter is a comprehensive and in-depth tutorial on all aspects of and issues pertaining to token-ring switching such as: *cut-through, store-and-forward, transparent switching, adaptive cut-through,* and the potential need for SRB.

Following a brief overview of LAN switching in general, the chapter kicks off by comparing and contrasting token-ring switching to ATM switching. It then compares token-ring switching to Ethernet switching, given that the latter is a relatively well-established technology whereas token-ring switching is still somewhat nascent. There is a sidebar that explains how Ethernet can now be freely deployed in SNA environments, despite the apparent roadblock of 37xx communications controllers not supporting SNA-over-Ethernet.

Section 4.3 looks at the principles of operation of a token-ring switch, paying particular attention to the key role played by transparent switching. Transparent switching is in effect Ethernet's transparent bridging vis-à-vis LAN switching. In many scenarios, transparent switching could, and should, be used in preference to SRB as used in traditional token-ring LANs. Consequently, this section also examines when one scheme should be used in preference to the other.

Once the mechanics of token-ring switching have been dealt with, Section 4.4 focuses on the ways in which token-ring switches can be profitably deployed. There are three primary scenarios for deploying such switches: in stand-alone workgroups, for splitting up large LANs, and for interconnecting LAN segments to a backbone. The advantages of token-ring switching in each of these scenarios are described, and there are figures showing exactly where and how the token-ring switches fit in.

Section 4.5 deals with all the various features of LAN switching including ATM *up-links*, cut-through mode switching, and Virtual LANs. Table 4.1 summarizes the features available on some of the key token-ring switches that will widely be used in the 1996–1998 time frame. In addition, there is a sidebar that explains how 32Mbps full-duplex token-ring can be realized with token-ring switching.

Token-ring switching is a stimulating and innovative new technology that can be profitably used to dramatically improve the throughput and enhance the overall manageability of *existing* token-ring LANs. Token-ring switching is an extremely cost-effective way of providing additional network bandwidth to LAN-backbones, LAN workgroups, or individual PCs/workstations. Token-ring switching can deliver this additional networking bandwidth, even to individual PCs/workstations (i.e., to the desktop), without the need for any new NIC cards or wiring, and with a minimum of network reengineering. Token-ring switching also does not require any changes to established LAN administration and management processes. Neither does it rely on any changes being made to the existing configuration of LAN workgroups or campus backbones.

Token-ring switching per se is not an ATM-based, fixed 53-byte cell size switching scheme. Instead, it is a variable length packet-switching scheme that works by rapidly switching standard token-ring frames, in their entirety, from one port on the switch to another. Some token-ring switches do, however, employ a cell-based backplane architecture for forwarding frames from one port to another. The presence of a cell-based backplane should not in any way be construed to mean that token-ring switching is related to or uses ATM technology.

There are two distinct frame-forwarding modes that can be used by a token-ring switch. In one mode, known as *store-and-forward*, the switch waits until it has received a complete error-free token-ring frame before it starts forwarding the frame to the destination port. In the other mode, known as *cut-through*, the switch starts forwarding the bits that make up a frame as soon as it has determined the appropriate port by reading the 6-byte, MAC-Layer "Destination Address" that appears near the start (i.e., fourth byte) of every token-ring frame. The pros and cons of these two forwarding modes are discussed in detail in Section 4.5.3.

Token-ring switching, and for that matter LAN switching (e.g., Ethernet or Fast Ethernet switching), in general, like ATM, is a Layer 1 and 2 switching methodology. LAN switching and ATM, at Layer 2, operate at the MAC sublayer—that is, the lowest sublayer within Layer 2. Figure 3.1 illustrated how Layer 2 of the OSI Reference Model consists of two sublayers: the LLC sublayer on top of the MAC sublayer.

In common with campus ATM, the maximum bandwidth gains possible with LAN switching can only be derived by attaching individual PCs/workstations, LAN servers, and computers—via dedicated media (e.g., UTP)—to separate ports on a LAN switch. Figure 4.1 shows a token-ring switch with PCs/workstations and LAN servers attached to it via dedicated media.

Just as with campus ATM scenarios, LAN switches can also, however, be used in hybrid configurations where entire LAN segments with multiple devices can be attached to individual ports on the switch. Figure 4.2 shows a token-ring switch that is supporting both multiple device LAN segments as well as dedicated media-attached LAN servers and other devices. In shared-media configurations involving multidevice LAN segments as shown in Figure 4.2, LAN switches, rather than increasing the bandwidth at the disposal of individual devices, increase the bandwidth available for LAN-to-LAN or LAN-to-Server interactions.

The fact that token-ring switches work by purely switching unadulterated, traditional token-ring frames, as opposed to fixed size cells or a new type of frame, has two major and indisputable advantages. For a start, token-ring switching works with existing PC/workstation NICs

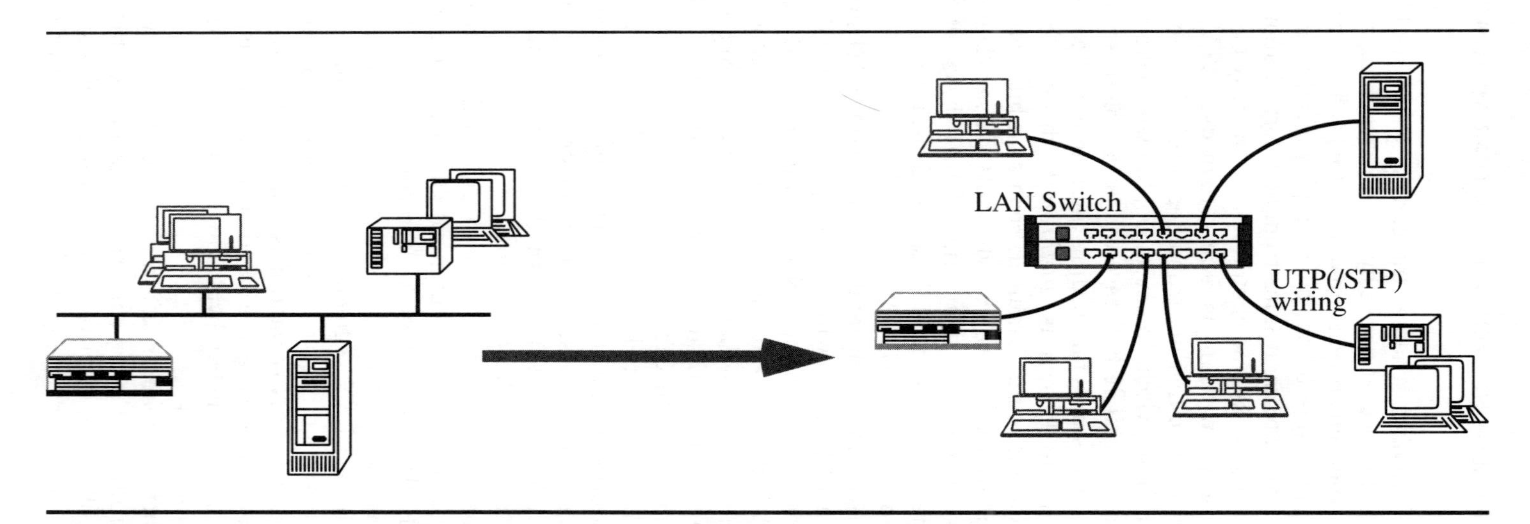

FIGURE 4.1 Moving from a shared media LAN to dedicated media LAN-switching.

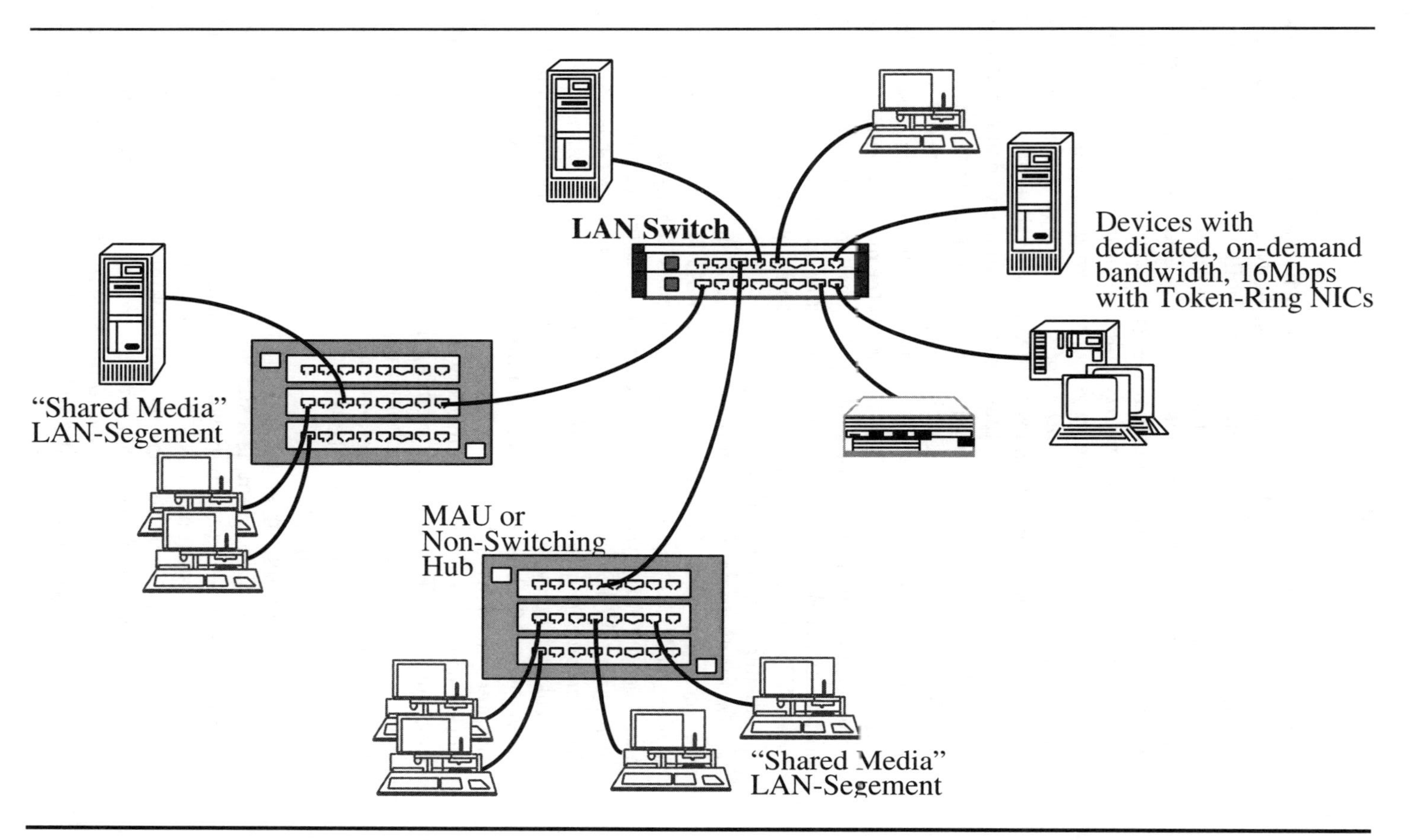

FIGURE 4.2 A LAN switch supporting both multiple device LAN segments, as well as devices such as servers attached via dedicated media.

and within existing LAN configurations, whereas ATM to the desktop is contingent on the replacement of existing NICs with ATM adapters. In addition, token-ring switches can be built using existing, proven, well-known, and relatively inexpensive token-ring technology—in particular standard token-ring chip sets. Consequently, token-ring switching tends to be a somewhat low-risk technological proposition.

Thanks to its low-cost, nearly nondisruptive, and low-risk traits, token-ring switching will end up being the campus bandwidth-enhancing precursor to ATM across much of the IBM networking community. Token-ring switching, as opposed to ATM, will thus provide this community with their first real taste of contemporary, dedicated media-oriented but shared-media-compatible, switching technology.

Most token-ring switches, by mid-1996, will support ATM *up-links*. A token-ring switch will use these ATM up-links (also referred to as *fat pipes*) to either realize 155 Mbps (or more) broadband connections with other intracampus token-ring switches, or gain access to a port on an ATM switch so that it can forward traffic from its LAN ports to an ATM network. Figure 4.3 depicts some of the ways that token-ring switches will use ATM fat pipes.

In many instances, members of the IBM networking community may get their first exposure to bona fide ATM switching when they deploy ATM fat pipes between intracampus token-ring switches. In addition, for many years to come token-ring switching will end up complementing ATM at the campus level, acting as a lower-cost, token-ring NIC-compatible adjunct for enhancing LAN workgroup bandwidth.

Token-ring switches work with existing PC, workstation, and LAN server 4- or 16Mbps token-ring adapters. For full-duplex operation the adapters used by LAN servers—but not those of the client PCs or workstations—need to be replaced with full-duplex, 16/32Mbps token-ring adapters. Token-ring switching is contingent on only one new piece of equipment—the token-ring switch itself. The token-ring switch will replace either a token-ring MAU (Multistation Access Unit wiring concentrator à la the IBM 8228), a nonswitching wiring hub, or a source-route bridge.

LAN switching in general, with token-ring switching being the token-ring-specific variant, is a simple, nonintrusive, and cost-effective means for increasing the bandwidth on LANs, as well as for providing certain devices, such as LAN servers, with full-duplex capability. Like ATM it also sets out to increase networking bandwidth. *Token-ring switching, however, delivers this increased bandwidth without the need for new adapters, wiring schemes, and management schemes as with ATM.*

The initial groundbreaking work on token-ring switching was done around the end of 1994 by Standard Microsystems Corp. (SMC),

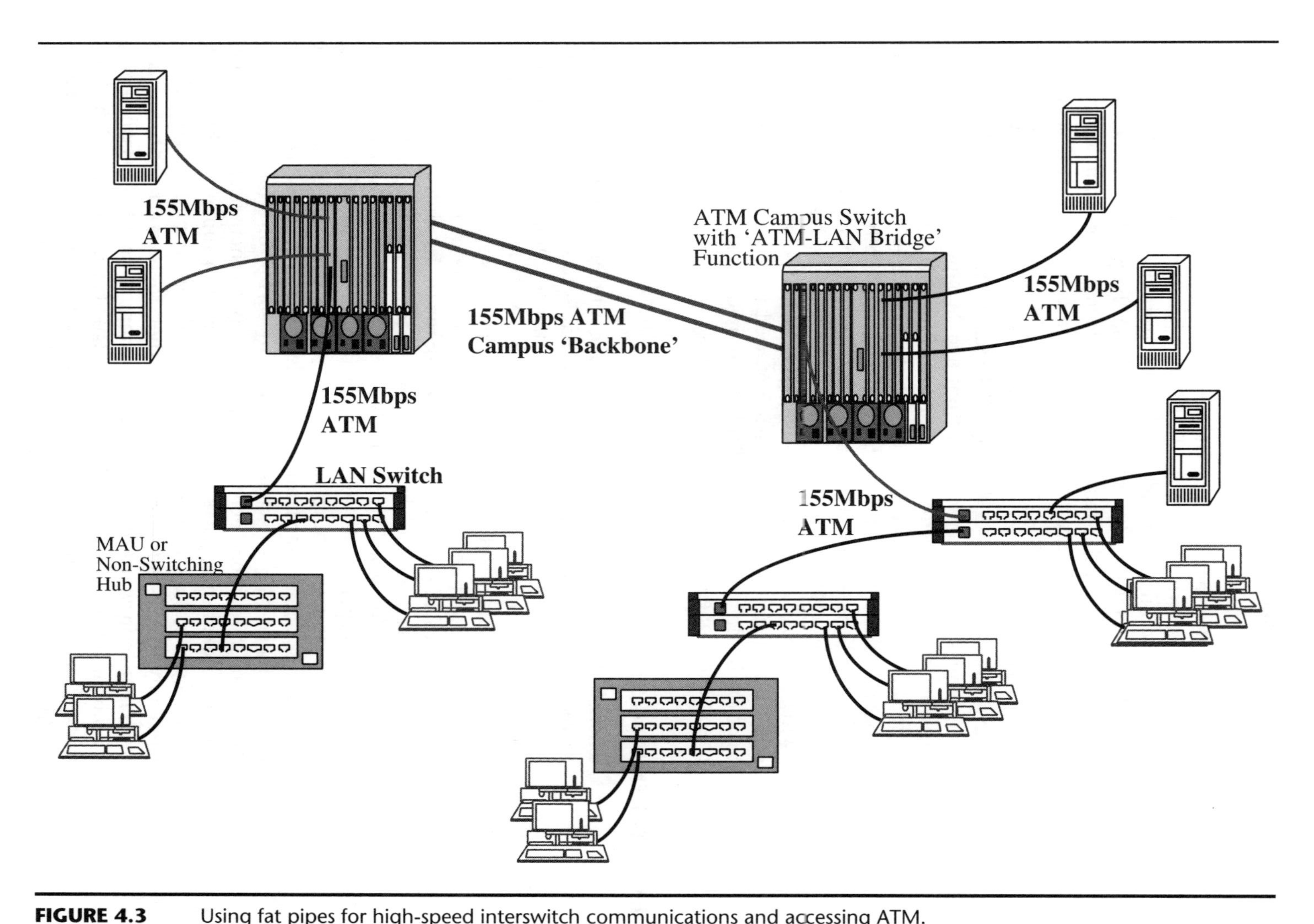

FIGURE 4.3 Using fat pipes for high-speed interswitch communications and accessing ATM.

Centillion (now acquired by Bay Networks), and Nashoba Networks (originally known as NetTrend). At that juncture, token-ring switching lagged behind Ethernet switches by around 18 months. There was a good reason for this: Token-ring LANs on the whole were not at that point experiencing the same levels of performance-sapping congestion that was by then plaguing comparably sized Ethernet LANs. As was discussed in Chapter 3, there are three tangible reasons as to why token-ring LANs fare so much better than comparably sized Ethernet LANs when it comes to dealing with heavy traffic loads. These are: token-ring's deterministic access that avoids bandwidth blackouts due to collisions; the short-duration nature of much of the traffic on token-ring LANs; and the fact that a 16Mbps token-ring offers at least 2Mbps of additional bandwidth compared to a 10Mbps Ethernet.

Toward the end of 1995, token-ring switches were available from: Bay Networks/Centillion, Cabletron who OEM the Nashoba Networks switch, Cisco, CrossComm, Madge, and SMC. 3Com, IBM, LanOptics, and Proteon, among others, have announced switches that are expected to be available by early 1996. Table 4.1 provides a summary of features of some of pioneering token-ring switches.

4.1 TOKEN-RING SWITCHING VIS-Á-VIS ATM

Though they are both Layer 1–2 switching schemes that can be used to increase the networking bandwidth available to individual PCs/workstations, LAN segments, or campus backbones, token-ring switching and ATM are indeed very different beasts. ATM for a start sets out to provide scalable broadband bandwidth to every portion of an entire far-flung network, from individual desktops on one campus all the way across a WAN to desktops at remote campuses. In contrast, the bandwidth increases provided by token-ring switching are limited to campus backbones, LAN workgroups, LAN servers, and individual PCs/workstations. In other words, token-ring switching is strictly LAN-oriented.

Token-ring switching technology per se does not have a native networking scheme for transporting data across a WAN. None of the original token-ring switches provided any sort of interface for switch-to-switch interactions across a WAN. Many in their first incarnation had only token-ring ports. A few then started to support either FDDI or ATM ports. If a token-ring switch has only token-ring or FDDI ports, then any WAN communications, whether to interact with another switch or to forward data to a remote campus, would be possible only by having a bridge/router, FRAD, or bridge attached to one of the LAN ports. Figure 4.4 shows how token-ring switches can communicate across a WAN using a bridge/router or FRAD. In this type of bridge/router or

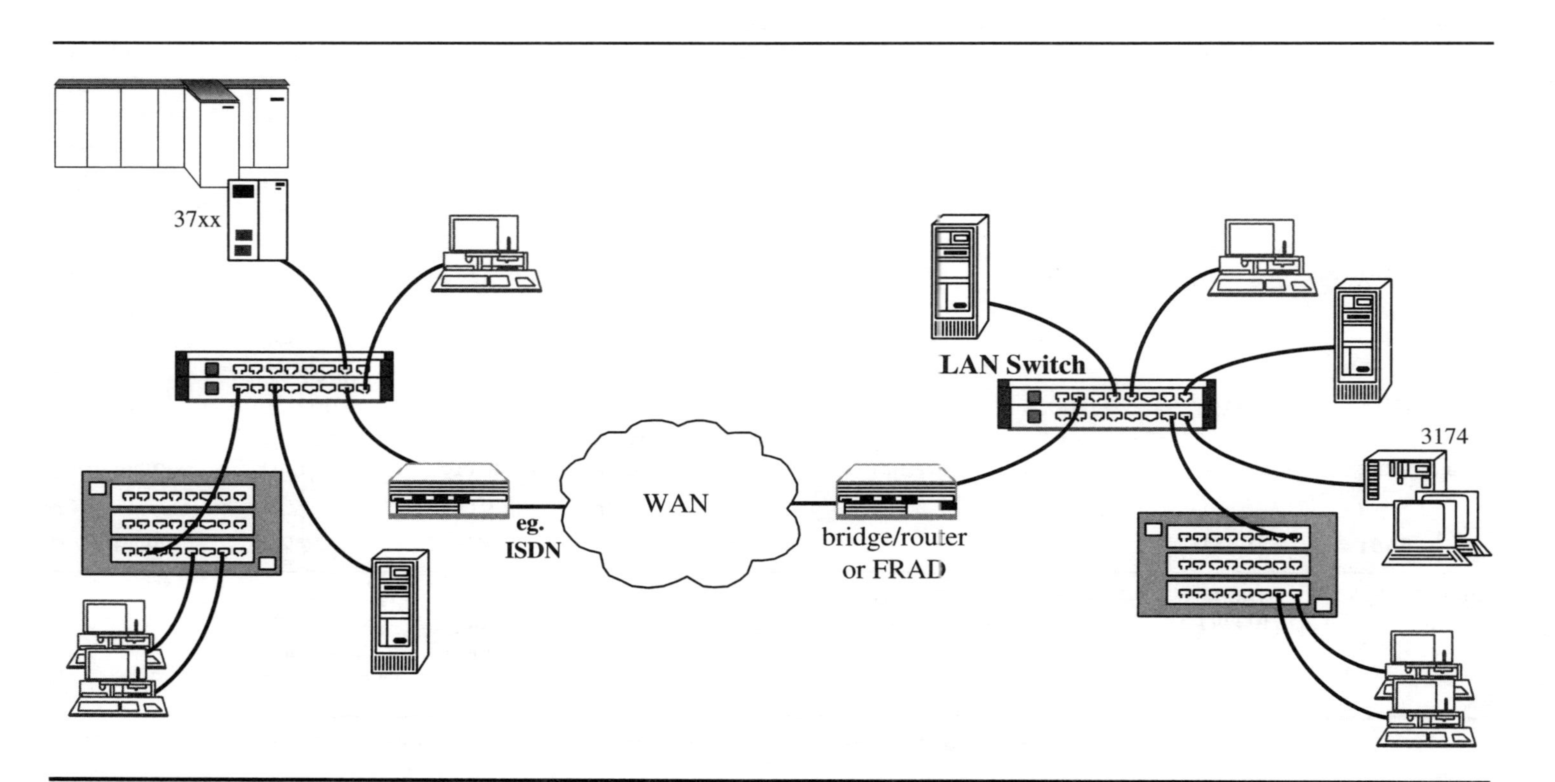

FIGURE 4.4 WAN communications between sits using LAN-switching.

TABLE 4.1 Summary of the Features Offered by Some of the Technology Pioneering Token-Ring Switches

	BayNetworks Centillion Speed Switch	*Nashoba Concord Cabletron TSX-1620*	*SMC ES/1*	*Cisco/ Madge ProStack RingSwitch*	*CrossComm XLT Module*	*IBM 8272 LANStreamer*	*3Com LinkSwitch 2000TR*	*LanOptics T-SwitchPro*
Availability	Now	Now	Now	Now	Now	1996	1996	1996
Ports	4/24	8/16	4/20	8	4	8/12/16	12	8
Form factor	Hub Chassis	Stackable	Hub Chassis	Hub Chassis	Router Chassis	Stackable	Stackable	Stackable
Price-per-port	$1,800 ->2,500	$1,500	$3,150 -> 5,800	$2,000	$2,000 -> 2,500	$700	$700	$1,500?
Store-and-forward	Yes	Yes	Yes	Yes	Yes	Yes	Yes	Yes
Cut-through mode	No	No	No	Yes	No	Yes	Yes	Yes
Transparent bridging/ switching	Yes	Yes	Yes	4Q95	Yes	Yes	Yes	Yes
Source-route bridging	Yes	Yes	Yes	Yes	Yes	2H	Yes	Yes
Fat pipe (possibly in future)	FDDI & ATM	FDDI & ATM	FDDI	FDDI & ATM	Through Edge-Route Module	FDDI & ATM?	None Announced	None Announced

Management	SNMP	SNMP and RMON	SNMP	SNMP	SNMP and RMON	SNMP	SNMP	SNMP and RMON
Address filtering/ NetBIOS caching	Yes	Yes	No	Yes	Yes	Not announced	Not announced	Not announced
Support full-duplex Adapters	No	No	No	Yes	Yes	Yes	Yes	Yes
Redundant power supply	Yes	Yes	No	Yes	Yes		Not announced	Not announced
Activate/ seactivate ports on time-of-day	No	Yes	No	No	No	No	No	No –
Layer 3 IP and IPX Routing	No	No	No	No	Yes	No	No	No

FRAD-based WAN configuration, the token-ring switches appear and act as MAC-layer (i.e., Layer 2) bridges.

By 1996, most token-ring switches will offer one or more optional ATM ports. Consequently, ATM—especially when public ATM networks get around to offering tempting tariffs—will become, in time, the preferred means for WAN interactions between token-ring switches. Public WAN ATM, in theory, has the potential to provide for the first time the cost-justifiable 16Mbps to 32Mbps bandwidth necessary to ensure that token-ring switches can interact with each other at "full bore."

In addition to being LAN-centric, token-ring switching in theory is restricted to data applications. Token-ring LANs per se have never been seriously touted as a means for multimedia interactions. ATM, on the other hand, was also designed from day one and ground up to support voice, video, and data traffic. A token-ring switch when forwarding frames from one port to another does not slice-and-dice the data traffic into ATM-like cells in order to accommodate voice and video traffic. (Token-ring switches with ATM fat pipes will, of course, slice-and-dice the data traffic when sending it across these ATM links—unless of course they support a PTM or Frame/Cell mechanism that precludes the need to slice-and-dice when dealing with data-only traffic.)

Given that they do not have any explicit mechanisms in order to recognize or support CBR traffic, token-ring switches cannot guarantee any level of quality of service vis-à-vis video or voice traffic. Trying to support video or voice traffic across ports that may also be handling data traffic would be problematic, to say the least. The voice or video traffic may end up having to await while one or more 4Kbyte data frames that arrived at the source port ahead of them were being forwarded. This level of delay would result in jerky and stalled video images and unacceptable gaps in voice conversations. Thus, in general, token-ring switching is not a feasible means for supporting crucial multimedia applications. In certain configurations, however, token-ring switching, particularly if used with dedicated ports that are not subject to any data traffic, may provide sufficient bandwidth gains to make it possible to support at least some amount of digitized, real-time, full-motion video.

Token-ring switching and ATM also differ in the amount of bandwidth they can offer. ATM offers full-duplex bandwidth in the 155Mbps (i.e., 310Mbps when used with full-duplex I/O) range, or even higher. The maximum LAN bandwidth available with token-ring switching with *standard* token-ring NICs is 16Mbps. Moreover, this 16Mbps bandwidth is strictly half-duplex given that standard token-ring NICs do not support full-duplex operation. Full-duplex token-ring NICs for PCs/workstations or LAN servers are now available and are currently priced in the $700 to $800 range. (Refer to the "Full-Duplex Token-Ring" sidebar.) In

FULL-DUPLEX TOKEN-RING

Full-duplex token-ring operation is a natural and logical extension to the dedicated bandwidth, over dedicated media, notion promulgated by token-ring switches. With dedicated media (i.e., only one LAN station per port on the switch) the needs for arbitrating access to the media are greatly reduced. A LAN station with dedicated media can at any time transmit data frames as soon as a frame is received at the NIC without having to lose any time while contending to gain access to the media. A LAN station can also receive frames at any time without incurring any delays because other stations are using the media.

With dedicated media it would thus be possible for a given LAN station—with a suitably modified NIC—to concurrently send and receive frames to/from the token-ring switch independent of the other functioning of the other stations attached to the switch. This is the basis for full-duplex token-ring operation. The key to this mode of operation, however, is special, full-duplex-capable NICs which are currently priced in the $700 to $800 range.

Token-ring NICs, till quite recently, were only designed for half-duplex operation. A fundamental precept of conventional token-ring is that each frame transmitted on the LAN travels through the token-ring NICs of every station attached to that LAN. To facilitate rapid frame propagation through each NIC, half-duplex token-ring adapters maintain a repeat path, at the MAC Layer, for all frames received. This repeat path precludes the need to de-queue received frames and then having to re-queue them so that they can be propagated through the LAN. In a full-duplex adapter this internal repeat path is eliminated.

Full-duplex adapters also do not perform conventional token-ring-related functions such as: token-ring capture for media access, token recovery, frame repeating (i.e., repeat path), or the updating of the *Addressed-Recognized* and *Frame-Copied* (A&C) bits in the Frame Status Field at the end of the MAC/LLC frame. In addition, a full-duplex adapter will use an internal clock within the NIC or LAN station. This clock will provide bit clocking for frame transmission since there will no longer be an Active Monitor, as is the case with conventional token-ring LANs, providing master clocking for the entire LAN.

Despite these changes, full-duplex adapters still use, without any modifications, the standard token-ring MAC/LLC frames—

which include the MAC addresses, the SAPs, the RIF, the LLC control field, and so on. Thus, full-duplex operation does not require any changes to existing token-ring applications. Just as with half-duplex NICs, full-duplex token-ring NICs will work over existing UTP/STP wiring.

Dedicated media is a prerequisite for full-duplex operation. It would, however, be customary that a full-duplex station—typically a LAN server—will invariably interact with two half-duplex stations at any one time (rather than with another full-duplex station). The full duplex station will receive frames from one half-duplex station while transmitting frames to the other. The half-duplex stations interacting with a full-duplex station could be on shared media (e.g., multiple stations on a LAN segment).

Throughput to/from, and the delays associated with accessing, a LAN server can be dramatically reduced if the LAN server could transmit data to one station while concurrently receiving data from another station. Hence, full-duplex adapters are targeted for use in LAN servers as opposed to every single LAN station on a LAN. Restricting full-duplex adapters just to LAN servers significantly constrains the overall cost of this solution.

addition, the ports on most token-ring switches can work in full-duplex mode without the need for any additional hardware or software upgrades. When a device with a full-duplex NIC is attached to a token-ring switch—via a *dedicated UTP/STP cable*—the switch will support full-duplex 32Mbps I/O to/from that device.

Given the cost of full-duplex NICs it is likely that they will typically just be installed on heavily accessed LAN servers. Individual PCs/workstations will continue to have half-duplex NICs. (Replacing the half-duplex NICs of all PCs/workstations with full-duplex ones would end up costing around 70%–80% of what it would cost to bring 25Mbps half-duplex, 50Mbps full-duplex, initially data-only ATM to each desktop.) A full-duplex LAN Server attached to a token-ring switch will accept data from one device attached at 16Mbps, while simultaneously transmitting data to another device at 16Mbps. Thus, it is just the LAN server and token-ring switch port to which it is attached that is running at 32Mbps. The PCs/workstations with standard half-duplex ports, as well as the switch ports to which these PCs/workstations are attached, continue to operate at 16Mbps (or even 4Mbps, if 4Mbps rather than 16Mbps NICs are being used). Figure 4.5 illustrates full-duplex operation à la token-ring switching.

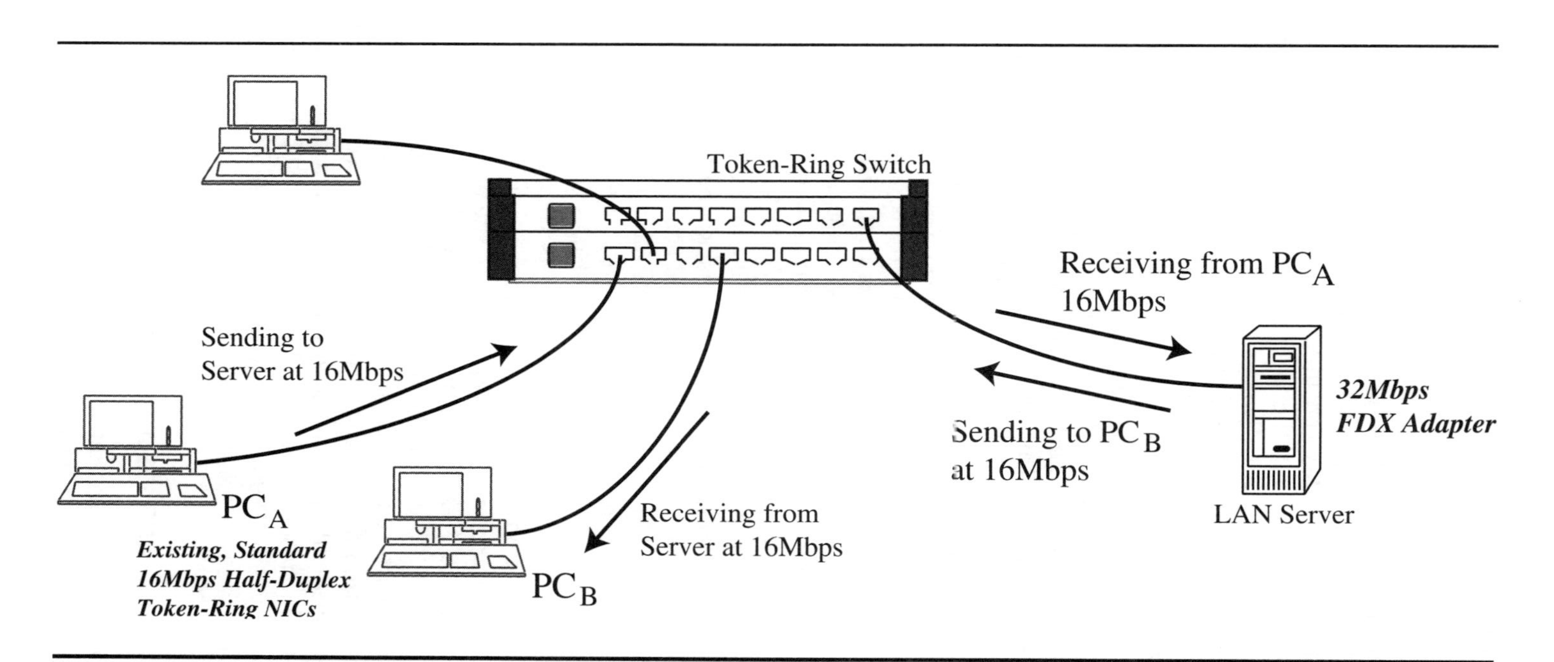

FIGURE 4.5 Full-duplex LAN server operation with token-ring switching.

Despite its limited bandwidth capability compared to that of ATM, token-ring switching does have the indubitable virtue that it works all the way to the desktop with existing NICs. ATM to the desktop is, on the other hand, totally contingent on, at a minimum, new adapters, and in some cases even the need to replace ISA bus-based PCs. Thus, price/performance—a bellwether indicator in networking—becomes a key differentiator between these two switching technologies. ATM offers significant amounts of bandwidth from the desktop all the way across WANs and intrinsic support for multimedia applications—albeit at a relatively high cost, especially in the case of ATM to the desktop. Token-ring switching, in comparison, only offers relatively modest increases in bandwidth for campus-level and data-only applications, but delivers this additional bandwidth for what is also a reasonably modest cost using existing NICs. Table 4.2 summarizes the pros and cons of ATM and token-ring switching.

Given their different price/performance curves, ATM and token-ring switching rather than being overtly competing technologies end up being complementary technologies that will address different segments of the market. During 1996 at least, token-ring switching will fulfill much of the demand for additional desktop, workgroup, and campus backbone bandwidth within the IBM networking community. ATM, will enter the picture by mid-1996, and will initially complement token-ring switching at the campus backbone level and then broaden its scope in time to embrace WAN and some multimedia applications.

By 1998, ATM is likely to be the first choice for campus backbone and LAN server bandwidth needs with such backbones and servers now running at least 155Mbps (and 310Mbps full-duplex). Token-ring switches, however, will continue to be used to provide additional bandwidth to desktops and LAN workgroups. Individual PCs/workgroups and LAN workgroups will gain access to the broadband ATM backbone as well as the ATM-based servers through token-ring switches. Figures 1.4 and 3.6 depict the overall configuration of an ATM campus backbone with token-ring switches acting as its feeder nodes. Given this migration and coexistence scenario, token-ring switching, in the case of data-only applications, has the potential of making 25Mbps ATM to the desktop obsolete. The availability of multimedia support, at least by mid-1996, could avert this possibility and give 25Mbps ATM a whole new lease on life.

4.2 TOKEN-RING SWITCHING VIS-Á-VIA ETHERNET SWITCHING

Ethernet switching and token-ring switching are indeed birds of a feather. They both share the same objectives: trying to provide cost-

TABLE 4.2 A Comparison of the Pros and Cons of ATM Switching with Those for Token-Ring Switching

ATM Token-Ring Switching	*Token-Ring Switching*
Pros:	**Pros:**
• Inherent support for voice and video	• Nondisruptive and low-risk
• Ability to deliver very high bandwidth, i.e.,155Mbps to 622Mbps and beyond	• Works with existing NICs and wiring
• Inherently full-duplex. All quoted bandwidth is full-duplex; i.e., 155Mbps is in reality 310Mbps	• Relatively inexpensive
• Applies to desktop, campus, and WAN sceanrios	• May satisfy the short- to mid-term campus-level bandwidth gains required by some enterprises—especially if coupled with 155Mbps ATM fat pipes between the token-ring switches
• Quality of service, such as latency, can be associated with a given user to customize the bandwidth to meet the demands of a particular application.	• Could be used to support some video applications
• Strategic	• ATM and FDDI fat pipes
Cons:	**Cons:**
• ATM to the desktop requires new adapters and possibly even new PCs.	• Maximum bandwidth is 32Mbps full-duplex—and even that with new full-duplex adapters
• A total solution even at the campus level is still relatively costly compared to 100Mbps LAN solutions.	• Restricted to campus environments. Not applicable in WAN scenarios.
• Supports data applications via LAN Emulation. Native ATM API for data is still in the process of emerging.	• Only half-duplex operation possible with standard NIC cards
• Standards, interfaces, and APIs for voice and video applications now readily available.	• No formal support for video or voice
• Quality of service, such as latency, can be associated with a given user to customize the bandwidth to meet the demands of a particular application.	

compelling bandwidth gains to individual desktops, LAN workgroups, and campus backbones. They both have the same limitations: being strictly LAN-centric (i.e., no explicit WAN support) and only being able to provide relatively modest boosts in bandwidth (i.e., 20Mbps in the case of full-duplex Ethernet) compared to that promised by ATM. In addition, both as is to be expected, share many of the same concepts, precepts, and port-to-port frame-forwarding methodology. So much so that from a high-level perspective these two technologies could even be thought of as being interchangeable. Consequently, nearly all, if not all, of the claims made in this book regarding token-ring switching also apply to Ethernet switching.

The fundamental difference between these two variants of LAN switching is the type of NICs installed in the PCs/workstations and LAN servers involved. If they have Ethernet NICs they will end up using Ethernet switching while if they have token-ring NICs they will end up using token-ring switching. This book, and this chapter in particular, focus on token-ring switching given the preponderance of token-ring LANs, to date, in IBM networking environments. Ethernet, however, is now becoming increasingly popular in IBM environments given its considerably lower cost compared to token-ring and the fact that supporting SNA over Ethernet all the way up to a mainframe is no longer an issue. (Refer to the "Ethernet LANs in SNA Environments" sidebar.) *Thus, it would not be uncommon to see Ethernet switching being used in IBM environments.* In some environments it could end up being all Ethernet switching with ATM stepping in to provide campus backbone and WAN bandwidth. It is also likely that there will be quite a few environments where some campuses end up using Ethernet switching while others use token-ring. In the larger campuses there is even the potential that both types of switching may be used within the same campus.

Bay Networks (who now owns token-ring switching pioneer Centillion) already has a chassis-based switch that supports both token-ring and Ethernet switching modules. Right now none of the token-ring switches or Ethernet switches, including the Bay Networks' combo switch, provides support for interoperability between token-ring-attached devices and Ethernet-attached devices. However, the translation required for this type of mixed-LAN interoperability is readily available within bridge/routers or bridges with the Translation Bridging feature (e.g., IBM 8209). If interoperability is required between Ethernet and token-ring switches this could be realized today by inserting a bridge/router or translation bridge between the disparate switches. Figure 4.6 shows how interworking between Ethernet and token-ring switches can be achieved by using a bridge/router between them.

There are, however, also some small differences between Ethernet and token-ring switches. The main ones are: price per-port; the way they are primarily deployed; the preferred method of frame-forwarding (i.e., cut-through or store-and-forward); and the applicability of source-route bridging.

Ethernet technology has always been less expensive than equivalent token-ring technology. For example, while it is now possible to get Ethernet NICs for under $100, the least expensive token-ring NICs are still in excess of $200. Two reasons are always cited for this price discrepancy.

ETHERNET LANS IN SNA ENVIRONMENTS

Even at the beginning of the 1990s, there was a marked dichotomy as to which enterprises used Ethernet LANs and which used token-ring LANs. Given that SNA-over-Ethernet was not widely supported by most networking devices, the SNA community had little choice but to standardize on token-rings. This despite the fact that token-ring technology (e.g., NIC or a bridge) was and still is generally more expensive than the equivalent Ethernet counterparts. This unfortunate demarcation whereby token-ring LANs were a prerequisite if a LAN had to support SNA traffic is no longer the case. IBM began supporting Ethernet and SNA-over-Ethernet (or the 802.3 variant of Ethernet) in earnest around 1991.

Today, all of IBM's strategic platforms support Ethernet LANs and SNA-over-Ethernet (or 802.3) is available on nearly all contemporary products including IBM 3174s, AS/400s, 3172s, RS/6000s, 2217s, 2210s, OS/2 workstations, and so on. The one conspicuous exception when it comes to SNA-over-Ethernet/802.3 is the venerable IBM 3745 communications controller, which happens to be IBM's flagship gateway for mainframe access. Though the 3745 has had an Ethernet/802.3 adapter since September 1992, IBM still only permits TCP/IP to be used with the Ethernet adapter. This limitation is now unlikely to ever be rectified.

The good news, however, is that the 3745 impediment to SNA-over-Ethernet should not in any way preclude the SNA community from being able to profitably use Ethernet LANs. Ethernet NICs for PCs are now available for under $100—at least half the price for comparable token-ring NICs! Given this significant price differential enormous cost savings can be realized by opting for Ethernet rather than token-ring—particularly when addressing remote branch office scenarios that may involve tens of thousands of PCs.

With SNA-over-Ethernet, the 3745 roadblock can be circumvented in at least two different ways. Routing technology, for a start, can provide transparent and effortless interoperability between token-ring and Ethernet LANs. By using a TCP/IP or RFC 1490-based encapsulation technique for SNA traffic, remote Ethernet LANs can be seamlessly and effortlessly interfaced with a token-ring adapter on a 37xx. The other option would be to use an IBM 3172, a channel-attached bridge/router (e.g., Cisco) or a channel-attached intelligent hub (e.g., Cabletron) as the Ethernet-to-mainframe gateway given that all these devices support SNA-over-Ethernet.

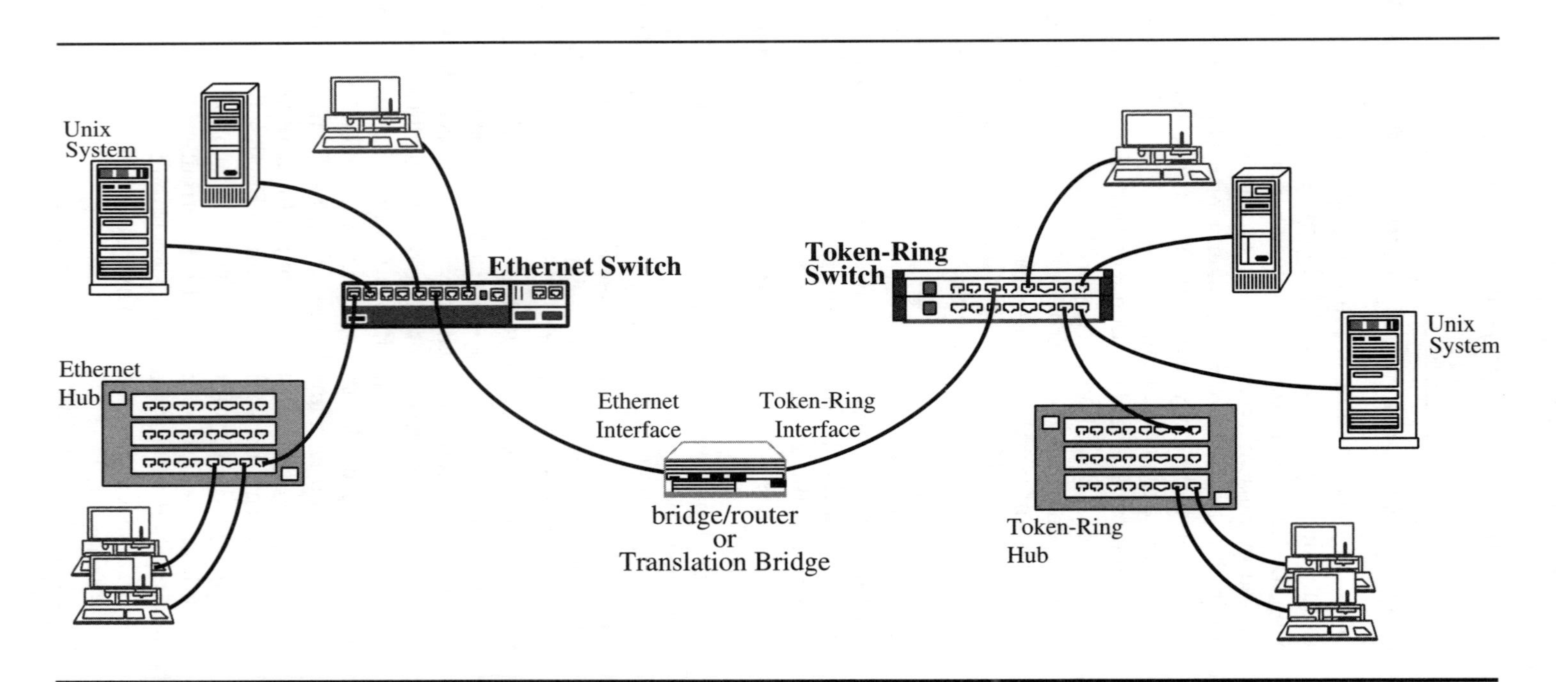

FIGURE 4.6 Interworking between ethernet and token-ring switches via a bridge/router or translation bridge.

Ethernet has always had a significantly larger market share and following than token-ring, whose popularity has been restricted to IBM environments. This larger market share, which is also continually expanding thanks to the lower price, leads to a better economy of scale, which helps lower prices. The larger market also attracts more competition, which in turn also lowers price. In addition, IBM and others have always maintained that token-ring technology, in particular the token-ring chip set, is more complex than Ethernet and consequently costs more to develop and produce. Then there is also the issue of various royalties associated with token-ring technology, which also contribute to the cost.

This traditional price differential between Ethernet and token-ring is also understandably reflected in the price of the switches. While the per-port cost for an Ethernet switch is already down to around the $300 mark, the per-port costs of currently shipping token-ring switches are still around $1,500. The fact that Ethernet switches have been on the market for at least 18 months prior to token-ring ones, the much higher demand for these switches, and the much larger number of vendors vying for a share of the market have all contributed to this rather attractive per-port price. IBM has claimed that the per-port cost for token-ring switches will be around $500 by the end of 1996. At that juncture, the typical per-port price for Ethernet switches is likely to be around $200.

The deployment of Ethernet switches, at least for the time being, tends to differ from that for token-ring switches. Token-ring switches are primarily being used at present for enhancing campus backbone bandwidth. The token-ring switch itself acts as a backbone, or it provides a high throughput interconnection between various LAN segments and a high-speed (e.g., FDDI) backbone. In some instances, token-ring switches are also used to segment large LANs into a series of smaller ones or to ensure that LAN servers have access to full-duplex, dedicated 32Mbps bandwidth. Today, token-ring switches are very rarely, if ever, used to enhance the bandwidth to individual PCs/workstations or of a standalone LAN workgroup. (These various deployment scenarios for token-ring switches are described in detail, along with figures, in Section 4.4.) In contrast, Ethernet switches tend to get equally and freely deployed in all three of these scenarios. Many Ethernet environments now consist of a hierarchy of switches–workgroup switches providing dedicated bandwidth to individual PCs/workstations feeding into departmental switches, which in turn are connected to switches serving the campus backbone.

There are two reasons why these two switch types tend to get deployed in different configurations. For a start, as discussed earlier, token-ring LANs in general are in less need of bandwidth assuagement than comparably sized Ethernets. Thus, there is less of a justifiable need for token-ring switches for desktop and workgroup level applications. This is, however, not the case with Ethernet LANs. They can profitably benefit from having switches at this level. Then there is also the issue of costs. Ethernet switches with their much lower per-port costs are more affordable even for desktop or workgroup applications. This as yet is not the case with token-ring switches. At around $1,500 per-port, deploying switches to provide dedicated bandwidth for each desktop is price prohibitive. However, as the per-port costs start to drop starting in 1996, token-ring switches will also gain more popularity at the workgroup and desktop level.

The other two differences between Ethernet and token-ring switches have to do with the preferred mode for frame forwarding and the availability of SRB. Ethernet switches, in general, support cut-through and most switches are used in this mode. On the other hand, cut-through mode, as shown in Table 4.1, is not available on all of the current crop of token-ring switches. To support cut-through mode, a token-ring switch vendor has to have unencumbered access to the MAC-layer code being used with the token-ring switch. Due to a combination of technical, fiscal, license, and support-related issues, many of the token-ring pioneers decided to forgo trying to obtain and modify MAC-layer code in order to be able to offer cut-through.

Cut-through mode, as is explained in Section 4.5.3, does have a drawback in that it does not preclude the forwarding of corrupted frames. Token-ring environments eschew corrupted frames. The basic token-ring architecture provides a 4-byte Frame Check Sequence plus some control bits at the end of the frame to ensure that corrupted frames can be quickly identified and rejected. The store-and-forward mode works on this principle. It does not begin to forward a frame until it has determined that the frame is uncorrupted and error-free. Thus, there is some valid justification for just offering store-and-forward mode in token-ring environments. Nonetheless, most of the newer switches are now beginning to offer both cut-through and store-and-forward along with a variation known as adaptive cut-through. With *adaptive cut-through* the switch will automatically revert to store-and-forward if it encounters too many corrupted frames while working in cut-through mode.

SRB is an integral feature of token-ring networking. Consequently there are situations—for example, when alternate data paths are desired within a token-ring environment—that token-ring switches need to know about and explicitly support SRB. This is described in detail in Section 4.3.2. Ethernet LANs do not use SRB, thus, SRB is not a relevant feature in Ethernet switches.

The bottom line here is that Ethernet switches and token-ring switches are very similar beasts. With Ethernets becoming increasingly popular in IBM environments Ethernet switching could end up playing as much a part in these environments as their token-ring brethren. The key differences between these two types of switches—such as price, the availability of cut-through, and the way in which they are deployed—will begin to narrow as token-ring switching matures.

4.3 THE BASIC PRECEPTS OF TOKEN-RING SWITCHING

Though rarely thought of as such, conventional LANs, whether Ethernet, token-ring, FDDI, or token-bus are by design *shared* bandwidth transmission schemes. A given LAN station only has use of the bandwidth offered by that LAN every time it manages to gain access to the LAN media in order to transmit a message frame.

All LAN types provide a mechanism whereby LAN stations can in effect bid to gain access to the LAN bandwidth. In token-ring LANs this access is governed by a LAN station being able to capture the token that circulates around the ring. With Ethernet, this access is dictated by a contention scheme-based around the collision detection and backoff conventions of CSMA/CD. When a LAN station does not have access to the media—as is the case when a LAN station on a token-ring does not have control of the token—it has zero bandwidth.

A PC/workstation or LAN server on a conventional 16Mbps token-ring thus does not have at its disposal a constant and dedicated bandwidth of 16Mbps. Instead, it only has that 16Mbps bandwidth for certain, relatively short durations of time. As the number of stations on a LAN increases, each station has to wait for longer periods before it can gain access to the media. This increased delay in gaining access to the media has the effect of further reducing the overall usable bandwidth available to a given LAN station over a given period of time. Hence, the degradation in performance, particularly so in Ethernet environments, being experienced by users as the size of LANs has started to increase.

It is, however, no longer necessary to have to share LAN bandwidth. Bandwidth sharing was predicated on the fact that early LAN technology was based around shared LAN media—a continuous coax cable in the case of Ethernet, or unpowered, dumb concentrators in the case of token-ring. The use of shared LAN media has already become passé. Highly manageable and fault-resilient intelligent hubs (e.g., IBM/Chipcom 8250) that support hub-and-spoke star-wiring schemes have been the strategic means for implementing LANs for the last few years. With such hubs individual UTP/STP wires go out to each LAN station. With hub-based star-wired LANs, each LAN station has dedicated LAN media (i.e., UTP/STP wiring) between itself and individual ports on the intelligent hub. Switched LANs, whether token-ring or Ethernet, are based on this notion of dedicated LAN media.

With a token-ring switch each LAN station attached to one of its ports effectively has dedicated bandwidth, corresponding to the LAN speed (e.g., 4 or 16Mbps) of its NIC. When a LAN station has data to transmit, it can proceed to do so at once without having to first wait to gain access to the LAN media—that is, waiting to capture the token. The impact of having this dedicated LAN bandwidth would be very apparent, for example, if a large file transfer—say of a 10MByte file—is conducted between two PCs. Provided that there is no traffic destined to these two PCs from other ports, the file transfer would take place at more or less line-speed bandwidth, with no delays caused through having to wait to capture the token after each frame has been transmitted.

The mechanics required to realize a switched token-ring LAN or a switched Ethernet LAN are contained entirely within the LAN switch. These LAN switches provide port-to-port, cross-connect switching—across a high-speed backplane—between any two pairs of ports. In some instances this high-speed backplane may use a cell switching scheme à la ATM to forward data frames from one port on the switch to another.

With LAN switching, frames received at one port are dynamically forwarded to the relevant destination port-based on the destination MAC address that appears near the start of each frame. The forwarding of frames from the input port to the destination port may be performed either "on the fly" in cut-through mode, or once the entire frame has been assembled and error-checked, in store-and-forward mode. The switch maintains a directory with entries specifying which port number serves either a particular destination MAC Address, or a set of destination MAC addresses in order to perform the port-to-port forwarding of frames.

Token-ring switches provide a limited amount of buffering for frames that cannot be immediately delivered to the destination port. Such delays in delivery would happen if the destination port is already involved in a data transfer operation (i.e., it is busy). This might be the case when two PCs are simultaneously trying to send data to the same LAN server. Frames from one of the PCs will have to be buffered while frames from the other PC are being forwarded to the LAN server.

4.3.1 Transparent Switching

The entries in a LAN switch's directory are typically created dynamically. LAN switches, whether token-ring or Ethernet, have a MAC address learning process identical to that employed by the Transparent Bridges used in Ethernet environments. With this learning process the switch reads and notes the source MAC addresses of frames it receives on each port. This information is then used to dynamically create and update directory entries. For each address learned an entry is created specifying the port number on which that MAC address was received. (In other words, each entry specifies the port through which a given MAC address can be reached.) Now when trying to forward a frame received on one port to its intended destination the switch can check its directory to see if there is an entry corresponding to the destination address. If the destination address had been previously learned, as a result of the switch encountering a frame that had that address as its *source* address, the directory entry will specify the necessary port number.

If the destination address for a frame to be forwarded is not found in the directory the switch will automatically forward that frame to all the potential destination ports on the switch. (The destination ports in this instance would typically be all the other ports bar the one on which the frame was received. However, most switches provide various partitioning and Virtual LAN schemes when certain sets of ports are segregated from others. In such scenarios the destination ports to which the original frame is propagated will be restricted to the ports within a given partition or Virtual LAN.) This process of propagating frames whose destination port has yet to be determined to all potential destination ports is sometimes referred to as *flooding the ports*. It is in effect nothing other than an all-routes broadcast search, which is really no different from, what a Transparent bridge or SRB does when it encounters an unknown destination address.

The flooding process does ensure that the frame has the maximum possibility of reaching its destination. When the frame eventually reaches its destination there is a probability that the recipient would at some point issue a response. That response would have the sender's MAC address as the source address. This source address would be what was the unknown destination address in the original frame. The switch, on learning this new source address, will now create a directory entry for it. The next time this address appears as a destination address there will be no need to flood the ports.

This continual learning process, with its built-in directory update mechanism, ensures that a switch only has to do a broadcast search the very first time it encounters an unknown address—as opposed to traditional SRBs that do not cache the routes they learn. Today's token-ring switches have directories that can hold between 5,000 to 10,000 entries. Most switches also enable directory entries to be manually created to minimize the number of broadcast searches that have to be performed.

The port flooding search process described above is referred to as *Transparent Switching* or *Transparent Bridging*, given that this process is exactly the same as that done by a transparent bridge. Transparent switching is supported by nearly all, if not all, token-ring switches. It tends to be the preferred and default mechanism for locating unknown destination addresses. It has the indubitable virtue of being a simple, intuitive, and proven scheme that will work in any networking scenario. It is also a fail-safe technique that ensures that every effort is made to make sure that a copy of the frame has a chance of reaching its intended destination.

4.3.2 Source-Route Bridging and Transparent Switching

Transparent switching in token-ring switches works totally independent of SRB. (The mechanics of SRB are described in detail in the next chapter.) Transparent switching does not know, and does not need to know, anything about SRB.

Source-Route Bridging uses and relies on a field within the MAC header known as the *Routing Information Field* (RIF) to maintain a description of the route being followed by a particular token-ring frame. During a SRB broadcast search, each bridge before forwarding a frame will update the RIF of that frame. The update will denote the bridge and LAN segment that it has just traversed. When a frame eventually reaches its intended destination during such a search, its RIF will contain a complete audit trail of every LAN segment and bridge that was traversed in order to get there.

The recipient of the search frame will return a positive acknowledgment to that frame to indicate its safe delivery. This acknowledgment would reflect the RIF that was found in the search frame. From then on all frames flowing, in both directions, between the original sender of the search frame and its recipient will include this RIF. Since the route to the destination is now known, a bit at the start of these RIFs will indicate that these data frames are not involved in a search process.

Source-Route Bridges, on receiving a frame with a nonsearch RIF, will read the RIF to determine to which destination port it should now forward that frame. The SRB will determine the destination port by first finding its name within the RIF and then reading the name of the LAN segment that follows it. The required destination port would correspond to the port to which that named LAN segment is attached.

Transparent switching is unaware of RIFs. When trying to locate an unknown destination, transparent switching does not look at, update, or create a RIF, or any entries within a RIF. The destination MAC address is the only criteria and MAC header field of consequence vis-à-vis transparent switching.

By mid-1996 all token-ring switches will support SRB in addition to transparent bridging. *Source-Route Bridging, however, is not a mandatory requirement for token-ring switching.* This was highlighted by IBM with the announcement of their original 8272 token-ring switch, which incongruously did not support SRB. A token-ring switch can use transparent switching in most scenarios. There are, however, two situations where the lack of SRB can be an issue. Without explicit support for SRB, a token-ring switch will not be able to deal with alternate paths through the network. Neither will it be able to deal with the notion of token-ring *ring numbers.*

Source-Route Bridges name LAN segments by assigning them ring numbers. A Network Administrator when installing a bridge will assign each of its ports a ring number. The LAN segment attached to each port assumes that ring number as its LAN name. Ring numbers only have meaning to and are only used by SRBs. Token-ring NICs do not care about or in any way deal with ring numbers.

In most large token-ring environments, however, network administrators and even users have gotten into the habit of using ring numbers to identify various LAN segments. When dealing with LAN issues, for example when reporting a possible LAN failure, it is not uncommon for users and administrators to refer to the LAN in question by its so-called ring number. Thus, many token-ring network management processes that require human intervention are likely to involve ring numbers.

While it is indeed possible to have a transparent switching-based token-ring environment that does not have ring numbers, this could disrupt existing network management procedures, as well as disorient users and administrators. Not using SRB also precludes redundant (or alternate) paths from being available within the network and the possibility of using duplicate MAC addresses on separate LAN segments. Both of these are features are often used in IBM SNA environments to realize a degree of alternate routing. Networks that rely on ring numbers for network management purposes or have a need for redundant paths should opt for using token-ring switches with SRB activated. Using external SRBs with just transparent switching on the switches would also work. However, with SRB now universally available on all switches it would be somewhat pointless to resort to external bridges. Figure 11.4 in Chapter 11 shows how a token-ring switch with SRB can be used to provide redundant LAN paths to a 37xx communications controller.

If ring numbers and redundant paths are not required, SRB should not be used on token-ring switches. Using SRB in scenarios where it is not required will just unnecessarily increase the amount of processing, per frame, that has to be done by the token-ring switch, thus impacting its overall throughput—albeit only by a hardly discernible factor.

4.4 WAYS OF DEPLOYING TOKEN-RING SWITCHES

Token-ring switches can be successfully and cost-effectively used to increase LAN throughput in a variety of campus-level configurations. Invariably these configurations tend to fall into one of three networking scenarios, or be variations of one of these three. The three primary networking scenarios (or applications) for token-ring switching are as follows:

1. Stand-alone LAN workgroups
2. Splitting up of large LANs into smaller LAN segments
3. Interconnecting LAN segments to a backbone LAN

In a stand-alone LAN workgroup configuration, a single PC/workstation, LAN server, or computer is attached to each port of the token-ring switch via dedicated UTP/STP wiring. See Figure 4.1. In this type of one-device-per-port configuration, each device is able to enjoy dedicated 4 or 16Mbps bandwidth when sending data to, or receiving data from, its *mother* port on the switch. (the “mother” port being the port that a given device is attached to). Figure 4.7 illustrates how a conventional token-ring workgroup built around a hub or MAU can be converted to a token-ring switch-centric workgroup.

The speed at which a token-ring device can operate in this type of switching configuration will only be dictated by the speed of its NIC. Devices with 4Mbps NICs will be able to function at 4Mbps, while those with 16Mbps NICs will be able to run at 16Mbps. Token-ring switches can support *mixed-speed* workgroup configurations where devices attached to some ports have 4Mbps NICs while the devices on other ports have 16Mbps. Such mixed-speed configurations do not in any way force all the ports to work at the slower speed. Each port continues to work at its maximum speed. The switch will also permit frames to be freely exchanged, in both directions, between the 16Mbps and 4Mbps devices despite the difference in speed.

This ability to support mixed-speed environments and permit data transfer between devices with different speed NICs is another key advantage of using a token-ring switch in this type of configuration. When a conventional hub or MAU is used to realize a token-ring workgroup, the maximum speed that the resulting LAN will be able to function at will be gated by the slowest speed NIC on the LAN. Thus, if the

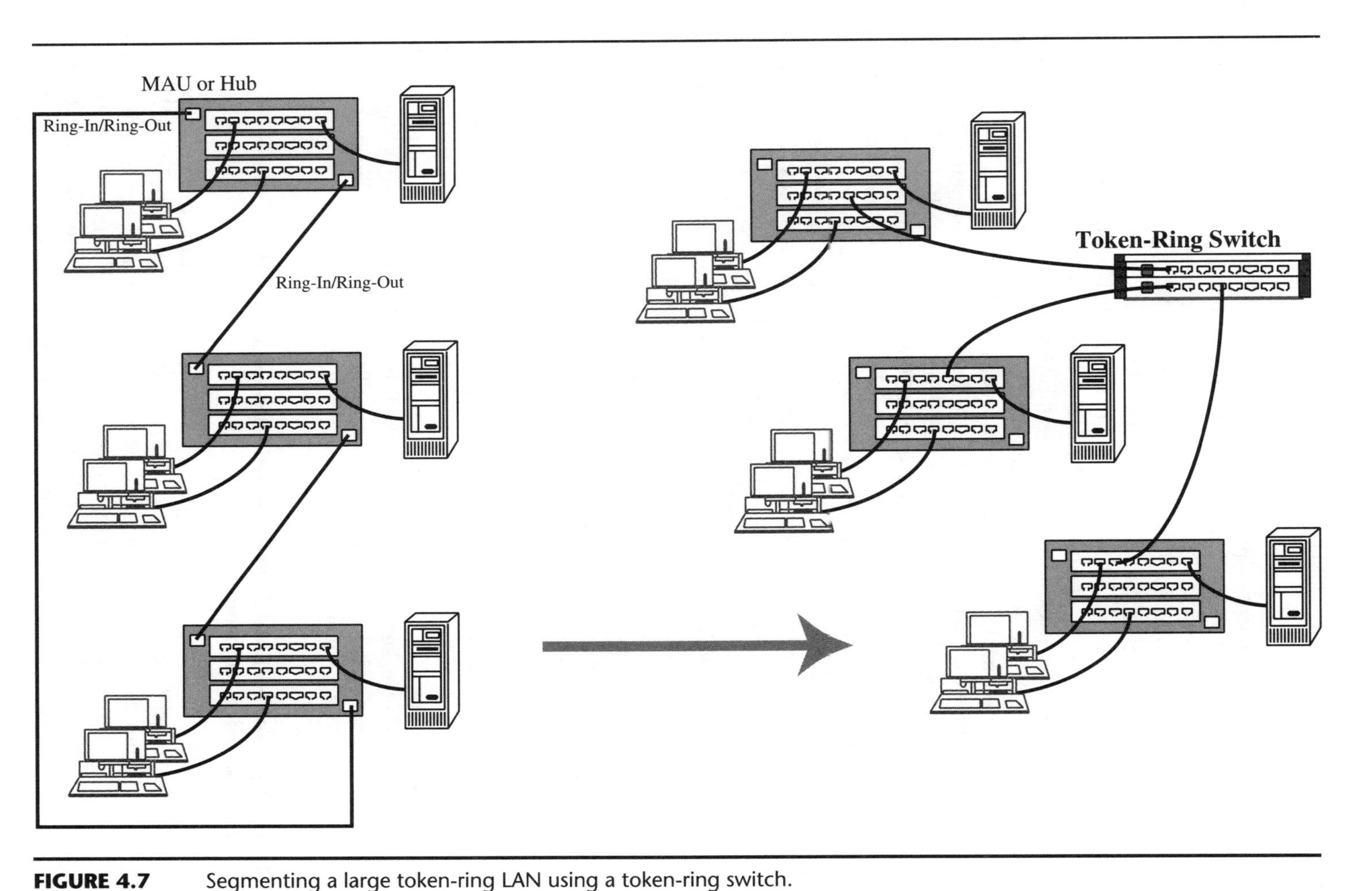

FIGURE 4.7 Segmenting a large token-ring LAN using a token-ring switch.

workgroup consists of devices with both 4Mbps and 16Mbps NICs, the LAN will only be able to operate at 4Mbps. (16Mbps token-ring NICs are always capable of working in 4Mbps mode to cater to exactly this type of situation. Consequently 16Mbps NICs are often marketed as being "16/4Mbps" to highlight this fact.)

Devices with the faster and more expensive 16Mbps NICs are forced to operate at one quarter their true capacity in mixed-speed LANs. The need to operate such mixed-speed LANs at 4Mbps is dictated by the fact that, in the absence of LAN switching, the LAN can only operate in shared media mode. In shared media mode, tokens as well as all the frames transmitted on the LAN have to be propagated through the NICs of all the attached devices. Since the 4Mbps NICs will not be able to keep up with data being sent at 16Mbps, the entire LAN has to work in 4Mbps. With enterprises invariably having PCs and workstations of different vintages, mixed speed LANs are not uncommon.

In LAN workgroup configurations, certain devices such as LAN servers or computers could be equipped with full-duplex NICs as shown in Figure 4.8. The devices equipped with full-duplex NICs will be able to operate at a burst mode equivalent to 32Mbps when receiving data from a device attached to the switch at 16Mbps while simultaneously transmitting data to another device at 16Mbps.

LAN workgroup configurations, particularly with full-duplex NICs on LAN servers, permit token-ring switches to deliver maximum potential bandwidth to individual PC/workstation users and LAN servers. This type of configuration, however, as discussed earlier, is as yet rarely used with token-ring switches. For one, the approximately $1,500 per-port cost of token-ring switches is still too high to be conducive to this type of application. Even more importantly, few token-ring environments, at present, have a genuinely justifiable need for this amount of bandwidth. Over the next couple of years this type of token-ring-centric LAN workgroup configuration will begin to proliferate as the prices of these switches starts to drop and ATM-based campus backbones stimulate the appetite for more bandwidth to the desktop.

4.4.1 Using a Token-Ring Switch to Split Up a Large LAN into Smaller Segments

A token-ring switch can also be profitably used to split up a large LAN consisting of 60 or more PCs/workstations, and multiple LAN servers, into a series of smaller LAN segments. In such LAN segmentation configurations each of the newly created smaller LAN segments are connected to individual ports on the token-ring switch. Figure 4.8 illustrates how a large LAN can be segmented into a series of smaller ones—in this illus-

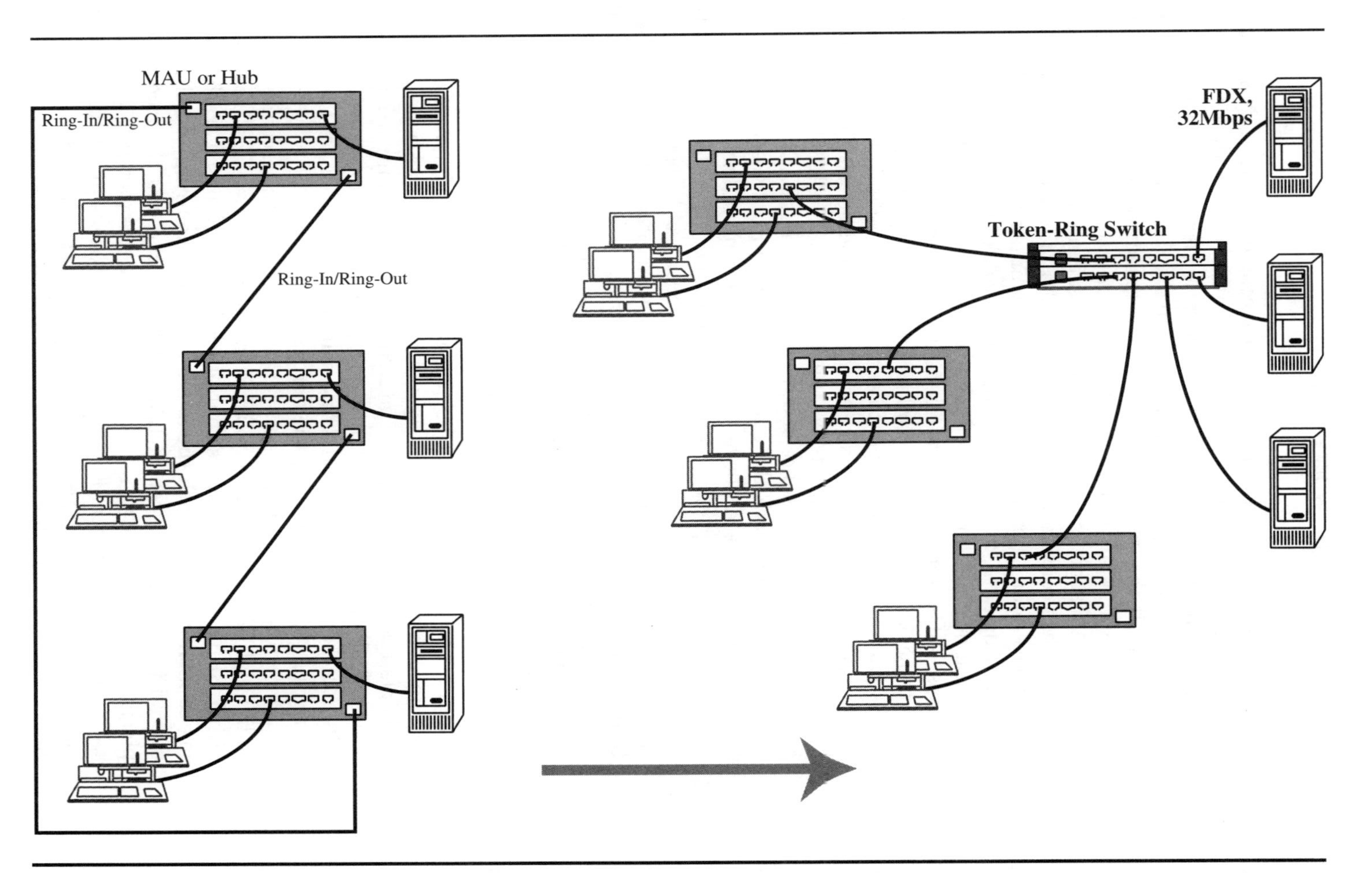

FIGURE 4.8 Segmenting a large token-ring LAN and at the same time making all the servers operate at full-duplex 32 Mpbs.

tration there happens to be three. In such configurations, LAN throughput enhancement is at the segment level, rather than at the device level as in the standalone LAN workgroup scenario discussed above.

Each new LAN segment can now have a dedicated 16Mbps (provided that all the devices on the LAN segment have 16Mbps token-ring adapters) relative to the switch port for interactions with devices on other LAN segments or with the servers. Since each new LAN segment now has considerably fewer devices than the original large LAN, each device on a segment will be able to gain access to the LAN, for transmitting data, at a much higher frequency than before.

If the original large LAN included devices with both 4Mbps and 16Mbps NICs, token-ring switch-based segmentation could also be used to neatly segregate the 4Mbps devices onto their own LAN segments and the 16Mbps ones onto separate segments. This NIC speed-based LAN segmentation ensures that the devices with the 16Mbps NICs can run at full speeds on the 16Mbps LAN segments without being slowed by the presence of devices with 4Mbps NICs.

A variation for both the same-speed and mixed-speed segmentation scenario discussed above would be to equip all the LAN servers with full-duplex adapters and have them connected to dedicated ports on the switch. Figure 4.8 illustrates this variant of LAN segmentation where all the LAN servers have full-duplex adapters and are attached to dedicated ports on the switch. This way, each LAN server could work at 32Mbps receiving data from a device on one LAN segment while transmitting data to a device on another LAN segment.

4.4.2 Interconnecting LAN Segments to a Backbone LAN

Token-ring switching can also be used to significantly improve the performance and the overall manageability of campus backbones, where multiple token-ring LANs are interconnected to a central backbone. This so-called Backbone Interconnect scenario, as discussed before, will in most cases be the initial and primary application for token-ring switching technology during the early part of 1996.

Prior to the advent of token-ring switches, the interconnection of individual LAN segments onto a central backbone was realized using either SRBs, or LAN servers that offered software-based traffic routing capabilities. Now with the availability of token-ring switches, the overall throughput to/from a LAN backbone can be enhanced by replacing the SRBs used to interconnect LAN segments to the backbone with a token-ring switch. Figure 4.9 depicts a token-ring switch-centric backbone interconnect scenario where a switch replaces a series of SRBs. In this type of interconnect scenario if only one switch is involved

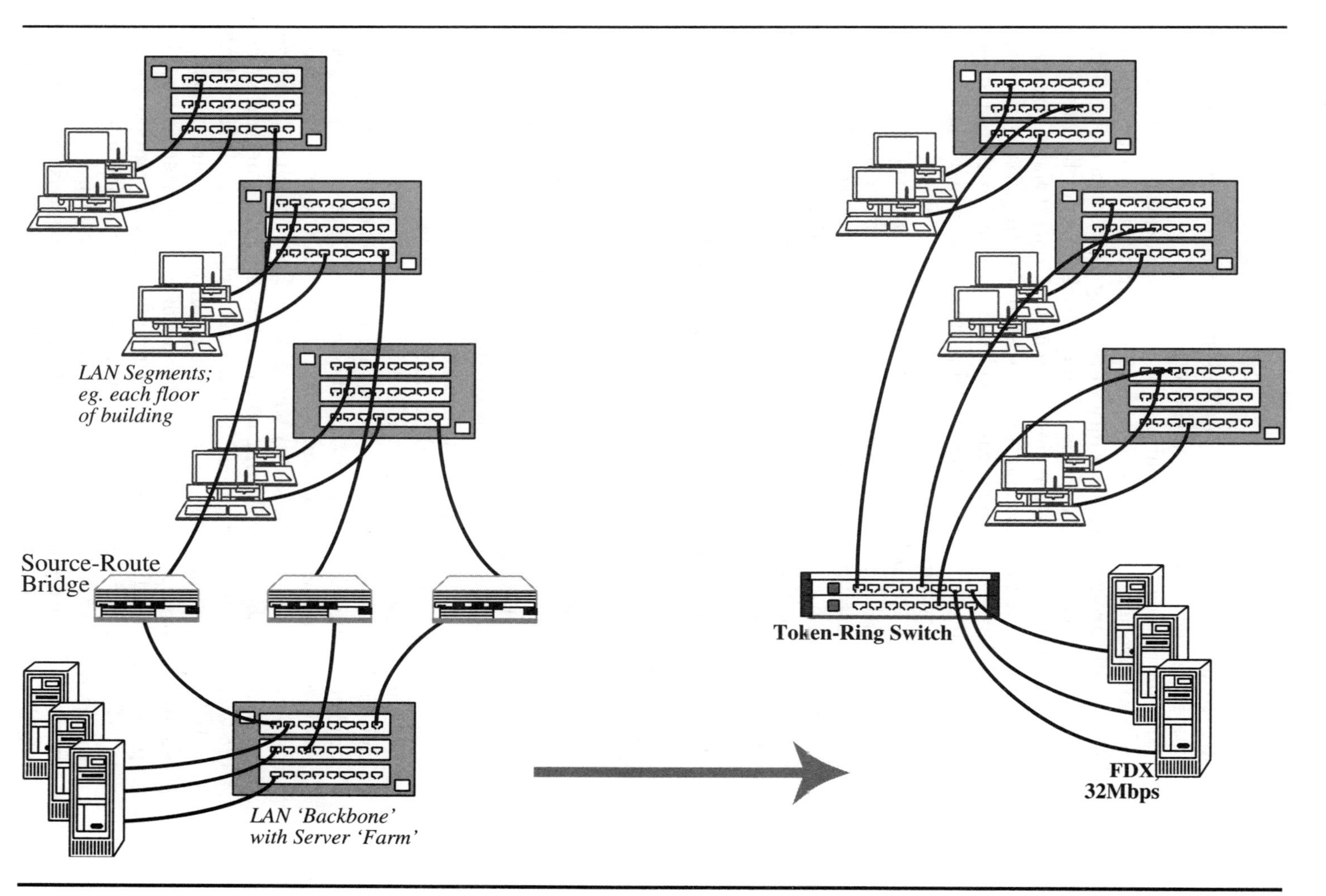

FIGURE 4.9 LAN backbone interconnect scenario where a token-ring switch is used to replace a series of source-route bridges.

it becomes the backbone per se. This type of configuration is sometimes referred to as a *collapsed backbone*.

A token-ring switch could also be used for backbone interconnect in configurations where LAN servers (e.g., Novell NetWare Server) with a software-routing capability were being used, rather than SRBs, to interconnect the LAN segments to the backbone. In this latter scenario, however, the token-ring switch will not replace the LAN servers—just the backbone interconnect function they were providing. The LAN servers will now most likely be equipped with full-duplex adapters and connected to individual ports on the switch à la the configuration shown in Figure 4.9.

Moreover, if each of the LAN Servers in such a configuration is associated with a particular LAN segment, the Virtual LAN and firewalling features available on some token-ring switches could be used to directly map each LAN server with the LAN that it serves. This Virtual LAN capability can further improve performance by minimizing any broadcast traffic that flows between the clients and the server. It can also ensure that intra-LAN traffic between the LAN clients and the LAN Servers is contained to each Virtual LAN.

This traffic segregation into Virtual LANs is also seen by some as a powerful security feature in that traffic does not go across different LAN segments. A basic precept of LANs is that all LAN traffic passes through the NICs of all the devices attached to that LAN. The functioning of LANs is invariably contingent on the fact that all NICs see all the traffic flowing across the LAN. This characteristic of LANs can be abused in that any device with the appropriate, but relatively standard, software can eavesdrop on all the traffic flowing on a LAN by "promiscuously" reading all the frames passing through its NIC. This type of promiscuous mode reading of all data frames on a LAN is actually the standard modus operandi of bridge and bridge/routers since this is the only way that they can intercept frames that have to be bridged to other LANs. Obviously the so-called Data Sniffers or Data Analyzers that permit any and all traffic on a LAN to be captured and analyzed also rely on the ability whereby any LAN-attached device can intercept all the traffic flowing on that LAN. By containing LAN traffic within tightly segregated Virtual LANs where each Virtual LAN only has users that belong to the same group or department, network administrators can minimize the danger of unauthorized eavesdropping of LAN traffic by users in other groups or departments.

Throughput increases in token-ring-centric backbone interconnect configurations are derived in three distinct manners. For a start, token-ring switches are equipped with fast, typically RISC, processors and run highly optimized software. Consequently they can forward packets to/from the backbone faster than most conventional SRBs, the majority of which are based on PC technology from the late 1980s. In other words, token-ring switches have a lower latency, or processing delay, than bridges when it comes to forwarding frames to/from the backbone.

In addition, by connecting servers to dedicated ports on the switch, typically with full-duplex adapters if the backbone is also token-ring, all interactions involving the servers can be greatly expedited compared to the original backbone configuration where the servers had to contend for bandwidth and were only capable of working in half-duplex mode. The third performance-enhancing factor relates to SRB. When SRB is not required, the switch can use transparent switching for forwarding packets from their source port to the relevant destination port. Transparent switching involves less processing than SRB in that it does not have to read and act upon the routing information contained in the RIF. Thus, transparent switching further reduces the latency of a switch compared to that of a SRB.

Further performance increases can be gained in token-ring switch-centric backbone interconnect scenarios by replacing the hubs or MAUs used to implement the various LAN segments with token-ring switches. Figure 4.10 shows a configuration where token-ring switches are used both to create LAN workgroup segments as well as to interconnect these segments and provide them with access to the LAN servers. Collapsed backbones are not the only backbone interconnect configurations that can be used with token-ring switches. A campus backbone supported by token-ring switches may consist either of multiple switches interconnected via FDDI or ATM fat pipes, or an 100Mbps FDDI backbone where the switches interconnect the token-ring segments to the FDDI backbone. Figure 3.6 in Chapter 3 depicts a campus backbone consisting of multiple token-ring switches interconnected via ATM.

4.4.3 The Difference Between Backbone Interconnect and LAN Segmentation

In the case of token-ring switch-centric collapsed backbones as shown in Figures 4.9 and 4.10, the backbone interconnect scenario may appear at first sight to be just another variation of the LAN workgroup segmentation scenario described in Section 3.4.1 and depicted in Figures 4.7 and 4.8. There is, however, a crucial difference between these two scenarios.

With large LAN segmentation, what was a single LAN is broken up into multiple LANs. Given that there were no SRBs in the original LAN there were no ring numbers. Ring numbers, as described earlier, are only introduced into token-ring environments when SRBs

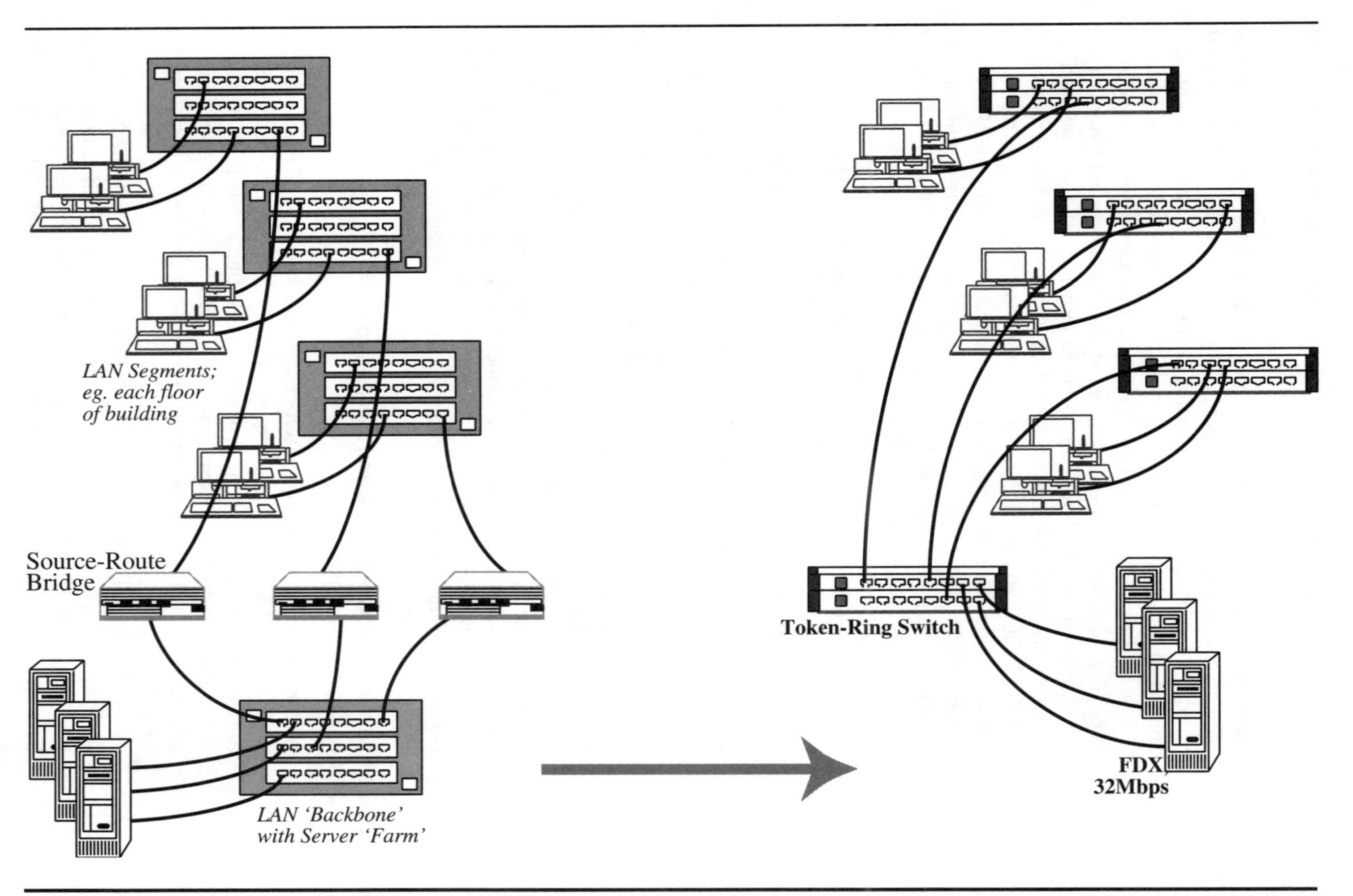

FIGURE 4.10 Segmenting a large token-ring LAN and at the same time making all the servers operate at full-duplex 32Mpbs.

are installed. Given that there were no ring numbers involved to begin with, token-ring switches do not insist that the new LAN segments are each given a separate ring number. Instead, in such configurations a token-ring switch works totally independent of ring numbers. The switch will also rely on transparent switching rather than SRB to locate unknown destination MAC addresses. The bottom line here is that LAN segmentation can be realized without recourse to SRB.

On the other hand, in backbone interconnect configurations, where a switch replaces multiple SRBs as shown in Figure 4.9, ring numbers would have been present and consequently have to be dealt with. To deal with separate ring numbers, a token-ring switch has to be able to support SRB.

It is theoretically possible not to use ring numbers or SRB in backbone interconnect situations. Not using SRB, and hence ring numbers, in backbone interconnect configurations, does have two drawbacks, as discussed earlier: It disrupts existing LAN network management procedures and precludes the ability to define and use redundant paths, or duplicate MAC addresses.

Of the three scenarios for beneficially deploying token-ring switches discussed above, backbone interconnect followed by workgroup segmentation—in some cases both being used in parallel as shown in Figure 4.10—will prove to be the most popular during 1996–1997. Given token-rings inherent, token-capture-based, polling-like deterministic media-access mechanism, noticeable performance degradation of token-ring LANs does not occur until LAN utilization begins to creep past 70%. In marked contrast, Ethernets start experiencing performance hits even at utilization rates lower than 50%.

4.5 KEY DIFFERENTIATING FEATURES OF TOKEN-RING SWITCHES

The key differentiating features of token-ring switches are as follows:

1. The fat-pipe technique used either to interconnect the switch to a high-speed backbone (e.g., FDDI or ATM), or to interconnect multiple switches together.
2. Availability of SRB, as discussed above relative to backbone interconnect scenarios
3. Cut-through, as opposed to store-and-foreward, mode of switching used to forward packets between ports
4. Filtering and security features for minimizing the amount of repetitive broadcast traffic forwarded by the switch

5. Virtual LAN facilities available
6. Network management features—particularly Remote Network Monitoring (RMON) for in-depth session trace diagnostics and problem determination

4.5.1 Fat-Pipe Alternatives

Token-ring switches use fat pipes to gain access to high-speed backbones or to realize connectivity between multiple switches. The need to interconnect multiple switches together may arise to either obtain more ports than are available on a single switch, or to realize LAN-to-LAN interactions between dispersed LANs within the same campus or possibly even across a WAN.

FDDI, given its prevalence as the technology of choice for high-speed backbones over the last few years, and predictably ATM, are the two fat pipes invariably now being offered by token-ring switches—with most concentrating on first providing FDDI. With ATM-based backbones currently being rare, particularly in IBM environments, ATM in general is aimed at switch-to-switch interconnection. Ironically, ATM per se, may, however, not be the optimal technique for interconnecting switches in every instance.

The inefficiency of using ATM just for data, as opposed to some form of frame relay like variable length packet transfer, was discussed at length in the previous chapter. The ideal solution would be to always include an optional packet switching mode, à la PTM or Cell/Frame, with the ATM fat pipe so slicing-and-dicing is only active when there is CBR traffic present. As yet none of the switch vendors that plan to offer ATM fat pipes have indicated their intent to offer a Cell/Frame option such as a PTM-type option. IBM, however, has made some claims that it would offer 160Mbps PTM-based fat pipe just for data-only applications. This, though not the optimal solution, may get IBM and others thinking about the possibility and desirability of offering ATM fat pipes with a built-in PTM option.

4.5.2 Source-Route Bridging, Layer 2 Filtering, and Virtual LANs

SRB is a key value-added feature provided on token-ring switches. While a token-ring switch could be made to work in certain configurations without SRB, as discussed above, having an SRB capability that can be activated on a per-port basis can enhance the flexibility and reduce the overall cost associated with implementing token-ring switching solutions.

If SRB is not available, network administrators who wish to either use separate ring numbers to identify specific LAN segments or implement redundant paths across the network have no option but to use external SRBs. Having to rely on external bridges could prove to be regressive and expensive. Not only does it increase the overall costs it would also compromise some of the throughput gains realizable with the switch given that improving the throughput possible via SRBs is one of the rationales for deploying switches!

Just because SRB is available on a switch does not mean that it should be used indiscriminately. Having to read and update SRB-related RIFs will increase the amount of processing that has to be done by the switch—and will thus have a small impact on latency. *So if throughput is the paramount goal, which will be the case in most instances, SRB should only be used if ring numbers are required to satisfy network management or redundant path requirements.*

Broadcasts, and the potential for bandwidth-sapping and traffic-disrupting broadcast storms, are always a concern in token-ring environments with SNA and NetBIOS, and today even TCP/IP, using repetitive broadcasts to locate addresses and names (e.g., LAN servers or an SNA gateway such as IBM 3745). Having seen the problems that can occur if broadcasts are not controlled, most of the token-ring switch vendors already provide filtering and address/name caching options to minimize the amount of broadcast traffic that has to be propagated between ports. Given that switches operate at Layer 2 the filtering provided is restricted to Layer 2, entities such as MAC addresses.

With some protocols, for example TCP/IP and IPX, comprehensive filtering can only be provided by using routing techniques that span both Layers 2 and 3. Whether a switch should provide Layer 3 routing functions is debatable—and also begs the question whether a switch with routing is a bona fide switch or just a fast router masquerading as a switch given that routing technology is now considered by some to be passé. At present some of the switch vendors, with CrossComm to the fore, have stated an intent to provide Layer 3 routing functions. *In the absence of Layer 3, networks that require the level of filtering and firewalls only possible with Layer 3 routing technology will have to think about judiciously complementing their token-ring switches with strategically placed external routers.*

So-called Virtual LAN capabilities are available on nearly all token-ring and for that matter Ethernet switches. There is, however, at present no commonality or consistency as to what is referred to as a VLAN facility. The term *VLAN*, in general, is used by switch vendors to describe any type of LAN user segregation facility that permits broadcast and data traffic to be somehow restricted to a predefined set of LAN segments. The notion of using this type of VLAN to restrict LAN traffic to individual LAN segments for the purposes of security has already been discussed above.

At present the VLAN capabilities of most, if not all, token-ring switches consist of the ability to group together either a set of ports on the switch, or if SRB is being used, a set of ring numbers. With this type of grouping, no traffic to/from these ports or rings, including any broadcasts, is ever forwarded to ports or rings that are not a part of that closed group VLAN. Some switches, in particular Ethernet switches, permit VLANs to be created whereby LAN users are grouped together either by the Layer 3 protocol they are using (e.g., IPX or TCP/IP), or by the Layer 3 IP-like subnetwork addresses assigned to them. Again the goal is to restrict traffic, and in particular broadcast traffic, just to the users that belong to a given VLAN. This type of Layer 3 attribute-based VLANs should be available on token-ring switches in 1996—particularly when they start to offer optional Layer 3 routing functions to enhance their filtering and fire-walling capabilities.

The ultimate level of VLANs would be the ability to group together sets of users-based on their organizational (e.g., all the members of the Human Resources department) or functional responsibilities (e.g., all product planners involved in a certain project) irrespective of where these users are located. They could be located on different LAN segments within the same campus or be physically dispersed across a far-flung network consisting of myriads of LANs interconnected via a WAN—providing such sophisticated VLANs not only require a considerable amount of specialized software but also interoperability schemes that will span LAN switches, routers, and even ATM switches. This type of VLAN is unlikely to be readily available on token-ring switches at least till mid-1996.

4.5.3 Cut-Through Versus Store-and-Forward

Cut-through mode packet forwarding is a technique to *potentially* reduce the time taken by a switch to deliver a complete packet from one port to another. In cut-through mode, the switch starts to forward bits that make up a frame from the source port to the destination port, as soon as it has read the destination MAC-address that appears close to the start of the frame and determined the required output port.

Due to this on-the-fly bit forwarding from the source to destination port, all the bits comprising a complete frame will have passed through the destination port shortly after the final bit was received at the source port. The only delays, or latency, encountered in forwarding the complete frame from source to destination are the time taken to determine the output port and the processing time to move bits from one port to the other across the switch. With cut-through mode the total delay, or latency, through the token-ring switch for a given frame would be in the 30–50 microseconds range.

The alternate approach to cut-through is store-and-forward. With this mode, the switch waits until it has received a complete frame—free of errors as denoted by the FCS near the end of the frame—before it sets out to forward that frame over the output port. With this mode of forwarding, where the first bit does not hit the output port until the whole frame has been assembled and checked, the latency through the switch for a given packet could be around 100 microseconds.

The higher latency associated with store-and-forward does not mean that cut-through is always more desirable. For a start, with cut-through there is always a danger that erroneous frames may be forwarded—wasting bandwidth on the output port and LAN, as well as possibly delaying the forwarding of other error-free packets to the destination port. One way to overcome this problem is to provide the so-called adaptive cut-through mode. With this mode, a packet error-rate threshold can be assigned to each port on the switch. If this threshold is exceeded on a given port the switch will automatically revert from using cut-through mode to using store-and-forward for that port until the error rate falls below the threshold.

The throughput gains possible with cut-through are also dependent on congestion and the nature of the traffic being forwarded. If the output port required to forward a packet is busy, either receiving or forwarding another packet, cut-through mode cannot work—and essentially reverts to store-and-forward. As traffic volumes increase the chances of cut-through encountering busy ports also increase. Cut-through being foiled by the output port being busy can also occur if a stream of packets is being sent from one port to another—as would be the case with a file transfer to/from a server. While the first packet in such a stream may, with luck, get out on the destination port in cut-through mode it is likely that subsequent packets will end up getting queued waiting for the previous packets to be forwarded.

Thus, even though a switch may be working in cut-through mode it will be difficult to determine whether a given packet was forwarded in cut-through mode, without delay, or whether the forwarding was done via store-and-forward because the output port was busy. With the difference in latency between the two modes being in the microsecond range a workstation user is not going to be able to tell or perceive a difference.

4.6 REFLECTIONS

Token-ring switching is a compelling and cost-effective way of providing additional network bandwidth to LAN backbones, LAN workgroups, or individual PCs/workstations. Token-ring switching can deliver this additional networking bandwidth, even to individual PCs/workstations (i.e., to the desktop), without the need for any new network adapters (i.e., NICs) or wiring, and with a minimum of network reengineering. Token-ring switching also does not require any changes to established LAN administration and management processes. Neither does it rely on any changes being made to the existing configuration of LAN workgroups or campus backbones.

By being a low-cost, nearly nondisruptive and low-risk technology, token-ring switching will end up being the campus bandwidth-enhancing precursor to ATM across much of the IBM networking community over the next couple of years. Token-ring switching, as opposed to ATM, will thus provide this community with its first real taste of contemporary switching technology.

Most token-ring switches, by mid-1996, will support ATM fat pipes. A token-ring switch will use these ATM fat pipes to either realize 155Mbps broadband connections with other intracampus token-ring switches, or gain access to a port on an ATM switch so that it can forward traffic from its LAN ports to an ATM network. ATM will debut in many IBM-centric networks when they deploy ATM fat pipes between their intracampus token-ring switches. In addition, for many years to come token-ring switching will end up complementing ATM at the campus level, acting as a lower-cost, token-ring NIC-compatible adjunct for enhancing LAN workgroup bandwidth.

CHAPTER 5

The First Steps in Building a Multiprotocol Network

A QUICK GUIDE TO CHAPTER 5

Cost-intensive, dreaded parallel data networks—one for SNA/APPN and the other for non-SNA—have unfortunately been a cross that the IBM-centric networking community has had to bear for much of the last ten years. Now at long last, stable, proven, and reliable parallel network consolidation technology is at hand. Consequently the first and most decisive step in the IBM network reengineering process is that of converging the parallel data networks into an effective and efficient SNA/APPN-capable multiprotocol LAN/WAN network.

There are at least eight very different technologies that can be called upon just to integrate SNA/APPN and non-SNA LAN traffic as mentioned Chapters 1 and 3. In addition there are four viable techniques that can be used to dovetail traffic from link-attached SNA and BSC devices into a multiprotocol network.

This is a somewhat technical, "bits-and-bytes" chapter, which sets out to describe seven of these network consolidation technologies: three LAN integration techniques and all four of the link traffic integration methodologies. (SNA Session Switching, though described as a link traffic integration scheme, can be used to consolidate both LAN- and link-based SNA and non-SNA traffic.) All three of the LAN integration techniques addressed fall into the category of being non-SNA technologies that strive to accommodate SNA/APPN traffic. Hence these three technologies are best suited for multiprotocol networks where the SNA/APPN component of the WAN traffic mix is 50% or less. The SNA/HPR-oriented network consolidation technologies that are suited for networks with a preponderance of SNA/APPN WAN traffic are described in Chapters 7 and 8.

Detailed principles of operation, usually replete with at least one illustration representing its characteristic modus operandi, are included for each of the seven technologies covered. The strengths and weaknesses of each are also discussed at length. The seven technologies are as follows:

- Three LAN Integration Technologies:
  1. Bridging: Source Route, Transparent, Source-Route Transparent, and Translation (Sections 5.1–5.4).
  2. TCP/IP Encapsulation of SNA/APPN traffic à la DLSw (Section 5.5).
  3. Protocol Independent Routing (Section 5.6)
- Four Link-Traffic integration Technologies:
  1. Synchronous Pass-Through (Sections 5.7.1)
  2. Remote Link Polling (Section 5.7.2)
  3. SDLC-to-LLC:2 Conversion (Section 5.7.3)
  4. Remote SNA Session Switching with Host Passthrough (Section 5.7.4).

This chapter also includes a sidebar that describes the notion of "Local LLC:2 Acknowledgment."

Attempting to consolidate their dreaded parallel networks to contain costs, minimize the operational burden, and set the stage for the eventual migration to ATM is now the major preoccupation taxing most of the IBM networking enterprises. Fortunately for these enterprises there is now an abundance of technology, some of it already proven and established, at their disposal for successfully realizing this consolidation. This technology ranges from straightforward MAC-level bridging to AnyNet-based protocol conversion of all the protocols relevant to HPR—the heir apparent to both SNA and APPN. As discussed earlier, there are at least eight viable and primary technologies available just for integrating the LAN traffic. Of these, six are still strategic. Enterprises should evaluate all six, at least at a high level to begin with, since they are all very different from each other and have their own set of unique strengths and weaknesses. This and the following three chapters will furnish all the pertinent data required to make this evaluation.

In Chapter 3 a scheme was put forward for selecting a parallel network consolidation technology based on an enterprise's envisaged traffic mix in 1997–1998. This "vendor neutral" scheme, which was created by the author as a means of easily differentiating between the non-SNA and SNA-based technologies for network consolidation, provides enterprises with a concrete metric for making an unbiased decision. Table 5.1 augments this technology selection scheme and builds upon the data that was presented in Table 1.1 by summarizing the best-fit selection criteria for all eight network consolidation technologies. Table 5.1 in essence provides a bird's-eye view of the technologies described in the next three chapters.

Much of this parallel network consolidation technology, in particular the most pertinent and popular ones (e.g., DLSw or RFC 1490), is available not from just from one or two vendors but from a range of aggressively competing vendors. For example, Cisco, Bay Networks, 3Com, and IBM all offer bridging, DLSw, RFC 1490, and APPN NN routing side by side on their bridge/routers. Moreover, multivendor interoperability is even theoretically possible if any one of these four technologies is selected. Consequently, much of this consolidation technology has become vendor neutral. Thus, selecting a given technology, with the possible exception of AnyNet, does not unduly restrict the number of potential vendors, nor does picking a favored vendor limit the technologies that may be used.

Ironically, this freedom of choice can be a problem. With all the major networking vendors in this arena now offering multiple solutions for parallel network consolidation the onus has fallen on network administrators to select the best one of their particular needs. This book endeavors to provide as much "apple-to-apples" comparative analysis as possible to help network administrators make the right choice. Having such a plethora of choices when it comes to technology to address a given requirement is a relatively new phenomenon to the SNA community. For nearly two decades IBM set the technical agenda for SNA networking and dictated the parameters of the technology that could be used within these networks. Though various degrees of SNA technology per se are available from a multitude of vendors these vendors are best characterized as offering "plug-compatible" solutions. *The problem when it comes to parallel network consolidation is that there is no overall architecture, master plan, or even a universally accepted approach for realizing SNA/APPN-capable multiprotocol LAN/WAN networks.*

TABLE 5.1 How to Decide Between the Eight LAN Traffic Integration Technologies (Part 1)

	Bridging	*TCP/IP Encapsulation of SNA/APPN Traffic*	*Frame Relay RFC 1490*	*APPN/HPR NN Node Routing*
Consider using in these scenarios:	1. Relatively small networks; e.g., 50 nodes or less, where the bulk of the traffic is SNA and NetBIOS.	1. IP-oriented networks where SNA/APPN consitutes less than 50% of the overall WAN traffic mix. 2. When the WAN fabric is not frame relay. 3. When there are four or more mainframes (or NetBIOS destinations) in the network.	1. When the WAN fabric is frame relay and there are only one or two mainframes in the network (also not too many remote NetBIOS destinations). 2. When the WAN fabric is frame relay and there is a need to support non-SDLC link traffic (e.g., BSC, Async.) from branch offices. 3. When the WAN fabric is frame relay and there is a need to support some level of voice traffic alongside the data traffic.	1. In AS/400 networks currently using APPN. 2. If the network has many mainframes and direct SNA Routing is sought for switching between the mainframes.

Avoid using in these scenarios:	1. Large networks. 2. Networks where the bulk of the traffic consists of the so-called routable protocols such as IP, IPX, or DECnet. 3. In tightly interconnected mesh networks where there can be multiple routes between any given source and destination. (Can result in broadcast storms.)	1. When there is little or no native (i.e., not related to TCP/IP encapsulation) TCP/IP traffic. 2. When the WAN fabric is frame relay and there are only one or two mainframes in the network (also not too many remote NetBIOS destinations). 3. Large, 500+ remote node networks.	1. Non-frame-relay-based networks. 2. When there are four or more mainframes in the network.	1. When the WAN fabric is frame relay and there are only one or two mainframres in the network. 2. Networks where SNA/APPN consitutes less than 40% of the overall WAN traffic mix.

TABLE 5.1 How to Decide Between the Eight LAN Traffic Integration Technologies (Part 2)

	AnyNet non-SNA to SNA/HPR Gateways à la the IBM 2217	*LAN-over-SNA*	*Remote SNA Session Session Switching*	*PIR*
Consider using in these scenarios:	1. SNA-oriented networks where SNA/APPN consitutes more than 80% of the overall WAN traffic mix. 2. If the network has many mainframes and direct SNA Routing is sought for switching between the main-frames, provided that SNA/APPN traffic account for at least 50% of the WAN traffic mix. 3. Desire to standardize on an HPR-based WAN as the means for multiprotocol LAN/WAN networking.	1. SNA-oriented networks where SNA/APPN consitutes more than 80% of the overall WAN traffic mix. 2. If the network has many mainframes and direct SNA Routing is sought for switching between the main-frames, provided that SNA/APPN traffic account for at least 50% of the WAN traffic mix.	1. IP-oriented networks where SNA/APPN con-situtes less than 50% of the overall WAN traffic mix, *and there are four or more mainframes attached to the network.*	1. Relatively small net-works, e.g., 100 nodes or less, where the the bulk of the traffic is SNA and NetBIOS

Avoid using in these scenarios:	1. Networks where SNA/APPN consitutes less than 50% of the overall WAN traffic mix. 2. When the WAN fabric is frame relay and there are only one or two mainframes in the network (also not too many remote NetBIOS destinations). 1. Networks where SNA/APPN consitutes less than 50% of the overall WAN traffic mix. 2. When the WAN fabric is frame relay and there are only one or two mainframes in the network (also not too many remote NetBIOS destinations).	1. Networks where SNA/APPN consitutes less than 50% of the overall WAN traffic mix. 2. When the WAN fabric is frame relay and there are only one or two mainframes in the network (also not too many remote NetBIOS destinations). 1. Networks where SNA/APPN consitutes less than 50% of the overall WAN traffic mix. 2. When the WAN fabric is frame relay and there are only one or two mainframes in the network (also not too many remote NetBIOS destinations). *No longer strategic.*	1. If there are only one or two mainframes in the network.	1. Large networks. 2. Networks where the bulk of the traffic consists of the so-called routable protocols such as IP, IPX, or DECnet. *No longer strategic.*

Thus, in marked contrast to multivendor SNA networking, multiprotocol LAN/WAN networking, irrespective of the number of vendors involved, cannot be tackled just in terms of putting together a plug-compatible solution since there is no one overriding solution that one can be plug-compatible with. The popularity of a given consolidation technology as the best solution around has, as is to be expected, depended on when that technology was available and what other alternate technologies were also available at that time. There has never, however, been an outright favorite technology that has been able to genuinely claim indubitable superiority over the rest. For a while, up until around late 1994 it really was just a two-horse race between SRB and TCP/IP encapsulation of SNA/APPN traffic (e.g., DLSw). CrossComm's PIR was looked upon as a sophisticated variant of SRB. Today, this race involves at least six horses with at least two—DLSw(+) and RFC 1490—vigorously jostling to stay at the front of the pack with APPN NN and AnyNet gateways valiantly trying to secure third place.

Token-ring LANs, and thus the genesis for multiprotocol networking, entered the IBM world in 1986. For the next six years, IBM's recommended and only real solution for implementing multiprotocol LAN/WAN networking was SRB. In the early 1990s Cisco came up with the very first scheme for encapsulating SNA/APPN traffic within TCP/IP message units. This scheme was referred to as Remote Source-Route Bridging (RSRB). With the advent of RSRB, the SNA community was faced with a choice. Now they could use SRB, which had some irritating shortcomings, or opt for RSRB, which involved making SNA traffic subservient to TCP/IP, which was relatively new entity to the IBM world. (One of the key shortcomings of SRB is its insistence on doing repeated broadcast searches for the same destination address.)

Around this time, Sync Research and Netlink popularized SDLC-to-LLC:2 conversion, which could be used with either SRB or RSRB, as the strategic means for incorporating SDLC traffic into a multiprotocol LAN/WAN network. In mid-1991, IBM made a tentative stab at LAN-over-SNA technology with its LAN-to-LAN Wide Area Network Program (LTLW). LTLW encapsulated NetBIOS traffic within LU 6.2 message units so that NetBIOS traffic could be transported from one LAN to another across an SNA network. By 1993 LTLW had been enhanced to support TCP/IP and Novel NetWare IPX traffic in addition to NetBIOS. Until the mid-1995 debut of the IBM 2217 MultiProtocol Concentrator, which is based on an amalgam of both AnyNet and LTLW technology LTLW along with its SNA Links counterpart from Novell were the epitome of LAN-over-SNA technology.

In mid-1992, IBM's first bridge/router, the star-crossed 6611, entered the fray. The 6611 offered SRB, DLSw, and APPN Network Node as three very different ways of implementing an SNA/APPN-capable multiprotocol network. DLSw was a family name for a core set of four SNA/APPN traffic integration facilities. These four facilities are: a TCP/IP encapsulation scheme for SNA/APPN and NetBIOS traffic, which, though similar to Cisco's RSRB, was not compatible with it; integrated SDLC-to-LLC:2 conversion, which would obviate the need for an external conversion box à la Sync Research or Netlink; local LLC:2 acknowledgment, which eliminated the need to transport Layer 2 acknowledgments across the WAN; and a caching facility that precluded the need to repeatedly search for destination MAC addresses or NetBIOS server names.

The APPN NN scheme was of limited consequence at this juncture since the Dependent LU technology that was required to support non-LU 6.2 traffic (e.g., the predominant 3270 traffic) within the context of APPN was not available. Though the 6611 was and still is a lackluster product, its introduction served as a watershed in this arena. With the 6611, IBM endorsed TCP/IP encapsulation of SNA/APPN traffic as had been originally advocated by Cisco with RSRB, and also postulated APPN as a possible long-term solution. What IBM did not do, and has yet to do, was to articulate when and how an enterprise should decide between SRB, DLSw, APPN, or its LAN-over-SNA (e.g., LTLW). To exacerbate this situation, RFC 1490-based encapsulation, which is a direct alternative to bridging, and TCP/IP encapsulation came into being in 1994. To date, the vendor community as a whole continues to be reticent in providing nonpartisan, objective and incontestable guidelines for choosing between these technologies.

The remainder of this chapter will concentrate on the four original non-SNA convergence technologies—bridging, TCP/IP encapsulation, PIR, and session switching—as well as the major technologies for incorporating SDLC and BSC traffic into multiprotocol network. Session switching is described in Section 5.7.4 under the theme of link traffic support given that unlike the other techniques it can be used to support both SNA LAN and link traffic.

5.1 SOURCE-ROUTE BRIDGING

SRB is the original, and in essence the default, method for interconnecting multiple token-ring LANs. SRB has been a standard and key feature of the IBM/IEEE 802.5 token-ring architecture from day one. It permits two or more token-ring LANs to be uniformly and transparently interconnected, such that they appear to form a single, seamless, consolidated LAN.

SRB may be used to interconnect LANs that are either adjacent to each other (e.g., within the same building) or geographically dispersed. The former scenario, referred to as *local bridging,* does not involve a WAN connection, while the latter, which does require WAN connections, is referred to as *remote bridging.* (A bridge performing remote bridging is sometimes called a split bridge by IBM, to highlight that the bridging function has essentially been split in two and is being performed at either end of a WAN connection.)

SRB permits a device (e.g., PC) on one LAN to interact with a device on another LAN (e.g., IBM 37x5 mainframe gateway) as if both devices were directly attached to the same physical LAN. That is the beauty of SRB, or for that matter any other type of LAN bridging. Bridging creates a single LAN image of a multi-LAN configuration and permits free and consistent interactions between LANs transcending any physical separation between the various LANs. It is this single LAN image of physically dispersed LANs provided by SRB that is exploited by bridge/routers when having to transport SNA/APPN traffic between remote sites and a central mainframe or AS/400 location.

Bridge/routers, when supporting SNA/APPN LAN traffic on multiple dispersed LANs, use SRB to logically interconnect the various LANs such that they form a single, virtual, SNA super-LAN. This SNA super-LAN would then be interfaced to a mainframe or AS/400-centric core SNA environment, through one or more SNA LAN gateways, such as a channel-attached 3745, 3174, 3172, or an AS/400's native token-ring interface.

Thanks to SRB's single LAN image, all the devices on the SNA/APPN super-LAN, including the SNA LAN gateway, all think and act as if they were attached to the same physical LAN. Thus, host system control software such as ACF/VTAM, ACF/NCP, or AS/400 OS/400 and consequently the host-resident SNA application programs can view and treat the various SNA devices on the dispersed LANs as if they were all on one LAN.

SRB is a straightforward LAN interconnection technique that can be used with impunity to handle most, if not all, SNA/APPN LAN traffic consolidation scenarios in relatively small (i.e., 50 or less remote sites) networks where SNA/APPN and NetBIOS traffic account for most of the traffic. Like all other bridging techniques, SRB is a Data Link (i.e., OSI Layer 2) level process. Thus it operates independent of, and is totally transparent to, higher level (i.e., Layer 3 and above) networking protocols such as SNA, APPN, NetBIOS, or Novell IPX/SPX. SRB works independently and outside the scope of bona fide SNA routing. Consequently, it does not support SNA or APPN routing between multiple mainframes or AS/400s.

5.1.1 Discovering the Mechanics of Source-Route Bridging

In order for SRB to be able to provide seamless interconnection between two devices on two different LANs, a path between these two LANs—in terms of a chain of intermediary bridges and LANs that need to be traversed—needs to be established. A dynamic, broadcast search technique is used to establish this path. The technique used is similar in concept to that employed by SNA since 1984 to locate undefined resources in multihost environments (i.e., the Default SSCP Selection feature of ACF/VTAM), and what is now available in APPN for dynamically locating remote resources.

The SRB broadcast search technique is initiated by the *source device* (e.g., PC), rather than by one of the bridges, hence the name *Source-Route*, denoting that it is the source that determines the eventual route across the network. A source device will initiate an SRB search by issuing a token-ring Layer 2 TEST or XID (i.e., Exchange Identification) command. TEST and XID are a part of the Layer 2 LLC sublayer repertoire of commands and responses. The LLC sublayer commands and responses used by SNA/APPN and NetBIOS traffic are virtually identical to the commands and responses available with SDLC. The LLC variant that supports this SDLC-like repertoire of commands and responses is referred to as LLC Type 2 (LLC:2). LLC:2 provides a *connection-oriented* (i.e., session-oriented) mode of operation where all data frames carry a LLC:2 sequence number and there is a mechanism for either acknowledging the successful receipt of frames or requesting the retransmission of corrupted or lost frames. The term *LLC:2* used in expressions such as *SDLC-to-LLC:2 conversion* and *Local LLC:2 Acknowledgment* refers to this SDLC-like, session-oriented mode of token-ring operation.

The source device initiating an SRB search will identify the destination device being sought via its unique, Layer 2, 6-byte, MAC address. The MAC address in this instance could be the globally unique, IEEE-administered Universal Adapter Address burned into each token-ring interface adapter. It is also possible to use an enterprise-specific locally administered address MAC address that only has to be unique relative to a given interconnected LAN configuration. Since the SRB search process is restricted to Layer 2, it is totally oblivious to, and is conducted entirely independent of, any higher-level addresses or names, such as SNA/APPN LU Names, SNA Network (/Local) Addresses, or NetBIOS Names. Figure 5.1 illustrates the basic structure of a token-ring MAC/LLC frame.

The source device will initially issue the TEST or XID command without an indication that the command should be propagated (i.e.,

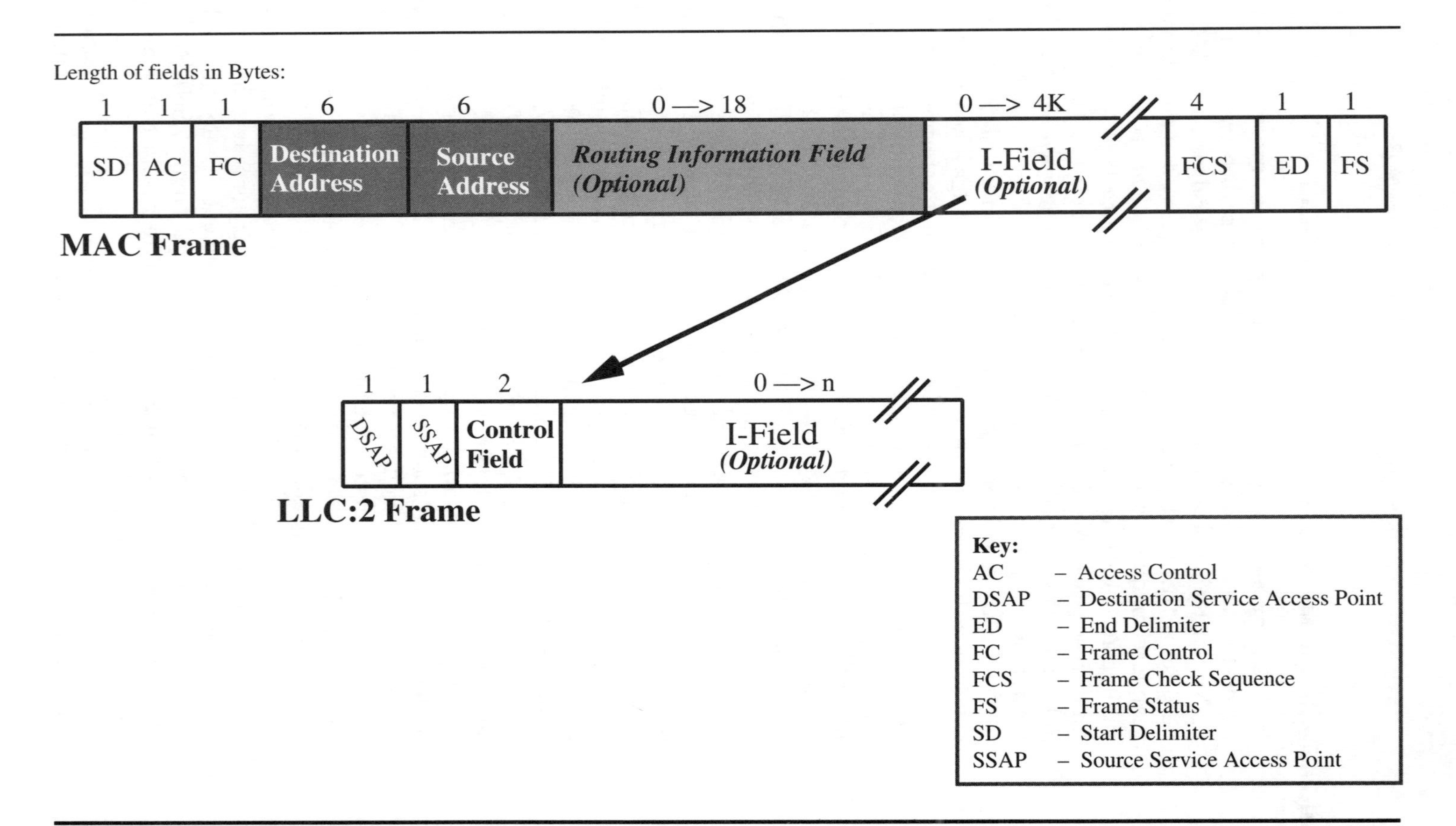

Length of fields in Bytes:
1
1
1
6
6
0 —> 18
0 —> 4K
4
1
1
SD
AC
FC
Destination Address
Source Address
Routing Information Field (Optional)
I-Field (Optional)
FCS
ED
FS
MAC Frame
1
1
2
0 —> n
DSAP
SSAP
Control Field
I-Field (Optional)
LLC:2 Frame
Key:
AC – Access Control
DSAP – Destination Service Access Point
ED – End Delimiter
FC – Frame Control
FCS – Frame Check Sequence
FS – Frame Status
SD – Start Delimiter
SSAP – Source Service Access Point

FIGURE 5.1 Format of a token-ring MAC/LLC frame.

broadcast) outside the local LAN (i.e., LAN-segment) to which it is attached. This initial search is known as an *on-segment search*. The source device can determine if the TEST or XID command reached its intended destination (i.e., the device identified by the destination MAC address) in one of two ways. One way is to check the so-called A&C bits that appear in the Frame Status field that appears at the very end of each token-ring frame. (See Figure 5.1.) The other option is to wait for a response.

The A bit in the A&C combo is the Address-Recognized Bit, while the C bit is the Frame-Copied bit. Both these bits are set to zero when the TEST or XID command is issued. If another device on the LAN recognizes the destination address as its own it is supposed to turn the A bit to a b'1'. If in addition to recognizing its own address the destination device also copied the frame into its input buffer (i.e., read the frame), it is supposed to set the C bit to a b'1'. With token-ring LANs, each transmitted frame traverses the LAN and eventually comes back to the source device—irrespective of whether the frame was read by its destination or not. The source device is then responsible for removing that frame from the LAN. When removing the TEST or XID command, the source device can check the A&C bits. An A&C setting of b'11' would immediately indicate that the intended destination was not only found but it read the frame. An A&C setting of b'00' would indicate that the intended destination was not found on the ring.

Independent of analyzing the A&C bits, the source device could wait to receive a response to the TEST or XID command per the conventions of the LLC:2 protocol. If the source does not receive a response to the TEST or XID command from the destination being sought prior to the expiration of a prespecified time-out period, it assumes that the destination device must be on another LAN. (Note, however, that the destination device, though on the same LAN, could have been temporarily off-line (e.g., powered off) while the on-segment search was being conducted, thus causing an unnecessary broadcast search.)

If the on-segment search is unsuccessful, the source device immediately resorts to an *off-segment search* in order to contact the destination, which is now assumed to be on another LAN. It does this by again issuing a TEST or XID command. This time around, however, the command will include an RIF in the MAC header prefixing the command. Two flags (represented by 4 bits) in the header will indicate the presence of the RIF, as well as the need for a broadcast search to be conducted. Figure 5.2 shows the format of a token-ring RIF.

The broadcast search flag can denote that the search is to be conducted in one of two distinct methods. These two methods are: *all-routes broadcast* or *single-route broadcast*. The method requested is selected by

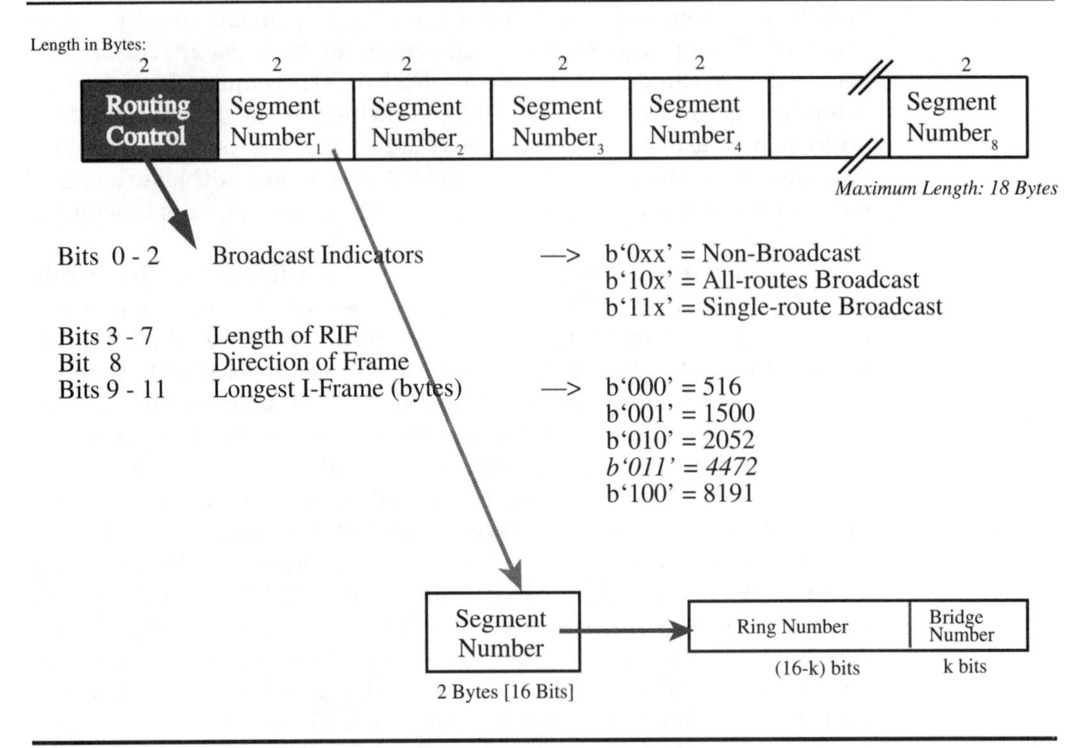

FIGURE 5.2 Format of a token-ring routing information field (RIF).

the software on the source device. In general SNA software will request all-routes searches while NetBIOS opts for single-route.

Discovering all the potential alternate routes within the network between the source and the destination is a fundamental precept of SRB. Ironically, though it goes to the trouble of discovering all the routes, SRB does not exploit this acquired knowledge by supporting dynamic alternate routing. Nonetheless, given this innate need to discover all potential routes, both the all-routes broadcast and the single-route broadcast involve a search process that traverses all available routes within the network. However, these two methods conduct the all-routes search differently. One option is to do the search on the outbound trip with the TEST or XID command. The other is to do it on the way back using the response to the TEST or XID command. With an all-routes broadcast the search is done on the way out. With a single-route broadcast the all-routes search is done on the way back using the response to the original TEST or XID.

When an all-routes search is specified on a TEST or XID command, each bridge encountering that command will make a copy of it and forward that copy to every other token-ring LAN to which it is attached (i.e., propagate it on all other ports). In the case of remote bridging the copy of the TEST or XID command will be sent over the WAN connection(s) to the remote bridge(s).

5.1.2 Exploring All the Routes with Source-Route Bridging

When forwarding a TEST or XID command to another LAN or WAN port the bridge will update the accompanying RIF to reflect the identity of the bridge that it has just crossed. The identification will be in the form of a 16-bit segment number, which consists of a unique number denoting the token-ring from which the command was copied, and a unique number identifying the bridge. Identifying both the bridge and the token-ring permits multiple bridges to be used between a pair of token-rings.

The current token-ring architecture does not permit an RIF to exceed 18 bytes. Hence an RIF, which must always begin with a 2-byte routing control field, can at most contain only eight segment numbers. Since the identification of the LAN containing the source device has to be the first entry in the RIF, SRB only permits LAN interconnections that do not span more than seven intermediary LAN segments. This drawback is referred to as the *seven-hop limitation* of SRB. A way to overcome this limitation is described in Section 5.1.4. (The routing control field at the start of the RIF, as shown in Figure 5.2, in addition to denoting the type of broadcast search being conducted—i.e. all-routes or single-route—also has fields to indicate the number of LAN segments currently present in the RIF, the longest information frame that will be supported once a session is established, and a flag to indicate whether the command containing the RIF is flowing from the source to the destination, or vice versa.)

With all-routes broadcast searches the destination device will receive as many copies of the TEST or XID command as there are available routes between the source and destination LANs. For example, if there are four different routes that can be used to get from the source to the destination, the destination device will receive four copies of the TEST or XID command that was issued by the source device. Each command received will reflect in its RIF the exact path that it has traversed—in terms of the intermediate LANs and bridges—in the order in which they were crossed.

The destination device will return each command it receives—replete with its RIF—back to the source device. The reversal of the direction of flow of the command is denoted by setting the direction flag in the routing control field to indicate *Destination to Source*. In addition the destination device will change to all-routes flag at the start of the routing control field to now indicate *Non-Broadcast*. The non-broadcast status ensures that bridges encountering the subject command route it back to the original source device, using the exact route—but in reverse—as specified in its RIF. Thus, the response traverses the same path as that taken by the original command.

The source device will now start to receive the responses being returned by the destination device. Theoretically the source could then determine the optimal route by evaluating the route structure found in the RIF of each response. For example, the source could decide that the route containing the least number of intermediary bridges (i.e., the least hop route) is the optimal path. It could also record the routes specified in the other RIFs to use as potential alternate routes if the initially chosen route fails during the dialogue. Unfortunately, token-ring implementations do neither.

Instead of trying to determine an optimal route based on the information in the RIF, the source blindly assumes that the route traversed by the first response received by it must be the best route, given that this route was obviously the fastest for the round-trip involved in this search process. Figure 5.3 depicts the overall dynamics of an all-routes broadcast search. The assumption that the route traversed by the first response is the best path between the source and the destination may not be valid. A temporary aberration (e.g., a WAN retransmission) on what would normally have been the fastest path may have caused an inferior path to be selected instead of what should have been the best path.

The source ignores and discards the routes specified in subsequent responses. The RIF in the first response will then be inserted in the headers of all frames interchanged between the source and destination devices for the duration of that dialogue. This RIF will provide all the bridges with the routing information required—in the form of ordered lists of intermediary LANs and bridges—to forward the frames from the source to the destination or vice versa. (When frames are being sent from destination to the source, the reverse direction bit will be set in the routing control field that appears at the start of the RIF.) Thus, SRB, like traditional SNA and even APPN, is an inherently fixed-path routing scheme.

The source device discarding RIFs specifying potential alternate routes is not the only inexcusable and incongruous anomaly in SRB. *There is also the issue that an SRB does not maintain any type of directory (e.g., cache directory) of previously discovered MAC addresses.* Thus an SRB without any compunction will conduct repeated broadcast searches for the same destination MAC address—even within a few

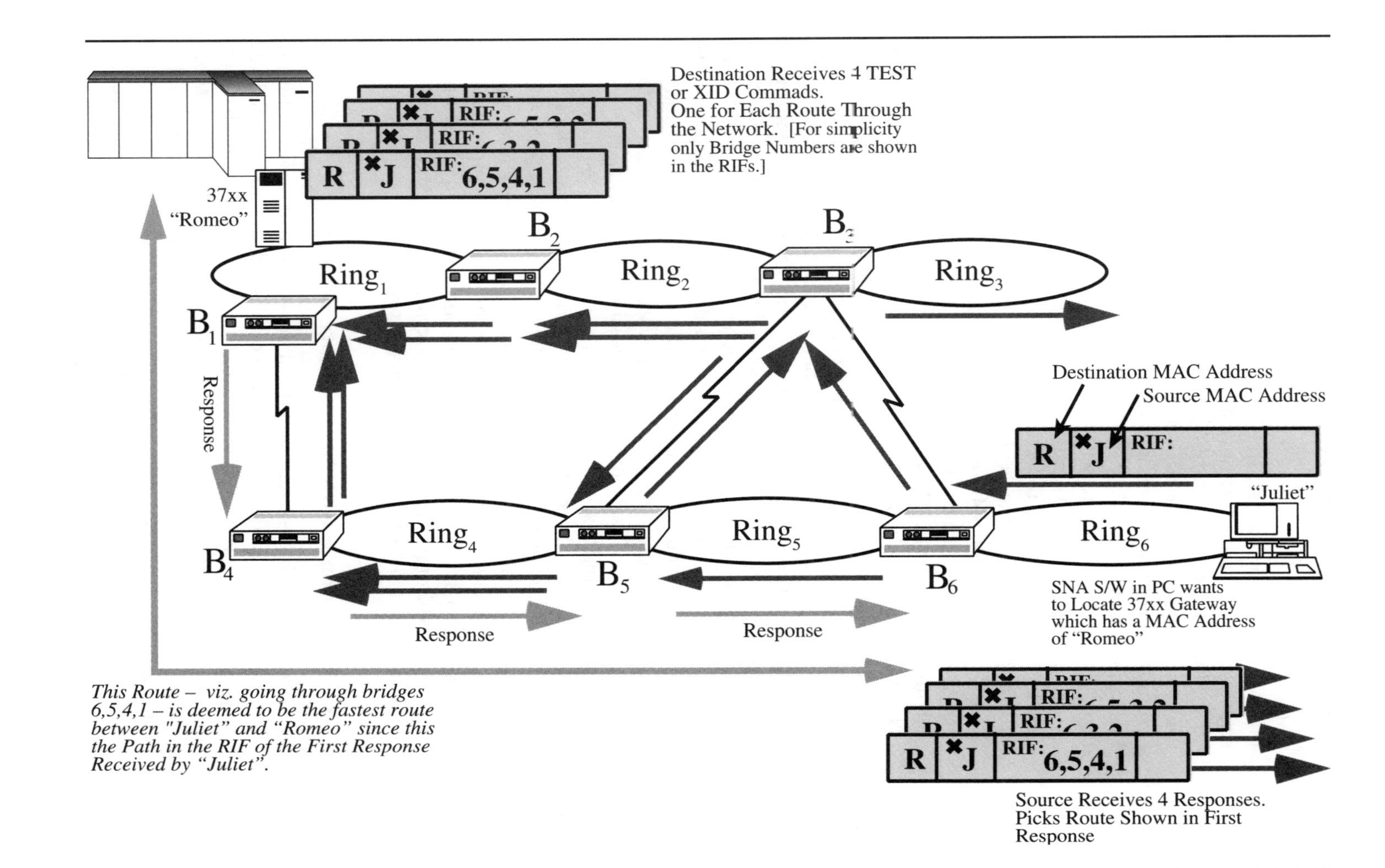

FIGURE 5.3 The overall dynamics of a token-ring all-routes broadcast search.

milliseconds of each other. In SNA/APPN environments such repeated searches for the same destination MAC address may occur on a regular basis. Each PC or workstation that needs to access applications running on a mainframe or AS/400 will contain some type of SNA/3270 emulation software, which will include the MAC address of the SNA LAN gateway (e.g., a 37xx in the case of a mainframe environment) through which this access will be realized.

When the SNA/3270 emulation software is activated it will issue a TEST or XID command to locate this SNA LAN gateway. This process will happen each time a PC or workstation user activates the SNA/3270 emulation software in order to access an SNA/APPN application. At certain times, for example around 9 A.M. each work day morning, quite a few PC/workstation users may have to log on to SNA/APPN applications. This would result in each PC/workstation initiating a broadcast search to locate the relevant SNA LAN gateway—which in the case of mainframe environments will invariably be a 37xx communications controller. (See Figure 5.3.)

At certain times of day—especially first thing in the morning or after lunch—it may not be unusual for a few hundred or possibly even a thousand PC/workstation users to try to log on to SNA/APPN applications around the same time. This could unleash hundreds or even thousands of broadcast searches—for the same destination address—taking place one after the other. Since a standard SRB does not cache the location of previously discovered destination addresses—in terms of one or more RIFs that will lead from the source LAN to the necessary destination LAN—there is no way to circumvent these bandwidth-sapping repeated searches that will disrupt the flow of actual data. The additional traffic and the disruption caused by these repeated broadcast searches, particularly when a lot of them happen together, are referred to in token-ring circles as *Broadcast Storms*. Section 5.1.4 describes a technique, referred to as *Proxy Responder*, that can be used to minimize the possibility of broadcast storms.

5.1.3 Sticking to a Single Route with Source-Route Bridging

In the case of a single-route broadcast search a bridge will determine the route over which the TEST or XID command should be propagated based on a single-route broadcast path maintained by each bridge. This broadcast path is constructed using a *spanning tree algorithm*. A spanning tree can be thought of as a logical graph that connects all the nodes in a network using the least number of connections, and avoiding duplicate paths between nodes. The graph is constructed relative to a particular root node that is selected on the basis of some predefined criteria such as the bridge with the numerically lowest bridge number. With a spanning tree the route from the root node to any other node is always optimal (i.e., the shortest), but this is not the case for the routes between the other nodes. The spanning tree algorithm is an integral component of the transparent bridging scheme used to interconnect Ethernet LANs.

With single-route broadcast searches the destination device receives one copy of the TEST or XID command. In marked contrast to the all-routes method, the RIF in the command received by the destination will contain only a routing control field and will not indicate the route traversed by the command. Just as with all-routes, the destination device will proceed to return the command back to the source by toggling the direction flag.

The destination device, however, now sets the broadcast flags at the start of the routing control field to denote All-Routes Broadcast. This causes the response to be returned to the source via all available paths between the destination and source. The route taken in each case is recorded in the RIF as in the case of a all-routes search. Thus, in the case of a single-route search the routing information is collected on the return trip, as opposed to the destination location trip.

The source will receive multiple responses, and as with an all-routes broadcast, will decide to use the route indicated in the first response received as the best possible path.

5.1.4 The Limitations of Source-Route Bridging

The greatest virtue of SRB is that it is pretty close to being a plug-and-play networking scheme. With SRB, interworking between devices on different LANs can be realized with a minimum of manual predefinition. Typically the only definitions that are required by an SRB are a numeric bridge number for that bridge and another numeric LAN segment number for each LAN attached to each of its ports.

SRB does, however, suffer from some major limitations. These are: fixed-path routing with no dynamic, alternate rerouting in the event of path failure; the additional overhead of the traffic generated during the broadcast searches, the potential for broadcast storms and the inherent seven-hop limitation on the number of intermediate LANs/bridges that may occur between the source and destination. Most of the leading bridge/router vendors (e.g., Cisco, Wellfleet, CrossComm) devised a variety of innovative facilities to circumvent these limitations of SRB during 1991 to 1992.

To alleviate the seven-hop limitation most vendors now offer an extended SRB facility, whereby a bridge/router subnetwork from a given vendor, irrespective of the number of SRB bridges/routers and LANs involved, always appears as a single hop to other bridge (/routers) in the overall network. The technique is totally transparent to other standard SRB bridges/routers, which see a single RIF LAN Segment entry that happens in reality to correspond to a multisegment subnetwork. In practice most enterprises have not as yet encountered the seven-hop ceiling of standard SRB. It is, however, comforting to know that extended SRB could fix this problem if it occurs, and that future networks can be contemplated without constant checkpoints to determine if the seven-hop count has been exceeded.

The additional traffic generated and the disruption caused by broadcast storms can be a source of major concern particularly in remote bridging configurations that use relatively low-speed WAN connections. The potential of such storms has been a perennial criticism of SRB. The threat of potential broadcast storms can be minimized by the use of SRB proxy responder agents, which are a vendor-specific extension to SRB available on certain bridge/routers. Such agents provide a local cache directory scheme where the addresses of remote destination devices can be saved following an initial SRB search. Before performing any broadcast searches an agent checks its cache directory to see if it already knows of a route to the destination device being sought. If a route is found in the cache directory, it will insert that route in the RIF of the TEST or XID command issued by the source device, and return the command back to the source, thus short-circuiting the need for an exhaustive broadcast search.

5.2 TRANSPARENT BRIDGING

Transparent bridging is to Ethernet LANs what SRB is to token-ring LANs. Transparent bridging is the native LAN interconnection mechanism for Ethernet LANs. As with SRB it can be used in both local (i.e., no WAN) and remote (i.e., across WAN) configurations. Transparent bridging, which predates SRB, is a straightforward and intuitive scheme that is fundamentally different from SRB. In transparent bridging the bridges do all the work necessary to discover a route between the source device and the destination device. With SRB the source device has to do the bulk of this work.

Transparent bridging, as denoted by its name, tends to be totally transparent to the LAN devices. With transparent bridging a LAN device never needs to know whether a destination device it wishes to interact with is co-located on the same physical LAN as itself or is attached to a faraway LAN. Transparent bridges ensure that LAN devices can interact with each other irrespective of their locations as if they are co-located on the same LAN. With transparent bridging a source device does not have to do anything special to invoke the bridging operation and locate a distant destination. (Note that with SRB, the source device first does an on-segment search followed by an off-segment search.)

The basic workings of transparent bridging are very similar to the transparent switching process described in Chapter 4. Transparent wwitching is in essence transparent bridging performed by a faster bridge, which now happens to be called a switch. Thus, as with transparent switching, a transparent bridge goes through a learning process during which it promiscuously reads the *source* MAC addresses of all frames it encounters on each of its ports. These source MAC addresses are stored in a cache directory along with the port number on which they were received. As a result of this learning process the bridge ends up with a cache directory that tells it the port number through which it can reach a MAC address that had been previously encountered as a source address of a frame. This directory entry creation process is exactly the same as that used by transparent switching.

Each time a transparent bridge encounters a frame it checks the destination MAC address of that frame against its cache directory. If it does not have an entry for that destination address (i.e., the bridge has yet to encounter a frame with that address as its source address) it propagates copies of that frame through all of its ports except the port on which the subject frame was received. If on the other hand, the bridge does find an entry for that destination address in its cache directory, it checks to see if the port indicated in the directory is the same as the one on which the frame was received. If they are the same it discards the frame since the frame would have been received over the same LAN containing the destination device. Thus, the destination device would have already seen that frame more or less at the same time the bridge saw it. If, however, the frame was received on a different port from the one specified in the cache directory the bridge will now forward the frame over the port specified in the directory. This will ensure that the frame will eventually reach its destination even if it has to traverse other bridges along the way.

5.2.1 Spanning Tree Algorithm vis-à-vis Transparent Bridging

Transparent bridging per the cache directory lookup and forward process described above is obviously very uncomplicated, easy to fathom, and even easy to implement. Unfortunately, the process described above has a fatal flaw if used in a network that has alternate paths (i.e., loops). In networks with alternate paths, frames could loop indefinitely through the network without ever getting to their destination. Moreover, each time a frame loops it could end up proliferating more copies of itself resulting in the network getting inundated with copies of the same frame. The reason for this is that transparent bridges can get confused as to where a given frame originated. Due to the alternate paths, a bridge may encounter a frame with the same source address on two different ports. Given that the MAC addresses of LAN devices are meant to be unique, this "peek-a-boo" appearance of the same address on two different ports throws the bridge into disarray, and it continually updates its cache directory to reflect where it thinks that address is currently resident.

Figure 5.4 illustrates a very simple two-LAN environment with three alternate routes between them to demonstrate the problems transparent bridging has with this type of network. Let's assume that LAN device A on LAN_0 sends a frame to device G on a LAN not even shown in the figure, and none of these three bridges has a cache directory entry for G. Since these are transparent bridges none of them will know about the existence of the others since transparent bridges are transparent to each other as well as to the LAN devices. Each of the three bridges will see the frame issued by A and learn that MAC Address A is reachable via their ports connected to LAN_0.

Each of these bridges will then proceed to forward this frame to LAN_1 per their forwarding algorithm given that they do not know where G is resident. Given that all three cannot transmit at the same time, one bridge will succeed in forwarding this frame to LAN_1 ahead of the other two. Now suddenly the other two bridges see a frame on their ports connected to LAN_1 that claims to have originated from device A. Transparent bridging, unlike SRB, does not make use of a field such as the RIF to indicate that bridging is taking place and the route that has been traversed by a frame. Thus a transparent bridge encountering a frame cannot determine whether that frame has just been originated by the source or whether it has been bridged across from another LAN by another bridge. This is a problem.

The two bridges that see the subject frame on their LAN_1 ports now have no choice but to assume that device A must have somehow moved and is now reachable via LAN_1 as opposed to LAN_0 as was the case before. Given that they still do not know where G is, these two bridges are honorbound by their forwarding algorithm to forward copies of the subject frame back to LAN_0. The general drift of this continuous to-and-fro frame-forwarding process now becomes clear. In essence, just like a program that is in a loop, these bridges will keep on proliferating new copies of the original frame and looping these frames from one LAN to the other.

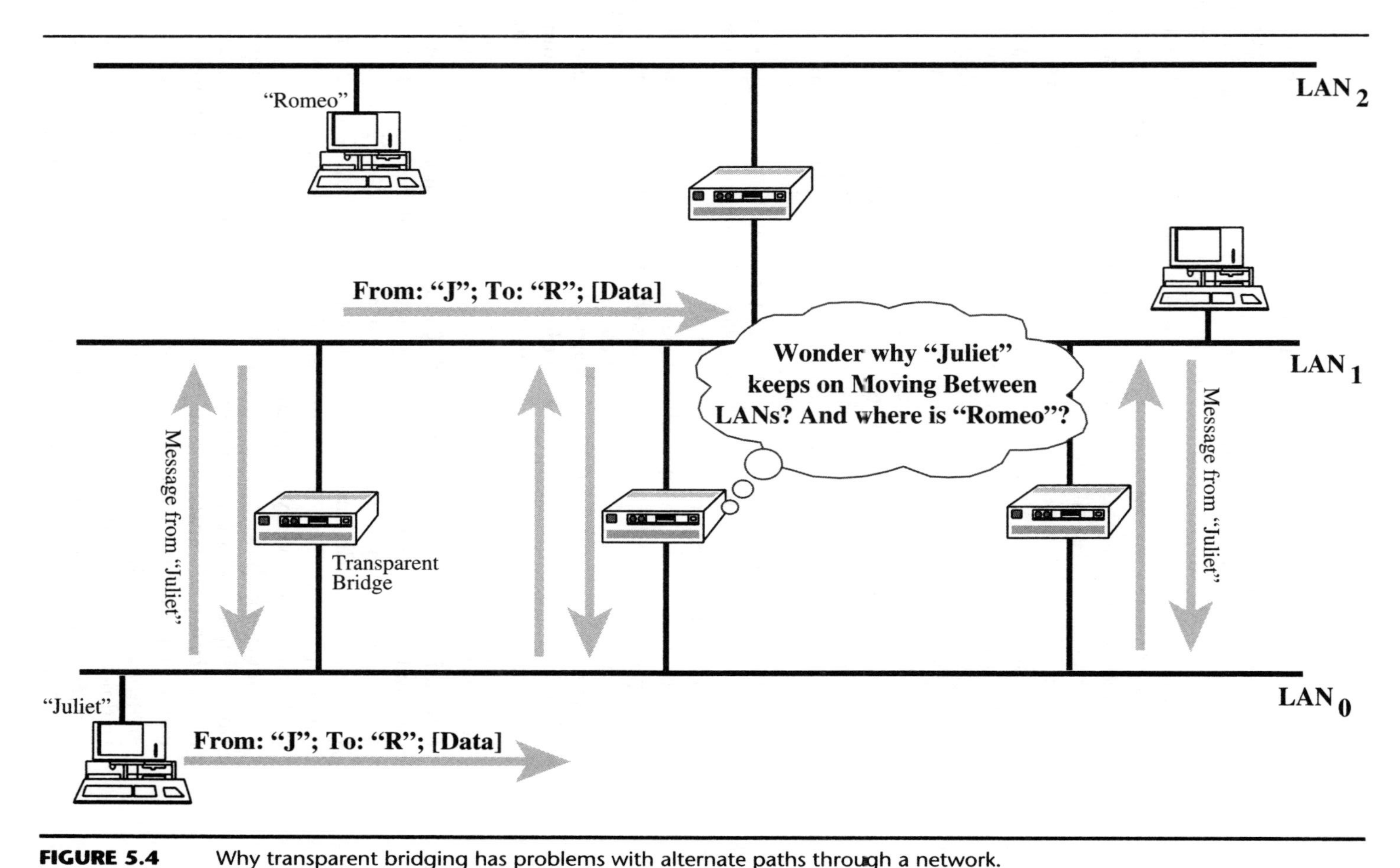

FIGURE 5.4 Why transparent bridging has problems with alternate paths through a network.

The bottom line here is that transparent bridging cannot be used within networks that contain active loops (i.e., alternate paths). Consequently, transparent bridging resorts to using the spanning tree algorithm to eradicate any active loops that may be present in a physical network and to create an instance of a logical loop-free network that can be used to transport traffic from one LAN to any other LAN. This leads to another fundamental difference between SRB and transparent bridging. SRB encourages alternate paths within its networks. It even goes to great lengths to discover all the potential paths between a given source and destination even though in the end it does not exploit this alternate path facility by supporting dynamic alternate routing. In contrast, transparent bridging cannot work if there are alternate paths within the network. It eradicates alternate paths using the spanning tree algorithm. Consequently, it also cannot support dynamic alternate routing if the original path fails. What it can do is to recalculate another spanning tree graph for the network using a previously blocked alternate path for the one that failed. (It should, however, be noted that certain bridge/routers offer proprietary extensions to transparent bridging that permit SNA TG-like parallel links, typically known as *circuit groups*, to be used with transparent bridging. Such extensions are mainly targeted for providing load balancing across WAN links.)

Recalculating the spanning tree results in connections being disrupted and traffic getting discarded. Ironically much the same thing happens with SRB in the event of a path failure. The devices affected will initiate a broadcast search to locate another route to the destination. This search process could take some time depending on the size of the network and the number of WAN links involved. During this time many of the SNA/APPN sessions could get disrupted since the SNA Path Control Network software is likely to detect that the original path appears to have failed. Thus, even though SRB can locate alternate paths, the process used, by not being dynamic, will result in session disruption and the need to relog on to SNA/APPN applications. Consequently, SRB and transparent bridging end up sharing a rather unfortunate common trait—neither, for very different reasons, supports nondisruptive dynamic alternate routing.

Figure 5.5 shows how the spanning tree algorithm will prune a network with many interconnections between its nodes into an instance of a loop-free configuration that still provides unrestricted any-to-any connectivity between the nodes. Figure 5.6 goes on to show how the spanning tree algorithm can be used to create a loop-free instance of a real network by disabling certain ports on some of the bridges. If there is a path failure within the active loop-free configuration shown in Figure 5.6 the bridges will rerun the spanning tree algorithm to come up with a new configuration where some of the previously disabled ports may now be activated to support the new configuration.

The spanning tree algorithm is a standard and integral component of transparent bridges. Bridges within the same interconnected network will exchange network topology-related message units—formally referred to as *Bridge Protocol Data Units* (BPDUs)—with each other that enable them to dynamically compute and activate an instance of a loop-free bridging configuration for that network. The destination MAC address on all BPDUs is a bridging-specific, universal, and *reserved multicast* or *functional address*. All transparent bridges support and recognize this *multicast* or *functional address*. Thus a BPDU issued by any bridge in the network is automatically intercepted, read, and acted upon by all the other active bridges in the network.

Transparent bridges do not require any human intervention in computing such a spanning tree-based loop-free configuration. However, certain parameters need to be assigned to each bridge in order for the spanning tree algorithm to determine a root bridge and then compute an active loop-free configuration around that root bridge. The minimum set of parameters that need to be specified for each transparent bridge are: a unique bridge identifier, port identifiers for all the ports on the bridge, and a path cost value representing the least cost path to the current root from the subject bridge.

The bridge with the highest priority bridge number (which happens to be that which is numerically the lowest) at any given juncture is automatically designated during the spanning tree computation process as being the current *root bridge*. If there are multiple bridges attached to the same LAN, the one closest to the root (i.e., the one with the least path cost to the root) is also automatically chosen during the computation process as the designated bridge for that LAN. The *designated bridge* will be responsible for forwarding frames from that LAN to the root bridge. (This notion of having a designated bridge precludes the possibility of having simultaneously active bridges that are acting in parallel to each other between two LANs.)

Each bridge transmitting BPDUs to the other bridges does not cease once a loop-free configuration has been established for the network. All the bridges will continue to transmit topology information bearing BPDUs every 1 to 10 seconds. The exact time interval to be used can be specified by a network administrator using *hello time* timer. This regular exchange of BPDUs enable the bridges to ensure that the current root bridge is still active and that the current topology is intact. If a BPDU exchange sequence indicates that the root bridge has failed or that the topology has changed the bridges will immediately recompute another loop-free configuration. The bridges will

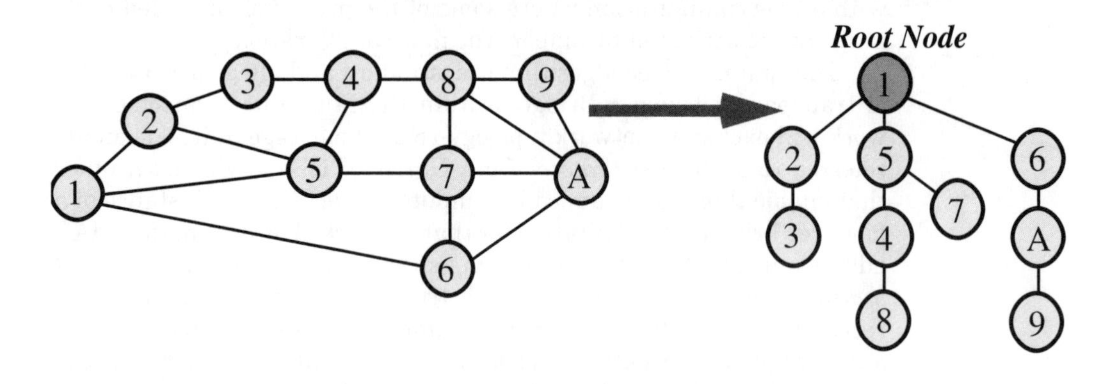

FIGURE 5.5 Pruning a network using the spanning tree algorithm to create a loop-free network.

cease forwarding frames while they are involved with computing a new configuration. This blackout period is referred to as a *convergence delay* since recomputing a new configuration is equivalent to reconverging the network.

Table 5.2 highlights the differences between SRB and transparent bridging.

5.3 SOURCE-ROUTE TRANSPARENT AND TRANSLATION BRIDGING

Within large enterprises it is not uncommon to have both token-ring and Ethernet LANs, possibly even co-located within the same floor of a building or within a campus. Neither is it unusual to have software (e.g., a LAN operating system or an e-mail package) running on a token-ring-attached PC/workstation that prefers to use transparent bridging rather than SRB. There could even be situations where the software is capable of working only with transparent bridging. There are multiple and obvious reasons as to why software developers may prefer transparent bridging to SRB.

For a start, transparent bridging does not require PC/workstation software to perform any bridging-related functions whatsoever. The bridge does all the work. Transparent bridging is indeed transparent to the software. In marked contrast, with SRB the PC/workstation software has to do a fair amount of bridging-specific functions. These include that of creating and inserting RIF fields, initiating on-and

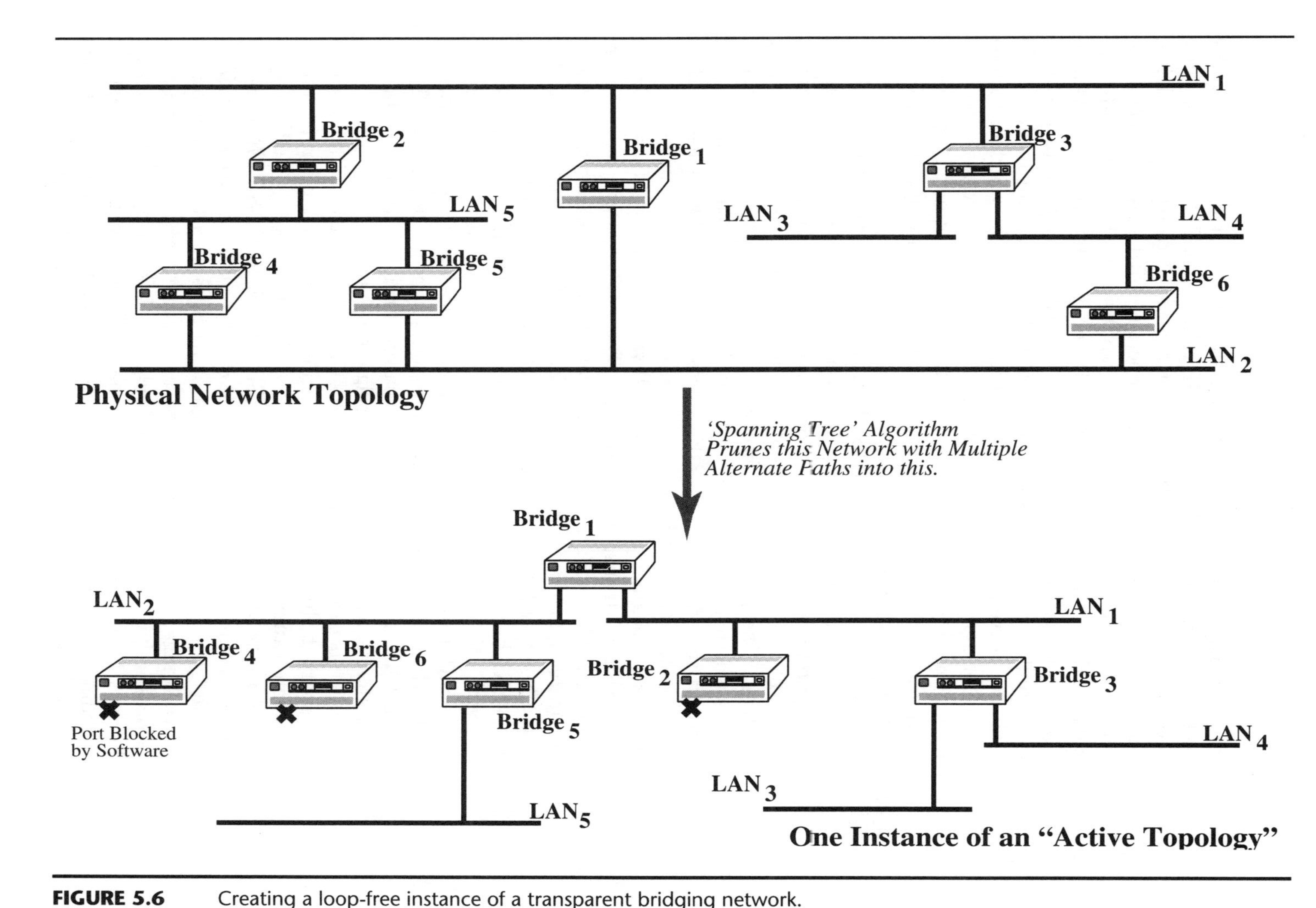

FIGURE 5.6 Creating a loop-free instance of a transparent bridging network.

off-segment searches, and selecting the path to the destination by picking the route specified in the RIF of the first TEST/XID response received, and so on. The transparent aspect of transparent bridging, compared to the work involved in supporting SRB, can alone persuade software developers to opt for transparent bridging instead of SRB. (To balance this, it is, however, worth pointing out that SRB with its support of multiple active routes within a network may provide more efficient route selection than is possible with transparent bridging.) In addition, there is also the factor that Ethernet LANs, which use transparent bridging rather than SRB, are much more pervasive and have been around longer than token-ring LANs. Hence, there is more software that intrinsically supports Ethernets and transparent bridging than token-rings and SRB.

Source-route transparent (SRT) bridging was developed as a way of ensuring that devices using SRB could *coexist* with devices using transparent bridging on the same network as well as on the same LAN. An SRT bridge concurrently supports both SRB and transparent bridging. The presence or absence of a RIF within the frame determines whether an SRT bridge will use SRB or transparent bridging to forward that frame. If an RIF is present it will always use SRB. If there is no RIF it will use transparent bridging. Thus, with SRT bridges it does not matter if the software in a LAN-attached device uses SRB or transparent bridging. The SRT bridges will support either. *An SRT bridge, however, will not allow a device using transparent bridging to talk to one using SRB nor will it allow a device attached to a token-ring LAN to interact with one attached to an Ethernet LAN.* With SRT bridging, devices using translation bridging can only interact with other devices also using translation bridging, while devices using SRB can only interact with other devices using SRB. Thus, SRT is strictly a means of providing coexistence between devices and LANs using SRB and those using transparent bridging. SRT is a facility available today on most leading bridge/routers.

Interoperability between LAN devices using SRB and those using transparent bridging, and consequently generic interoperability between Ethernet and token-ring LANs, can be realized using a *source-route translational bridge* (SR-TB). *With an SR-TB, a device using translation bridging can freely interact with a device using SRB.* An SR-TB is seen by LAN devices using SRB as a bona fide source-route bridge, while it appears as a standard translation bridge to devices relying on translation bridging. Token-ring and Ethernet Frames have different header formats. They even use different bit encoding schemes to represent the bits that make up a byte; token-ring transmits the most significant (i.e., leftmost) bit of a byte first, while

TABLE 5.2 Source Route Bridging Versus Transparent Bridging

Source-Route Bridging	*Transparent Bridging*
• End Stations Locate & Determine Route	• Transparent to End Stations
• No Central Control of Routes	• Routes Relative to Root
• Multiple Paths	• No Loops
• Overhead of Broadcast Traffic	• BPDU Exchanges
• Bridge Has to Be Configured	• Potential Plug-&-Play
• Simpler Bridges, but not Less Expensive	• Bridges are More Complex
• End Stations Must Reestablish New Route if Original Path Fails	• Learning, Setup, & Convergence Delays
• No Longer a Standard; Must Be SRT	• IEEE 802.1d Standard

Ethernet and 802.3 transmit the least significant bit of a byte first. An SR-TB converts between the different header formats and the bit ordering schemes. Figure 5.5 shows the type of header format conversion performed by an SR-TB.

Source-Route Translation Bridging was pioneered by IBM in the late 1980s with the IBM 8209 LAN bridge. SR-TB is now available on certain bridge/routers, particularly from Cisco and IBM. On a bridge/router, SR-TB is only required if MAC-layer bridging, opposed to IP or IPX type Layer 3 routing, needs to be performed between devices using SRB and transparent bridging. There is no need for translation bridging if traffic between devices on a token-ring and those on a Ethernet is being routed. The routing function will take care of the header format and bit ordering differences between these two types of LANs. Thus, if a bridge/router network is using TCP/IP encapsulation (e.g., DLSw) to support SNA/APPN and NetBIOS traffic there is no need to worry about SR-TB even if there is a need for interactions between Ethernet-attached devices and token-ring-attached devices.

This scenario for Ethernet to token-ring interworking is likely to occur fairly often these days in enterprises that have IBM 3745 communications controllers. Though the 3745 has had an Ethernet interface since late 1992 this interface does not support SNA/APPN traffic over Ethernet. This Ethernet interface can only be used to transport TCP/IP traffic to and from a mainframe. The inability of the 3745 to support SNA/APPN across its Ethernet interface does not, however, preclude enterprises from deploying SNA/APPN traffic bearing Ethernet LANs. SR-TB or TCP/IP encapsulation of SNA/APPN can be gainfully used in such scenarios to provide transparent and seamless interoperability between a token-ring interface on the 3745 and all of the Ethernet LANs.

5.4 OVERCOMING THE LIMITATIONS OF SOURCE-ROUTE BRIDGING

Bridging in general, whether source-route or transparent, has the infuriating weakness that it cannot offer nondisruptive, dynamic alternate routing in the event of a path failure. Bridging a path failure due to a problem with a WAN link or bridge will invariably result in SNA/APPN sessions getting disrupted and users having to relog on to applications. This lack of dynamic alternate routing can be particularly galling if the LAN/WAN network contains multiple alternate paths that could have been used to reroute traffic around the link or bridge experiencing problems.

In addition to the lack of dynamic alternate routing, SRB has other frustrating limitations not conducive to contemporary networking. These additional limitations of standard SRB include:

1. The repeated bandwidth-sapping and traffic-disrupting broadcast searches for previously discovered destination addresses (e.g., 37xx mainframe gateway) since SRB do not maintain a cache directory of known destination address. For example, an SRB bridge on a LAN with say 30 PCs that need to interact with a mainframe will unashamedly do 30 broadcast searches, one after the other, for the mainframe LAN gateway. In addition, this repetitive search for the mainframe LAN gateway will in turn be done by every bridge on the network that is supporting PCs that need to interact with the mainframe.
2. The potential for setting off broadcast storms by having multiple, far-flung broadcast searches taking place at the same time
3. No capability for load-balancing traffic across multiple parallel links (à la SNA Transmission Groups) given that SRB is a fixed-path routing scheme that does not support the notion of traffic from the same session being sent across parallel WAN links or multiple paths through the network
4. Inability to select the optimal path between the source and the destination based on criteria such as the least number of intermediate hops or the lowest cost
5. Only being able to cater to a maximum of seven intermediate hops between the source and destination

Around 1991, Cisco and CrossComm set out to develop ways to eradicate the limitations of SRB when it came to supporting SNA/APPN traffic within LAN/WAN networks. Cisco came up with the notion of encapsulating the SNA/APPN traffic within TCP/IP message units and then routing these TCP/IP message units from the source LAN to the destination LAN. Cisco called this TCP/IP encapsulation scheme for SNA/APPN traffic *remote source-route bridging* since it was essentially a TCP/IP-based version of source-route bridging.

CrossComm's solution was Protocol Independent Routing (PIR). PIR unlike Cisco's approach was not based on TCP/IP. Instead, CrossComm with PIR built upon the basic and original infrastructure of standard SRB to come up with a scheme that retained the plug-and-play simplicity of SRB while at the same time eliminating all of its major weaknesses. In a sense, PIR became what SRB should have been in the first place if the original implementors of SRB at IBM had taken the time and trouble to add a bit of ingenuity and pizazz to the basic and somewhat open-ended SRB routing scheme outlined in the token-ring architecture.

TCP/IP encapsulation and PIR both support dynamic alternate routing in the event of path failure. PIR can also optionally provide dynamic alternate rerouting in the event of link congestion. In addition, they both possess powerful cache directory mechanisms that preclude the need for repeated searches for previously discovered destinations. PIR and some TCP/IP encapsulation schemes (e.g., DLSw) also provide a facility whereby a router that already knows the whereabouts of a destination being sought by another router can provide the latter with this destination location information. This data sharing process short-circuits the time that it would take for a complete broadcast search to be completed.

TCP/IP encapsulation and PIR also support load balancing in certain situations. In the former, however, the availability of this feature is contingent on whether it is supported by the IP routing protocol in use. Some IP routing protocols such as the widely used Routing Information Protocol (RIP) will not support load balancing. Others such as Open Shortest Path First (OSPF) and Cisco's proprietary Interior Gateway Routing Protocol (IGRP) do. PIR's load balancing is typically realized between two or more pairs of parallel bridge/routers (e.g., two bridge/routers on the source LAN and two on the destination), as opposed to across parallel links between two end-to-end bridge/routers.

The dynamic alternate routing of SNA/APPN traffic encapsulated within TCP/IP is realized as a byproduct of the standard IP dynamic alternate routing capability. In the case of PIR dynamic routing is a function built, into PIR itself, given that PIR is not built upon another protocol such as IP. In any router-based network a path failure that necessitates alternate routing will also require that the routers update their network topology data and recalculate the valid and active paths across the network. This topology and route recalculation process is analogous to that described above for transparent bridging and involves the exchange of the router equivalent of BPDUs.

The time taken to do this recomputation of the new network layout is referred to as a *convergence delay*. Typically data traffic routing is disrupted during the convergence delay since the routers are preoccupied with exchanging updated topology information with each other and recomputing the new topology. The exact length of a convergence delay will depend on the size of the network as well as the speeds of the WAN links between the routers. In mid- to large-size networks (i.e., 50 or more routers) it is not unusual for convergence delays to be in excess of 20 seconds. Dynamic alternate routing will not come into play during the convergence delay since the routers are still determining what the active and valid routes through the network are. Dynamic alternate routing will thus only kick in after the network has been reconverged. If the convergence delay is excessive there is a possibility that some SNA/APPN sessions may get disrupted due to timeouts before dynamic alternate routing has a chance to kick in.

In the event of a route failure PIR networks are also subject to a convergence delay. PIR, however, goes to inordinate lengths to minimize the length of this delay. Early detection of link failures is an important factor in keeping this delay small. CrossComm routers employ a special Layer 1 hardware interface that permits them to detect the failure of a WAN link within two seconds. (In other bridge/routers WAN link failures are typically only discovered when a retry timer expires.) This head-start in detecting a link failure usually permits PIR to converge a network in a shorter time than that which would be required by non-PIR routers to converge an equivalent network. This reduction in convergence delay reduces the possibility of SNA/APPN sessions timing out and thus getting disconnected.

Cisco's RSRB was and still is proprietary. So is PIR. Other early TCP/IP encapsulation schemes such as that initially used by 3Com were also proprietary and vendor-specific. The proprietary nature of TCP/IP encapsulation schemes changed with the advent of DLSw. IBM's attempt at TCP/IP encapsulation appeared with the 6611 bridge/router in mid-1992. This TCP/IP encapsulation scheme was included within a set of SNA/APPN and NetBIOS integration facilities that were generically referred to as DLSw. In January 1993, IBM submitted a specification of DLSw as implemented on the 6611 to the IETF. This *DLSw* specification covered the message unit formats used by DLSw, the finite-state machines for DLSw state transitions, and the broadcast search protocol used by DLSw. This DLSw specification became an Informational RFC (I-RFC)—such as I-RFC 1434. An Information RFC is not an official Internet standard per se in that it has not been ratified by the IETF.

Today, DLSw implementations, currently based on I-RFC 1434, are virtually ubiquitous on all leading bridge/routers. Cisco, Bay Networks, 3Com, Proteon, IBM, CrossComm, and ACC among others currently support DLSw. Cisco supports standard DLSw and RSRB side by side on its bridge/routers. The Cisco term *DLSw+* refers to the fact that Cisco routers offer both multivendor interoperable DLSw as well as the proprietary RSRB. DLSw in general is the only nonproprietary TCP/IP encapsulation scheme for SNA/APPN traffic. DLSw thus offers the promise of multivendor interoperability. The latest version of DLSw, known as *DLSw Standard Version 1.0* (or I-RFC 1795), further facilitates multivendor interoperability by including an DLSw Capabilities Exchange handshake protocol that enables different DLSw implementations to agree on a common set of supported features.

The last major difference of note between TCP/IP encapsulation schemes and PIR relate to the amount of effort required to implement a network based on one versus the other. TCP/IP encapsulation schemes in general work in conjunction with SRB. Thus, the parameters required by a source-route bridge—which are typically just a bridge number and a ring number for each LAN attached to the bridge—still need to be specified even when TCP/IP encapsulation is being used. Though additional parameters can be specified to optimize performance, PIR when working in default mode does not require any additional parameters to those specified for SRB. Given their dependence on IP addresses, TCP/IP encapsulation schemes in general also require some IP address administration and management. Consequently, it is fair to say that all TCP/IP encapsulation schemes require more human intervention in order to be set up than that required by PIR. At this juncture it is, however, worth noting yet again that PIR despite its pioneering role and its indubitable strengths such as its plug-and-play capability is now not a major contender for parallel network consolidation given that CrossComm no longer actively promotes it.

5.5 TCP/IP ENCAPSULATION OF SNA/APPN TRAFFIC

The goal of TCP/IP encapsulation is to provide an IP-centric, reliable, and value-added mechanism for effectively transporting SNA/APPN and NetBIOS traffic within a multiprotocol LAN/WAN network built around bridge/routers. With this approach, SNA/APPN and NetBIOS traffic that needs to be transported from one LAN to another is first encapsulated within TCP/IP message units and then routed from the source LAN to the destination, most likely across a WAN, as TCP/IP traffic. At the destination LAN, the SNA/APPN or NetBIOS traffic is extracted from the TCP/IP message units and put on the LAN as standard token-ring frames. Figure 5.7 illustrates the overall concept of encapsulating SNA/APPN or NetBIOS traffic within a TCP/IP message unit.

TCP/IP encapsulation as discussed above eradicates all of the annoying weaknesses and limitations of SRB. It also has the virtue that SNA/APPN traffic that is thus encapsulated can be routed from bridge/router to bridge/router using standard IP routing conventions. When not encapsulated in TCP/IP, bridge/routers can only support SNA/APPN traffic using either bridging or APPN NN routing, which is still a relatively new and expensive feature of most bridge/routers. Unencapsulted NetBIOS traffic can usually only be bridged. TCP/IP encapsulation in effect imbues SNA/APPN and NetBIOS traffic with the dynamic networking properties of IP (e.g., dynamic alternate routing) and permits them to enjoy "nonalien" status when being dealt with by bridge/routers.

With TCP/IP encapsulation SNA/APPN and NetBIOS traffic that needs to be forwarded from one LAN to another, typically across a WAN, is intercepted by a bridge/router. In the absence of TCP/IP encapsulation this LAN-to-LAN traffic forwarding would normally have been realized using SRB. Thus in order to intercept SNA/APPN and NetBIOS traffic the bridge/router masquerades as a source-route bridge. (Bridge/routers can identify SNA/APPN and NetBIOS traffic by examining the Source Service Access Point [SSAP] value that is included within the LLC portion of a MAC/LLC frame; see Figure 5.1. SNA/APPN traffic will have SSAP values such as X04, X05, X08, and X0C, while NetBIOS traffic will have a SSAP value of XF0.) TCP/IP encapsulation thus always works as a transparent adjunct to SRB rather than a stand-alone methodology independent of SRB and the MAC addresses used by SRB. This symbiotic relationship with SRB is a fundamental concept of all TCP/IP encapsulation schemes including DLSw.

The need for TCP/IP encapsulation to build upon SRB boils down to one fundamental issue. This is the need for total and absolute compatibility with SNA/APPN and NetBIOS software running on PCs, workstations, other LAN attached devices (e.g., IBM 3174 control units), LAN-attached minicomputers (e.g., AS/400s) and SNA LAN gateways (e.g., IBM 3745). SNA/APPN software developed for LAN applications invariably uses MAC addresses to denote destinations, for example, SNA LAN gateway such as a 37xx. (The SNA/APPN software being described here relates to full SNA node implementations, as opposed to emulators that will use a non-SNA protocol between the LAN device and the emulation server or gateway. If a TCP/IP-based SNA/3270 emulation scheme such as tn3270 is being used IP addresses will be used to designate the tn3270 server as opposed to a MAC address.) NetBIOS applications supplement MAC addresses with NetBIOS name-based protocols.

Given the variety, diversity, complexity, and the sheer volume of SNA/APPN and NetBIOS applications, TCP/IP encapsulation schemes cannot afford to request that this software is in anyway modified. Hence, all TCP/IP encapsulation schemes use destination MAC addresses or NetBIOS names as their means (or metric) of locating desired destinations (e.g., SNA LAN gateway or NetBIOS server). They rely on intercepting and supporting standard LAN MAC frames. Consequently, TCP/IP encapsulation schemes can be thought of as being TCP/IP-based LAN-to-LAN bridging schemes. Figure 5.8 illustrates the notion that TCP/IP encapsulation schemes are essentially a form of bridging.

5.5.1 Correlating Destination MAC Addresses with IP Addresses

Since TCP/IP encapsulation, when dealing with SNA/APPN traffic, relies on the destination MAC address to indicate where that traffic needs to be delivered to, it needs mechanisms to do the following:

- Locate the LAN containing the destination (e.g., 3745) being sought.
- Once the destination LAN is found, obtain the IP address of a bridge/router-attached to that LAN that can now act as the *destination bridge/router* for traffic being sent to the LAN-attached device corresponding to the subject destination MAC address.
- Correlate each destination MAC address with the IP address of the bridge/router acting as the destination bridge/router for that address.

TCP/IP encapsulation schemes used to differ in terms of how the above processes were conducted. If the number of destinations in a network are relatively small—for example, an SNA network with one mainframe and one 37xx acting as the LAN to SNA gateway—the easiest and most intuitive way to realize the MAC address to IP address correlation would be to do it manually. The original 3Com scheme used this approach. On each bridge/router supporting TCP/IP encapsulation, a network administrator would manually define the IP address of the destination bridge/router attached to the LAN containing a given destination MAC address. For small networks with just one or two destinations this manual scheme can be extremely efficient and pain-

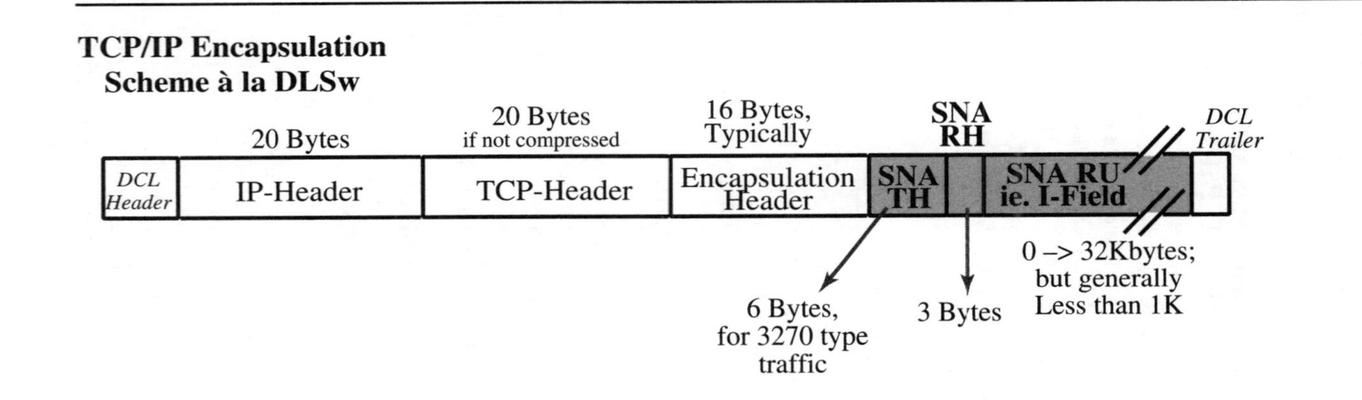

FIGURE 5.7 Encapsulating SNA message units within a TCP/IP packet.

less. Doing this MAC address to IP address correlation manually obviates the need to perform some type of broadcast search, à la SRB, to locate the destination LAN and the bridge/router serving it.

Manual correlation schemes are now out of favor. Apart from Cisco, all other major bridge/router vendors including 3Com have now standardized on DLSw as their TCP/IP encapsulation scheme for SNA/APPN and NetBIOS traffic. Cisco, in addition to supporting DLSw, also supports its original and proprietary RSRB scheme. Today, DLSw and RSRB are thus the only two TCP/IP encapsulation schemes of note. Given DLSw status as the de facto TCP/IP encapsulation standard for this industry this two-horse scenario is now unlikely to change. Any new entrants to this arena are most likely adopt DLSW rather than coming up with their own scheme.

DLSw and RSRB both use dynamic broadcast search mechanisms, which are both broadly based on SRB's search scheme, to locate the MAC addressed specified destination. They do not use manual correlation.

5.5.2 Overview of DLSw's TCP Connection Establishment Process

In order for DLSw to be used between any two bridge/routers, two TCP connections need to be established between them. This need for two TCP connections between each pair of bridge/routers is a prerequisite of DLSw. Between a given pair of DLSw bridge/routers, each bridge/router uses one TCP connection as its transmit path to the other bridge/router, and the other TCP connection as its receive path from its partner. Each bridge/router only establishes its transmit path TCP connections. The two connections between a pair of bridge/routers thus get established by each establishing just its outgoing transmit path.

In order to establish these TCP connections each DLSw-capable bridge/router in general needs to have a list of the IP addresses of the other bridge/routers it may have to forward SNA/APPN or NetBIOS traffic to. Each DLSw capable bridge/router when activated will set about trying to establish a transmit path TCP connection with all the other DLSw capable bridge/routers that it knows about. Each time a bridge/router successfully establishes a transmit path TCP connection it will receive another connection establishment from the other bridge/router to set up the second connection. DLSw bridge/routers set up their transmit path TCP connection using a DLSw specific TCP Port Number. There are many variations as to the amount of DLSw connections that need to be established by a given bridge/router and how it should go about establishing these connections.

In mainframe-centric traditional SNA environments it may not be

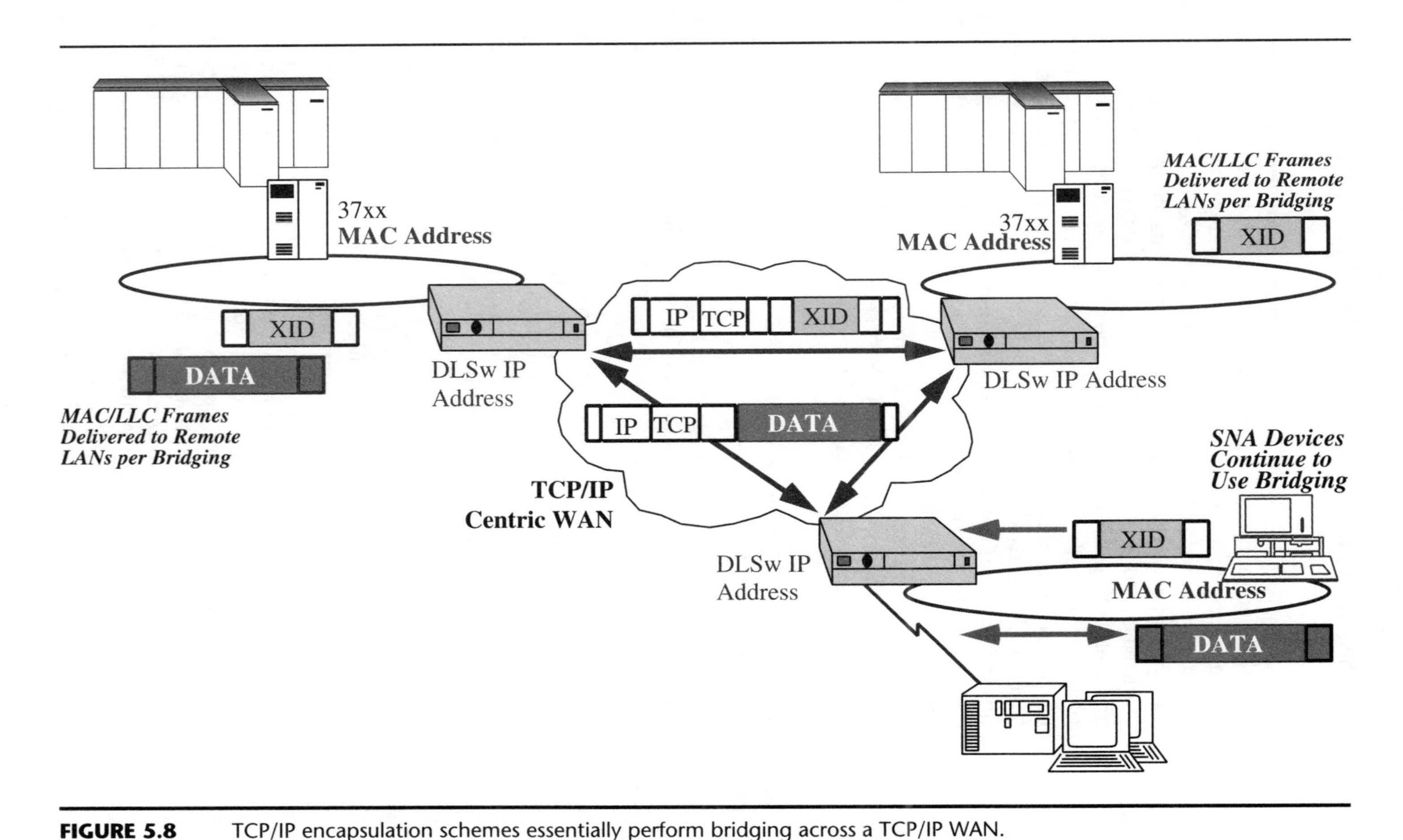

FIGURE 5.8 TCP/IP encapsulation schemes essentially perform bridging across a TCP/IP WAN.

necessary for the remote DLSw bridge/routers to establish TCP connections with all of the other remote routers. Given that all of the SNA traffic will be going to/from the mainframes it may be sufficient for the remote bridge/routers to establish TCP connections with just the central bridge/routers adjacent to the mainframes. In this type of network it would also be feasible for the central bridge/routers not to have an exhaustive list of the IP addresses of all of the remote bridge/routers. Instead they could wait to receive connection establishment requests from the remotes rather than trying to initiate these connections from the central bridge/router. This type of remote-initiated connection establishment will obviously minimize the amount of DLSw-related IP addresses that have to be defined at the central routers.

Figure 5.9 depicts the DLSw-related TCP connections between bridge/routers in a mainframe-centric SNA network. Proteon has an extension to DLSw that eliminates the need for the manual predefinition of any DLSw-specific IP addresses. Rather than relying on a list of IP addresses for the other DLSw bridge/routers, the Proteon scheme employs a DLSw OSPF multicast address. When a Proteon DLSw bridge/router is activated it automatically issues a "Hello. I am a new DLSw router" message using the DLSw multicast addresses. All active DLSw-capable bridge/routers will receive this multicast message. They will automatically learn the address of the new bridge/router since its address would be included as the source IP address. They will in turn respond to the multicast greeting. Each response will contain the IP address of its sender. With this scheme all the DLSw bridge/routers can dynamically learn each other's IP addresses. This Proteon multicast scheme, which is unfortunately still not a part of the DLSw standard, can make DLSw look like a plug-and-play scheme à la SRB or PIR.

Once the two TCP connections are established between a pair of bridge/routers they exchange DLSw Capabilities message units with each other. This DLSw capabilities exchange permits them to determine which DLSw facilities each supports and the version of DLSw implemented in each bridge/router. This capabilities exchange facilitates multivendor interoperability as well as backward compatibility with older versions of DLSw. Once the DLSw capabilities at each end have been determined the DLSw bridge/routers can now set about transporting actual data traffic.

5.5.3 Overview of the DLSw's Destination Location Process

Just as with SRB, DLSw's need to locate a remote destination is triggered when a DLSw-capable bridge/router receives an off-segment, broadcast search-required TEST or XID LLC command or a NetBIOS

NAME_Query request. The destination being sought would be indicated by the command's destination MAC address or a NetBIOS name. If the destination MAC address or NetBIOS name is not in the bridge/router's DLSw cache directory it would conduct a SRB-like broadcast search—but using DLSw-specific TCP/IP message units as opposed to the standard token-ring frames, replete with the RIF fields, as used by SRB. The two main TCP/IP-based message units employed by DLSw to conduct such searches are known as CANUREACH and ICANREACH. If the source bridge/router does have an entry for the destination address it will skip the broadcast search state and move to the next phase in establishing an end-to-end connection.

The first broadcast search message sent out by the source bridge/router initiating the search is known as a *CANUREACH Station-Explorer.* This is a 72-byte DLSw Control Message that will be included within a TCP/IP packet containing a 20-byte IP header and a 20-byte TCP header at its start. The destination MAC address being sought as well as the destination SAP address are extracted from the TEST or XID command and inserted within predefined fields of the CANUREACH Station-Explorer control message. Various data link correlators are also included in this message that will be returned back in the responses to this message. The source DLSw bridge/router will then send a copy of the CANUREACH Station-Explorer control message to every bridge/router with which it has previously established DLSw TCP connections.

On receipt of a CANUREACH Station-Explorer control message a target bridge/router will check its cache to see if it has an entry for the destination being sought. If it has an entry it will respond to the explorer message with an ICANREACH Station-Explorer control message. If it does not have an entry for the destination it will create an LLC TEST command with the destination MAC address and DSAP that was in the control message and transmit copies of that TEST message on all of its LAN ports. Now just as with a SRB search, this target DLSw bridge/router will wait to see if it receives a response to this TEST command. Every bridge/router that received a copy of the CANUREACH Station-Explorer control message would conduct this same search on all the LAN ports known to them using locally created TEST commands. If the destination is active it will eventually respond to one or more of these TEST commands. This response will then reach one or more of the DLSw bridge/routers that issued TEST commands.

A router receiving a TEST response will issue an ICANREACH Station-Explorer control message to the source bridge/router that issued the CANUREACH to indicate that it has access to the relevant destination device. This provides the correlation between the destina-

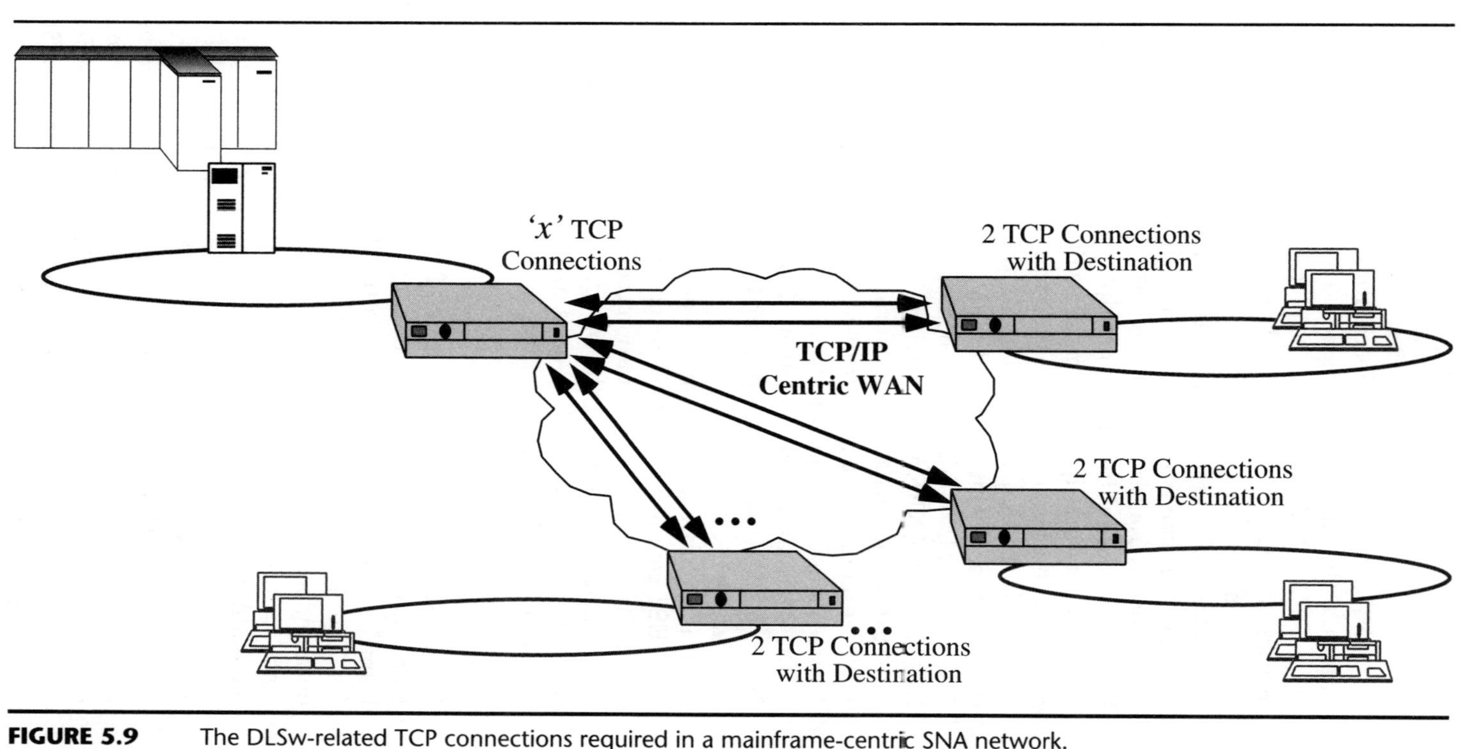

FIGURE 5.9 The DLSw-related TCP connections required in a mainframe-centric SNA network.

tion's MAC address and a destination bridge/router that can forward frames to that destination device. The source bridge/router on receiving an ICANREACH Station-Explorer control message responds to it by issuing a REACH Acknowledgment control message. At this stage a DLSw connection is not established.

There could be multiple routes—that is, multiple potential bridge/routers—through which to forward frames to the destination. The source bridge/router waits to see if it gets multiple, disparate ICANREACH responses indicating multiple options for reaching that destination. If it gets multiple responses it will use an implementation-specific metric to determine which destination bridge/router to use as its preferred path to that destination. At this juncture it has the option of caching these route(s) in its directory against an entry for the destination MAC address to obviate the need for subsequent CANUREACH–ICANREACH searches.

At this juncture the source bridge/router will construct and issue a TEST response that is addressed to the original source LAN device. This response will appear to the source as if it had been sent directly by the destination device. Given that DLSw does not update any RIF fields the destination will appear to the source as being on the same LAN as itself.

It is important to note at this point that no LAN frames—for example, TEST command or TEST response—have been transported end-to-end. Instead only the information required to recreate such frames has been exchanged between the bridge/routers. The bridge/routers have created the relevant LAN frames using the information that was contained within the TCP/IP control message. (All of these DLSw connection establishment control messages are 72-bytes long and are prefixed by a 40-byte TCP/IP header.) This is a very crucial concept of DLSw. It never transports LAN frames in their entirety between DLSw bridge/routers—either during connection establishment or data transfer. Instead it just forwards the data content with TCP/IP packets. The receiving bridge/router then creates a LAN frame around this data and sends it out on the appropriate LAN. This is in marked contrast to RSRB that does transport entire LAN frames, that is, complete MAC/LLC frames, LAN-to-LAN within TCP/IP packets.

Once the preferred destination bridge/router is selected the source bridge/router then sends it a CANUREACH Station-Circuit Start control message. This message essentially contains the same information that was sent out in the first message. The key difference is that this control message is only sent to one bridge/router—the designated destination bridge/router. On receipt of a CANUREACH Station-Circuit Start control message the destination bridge/router will set up the necessary control blocks to support a DLSw connection between itself and the source for traffic between the source LAN device (e.g., PC) and the destination LAN device (e.g., 3745).

The designated destination bridge/router will respond to the CANUREACH Station-Circuit Start control message with an ICANREACH Station-Circuit Start message which in turn will be acknowledged with a REACH Acknowledgment control message. When the source issues this last REACH Acknowledgment control message a bidirectional DLSw connection will be in place to transport traffic between source LAN device (e.g., PC) and the destination LAN device.

Cisco's DLSw+ has a scheme for optimizing the amount of TCP/IP-based broadcast searches that have to be performed in large, complex networks with many interconnections between bridge/routers. This optimization scheme is referred to as *peer grouping* or *peer groups*. DLSw+ peer groups set out to create a logical broadcast search hierarchy within the network based around the expected flow of broadcast searches within a network. Thus for example if a network has a hub-and-spoke structure where multiple branch office (or feeder) bridge/routers are connected to a regional bridge/router which is then connected to the core WAN, this regional collection of bridge/routers could be defined as a peer group. Figure 5.10 depicts the concept of DLSw+ peer groups.

Within a peer group typically one bridge/router is designated as the *border peer*. (More than one bridge/router per peer group may be designated a border peer to cater to requirements such as fault-tolerancy.) When a bridge/router within a peer group needs to conduct a search for an unknown destination it sends a search request only to its designated border peer. The border peer then propagates this search across the network and forwards the result of the search back to the source bridge/router. This border peer mechanism can cut down on the amount of broadcast search traffic that flows across the network.

5.5.4 Overview of the DLSw's Data Transfer Phase

When the source LAN device receives the TEST response it may respond with an XID command to initiate the establishment of the LLC:2 connection between itself and the destination. In the case of an SNA Type 2.1 peer-to-peer node or an APPN node, rather than there being just one XID in each direction, there could be a series of XIDs in each direction—known as an XID negotiation process. Such an XID negotiating process permits the two nodes involved to negotiate a variety of options. Key among these is which of the two nodes will be the Layer 2 Primary Station. DLSw bridge/routers transport XID information from source LAN to destination LAN using DLSw-specific XID Frame control messages.

Following the XID exchange, which may just be one XID in each direction or a protracted XID negotiation exchange, the LAN device containing the designated Layer 2 Primary Station will issue an LLC:2 connection establishing Set Asynchronous Balance Mode Extended (SABME) command. This SABME will trigger a DLSw CONTACT control message, which will result in a SABME being generated and issued at the other end. Figure 5.11 shows a typical DLSw connection establishment process leading all the way up to the data transfer stage.

Though LLC:2 connections are established via the SABME–UA exchanges at either end, there is no end-to-end LLC:2 connection between the source and destination LAN devices. The LLC:2 connections at each end are terminated at the DLSw bridge/router. This termination of the LLC:2 connections at each end is another fundamental precept of DLSw. The DLSw bridge/routers, using TCP's reliable transport mechanism, provides a TCP/IP-based virtual bridge between the two separate LLC:2 connections at each end. Not having an end-to-end LLC:2 connection also means that LLC:2 acknowledgments do not have to be transmitted across a WAN between the DLSw bridge/routers. LLC:2 acknowledgments are generated locally by the DLSw bridge/routers. This local LLC:2 acknowledgment facility is another well-known and quintessential feature of DLSw.

Once the end-to-end DLSw connection and the two local LLC:2 connections have been established, unconstrained bidirectional data transfer can take place between the original source and destination LAN devices. The DLSw bridge/routers intercept all the databearing LAN frames and extract the data portion (i.e., the Information Field) from these frames. This data is then placed within DLSw-specific TCP/IP-based INFOFRAME message units for transportation between the DLSw bridge/routers. An INFOFRAME message unit consists of a I-field data block extracted from a LAN frame (e.g., SNA Path Information Unit [PIU] message unit) prefixed by a 16-byte DLSw information message header. (There are proprietary extensions to DLSw that permit multiple SNA PIU message units from being placed within a single DLSw INFOFRAME message unit. This optional data blocking feature cuts down on the TCP/IP packets that have to be sent across the WAN and can expedite end-to-end packet delivery.)

This 16-byte header contains a DLSw identifier that serves as a DLSw session identifier between the source and destination DLSw bridge/routers. The INFOFRAME message unit, with its 16-byte header is then prefixed by a 40-byte TCP/IP header combination. Hence the total header

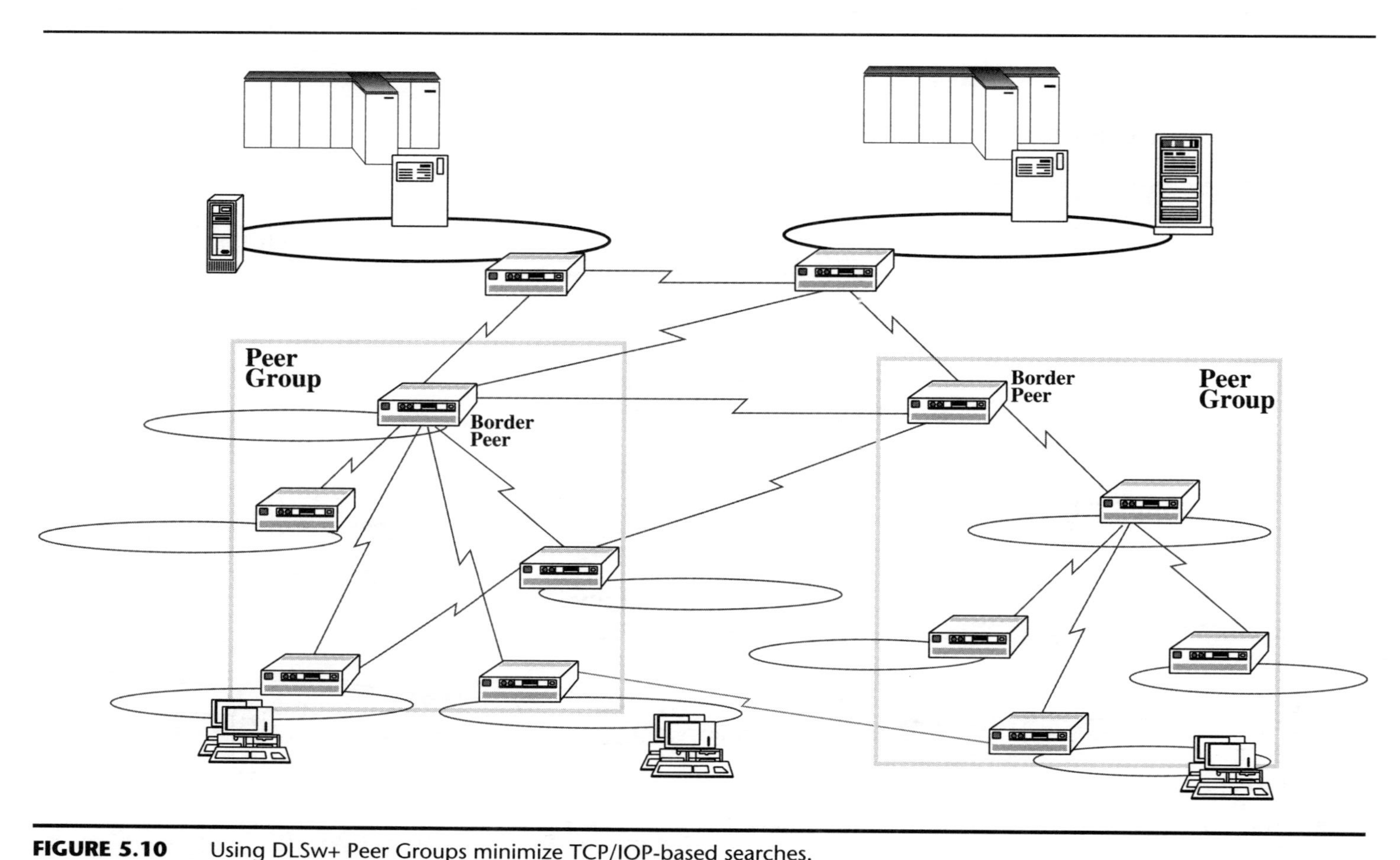

FIGURE 5.10 Using DLSw+ Peer Groups minimize TCP/IOP-based searches.

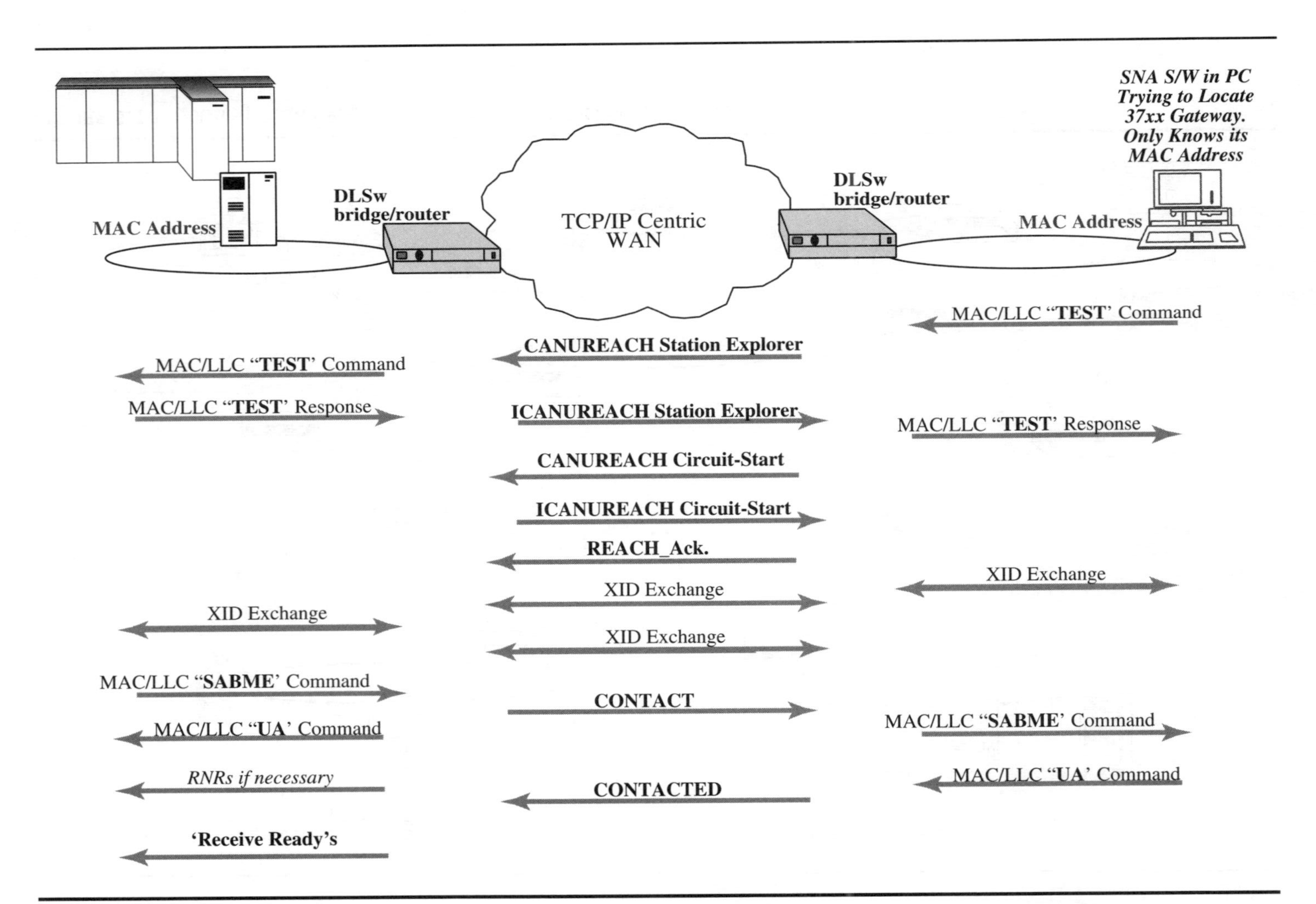

FIGURE 5.11 Overview of DLSs's end-to-end connection establishment process leading to the data transfer phase.

overhead of DLSw when transporting data traffic is 56 bytes—that is, 20-byte IP header, 20-byte TCP header, and the 16-byte DLSw header.

If frame relay is being used as the WAN medium between the DLSw bridge/routers, the DLSw INFOFRAME message unit replete with its 56 bytes of header will be encapsulated within a 10-byte frame relay frame. The 10 bytes that make up this frame relay frame consist of 6 bytes of frame relay framing fields (flags, address, and FCS) and 4 bytes corresponding to the RFC 1294 or 1490 frame relay encapsulation header for IP traffic. If the 10 bytes of frame relay header overhead is factored in, the total header overhead of DLSw when transferring data traffic over a frame relay WAN comes to 66 bytes. Figure 5.7 depicted the structure of a standard DLSw INFOFRAME message unit encapsulated within frame relay and compares it with the framing used by RSRB. The key difference between DLSw and RSRB is that RSRB transports a complete LAN frame—replete with MAC addresses, SAPs, and possibly even a RIF—end-to-end while DLSw only transports the I-field of a LAN frame.

5.5.5 Overview of the Feature Repertoire of the Latest DLSw Specification

DLSw from day one was always more than just a scheme for realizing TCP/IP encapsulation of SNA/APPN and NetBIOS traffic. It is best thought of as an integrated set of core facilities for effectively and efficiently dovetailing LAN- or SDLC-based SNA traffic, and LAN-based NetBIOS traffic, into multiprotocol LAN/WAN networks built around bridge/routers. The latest version of DLSw, known as *DLSw Standard Version 1.0* (or I-RFC 1795), was developed by a committee made up of members from all of the leading bridge/router vendors. This committee convened as the DLSw Related Special Interest Group within the overall auspices of the IBM-sponsored APPN Implementors' Workshop (AIW).

The repertoire of facilities in this multivendor-developed DLSw Standard Version 1.0 specification can be summarized as follows:

1. The encapsulation of SNA/APPN and NetBIOS traffic with TCP/IP packets and the routing of these packets, using IP routing methodology and protocols, as the means of transporting SNA/APPN and NetBIOS traffic between bridge/routers within a multiprotocol LAN/WAN network.
2. SDLC to LLC:2 Conversion that enables link-attached SNA/SDLC devices (e.g., 3274 control unit or an automated teller machine) to appear, and be treated, as if they were attached to a token-ring LAN to mainframe-resident software. This technology is described in detail later in this chapter.

3. Local LLC:2 Acknowledgment that precludes data link layer—in this case both LLC:2 and SDLC—receive-ready acknowledgments associated with SNA message units having to be sent end-to-end across the WAN. (This facility applies by default to SDLC traffic given that SDLC to LLC:2 Conversion makes SDLC devices appear as if they were LAN-attached devices using LL:2.) See "Local LLC:2 Acknowledgment" sidebar.
4. The caching of the exact location of destinations (i.e., what LAN are they resident on) of NetBIOS Names and MAC Addresses to preclude the need to do repetitive broadcast searches to locate them each time a new session has to be established. Though invariably mentioned as an explicit feature of DLSw, this caching scheme as described above is really an integral part of the CANUREACH–ICANREACH protocol used by the TCP/IP encapsulation scheme.
5. A DLSw Capabilities exchange whereby DLSw capable bridge/routers can specify which DLSw features they support and the version of DLSw they are using (e.g., I-RFC 1434 version as opposed to I-RFC 1795).
6. A congestion control mechanism modeled around APPN's adaptive pacing. Pacing per se is an old and proven SNA protocol whereby a receiving component (e.g., LU) can control the rate at which it receives data by opening, and closing, a pacing-specific transmission window. SNA, from day one, has had a session-level pacing scheme that operates using a fixed window size. A dynamic pacing scheme, where the window size could vary between a minimum and maximum value depending on the rate of data flow, was introduced in 1978. LU 6.2 and then APPN included a variant of this dynamic pacing where the window size could be dynamically set each time the window was to be open (i.e., prior to each new cycle of message unit transmittals). This technique, known as *adaptive pacing*, provides the most powerful, and flexible, approach to pacing to date.
7. SNMP-based Management Information Base (MIB) for managing the DLSw subnetwork within a multiprotocol LAN/WAN network.
8. A DLSw circuit prioritization scheme whereby different sessions using DLSw can be prioritized relative to each other.

5.5.6 Limitations of DLSw

In addition to not being a formally ratified industry standard and its relatively high header overhead (i.e., minimum of 66 bytes of header for each block of data), DLSw in general has two significant weaknesses. These two weaknesses are its inherent inability to support

LOCAL LLC:2 ACKNOWLEDGMENT

LAN-based SNA and NetBIOS traffic use 802.2 Logical Link Control (LLC) Type 2 as their Layer-2 data-link control protocol. LLC Type 2, which in general is only used by SNA and NetBIOS, is a *connection-oriented* (i.e., session-based) protocol that provides for guaranteed Layer 2 delivery by the use of sequence numbers and acknowledgments—in much the same way as SDLC/HDLC.

Just as with SDLC/HDLC, LLC:2 acknowledgments can be either "piggybacked" on data frames being sent in the opposite direction, or sent in the form of explicit 23-byte Receive Ready (RR) LLC:2 supervisory commands. With a 7-bit sequence number field, LLC:2, again like extended-mode SDLC/HDLC, can have an acknowledgment window size of 128—that is, only one acknowledgment is required for every 127 frames received.

The initial IBM implementations of SNA software for token-ring applications, however, did not use the large acknowledgment window size available with LLC:2. Instead, initial SNA LAN implementations used a window size of 1, with explicit RRs for each message unit being sent in each direction. In essence, IBM had reverted back to the much derided acknowledgment scheme used by BSC in 1967. Things have improved somewhat since then. Nonetheless it is still not uncommon for SNA LAN software to use relatively small acknowledgment window sizes such as 3 or 4.

This small window-based acknowledgment scheme is inefficient, squanders bandwidth, and gets in the way of real data transfer traffic. Given that most token-ring LANs invariably have excess bandwidth and work in the 4 to 16Mbps range it is, however, not unduly disruptive provided that these acknowledgments are restricted to a LAN. This is definitely not the case, however, with WANs. The explicit, RR acknowledgments for every handful of SNA frame—going in either direction—soak up WAN bandwidth and disrupt the flow of genuine data.

Theoretically, having to send RRs across a busy WAN, especially one involving multiple intermediate hops, could also result in Layer 2 time-outs, which are typically set to be about 20 seconds.

The Local LLC:2 acknowledgment feature precludes LLC:2 acknowledgments having to be transported across the WAN.

large networks (i.e., in excess of 500 remote sites), and its inability to dynamically reroute around a failed destination bridge/router.

DLSw's problem with supporting large networks arises from its inextricable reliance on TCP as its end-to-end transport scheme. The TCP protocol consumes relatively large amounts of bridge/router memory and processing power resources. This is due to TCP being a guaranteed delivery mechanism. Hence, bridge/routers at either end of each TCP connection have to maintain TCP-specific control blocks (with information on sequence numbers, etc.) for each connection, and deal with the acknowledgment and recovery protocols required to ensure guaranteed end-to-end delivery of TCP message units.

Even the most powerful and expensive of today's multiprotocol bridge/routers *at best* can support only around 1,000 concurrent TCP connections. Most run out of steam after a few hundred TCP connections.

In a mainframe-centric network such as that shown in Figure 5.9, DLSw requires, at a minimum, two TCP connections between every remote bridge/router and the central site bridge/router(s) that are adjacent to the mainframe(s). All the SNA session traffic between a given remote site and a central site is then multiplexed across these two TCP connections. Consequently central site bridge/routers need to be able to support at least two TCP connections for every remote site in the network. Thus, if a network has 50 remote sites, the central site bridge/routers have to be able to support at least 100 DLSw-related TCP connections. However, most SNA networks are considerably larger. A 300-remote-site SNA network is invariably going to be classed at best as being on the low side of a mid-size SNA network. Nonetheless, a 300-remote-site network would require that the central site bridge/router adjacent to the SNA mainframe is capable of maintaining 600 or more TCP connections. Most of today's bridge/routers will not be able to successfully sustain 600 TCP connections!

One obvious, but costly, work-around would be to deploy multiple bridge/routers at the central site and split the remote sites across these bridge/routers to ensure that no bridge/router has to maintain more than a couple of hundred TCP connections. This work-around could also have some scalability issues. As the number of remote sites increases, so do the number of bridge/routers required. Five or more bridge/routers may be required to support a 1,000-remote-site network—which for SNA is still but a mid-size network. Another option, offered by Cisco, is to automatically disconnect DLSw TCP connections that have been inactive for a predesignated amount of time. Cisco with RSRB, as opposed to DLSw, offers another way to circumvent this problem. It offers a nonguaranteed IP-based transport scheme referred to as *Fast Sequenced Transport* (FST) that does not use TCP connections. The downside to FST is that without TCP's LLC:2 or SDLC-like reliable transport mechanism, end-to-end SNA message units could get lost in transit or arrive out of sequence. Though SNA has high-level protocols that will recover from such transport layer vagaries, these protocols can disrupt end-to-end traffic flows and severely impact response times.

DLSw also cannot reroute around a failed destination bridge/router. In typical mainframe-centric SNA networks, the destination DLSw bridge/router(s) will be the bridge/router(s) located in the central site data center in front of a 37xx or 3172 LAN-to-mainframe gateway. Such central-site bridge/routers are the focal point of all SNA traffic going in and out of the mainframes. Realizing that such a bridge/router can be a vulnerable single point of failure, many networks implement a redundant configuration with two or more central site bridge/routers—in many instances even having each backup bridge/router-attached to a different LAN on the 37xx or 3172. DLSw cannot support this widely used and practical "hot" backup configuration. If the active destination router being used by DLSw fails, all of the SNA sessions are disrupted—even if the network went to the cost and trouble of installing a "hot" backup configuration for the central site destination bridge/routers. Figure 5.12 illustrates DLSw's inability to reroute around failed destination bridge/routers. In this respect DLSw is unfortunately

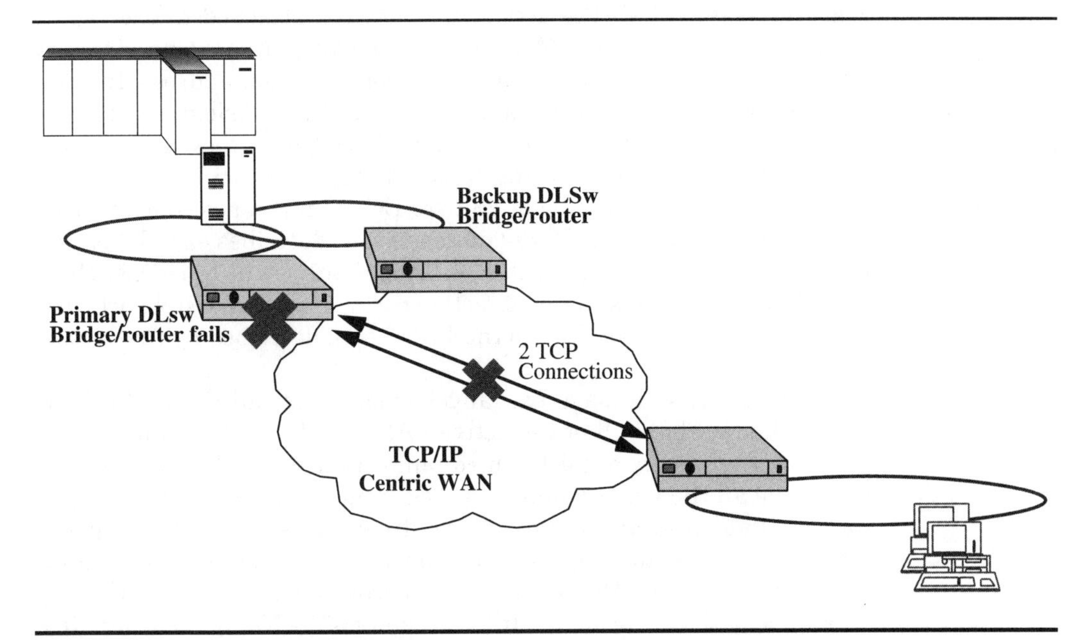

FIGURE 5.12 DLSw's inability to cope with the failure of a destination bridge/router.

no better than SRB. In reality, DLSw can only dynamically and nondisruptively reroute around failed intermediate nodes or links.

5.6 PROTOCOL INDEPENDENT ROUTING

Protocol Independent Routing was CrossComm's proprietary solution for overcoming the limitations of SRB. PIR, as discussed earlier, possesses all of the value-added attributes of TCP/IP encapsulation—such as dynamic alternate routing, address caching, load-balancing, and so on. Moreover, it is a plug-and-play networking scheme that is not based on TCP/IP. PIR does all of its routing, using Layer 2 MAC addresses, even in complex WAN topologies involving mesh connections.

Much of PIR's routing prowess can be directly attributed to the powerful, hardware-based *address processor* (also known as ALOGIC) mechanism that is an integral component of all CrossComm ILAN and ILAN XL bridge/routers. This ALOGIC permits these bridge/routers to dynamically learn the MAC addresses of LAN devices. It will then automatically index these addresses into a database that can maintain extensive entries about each MAC address starting with a description of where that LAN device is located within the network.

In much the same way as a transparent bridge, a PIR bridge/router, using its ALOGIC, sets about to automatically learn all the MAC addresses of all the active LAN devices attached to each of its LAN ports. Each learned MAC address is stored in a cache directory along with a reference as to which LAN contains that address. In addition to learning local MAC addresses, each PIR bridge/router also maintains a current routing table. PIR bridge/routers dynamically build and update these routing tables through the periodic (typically every 18 seconds) exchange of 28-byte ECHO messages. Each bridge/router issues a separate ECHO message that describes each LAN that is attached to that bridge/router. A bridge/router will broadcast these ECHO messages across all of its active ports, including WAN ports, bar the LAN port corresponding to the LAN being described by that actual message.

These ECHO message exchanges ensure that all the PIR bridge/routers know where all of the active LANs are located within the network, as well as enabling them to build up a network topology map indicating all the paths available to get from one LAN to another. It is this continuous awareness of the multiple paths across the network that enables PIR to support short convergence delay dynamic alternate routing. The ECHO messages also provide a mechanism whereby a realistic path cost between different pairs of LANs can be computed. This cost is defined in terms of a metric value.

The PIR metric value for a given link, which would typically be a WAN link, is dynamically ascertained by adding the actual time it takes for an ECHO message to traverse that link to the nominal cost of that link. The nominal cost of a link in PIR is determined by dividing the numeric value 100 million by the bits per second speed of a link. As an ECHO message passes through a network its metric value is continually updated as it crosses each link to reflect the total cost of the path it has traversed so far.

Once the topology and the path costs are known, each bridge/router will construct a spanning algorithm-based optimum path routing tree, with it as the root node. With transparent bridging there is only one spanning tree algorithm-based active topology for the entire network and one root node. With PIR there are multiple spanning trees, each optimized around a given source bridge/router. This notion of having multiple spanning trees optimized around a particular bridge/router is also used by conventional multiprotocol bridge/routers for routing protocols such as IP, IPX, and OSI.

When a PIR network is initially activated the participating bridge/routers will immediately start the process of building their routing tables and establishing the network topology by exchanging ECHO messages. In parallel, the ALOGIC in each bridge/router will be trying to learn as many of the local MAC addresses as possible. Once the current routing tables are in place and the network topology is known, each bridge/router sends a copy of its local cache directory to all of the other bridge/routers using PIR specific multicast address. At the conclusion of these directory content sharing exchanges each PIR bridge/router will have a fairly extensive MAC address cache along with where that MAC address is located.

This automatic network-wide MAC address sharing scheme obviously minimizes the amount of broadcast searches that have to be done in PIR networks to locate unknown destination MAC addresses. There will, however, still be instances when a bridge/router will encounter a destination address that is not in its cache directory. This is most likely to happen if the destination being sought has not as yet transmitted any frames. When confronted with an unknown MAC address the source bridge/router will conduct a broadcast search by sending copies of the original LAN frames containing the destination address to all the known LANs in the network.

The source bridge/router does this broadcast, not as a SRB-like all-routes broadcast, but as a single-router broadcast using its spanning tree-based optimum path routing tree to all the other bridge/routers in the network. Unlike with SRB, the responses to this single-router search is not returned as an all-routes broadcast since PIR bridge/routers do not rely on this scheme to learn about alternate routes within the network. They know about the alternate paths thanks to their periodic ECHO message exchanges. Once the source bridge/router determines the location of the hitherto unknown MAC address it will issue directory update messages to the other bridge/routers with this new information. Though ECHO messages are exchanged periodically, directory updates are only issued when there is an explicit change to a local directory such as the addition of a new MAC address entry.

PIR can thus be thought of as an intriguing amalgam of SRB and transparent bridging. It applies many transparent bridging concepts, such as MAC address learning, spanning trees, and periodic BPDU-type exchanges, to garner value-added support for SRB-based SNA/APPN and NetBIOS traffic. Given its reliance on transparent bridging techniques, PIR bridge/routers also support standard transparent bridging. Thus, PIR bridges are also to an extent SRT bridges. The key difference between PIR and SRT is that PIR does not use standard SRB per se but its brand of enhanced SRB replete with dynamic alternate routing, and MAC address caching. PIR bridge/routers can also be configured to perform PIR-independent, Layer 3-based bona fide IP, and IPX routing, as well as standard SRB. Thus it is possible to have PIR configurations where SNA/APPN and NetBIOS traffic is being transported using PIR, IP and IPX traffic is being conventionally routed, and DECnet traffic is being transparent bridged.

PIR, unlike DLSw and RSRB, is not an encapsulation technique per se in that it does not use any higher-level protocols, such as IP. PIR uses standard token-ring and Ethernet LAN frames. When transporting these LAN frames across WAN links PIR will use HDLC or frame relay. PIR, however, does include a PIR-specific 8-byte header into every LAN frame it transports. This header specifies frame type (e.g., ECHO message, data frame, MAC address data, etc.), the LAN number of the source LAN, and the number of the source bridge/router-specific spanning tree routing path being used to forward this frame. (This PIR header is not included in traffic that is being IP, or IPX, routed or being bridged using standard SRB or transparent bridging.)

5.7 DOVETAILING SNA LINK TRAFFIC INTO MULTIPROTOCOL LAN/WAN NETWORKS

Traffic from LAN-attached SNA devices can now be transparently, effectively, and relatively painlessly integrated with non-SNA LAN traffic in bridge/router of FRAD-based multiprotocol LAN/WAN networks using either bridging, TCP/IP encapsulation, RFC 1490-based frame relay encapsulation, or APPN NN routing. Unfortunately, support just for LAN-based SNA traffic will typically only be but a partial solution for most IBM networking enterprises. More than half of the current installed base of SNA devices, worldwide, rather than being LAN-attached, are connected to SDLC or 3270 BSC links. There are technical, historic, and fiscal reasons for this. Consequently, it is certainly not feasible in all situations to expect enterprises to be willing to readily upgrade or replace these link-attached devices to make them LAN-attached.

IBM only introduced support for token-ring LANs in 1986. In reality token-ring support did not become prevalent, even on IBM equipment, until around 1988. Even when token-ring support was available on a device—for example, the IBM 3174 control unit—many enterprises continued to opt for link-attached models while their LAN deployment strategies were being finalized. Thus, there is still a very large, heavily used, and not fully depreciated population of SNA devices that cannot be easily, or cost-justifiably, upgraded or replaced. In some instances, particularly in the case of some older minicomputers, automated teller machines, and RJE devices, LAN-attachment may not be supported. The only way to realize LAN-attachment would be to migrate to a new, and possibly, different system or model. This in many instances may not be an attractive or universally acceptable proposition. The repertoire of link-attached devices that may be encountered, today, at IBM-centric networking enterprises could include: IBM 3174s, 3274s, 3770s, 8100s, Series/1s, S/36s, S/38s, 3600s, 4700s, 5520s; automated teller machines, as well as minicomputers from other vendors such as DEC, Wang, Prime, and DG.

Thus, in order to deliver a universal, all-inclusive, multiprotocol LAN/WAN network that truly supports all the different types of traffic that may be encountered within an enterprise, bridge/routers of FRADs have to be able to comprehensively cater to link-based SNA traffic, in addition to LAN-based SNA traffic. Acutely cognizant of this requirement, all of today's leading bridge/routers and FRADs now offer at least one method for dovetailing SNA/SDLC link traffic with LAN traffic. If nothing else, DLSw that is now becoming somewhat ubiquitous on bridge/routers provides SDLC-to-LLC:2 conversion as a standard facility. Though FRADs typically do not implement DLSW as an option, link support not just for SDLC but for 3270, 2780/3780 RJE BSC, Async, and other protocols such as Burroughs Poll/Select is invariably their forte.

Today, there are four technologies that can viably be used by bridge/routers or FRADs to integrate SNA link traffic with multiprotocol LAN traffic. These four technologies are:

1. *Synchronous pass-through* (or *Synch pass-through* as it is often referred to) with or without value-added facilities such as guaranteed priority for link traffic over LAN traffic.
2. *Remote link polling*, which can also be gainfully utilized to support 3270 BSC traffic. SDLC-to-LLC:2 Conversion, listed directly below, though rarely thought of as such is in reality a variation and offshoot of Remote Link Polling.
3. *SDLC-to-LLC:2 conversion* as available with DLSw.
4. *Remote SNA session switching with host pass-through*. Though described in this section given that its genesis was in the arena of link support, this technique is now offered by CNT/Brixton as a part of its Convergence family of solutions to support both SNA LAN and link traffic with the added bonus of direct mainframe-to-mainframe access independent of SNA routing.

The first three technologies listed above are now relatively common with SDLC-to-LLC:2 conversion with its DLSw ties now rapidly becoming the de facto technology for SDLC support. SDLC-to-LLC:2 conversion, in addition to being offered by FRADs independent of DLSw, is also available on stand-alone conversion boxes from vendors such as Netlink and Sync Research.

Supporting link traffic is not the primary rationale for Remote SNA session switching. As implied by its name, its main claim to fame is its ability to directly switch SNA sessions between multiple mainframes. Though not based on SNA routing this technique provides an effective way to emulate SNA routing in many, but not all, instances. (For example, SNA session switching cannot be used as an alternative to SNI to realize SNA routing between mainframes resident in separate, autonomous SNA networks.) SNA session switching is a sophisticated and a relatively complex solution that is contingent on deploying a subset of SNA SSCP functionality—as now found in PU Concentrator Gateways—around the periphery of the WAN.

In addition to the above four technologies, there are other schemes that can theoretically be used to support link traffic within the context of multiprotocol networks. One such technique, that was rather publicly evaluated by Cisco in 1991 before it summarily rejected it as understandably impractical, was the implementation of a partial SNA Type 4 node (à la that found in an IBM 3745 communications controller) on a bridge/router. The lure of implementing a Type 4 node was that it would have provided bona fide, no-compromise support for SNA routing and would have enabled bridge/routers to interact with IBM 37xx's as peers. Today, APPN NN routing, now available on nearly all leading bridge/routers, has become the contemporary solution for providing a level of

SNA routing within bridge/router-based networks. APPN NN routing, though not exactly the same as SNA subarea-to-subarea routing, does nonetheless provide SNA application-based, session-layer routing between multiple destinations.

At this juncture, for the sake of gaining an accurate perspective it is worth noting that technologies 1, 2, and 4 listed above were successfully implemented in one form or another on many statistical multiplexors and X.25 packet-switches since the early 1980s. (The only reason for the omission of SDLC-to-LLC:2 conversion was that most of these multiplexors and X.25 switches did not directly support LANs.) This is yet another example of the pervasive *Plus ça change, plus c'est la même chose* phenomenon when it comes to IBM networking.

5.7.1 Synch Pass-Through—The Hands-Off Approach to Supporting Link Traffic

The most intuitive and theoretically the least risky way for bridge/routers or FRADs to support SNA/SDLC link traffic is via a synch pass-through. Understandably, it was thus the first, and for a while in the early 1990s the only, solution for incorporating SNA link traffic into a multiprotocol network. Unfortunately, from a technical perspective synch pass-through is also the least attractive of the schemes available for link traffic consolidation. In addition to being somewhat inefficient in the amount of spurious traffic transported across the WAN, synch pass-through also cannot generally be used to successfully support 3270 BSC traffic. It can, however, be used to support other SDLC-like protocols such as HDLC and in some cases 2780/3780 RJE-based contention mode BSC traffic.

With synch pass-through, each physical SDLC link emanating from a 37xx port and going to one or more (i.e., multidropped) remote SNA devices is replaced by a totally unobtrusive, clear-channel, port-to-port connection through the multiprotocol WAN. The 37xx port to which the original SDLC end-to-end .long-haul link was attached will now be connected to a serial port on an adjacent, upstream, local bridge/router or FRAD. This connection will most likely be done using a modem bypass cable or a short-haul modem configuration, as opposed to a true modem-to-modem connection, given the local nature (i.e., a few yards) of this connection. The remote device(s) are attached to a serial port on a downstream remote bridge/router. The two bridge/router or FRAD serial ports at either end of the configuration are then mapped to each other via a point-to-point route across the multiprotocol WAN. Figure 5.13 depicts the overall setup of a synch pass-through configuration.

With a synch pass-through configuration SDLC traffic arriving at

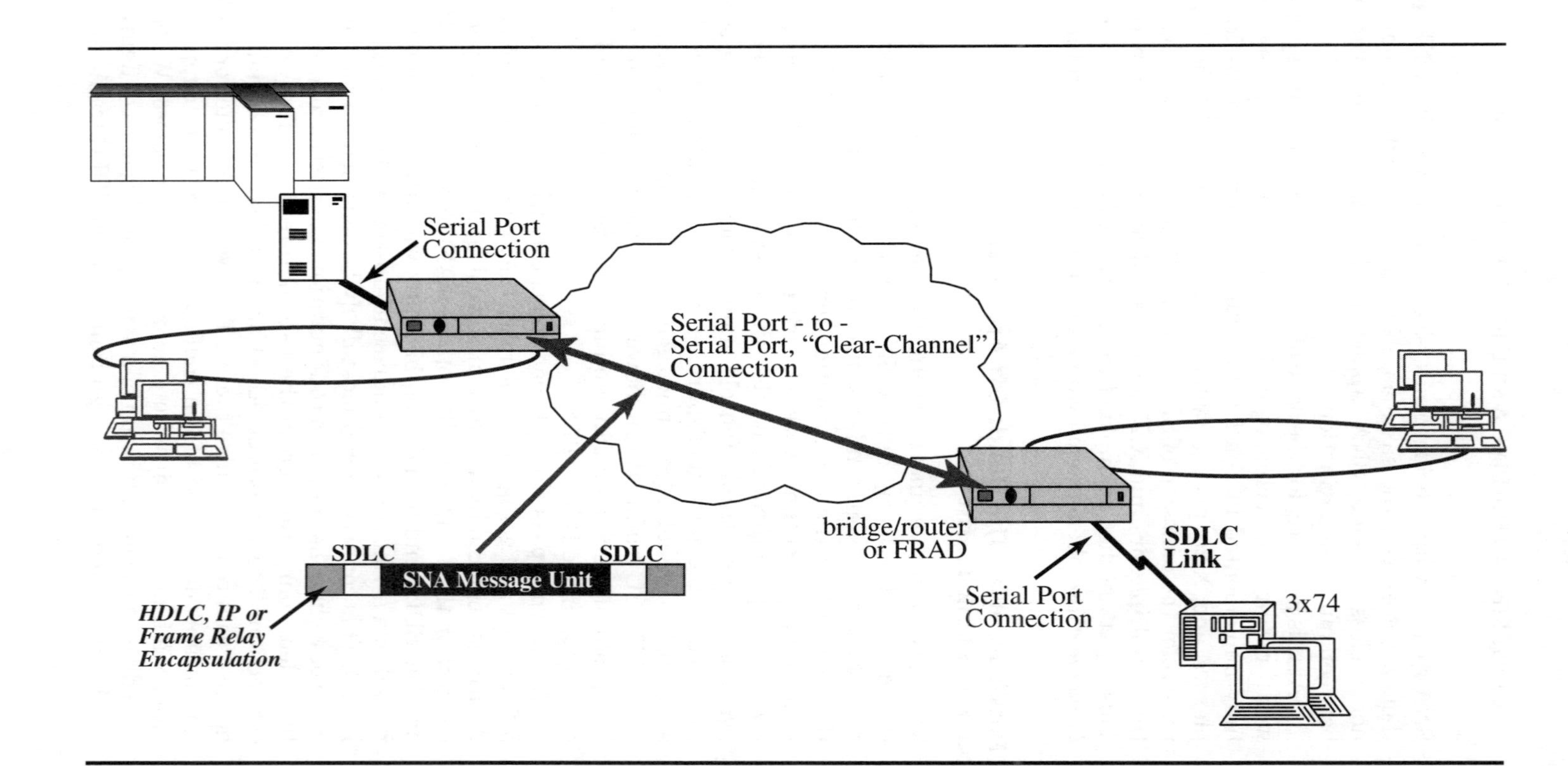

FIGURE 5.13 Overall concept of SDLC link support via synch pass-through.

either of the bridge/router or FRAD serial ports will be cleanly transported to the predefined partner port at the other end. This will be achieved by an encapsulating technique. Every SDLC frame, from the first bit of the start flag to the last bit of the end flag, is included as data I-field within either a TCP/IP packet, a HDLC frame, or a frame relay frame. The encapsulation protocol used depending on the WAN fabric between the end-to-end bridge/routers or FRADs. (For example, frame relay encapsulation is likely to be used if the WAN fabric is frame relay.) The bridge/routers or FRADs do not bother to read the SDLC frame, let alone modify it in anyway. What comes in at one serial port goes out unadulterated at the opposite serial port. synch pass-through is a zero intervention, totally hands-off approach for transporting link traffic across a multiprotocol network.

In general, synch pass-through is a safe-bet solution that permits SNA links to be cut over to a multiprotocol WAN with the minimum of fuss, and certainly without the need for either a new ACF/NCP host gen, or any software reconfiguration to the link-attached devices.

Since synch pass-through forwards unadulterated SDLC frames, which in turn may contain complete SNA message units, neither end (i.e., mainframe/37xx or link-attached SDLC device) sees any changes at the SDLC or SNA level. All they might notice is a possible change in the rate that the frames are being transported. Thus, both sides continue to work as if they were still connected to each other via a physical link, rather than through a complex multiprotocol WAN carrying other traffic. This nonintervention approach safeguards enterprises from any compatibility issues that may arise from mainframe/37xx software upgrades since what ever works on a leased line should also, in theory, work with synch pass-through. The only pertinent caveat to this is any ultra-time-sensitive functions that may require adjustments to various timers to compensate for any changes in link speed.

This hands-off approach also provides a high degree of compatibility with SNA-based network management systems like NetView. They will continue to have total end-to-end SNA control and visibility. They will, however, at least for the time being, not be aware of what is happening within the multiprotocol WAN. Thus, the SDLC links perceived by them will invariably appear to be extraordinarily clean links with little, or no, retransmissions, since any retransmissions required on the WAN will not be seen by them.

Ironically, the noninterventionist trademark of synch pass-through is also its biggest weakness. By keeping all of the SDLC frames intact, synch pass-through, in general, does not in any way reduce the amount of traffic associated with a given serial link. If anything it increases the traffic, since encapsulating SDLC frames within TCP/IP packets not only makes the transmitted packets longer, but also adds additional TCP/IP control interactions (e.g., acknowledgments) on top of the SDLC interactions. To compensate for this many vendors now offer some type of data compression capability, with some, such as Cisco, also offering the possibility of compressing the TCP header.

To provide comparable response times to those enjoyed previously with a physical end-to-end link, the multiprotocol network has to ensure that each end-to-end synch pass-through connection gets an allocation of bandwidth comparable to that of the original link. Normally this is not an issue. Most SNA links to peripheral devices, as opposed to those between 37xx's, are in the sub-19.2Kbps range. In contrast, the trunks in a bridge/router or the VCs in a FRAD-based WAN are unlikely to be less than 56Kbps. However, from the perspective of each SDLC link, the WAN has to be shared with high-speed, high-volume traffic from LANs, as well as with traffic from other SDLC links. Being able to assign a higher transmission priority to link traffic, at the expense of LAN traffic, is one way to alleviate SDLC link traffic becoming swamped by LAN traffic. This facility is now widely available.

The other unsavory feature of synch pass-through is the amount of spurious, nonproductive polling traffic that has to be continually transported across the WAN, consuming valuable bandwidth. Given that synch pass-through is a totally hands-off approach, the SDLC Receive Ready (RR) polls that occur whenever there is no active traffic on a link, and the corresponding RR response returned if there is no data that can be forwarded at that juncture, are always sent end-to-end across the WAN. End-to-end retransmission will also be required if one side requests a retransmission. Since SDLC does not support Selective Reject (i.e., the ability to request the retransmission of one particular frame) many retransmissions will result in multiple frames having to be resent even if some of these had previously been successfully received. The idle RR–RR polling—which in many SDLC links could account for over 60% of the traffic sent over a 24-hour period—as well as the end-to-end retransmissions squander WAN bandwidth that can profitably be used to support other non-SDLC traffic.

5.7.2 Remote Link Polling—The Precursor to SDLC-to-LLC:2 Conversion

Remote polling successfully overcomes the wasteful idle-poll and retransmission problems associated with synch pass-through with considerable panache. It can also be used to effectively and efficiently support 3270 BSC traffic. With this technique, only SDLC frames (or 3270 BSC blocks) containing bona fide data, that is, SDLC I-frames (or BSC Text

Block), which in turn would contain SNA message units, are transmitted across the WAN. Polling and retransmissions are performed, and responded to, at the periphery of the WAN by special SDLC (or BSC) remote polling link-driver modules. Figure 5.14 depicts the setup of a remote link polling configuration and highlights the location of the link-driver modules. This figure also illustrates the port-to-port and address-to-address mapping used to realize remote link polling.

The remote polling link-driver modules come in two distinct flavors: *primary modules* that issue polls and *secondary modules*, which respond to polls, The primary modules will be deployed in remote, downstream bridge/routers or FRADs to which the actual SNA devices (e.g., 3174, AS/400, or automated teller machines) are attached. The secondary modules will be used in the local, upstream, data-center-located bridge/routers or FRADs connected to the 37xx/mainframe ports.

Just as with synch pass-through there will be a predesignated port-to-port mapping, associated with each SNA link, between the router port-attached to the 37xx and the corresponding port at the other end to which the SNA devices are attached. However, in the case of remote polling, a link-address-to-link-address mapping (e.g., SDLC address at 37xx to SDLC address at remote device) is also required on top of the port-to-port mapping.

Once the port-to-port and address-to-address mappings have been established, the remote primary module starts to poll the devices attached to its port using a predefined *polling table*. This polling table specifies the link addresses to be polled, plus the order and frequency at which to poll the various addresses in the table. As each device becomes active, as indicated by a positive response to a poll (e.g., an Unnumbered Acknowledgment [UA] response to a Set Normal Response Mode [SNRM]), the primary will notify its partner secondary module that the device in question is now active and ready for data exchange using an implementation-specific handshake protocol.

In parallel to the primary module starting its polling cycle, the data-center-resident secondary module will enable its port and wait to receive activation messages (e.g., SDLC SNRM) from the 37xx port. When it detects that the bridge/router or FRAD port is active at the physical layer (e.g., detection of Data Set Ready) the 37xx will react by issuing activation polls. If the secondary module has received notification that the subject physical device corresponding to the address in the 37xx polls is now active, it will respond to the 37xx, as if it were the real device.

Once this initial activation sequence is complete, one side will start to receive data frames. It will accept them and extract the databearing I-fields from them. It will then forward just these I-fields to its partner. These I-fields will typically be transported end-to-end within either a

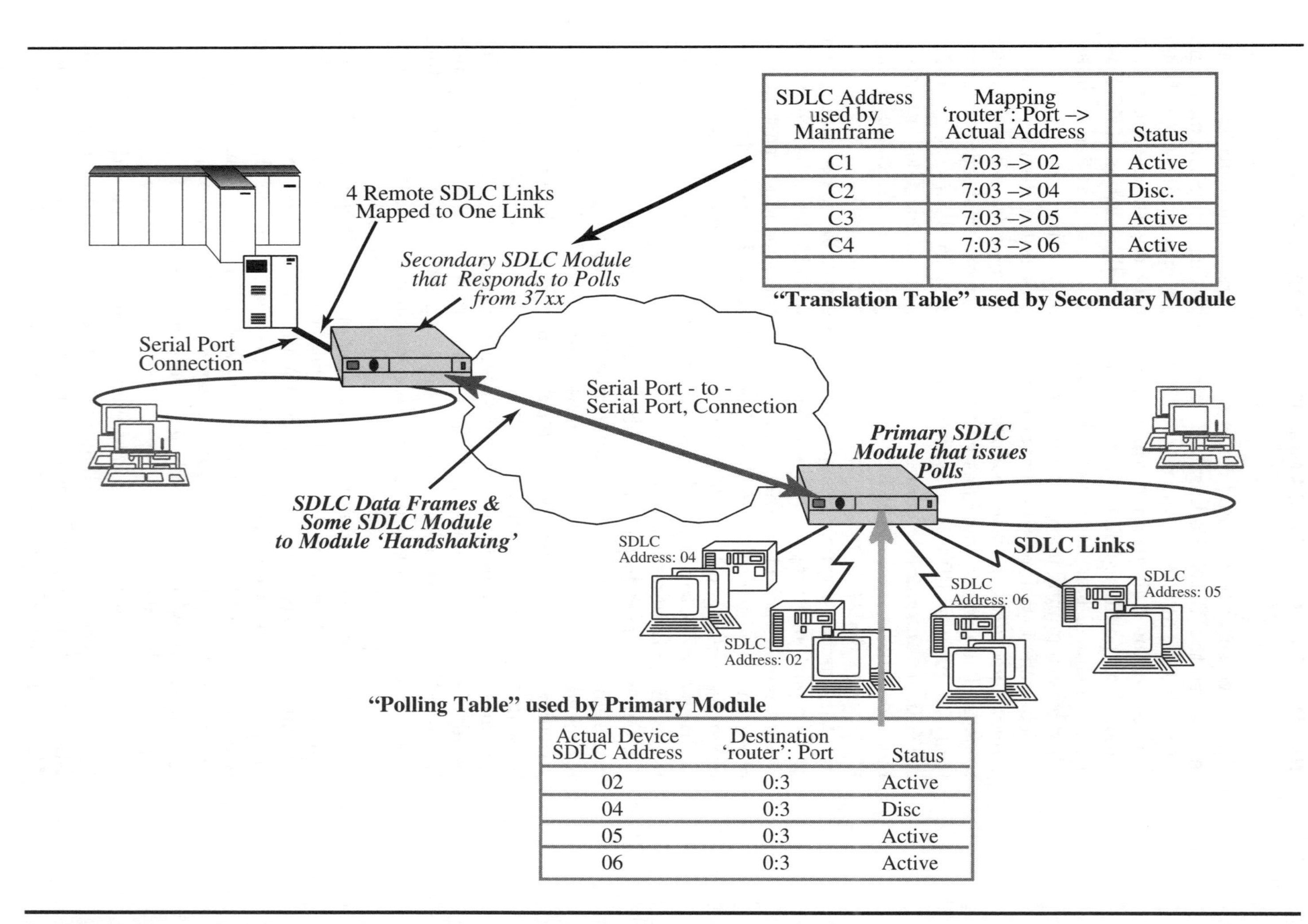

SDLC Address used by Mainframe	Mapping 'router': Port -> Actual Address	Status
C1	7:03 -> 02	Active
C2	7:03 -> 04	Disc.
C3	7:03 -> 05	Active
C4	7:03 -> 06	Active

"Translation Table" used by Secondary Module

"Polling Table" used by Primary Module

Actual Device SDLC Address	Destination 'router': Port	Status
02	0:3	Active
04	0:3	Disc
05	0:3	Active
06	0:3	Active

FIGURE 5.14 Remote link polling and the tables used to facilitate address remapping.

TCP/IP packet, a HDLC frame, or a frame relay frame depending on the WAN protocol being used. The partner module on receiving such I-fields will create SDLC frames, insert the I-fields within these frames, and ensure that they are delivered to the intended destination. The relaying of such data frames between the modules will be performed on a per-address basis. This process of receiving dataframes at one end and forwarding them to the other end for delivery to their rightful destination will then take place on a routine and regular basis.

Remote link polling, also referred to as *poll spoofing*, or *SDLC local acknowledgment,* is invariably considered by enterprises to be more desirable than synch pass-through given its indubitable virtue of reducing the volume of link traffic that has to be transported across the multiprotocol WAN. Remote polling will also eliminate the likelihood of Layer 2 timeouts occurring. Remote polling, just like synch pass-through, has the added merit that it in no way interferes with the SNA protocols. It is thus an SNA-transparent scheme, which will not be affected by any modifications made at the SNA level. It also provides NetView with full end-to-end SNA access and visibility, but now with slightly more distortion of its perception as to the real working of the underlying link. (For example, NetView will not know about any retransmissions that were required and performed at the remote end.)

Remote polling can provide an optional capability whereby link configurations can be transformed to achieve certain cost reductions and even possible polling efficacy. Given that remote polling functions on a link-address-to-link-address basis it is possible to transform and map link addresses of devices on multiple different links to addresses on a virtual, consolidated large link at the 37xx side. For example, as shown in Figure 5.14, the devices on four point-to-point links could now be presented to the 37xx as if they were four devices on a single, multipoint link. SDLC-to-LLC:2 conversion described below is an offshoot of this technique of mapping the physical link configuration at the remote end to a different link configuration at the 37xx end.

5.7.3 SDLC-to-LLC:2 Conversion: The Strategic and Popular Choice

This technique is in effect a variation of standard remote link polling as described above. It is based on remote link polling's inherent capability to map the physical link configuration at the remote end to a virtual configuration at the 37xx end.

In standard remote link polling the same data link control protocol (e.g., SDLC or 3270 BSC) is used both at the 37xx side as well as at the remote device end. With SDLC-to-LLC:2 conversion, different data link protocols are used at the two ends: SDLC on the link-attached device end, and token-ring Logical Link Control Type 2 (i.e., LLC:2) connection-oriented protocol at the 37xx side.

The primary rationale for SDLC-to-LLC:2 conversion is cost reduction. With SDLC-to-LLC:2 conversion enterprises can eliminate serial link ports at the central site, both on 37xx communications controllers, and on the bridge/routers or FRADs adjacent to those controllers.

Instead of transporting link traffic to/from the 37xx via serial ports, SDLC-to-LLC:2 conversion converts link traffic to token-ring LAN traffic, and conveys it to/from the 37xx over a standard 37xx token-ring interface (i.e., 37xx TIC). In other words SDLC-to-LLC:2 conversion converts SDLC link traffic to token-ring LAN traffic so that the link traffic can be dealt with purely as token-ring LAN traffic, vis-à-vis the 37xx and mainframe software. The link-specific token-ring traffic is converted back into link format, at the remote end, so that it can be transmitted over the actual physical links using SDLC.

Thus, rather than requiring serial ports for the link traffic and a token-ring LAN interface for LAN traffic, enterprises can now standardize on just a token-ring interface, at the mainframe end, for both LAN and link traffic. Figure 5.15 illustrates the overall set-up of a SDLC-to-LLC:2 conversion configuration. Compare this figure with Figures 5.11 and 5.12 to note how this approach eliminates the need for serial ports at the 37xx end.

SDLC-to-LLC:2 conversion also permits enterprises to support remote link-attached SNA devices using non-3745 SNA Gateways such as 3172 or channel-attached multiprotocol bridge/routers (e.g., Cisco) or intelligent hubs (e.g., Cabletron). Hence, in some instances SDLC-to-LLC:2 conversion will permit enterprises to displace 37xx's that were being used primarily to support link traffic in favor of a lower cost solution such as 3172s. Given these cost reduction possibilities and its association with DLSw, SDLC-to-LLC:2 conversion has now become the preferred option for link traffic integration.

As with remote link polling, SDLC-to-LLC:2 conversion is realized using two link-driver modules. The primary modules that are deployed at the remote end that issues the SDLC polls can be exactly the same as those used for remote polling. The secondary modules in this case act as bona fide token-ring LLC:2 stations. Just as with a real token-ring device, these secondary modules will have one or more (i.e., ideally one for each SDLC device being supported) MAC addresses assigned them. (If a secondary module can support only one MAC address it will use that one address with different Source Service Access Point [SAP] values to support multiple SDLC devices.) Whereas in remote polling there is an SDLC-address-to-SDLC-address mapping, with SDLC con-

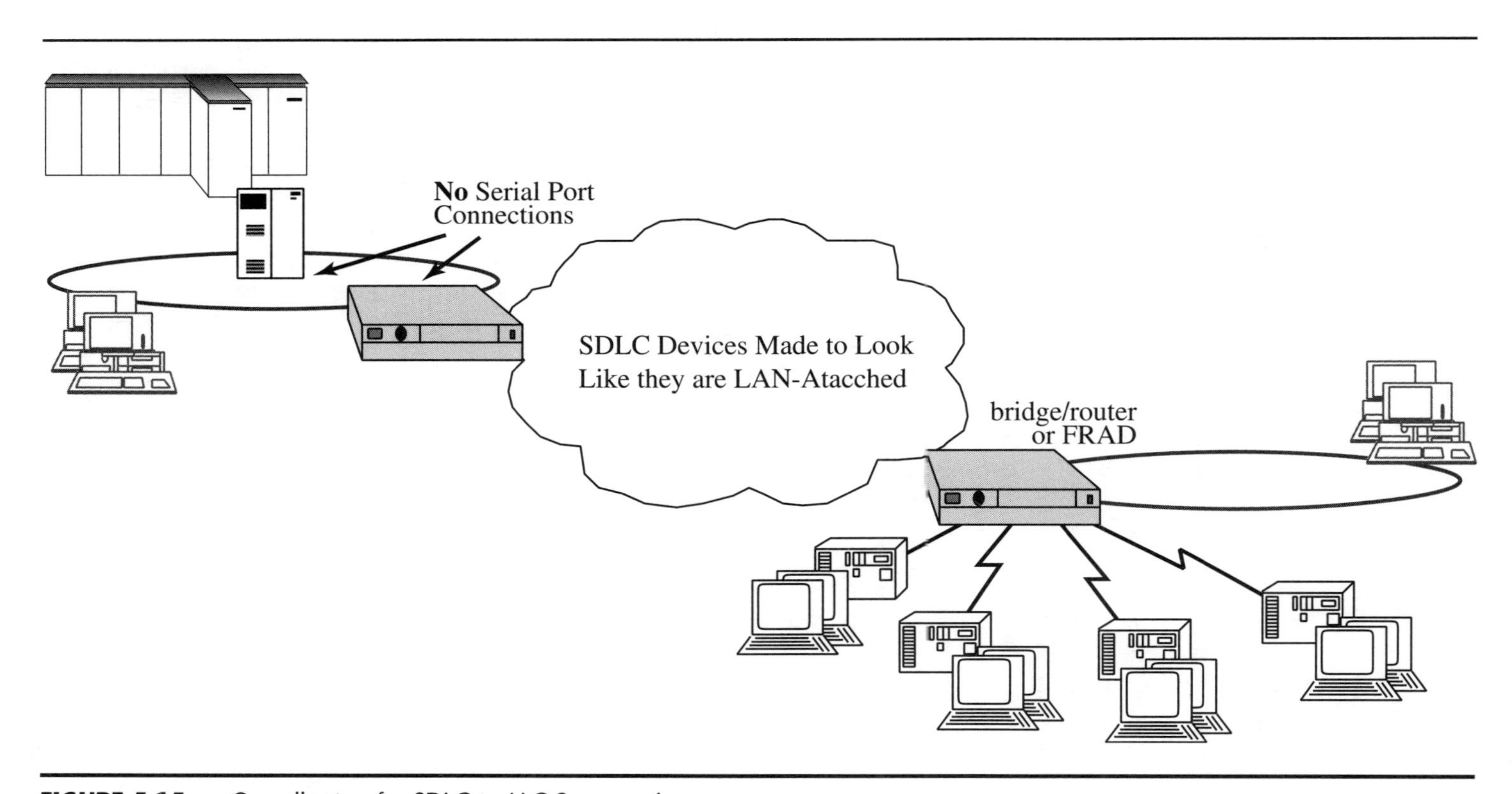

FIGURE 5.15 Overall setup for SDLC-to-LLC:2 conversion.

version there is either a SDLC-to-MAC address or SDLC-to-MAC-address cum SAP value mapping.

The secondary module will establish a LLC:2 connection with the 37xx by issuing and responding to appropriate TEST and XID commands and responses. Once this LLC:2 connection has been established, this module will receive MAC/LLC frames from the 37xx. In the case of databearing frames, it will extract the I-field from the LAN frame and forward it to its partner primary module. The primary module will insert this I-field within a SDLC frame and forward it to the appropriate SDLC device.

On receiving databearing SDLC frames, a primary module will extract the I-fields from these frames and forward them to its secondary module partner. The secondary module will insert these I-fields within LAN frames and forward them to the mainframe/37xx.

The primary modules will be deployed in remote, downstream bridge/routers or FRADs to which the actual SNA devices (e.g., 3174, AS/400, or automated teller machines) are attached. The secondary modules can actually be implemented in one of two places. One option would be to implement a secondary module co-located with the primary module within each remote bridge/router or FRAD supporting SDLC links. In this case, the secondary module will do the bidirectional SDLC-to-LLC:2 conversion locally. The converted SDLC traffic would then have to be forwarded to the mainframe/37xx side using a LAN traffic integration scheme such as bridging, TCP/IP encapsulation, or RFC 1490.

The other option in terms of deploying secondary modules is to implement them only in the upstream, data-center-located bridge/routers or FRADs that are actually connected to the 37xx/mainframe ports. If this scheme is used, the primary modules will have to forward I-field data to the secondary modules by encapsulating the I-fields within either a TCP/IP packet, a HDLC frame or a frame, relay frame depending on the WAN protocol being used. The advantage of this approach is that a single secondary module can support multiple primary modules (i.e., multiple remote bridge/routers or FRADs). It is, however, fair to note that most of today's SDLC-to-LLC:2 conversion schemes use the other option, namely, having a secondary module within each bridge/router or FRAD.

Just as with remote link polling, SDLC-to-LLC:2 conversion does not in any way interfere with, or modify, I-fields that contain actual end user data. Hence, SDLC-to-LLC:2 conversion, like remote link polling and for that matter synch pass-through, can be thought of as a technique that is totally transparent to all end-to-end SNA interactions.

This LLC:2 conversion scheme cannot usually be used to support

BSC traffic. The problem is that it is difficult, at best, to convince IBM mainframe and 37xx software that a BSC device is actually LAN-attached given that IBM never seriously envisaged the possibilities of LAN-attached BSC devices. Hence, BSC devices can invariably only be defined to ACF/VTAM or ACF/NCP as being link-attached. One way to overcome this limitation is to do a BSC-to-SNA protocol conversion at the same time as the link to LLC:2 conversion. *This dual conversion eliminates the need for any BSC serial ports at the mainframe end.*

At this juncture, it is worth stressing that all SDLC (or 3270 BSC) link support features offered on bridge/Routers and FRADS should be treated as short-term tactical migration aids, rather than as long-term, strategic offerings. They offer enterprises a cost-effective means of integrating their existing installed base of link-attached SNA devices into their new multiprotocol networks. Enterprises should not, however, treat this availability of link support as an open-ended excuse for continuing to acquire link-attached SNA devices, rather than equivalent LAN attachable SNA/APPN devices.

5.7.4 Remote SNA Session Switching with Host Pass-through

Remote SNA Session Switching with Host Pass-through is in effect remote link polling on steroids. It enables a level of nonstandard but nonetheless effective SNA routing to be realized in multiple mainframe environments. Whereas remote link polling and SDLC-to-LLC:2 conversion use Layer 2-based (i.e., SDLC or LLC:2) link drivers, this technique replaces those with primary and secondary SNA nodes. This scheme makes the quantum leap from mere dabbling at Level 2 to full-blown SNA emulation and intervention. It can be used to effectively integrate both LAN- and link-based SNA traffic into a multiprotocol network. This technique, which was developed for traditional mainframe-centric SNA environments, typically does not support APPN Network Nodes, though there are no technical reasons why it cannot be extended to also embrace APPN traffic.

Remote SNA session switching relies on an *SNA primary node* deployed at a remote downstream router emulating a subset of the SSCP functionality found in SNA Type 5 mainframe nodes. Using this SSCP functionality it actually captures and controls all the physical SNA devices (e.g., 3174s, AS/400s, PC, etc.) at the SNA *level* using bona fide SNA protocols (e.g., issuing SNA ACTPU and ACTLU requests). The outcome of this SSCP-based interaction is that the SNA devices will end up establishing the prerequisite SNA control sessions (e.g., SSCP-PU and SSCP-LU) that are required to be able to set up subsequent end user sessions, with the router resident SNA primary node. In the absence of this SNA primary node the end devices would have established these control sessions with mainframe resident ACF/VTAM.

The SNA devices are now entirely controlled by their adjacent SNA-speaking router though totally oblivious to it. Logon requests for establishing SNA sessions are now first received and processed by the SNA primary node rather than by ACF/VTAM. The SNA primary node approach used by remote SNA Switching is very similar to that used by PU Controller Gateways. Figure 5.16 depicts the overall setup for a remote SNA switching configuration and highlights the location of the SNA primary and SNA secondary nodes.

On receipt of a logon request the SNA primary node using a predefined *application routing table* can dynamically determine the appropriate mainframe to which the logon should be forwarded. It will then determine the optimal path through the multiprotocol WAN to reach that mainframe. With this technique all logons are directly forwarded to the mainframe in which the required application resides.

The secondary SNA nodes, attached upstream to 37xx ports, emulate SNA Type 2 (e.g., 3174) and present a series of SNA PUs and LUs, to the ACF/VTAMs resident in the mainframes. Rather than exactly corresponding to the PU and LU configuration of the actual downstream SNA devices, this virtual PU and LU configuration will typically reflect a pseudoconfiguration that would enable each host to believe that it owns all, or a subset, of the LUs in the actual network. PU Concentrator Gateways also use such real-to-virtual mappings of PU and LUs.

The main difference between a PU Concentrator Gateway and remote SNA switching is that the gateway does not bother trying to switch SNA sessions between multiple mainframes. Instead, each gateway is defined to a single mainframe. Consequently, gateways also employ a partner SNA secondary module at the mainframe end. A PU Concentrator Gateway is thus a subset of the remote SNA switching functionality.

Remote SNA session switching in essence bifurcates a physical SNA network and introduces a multiprotocol WAN between the two ends. The key, and in reality the only, advantage of this scheme is that it permits mainframe-to-mainframe routing to be performed directly by the multiprotocol WAN. In large, multiple mainframe environments this would result in much faster session establishment and more efficient routing of traffic across the WAN. Remote SNA session switching does, however, normally require that the entire SNA environment (i.e., all ACF/VTAMs and ACF/NCPs) is redefined to reflect the new virtual, bifurcated environment.

Remote SNA switching in multiprotocol LAN/WAN networks is at present supported only by CNT/Brixton. The CNT/Brixton approach can be thought of as an amalgamation of remote SNA switching

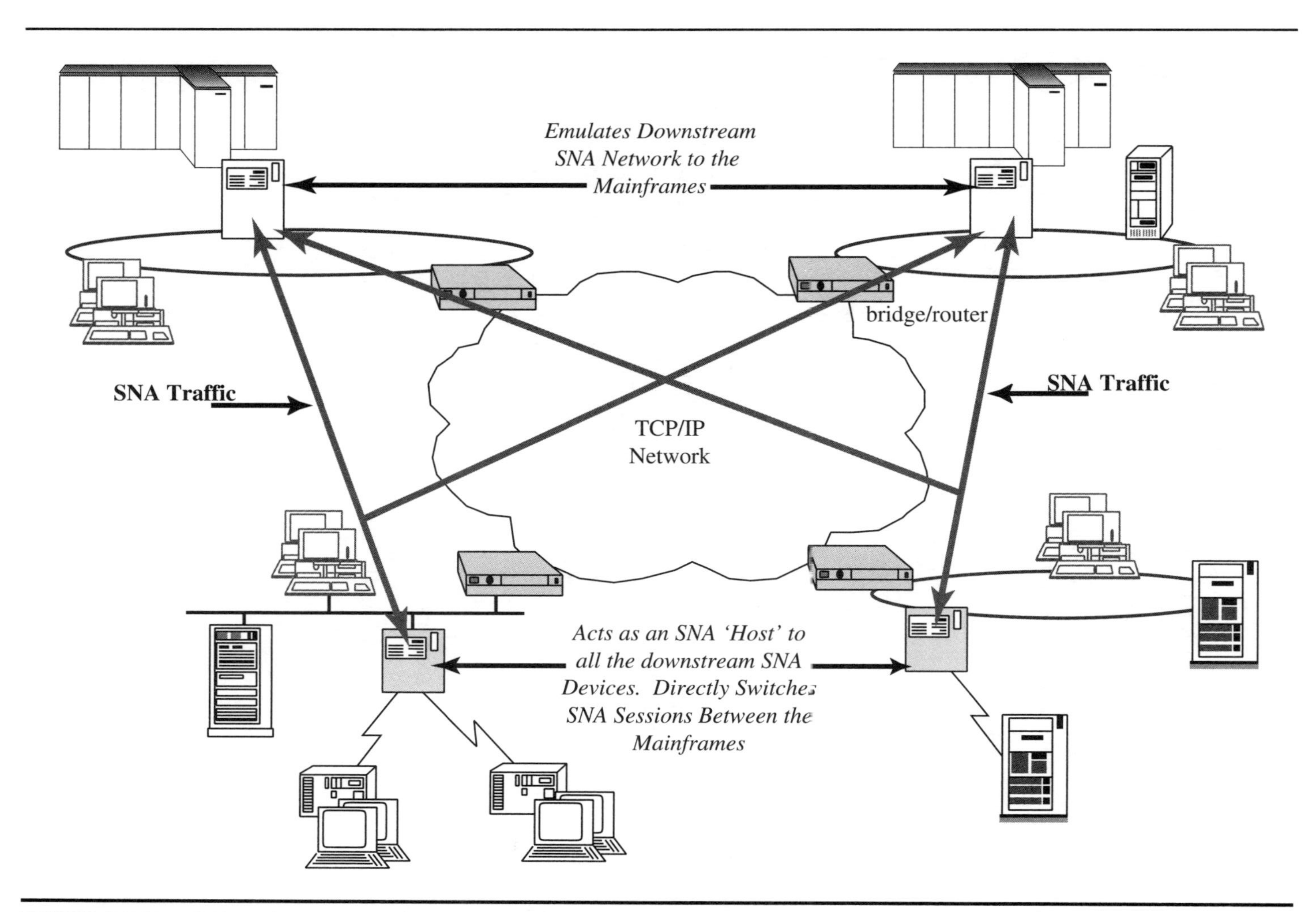

FIGURE 5.16 SNA session switching between mainframes across a TCP/IP WAN.

and TCP/IP encapsulation. The CNT/Brixton approach bifurcates the SNA network with a standard, plain vanilla TCP/IP WAN. SNA sessions are terminated at the SNA session layer at either side of the TCP/IP cloud using primary and secondary SNA nodes. Only actual SNA I-fields stripped of their SNA Transport Headers (TH) and Request/Response Headers (RH) are transported across the WAN—encapsulated within TCP/IP packets. Standard IP bridge/routers (e.g., Cisco or Bay Networks) could be used to implement the TCP/IP WAN and transport the SNA traffic end-to-end across that WAN.

5.8 REFLECTIONS

Dreaded parallel data networks no longer have to be an indelible fixture in contemporary networking. A plethora of technology, most now at third- or fourth-generation maturity and stability, is available from a range of vendors for integrating both LAN- and link-based SNA/APPN traffic into multiprotocol LAN/WAN networks. Of these, DLSw, given its near universal endorsement and support by the bridge/router vendors, is now gaining some momentum as a strategic means for realizing this SNA/APPN integration. DLSw does have some limitations. Its reliance on end-to-end TCP connections precludes it from supporting networks with large numbers of remote sites, at least at present. Though it supports dynamic alternate routing, it cannot reroute around failed destination (i.e., central site) bridge/routers. Given its use of both TCP and IP headers it also has a relatively high header overhead—unless proprietary vendor specific schemes are used to compress or eliminate these headers. RFC 1490-based frame relay encapsulation and APPN NN routing are viable alternatives to DLSw in certain scenarios—albeit the former. Unfortunately RFC 1490 is only an option if frame relay is being used as the WAN fabric. RFC 1490 is discussed in length in the next chapter while HPR/APPN NN-based routing is discussed in Chapter 7.

CHAPTER 6

Frame Relay: A Springboard to ATM

A QUICK GUIDE TO CHAPTER 6

Frame relay, without a doubt, is the best thing to happen to the SNA community since the advent of token-ring LANs a decade ago. The availability of native support for SNA over frame relay, in 1994, thanks to the so-called RFC 1490/FRF.3 standard, has opened up unprecedented options for slashing networking costs in IBM-oriented networking environments.

Today, frame relay (FR) can be used as a cost-compelling and bandwidth-rich alternative to leased, long-haul SDLC lines. Moreover, an IBM 3745/46 Front End Processor (FEP) will support up to 200 FR VCs on a single serial port. Thus, frame relay could also be leveraged as a means for reducing 3745/46 serial port costs.

Ironically, the considerable cost savings that can be realized by replacing SDLC with FR is only the tip of the iceberg in terms of FR's potential of minimizing cost. FR is also the ideal mid-term, mid-scale bandwidth (e.g., 4Mbps) WAN fabric for building SNA/APPN-capable multiprotocol, multivendor LAN/WAN networks. Bridging, DLSw(+), and APPN/HPR NN can all be profitably deployed across FR. In addition there is RFC 1490—FR's lean-and-mean native-mode encapsulation scheme for multiple protocols that can be at least four times more efficient than DLSw, vis-à-vis encapsulation headers, when it comes to supporting SNA/APPN traffic.

This chapter starts off by looking at every conceivable application for FR within an IBM-oriented network, including voice-over-FR and even the possibility of totally eliminating 37xx FEPs. It then provides a thumbnail sketch of FR and covers topics such as Committed Information Rates (CIRs), frame format, and congestion control.

Sections 6.3 and 6.5 offer in-depth and analytical examination of the exciting possibilities of using RFC 1490-based solutions for consolidating parallel networks. IBM's so-called BAN variant of RFC 1490 is also described. Section 6.6 compares and contrasts RFC 1490 against its nemesis—DLSw(+). In general, if FR is the WAN fabric and the network only has one or two mainframes, RFC 1490 is likely to be a more optimum solution than DLSw(+) or APPN.

Section 6.4 discusses the various ways in which a 3745 can be used relative to FR, including the possibility of implementing private, in-house FR networks built around 3745s masquerading as FR switches (i.e., Data Circuit-Terminating Equipments [DCEs]). This section also emphasizes why it is not cost-effective or practical to use a serial port to interface 3745/46 to an FR network. In general, a FRAD and LAN-based interface between an FR network and the mainframe is likely to be a more pragmatic and efficient solution.

The chapter concludes with a section that describes an innovative scheme for minimizing the possibility of SNA/APPN frames being discarded by an FR network—even if it experiences congestion. Given that discarded frames result in unacceptable performance fluctuations, this scheme, now being offered by some FRAD vendors, will be of great interest to the SNA community.

Frame relay is a new-generation, high-performance, low-overhead, connection-oriented *packet-switching standard* prescribed by both the CCITT and ANSI. Though originally developed in the early 1980s as a packet transport scheme for ISDN, frame relay today is a very popular, stand-alone WAN transport mechanism that is totally independent of ISDN. Cut to the chase: Frame relay can best be thought of as a slimmed-down and streamlined X.25. In X.25, data transport-related processing (e.g., flow control and error-checking) was done at both Layers 2 and 3. Frame relay does not do any Layer 3 processing. *The entire frame relay protocol is restricted to and contained within Layer 2.*

Frame relay can be used in SNA environments as a direct, packet-switching-based (i.e., resource-sharing) alternative to running SDLC over leased lines. Some SNA devices such as IBM 3174s, 3745s, AS/400s, and PCs running OS/2 can now directly run SNA over FR. Moreover, to make this switch to frame relay even more palatable, IBM 3745 communications controllers support FR connections in native mode (i.e., like SDLC). Hence, unlike with X.25, 3745s support FR without recourse to the resource-intensive and hard-to-configure NCP Packet Switching Interface (NPSI) software. In the case of SNA devices that do not directly support FR, there is now a thriving market in so-called mono-FRADs from vendors such as Motorola and Sync Research that provide very low-cost SDLC-to-FR conversion.

Providing a public packet-switching network-based solution for replacing costly SDLC leased lines, however, is not the only mouthwateringly compelling application for FR within IBM-centric networks. If anything, SDLC replacement is only the tip of the iceberg. Post 1995, enterprises should not be considering FR just as a means of eliminating SDLC links in SNA-only networks. Such SDLC replacement should only be thought of as a small part of a much larger and profound network rationalization and network consolidation process facilitated by FR. Rather than just doing SDLC-to-FR, it would make more sense to do SDLC-to-LLC:2 and then carry that SNA traffic across an FR WAN alongside SNA LAN and other multiprotocol traffic.

Frame relay, today, represents an unprecedented opportunity for dramatically slashing networking costs in IBM-centric networks. At the same time FR can increase the amount of bandwidth available to such networks. With FR enterprises can realize palpable cost savings in multiple ways. For a start, it provides a cost-effective means of obtaining low overhead, up to 2Mbps bandwidth that is ideally suited as the WAN fabric for consolidated, SNA/APPN-capable multiprotocol LAN/WAN networks. Such multiprotocol networks can be effectively realized in two distinct fashions. One would be to use bridge/routers with DLSw, APPN NN routing, or even straight bridging over FR. The other would be to use RFC 1490-based FRADs that would transport SNA/APPN traffic by directly encapsulating that traffic within FR frames.

A major cost-reduction opportunity is that of consolidating dreaded parallel networks using an FR WAN. This would permit an enterprise to dispense with all the long-haul leased lines being used for SNA, and possibly also to interconnect bridge/routers. Multiple vendors, with Hypercom and Motorola in the vanguard, now also offer reasonable-quality voice-over-FR. Consequently, further cost saving could be realized in branch office scenarios by combining data and voice traffic across an FR WAN. FR is thus not just a precursor to ATM but in a sense also a trendsetting adjunct to it.

In addition to eliminating the need for leased lines, the VC multiplexing capabilities of FR can be profitably used to reduce the number of serial ports (e.g., V.24, V.35) required on IBM 3745s. Devices and traffic previously supported via multiple SDLC serial ports could now be supported across a single FR port. Ten-to-one, or even 40-to-one port reduction is possible with the 3745 theoretically supporting a 200-to-one port reduction capability. The 3745 cost saving possible with FR, however, is not restricted to the elimination of serial ports. With a bridge/router or FRAD-based SNA/APPN-capable multiprotocol FR solution, an enterprise could replace a 3745 with a more cost-effective mainframe gateway such as an IBM 3172 or a channel-attached bridge/router. Figure 6.1 illustrates the so-called Gurugé Model for FR-based SNA/APPN-capable multiprotocol networking without the use of 3745s.

To be fair, the 3745 replacement scenario shown in Figure 6.1 is not contingent on using RFC 1490 and an FR WAN-centric solution. Effective non-3745-based LAN/WAN networks can also be realized using other parallel network consolidation technologies including DLSw(+), APPN NN, and even SRB. The 3745 replacement possibility is emphasized here since IBM and a few FRAD vendors promote RFC 1490 as a means of reducing 3745 ports. IBM's implementation of RFC 1490 for SNA/APPN, which is known as *boundary access node* (BAN) is unashamedly targeted at 3745-centric solutions with port reduction being the explicit inducement. The 3745 replacement scenario can be thought of as taking this port replacement argument to its ultimate conclusion!

Even if a 3745 is retained as the mainframe gateway, given its peerless record for fault-free, mission-critical networking, serial port reduction is not the only way to derive cost savings on this front. Rather than reducing the number of serial ports, serial ports can be eliminated altogether using a variant of the configuration shown in Figure 6.1. Instead of using serial ports to connect the 3745 to the FR network, an inexpensive FRAD-attached to a token-ring interface on the 3745 can be used as the means of accessing the FR network. Figure 6.2 depicts how a FRAD attached to a 3745 token-ring interface can provide access to an FR network and eliminate the need for using serial ports between the 3745 and the FR network. In general, using 3745 or 3746 serial ports to connect to an FR WAN is not recommended for both fiscal and technical reasons. These issues are discussed in detail in Section 6.4.1.

In such a token-ring interface-based FR configuration, whether a 3745 or 3172 is being used as the gateway, the remote SNA/APPN or IP devices will appear to the gateway and the mainframe as if they are directly attached to the gateway's token-ring. (This type of configura-

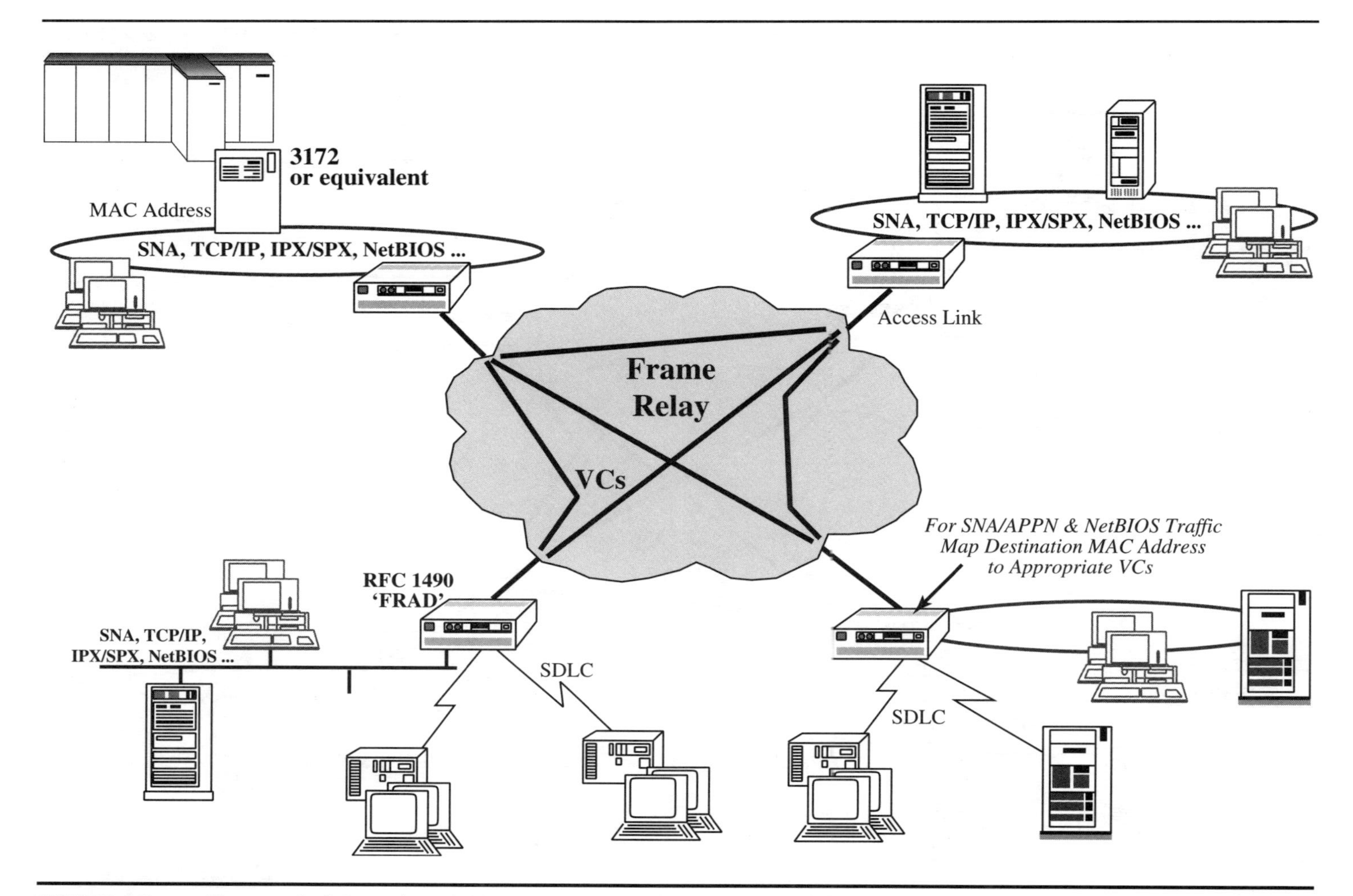

FIGURE 6.1 The so-called Gurugé model for frame relay-centric SNA/APPN-capable multiprotocol networking without 37xxs.

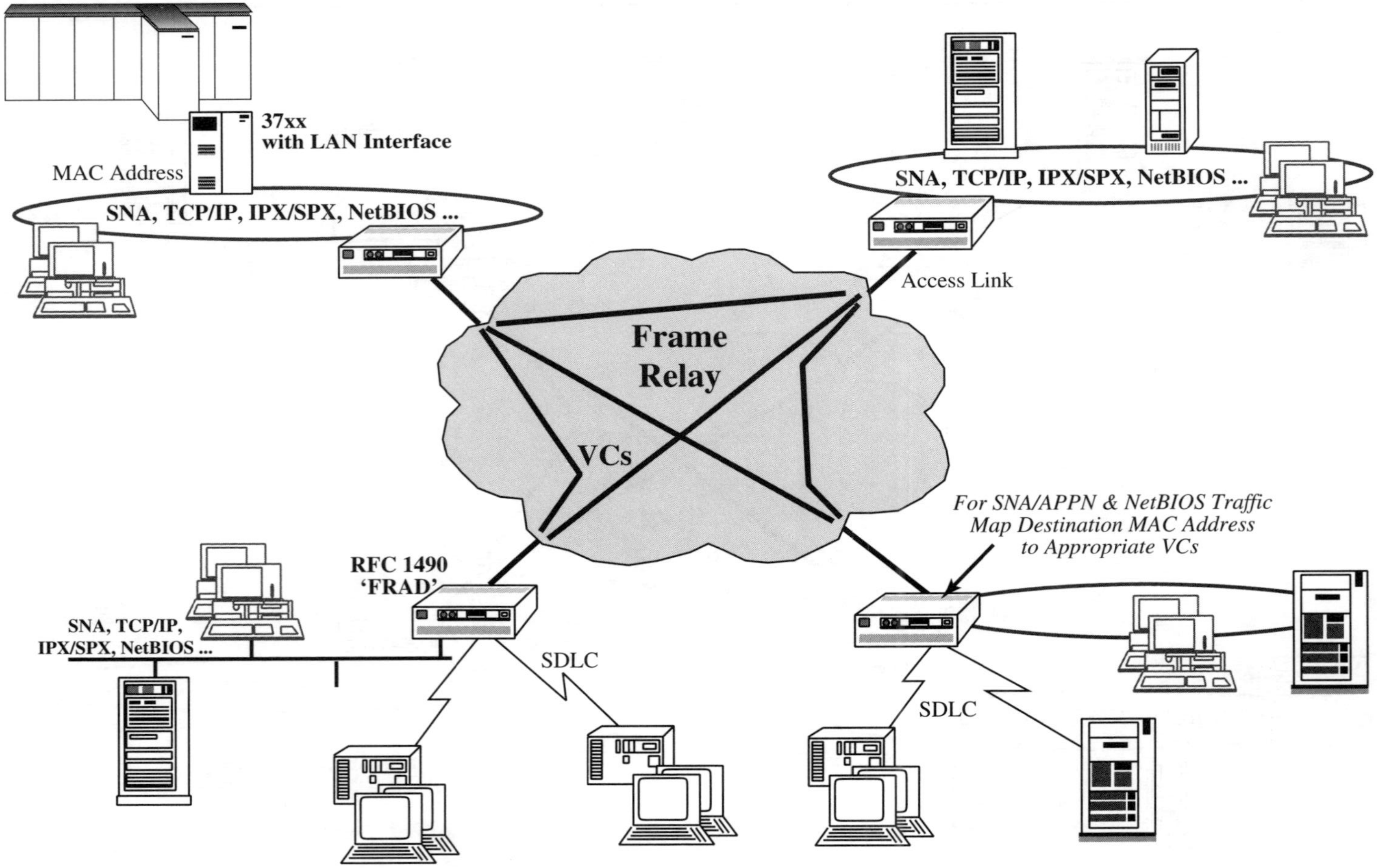

FIGURE 6.2 Supporting SNA-over-frame relay without using serial ports on a 37xx.

tion can also be used to transport IP traffic to/from a mainframe.) *The remote traffic is essentially bridged to the gateway's LAN across the FR WAN on a LAN-to-LAN or link-to-LAN basis by the bridge/routers or FRADs.* No FR-specific software is required on a 37xx or any other type of gateway since the traffic transported across the FR WAN appears to these gateways as unadulterated LAN traffic as opposed to FR traffic. (If a 3172 or a channel-attached bridge/router is being used, the gateway's LAN could be an Ethernet as opposed to a token-ring.)

With respect to SNA/APPN traffic, multiprotocol bridge/routers will achieve this bridging function across the FR WAN using either straight SRB-type bridging, TCP/IP-encapsulation à la DLSw, or APPN NN routing. SDLC-to-LLC:2 conversion will most likely be used to support SNA/SDLC devices. IP, IPX, Banyan Vines, and so on will typically be routed across the WAN. RFC 1490-based FRADs will transport the SNA/APPN traffic by encapsulating them within FR frames per the FRF.3 specification for supporting SNA/APPN traffic within the context of RFC 1490.

Just as with bridge/routers, most of these FRADs will usually resort to SDLC-to-LLC:2 conversion to support SDLC traffic. BSC, Async, and other link protocols (e.g., NCR Point-of-Sales protocol) will typically be transported across the WAN using either LLC:2, TCP/IP, or in some cases a modified Layer 3 X.25 as the encapsulation scheme and the end-to-end protocol. (If LLC:2 is being used, additional encapsulation and end-to-end protocol may not be necessary. The FRF.3 version of RFC 1490 provides a 4-byte LLC:2 field that can be used to realize an end-to-end LLC:2 protocol replete with sequence numbers and acknowledgments.)

With the exception of SDLC traffic supported through SDLC-to-LLC:2 conversion, the support for other types of link traffic is likely to be vendor-specific since there are no de facto standards, à la FRF.3, that specify how these somewhat esoteric traffic types should be handled vis-à-vis FR and RFC 1490. Thus, the transport of BSC and Async across FR is likely to be contingent on FRADs from the same vendor being deployed at each end on either side of the WAN. This lack of potential interoperability when it comes to supporting this type of link traffic should not in any way be construed to mean that RFC 1490 solutions cannot be used in multivendor scenarios. Far from it. Just as with bridging, DLSw or APPN NN multivendor interoperability is possible provided that the protocols are covered by RFC 1490 or supplementary standards such as FRF.3. In general, if the protocol repertoire is restricted to the widely used ones such as SNA/APPN, TCP/IP, IPX/SPX, NetBIOS, and SNA/SDLC, multivendor interoperability, at least in theory, is possible with RFC 1490-based solutions.

6.1 APPLICATIONS FOR FRAME RELAY IN IBM-CENTRIC NETWORKS

The primary applications for frame relay within IBM-centric networks can be summarized as follows:

1. *SDLC replacement within SNA-only networks:* FR can be used as a direct cost-compelling replacement for expensive SDLC leased lines within existing SNA networks. IBM 3745/46s, 3174s, AS/400s, and OS/2 PCs now have native, software-based support for SNA-over-FR. Thus, migration from SDLC to FR, provided that link speeds are not being greatly increased, will be possible with just a software upgrade. Hardware adapter upgrades may only be necessary if link speeds are being increased to exploit the cost-effective higher bandwidth rates typically available with FR.

The 3745 can theoretically support up to 200 FR VC per FR port. This VC multiplexing feature permits multiple downstream FR devices—for example, 3174s, 3174s acting as SNA LAN pass-through gateways to LAN-attached PCs, AS/400s, and so on—to interact with a 3745 over a single port. If a High Speed Scanner (HSS) with a T1 port is installed on a 3745, it would easily be possible to multiplex 40 to 60 56Kbps VCs across it, or even more in the case of slower VCs. Thus, in addition to offering cost-effective bandwidth, and possibly more bandwidth, FR provides a means for consolidating and eliminating 3745 ports.

Relatively inexpensive (e.g., $800 and falling fast) SDLC-to-FR mono-FRADs can be utilized to accommodate SNA devices that do not support FR. Today, these mono-FRADs can be used just in front of the remote SNA/SDLC device. A partner FRAD will not be required at the 3745 side since these FRADs can now "talk" the same protocol (i.e., RFC 1490/FRF.3) that 3174s and others use to interact with 3745/46s. Such, FRADs just at the remote site configurations are referred to as asymmetrical FRAD configurations to denote the absence of a partner FRAD at the opposite end.

Symmetrical configurations involving FRADs at either end of the WAN that end up doing a SDLC-to-FR-to-SDLC are now technically obsolete given that 3745/46s now have a native FR interface. Converting back to SDLC at the 3745 end will preclude the possibility of any port consolidation that would be possible with an FR connection. Figure 6.3 illustrates an SNA-over-FR network where some devices are using mono-FRADs while others are using native SNA-over-FR connections.

Despite the initially tantalizing possibilities of port consolidation, serial port-based interface FR networks are not recommended. For a start, FR is supported only on 3745s and 3746s. There is no FR support on 3720s and 3725s—given that these older models are no longer able to run the latest versions of ACF/NCP (e.g., Version 7 Release 1) required in order to gain FR access. There is still a fairly large installed base of 3720s and 3725s around the world. In addition, even on a 3745, 30 or more medium-speed (i.e., 32–56Kbps) VCs can only be accommodated across a single port if a HSS is being used. HSSs are expensive. A more cost-effective approach would be to LAN-attach an inexpensive FRAD to a 37xx and then have the FR access link-attached to the FRAD. This type of LAN-based solution will also work with 3720s and 3725s, in addition to non-37xx gateways such as 3172s.

Just SDLC-to-FR migration by itself is also not recommended. Rather than going through this intermediary upgrade step, enterprises should look at FR as a means of consolidating their dreaded parallel networks and getting rid of all their SNA/SDLC only links in favor of an FR-based SNA/APPN-capable multiprotocol WAN.

2. *Consolidation of dreaded parallel networks using a core FR WAN:* In reality there are multiple ways of realizing this consolidation per se even though they all involve using FR as the central WAN fabric. Some of the techniques that may be used to realize FR-centric parallel network consolidation include:

a. *Bridging across FR*, where for example source-route bridges are interconnected via an FR WAN, and SNA/APPN traffic is transported across the WAN by having the LAN frames containing that traffic encapsulated within FR frames

b. *DLSw(+) across FR*, where bridge/routers running DLSw(+) will interact with each other across an FR WAN, with SNA/APPN traffic being encapsulated within TCP/IP packets, which are in turn encapsulated within FR frames

c. *APPN NN routing across FR*, where bridge/routers will individually route SNA/APPN and non-SNA traffic across an FR WAN, with SNA/APPN message units encapsulated directly inside FR frames à la RFC 1490/FRF.3

d. *RFC 1490/FRF.3 encapsulation of SNA/APPN and other protocols*, where these protocols will be transported across an FR WAN encapsulated directly inside FR frames

e. *LAN-over-SNA or IBM 2217 MultiProtocol Concentrator-type solution across FR*, where SNA and non-SNA traffic will be transported across the FR WAN by having this traffic encapsulated within SNA message units, which are then encapsulated directly inside FR frames à la RFC 1490/FRF.3. The availability of ade-

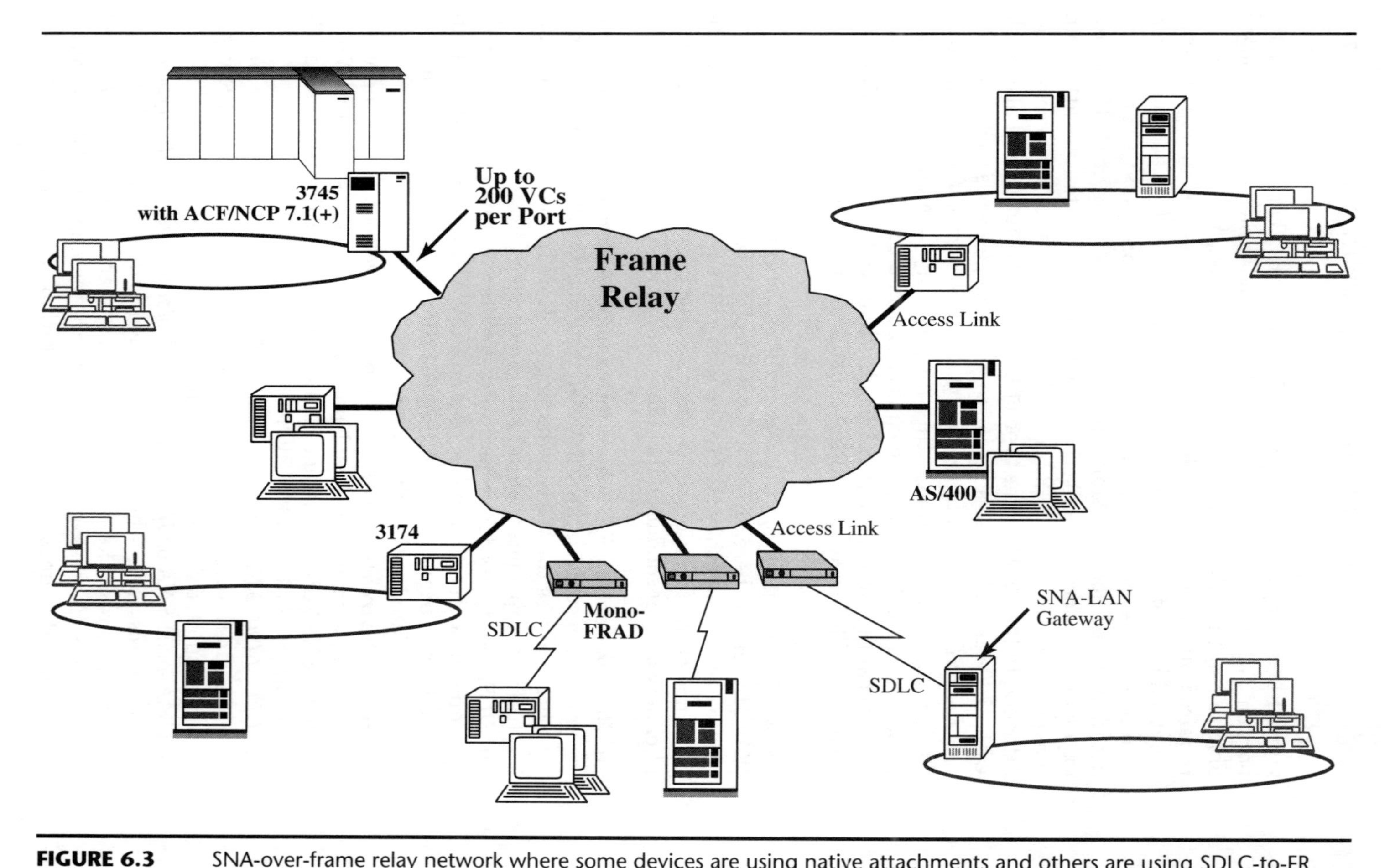

FIGURE 6.3 SNA-over-frame relay network where some devices are using native attachments and others are using SDLC-to-FR mono-FRADs.

quate bandwidth has always been a perennial concern with any LAN-over-SNA solution given that SNA-only networks with their 9600bps or 19.2Kbps SDLC links are not the epitome of bandwidth-to-spare networks. FR, obviously, now provides a cost-effective means of realizing 56Kbps range bandwidth for such LAN-over-SNA or AnyNet Gateway solutions.

f. *PIR across FR,* where CrossComm bridge/routers are interconnected via an FR WAN, and SNA/APPN traffic is transported across the WAN by having the PIR LAN-frames (i.e., LAN frame plus PIR header) containing that traffic encapsulated within FR frames

Note that though all six of the above parallel network consolidation solutions are based on FR they are indeed very different approaches. Thus when evaluating network consolidation techniques it is very important to remember two things. The first is that FR can play a role with all of them. The second is that the potential use of FR as the core WAN with all of these techniques does not in any way ensure that they all end up exhibiting the same overall characteristics. Far from it. Using DLSw(+) over FR is not the same as using RFC 1490 in native mode. As will be seen later, if nothing else, the header overhead of native mode RFC 1490 can be one quarter that of DLSw.

The first three techniques above, as well as PIR, can be thought of as the traditional bridge/router approaches to parallel network consolidation over FR. With all of these techniques, bridge/routers will typically use just one VC between each pair of bridge/routers–and forward all the traffic, independent of their protocol, across that VC. This, for obvious reasons, is referred to as the *single-VC* approach. This single-VC scheme, however, can be impractical if the "frowned upon" serial port connection approach is being used to interface a 3745 to the FR WAN. The 3745 will happily accept from the FR WAN any SNA or TCP/IP traffic being sent to a mainframe. Till recently it would not accept any other traffic. Thus, there was no option but to use a *separate-VC* scheme where traffic destined to the mainframe was kept on one set of VCs while the other traffic shared another set of VCs. This is another reason to discourage the use of 3745/46 serial port connections to FR. Today, the 3745 will accept other traffic on an FR serial port and will route it to a LAN port.

3. *Integration of the esoteric but essential link protocols used in bank branch office*—such as BSC from automated teller machines, Async from security/alarm systems, Burroughs Poll/Select—into a FRAD-based multiprotocol LAN/WAN network. FRADs, irrespective of vendor, tend to excel in providing proven support for such link protocols.

4. *Integration of voice into the multiprotocol LAN/WAN network:* Using the voice-over-FR facilities now being offered by certain FRAD vendors. Figure 6.4 shows an FR-centric FRAD-based LAN/WAN network supporting a variety of non-SDLC link traffic as well as voice.

5. *Realizing the equivalent of a fully interconnected mesh-network:* With direct point-to-point connections using VCs across a public FR network. Such an FR-based meshed network is likely to be significantly less expensive than one implemented using long-haul leased lines.

6. *Implementing a private, in-house FR network:* As described in Chapter 3, using FR/ATM switches such as IBM's 2220 Nways BroadBand Switches or Cisco's LightStream 2020 Switches

6.2 A THUMBNAIL SKETCH OF FRAME RELAY

X.25 has many sophisticated mechanisms to ensure error-free guaranteed in-sequence delivery of packets. Frame relay, on the other hand, eschews such measures and relies in the main on today's relatively error-free, typically fiber-based, digital transmission media to ensure that not too many frames get corrupted in transit. If and when a frame does get corrupted or discarded due to network congestion, frame relay per se does not bother to take any steps to have it retransmitted. Instead, just as with ATM, request of such retransmissions of corrupted or discarded frames is left to higher-level protocols being used by the end-to-end users.

Frame relay also does not check for corrupted frames at intermediate nodes; that is, there is no FCS processing for frame content integrity at intermediate nodes. Instead, frames are transmitted end-to-end with minimum processing at intermediate nodes à la ATM. This elimination of intermediate node processing can tangibly improve data traffic throughput across a WAN that has many intermediate nodes compared to the throughput possible with X.25, for example.

Frame relay, like X.25, is a packet-mode interface specification between end user devices (i.e., FR Data Terminal Equipments [DTEs]) and a network that is accessed via an FR DCE. Just as with X.25, FR users interact with each other across the network using VCs. Today, public FR networks offer VCs with capacities up to T1/E1 (i.e., 1.54Mbps/2.048Mbps). However, the FR specification already covers speeds up to T3 (i.e., 45Mbps). Thus, it is highly likely that public FR services, in the near future, will extend their bandwidth offerings to include at least 4 to 8Mbps VCs.

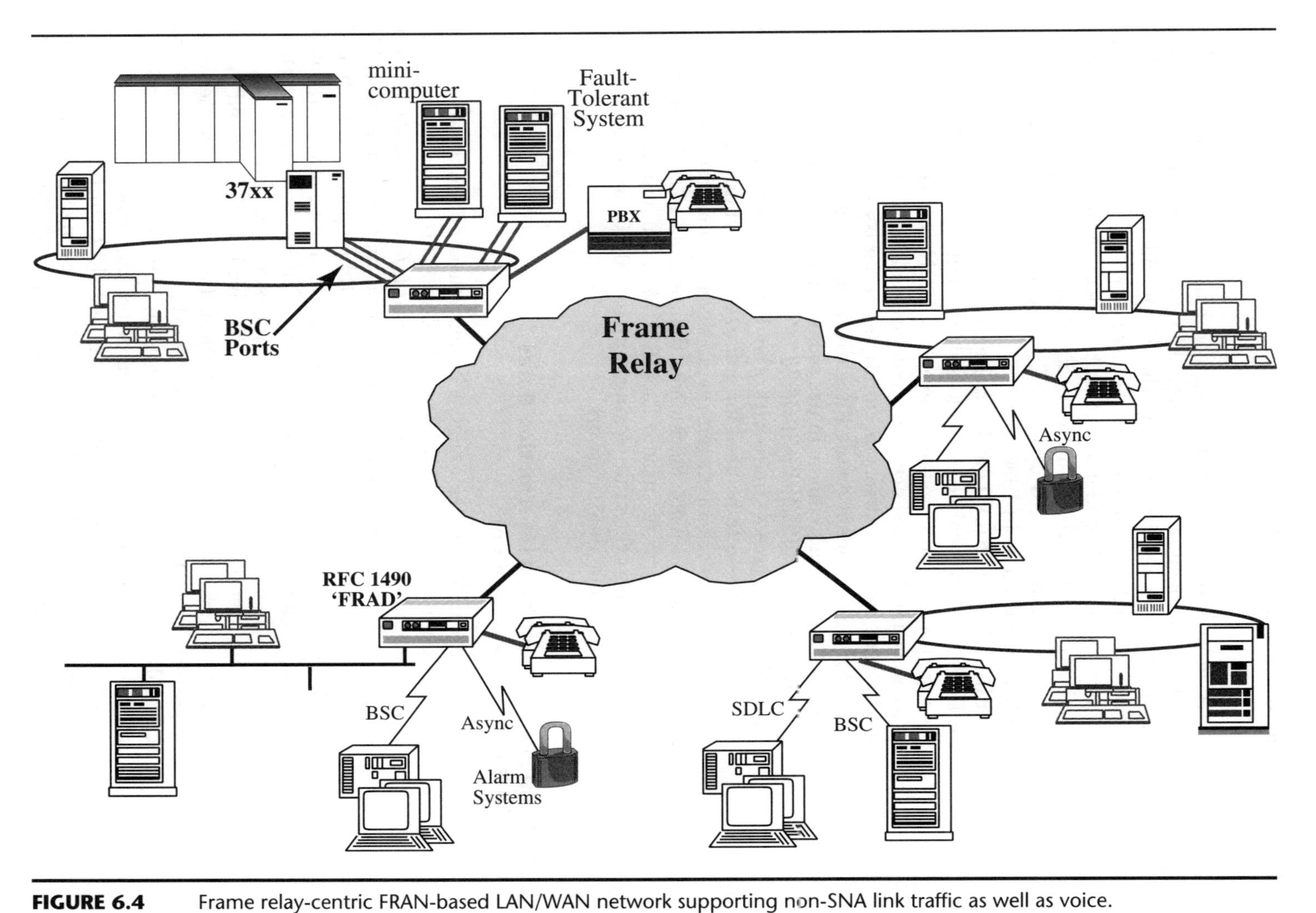

FIGURE 6.4 Frame relay-centric FRAN-based LAN/WAN network supporting non-SNA link traffic as well as voice.

At present public FR services support only permanent VCs (PVCs). A PVC is analogous to a leased line. PVCs are established per a customer's express requirements by the FR provider (e.g., AT&T, MCI, WilTel). A PVC once in place can be used at will at any time, in much the same way as a leased line, without the need to go through any type of call setup procedure. An industry standard for switched VCs (SVCs) is just now being worked upon. SVC support on public networks are likely to be available in 1996.

FR DTEs (e.g., FRAD, bridge/router, or 3745) are connected to the FR DCE via a physical access link—which would typically be a short-distance leased line or possibly even an ISDN connection. Multiple VCs can be multiplexed across a single-access link. The exact number of VCs that could in theory be multiplexed across a single physical access link is dependent on the FR addressing scheme being used. The prevalent addressing scheme in use today is based on a 10-bit address field. With this 10-bit scheme it would be possible to multiplex around 1,024 VCs per access line—with the FR specification permitting around 976 of these to be used for end user VCs while the others are reserved for network management and control purposes. FR has a built-in address extension mechanism that permits 16 and even 23-bit addressing fields to be used. (The IBM 3745 communications controller will support the multiplexing of up to 200 VCs on an access link between itself and an FR network.)

6.2.1 Frame Relay Committed Information Rates and Bandwidth-on-Demand

With FR each PVC can be assigned a guaranteed minimum amount of bandwidth—for example, 32Kbps or 56Kbps. This guaranteed bandwidth assignment is known as the *Committed Information Rate* (CIR) of a given PVC. Public FR network tariffs are usually based on the CIR required per PVC—with the cost being directly proportional to that of the CIR. In other words the higher the CIR (i.e., more guaranteed bandwidth) the higher the monthly cost. When a CIR has been assigned to a PVC, the network provider endeavors to ensure that frames that are sent within that allocation of minimum bandwidth are delivered to the other end without any frames being discarded to cope with any network congestion. Thus, if PVC has a CIR of 56Kbps, it should be possible to send 56Kbits per second worth of frames on that PVC with a relatively small chance of any frames being discarded by the network.

FR, however, does permit the CIR of a PVC to be temporarily exceeded provided the WAN is not experiencing congestion. This is the so-called bandwidth-on-demand capability of FR. To exploit bandwidth-on-demand the CIR of a PVC must be lower than the speed of the physical access link to the FR WAN. This is because the CIR applies only to the amount of bandwidth allocated within the WAN, as opposed to the amount of bandwidth available to access the WAN. If the WAN cannot be accessed at a bits-per-second rate higher than the CIR then it would not be possible to exceed the CIR.

For example, a 128Kbps Fractional T1 link could be used between an FR DTE and the FR WAN. If so, all transmissions across this physical access link between the DTE and the WAN will occur at 128Kbps. The DTE may have multiple PVCs multiplexed across this 128Kbps link. Let's assume that one of these PVCs has a CIR of 56Kbps. Any frames corresponding to this 56Kbps CIR PVC, when being transmitted across the physical link, will go across this link at 128Kbps, rather than at 56Kbps. In other words the speed at which data is sent across the physical access link is determined by the actual speed of that link rather than by any CIR-related value or factor. Hence, it would be possible to transmit 56Kbits across this access link to the network in .5 second. This 56Kbits pumped into the WAN in .5 second will still be within the CIR. The WAN should thus deliver this data to the other end without discarding any frames.

If the CIR is not to be exceeded, no more frames should be sent on this PVC for the next .5 second. Then another 56Kbits could be safely sent within the CIR limit over the next second. However, with bandwidth-on-demand the DTE may decide to try and push across more frames within the first second without waiting for another .5 second to elapse. Thus, after sending 56Kbits in the first .5 second it may send another 10Kbits over the next .5 second—given that the physical link per se permits up to 128Kbits to be transmitted over a period of 1 second. There is no guarantee that the additional frames that were sent after the CIR was exceeded will be delivered to their destination.

The frame relay provider reserves the right to arbitrarily discard any frames that have been transmitted once a PVC has exceeded its CIR for a given predefined time period—say every second. If the speed of the access link is the same as that of the CIR of a given PVC it would not be possible for that PVC to enjoy bandwidth-on-demand. It would also be pointless to have PVC CIR that is higher than the speed of the two physical access links—at either end of the WAN—that would be traversed end-to-end by that PVC.

FR bandwidth-on-demand is just like exceeding the speed limit on a highway. Staying at or below the speed limit will eliminate the risk of incurring any speeding violations. Exceeding the speed limit, every once in awhile, should not, with luck, result in a speeding violation each and every time. Hence, some exceed the speed limit and take a chance of getting a speeding violation. This is much the as same with bandwidth-on-demand. Staying at or below the CIR will minimize, but unfortunately not guarantee, that frames will not be discarded. When the CIR is exceeded there will be times when large numbers of frames get through, and other times when some get discarded. Just as with getting caught speeding it is a matter of chance. This analogy also highlights the need for a faster access link if the CIR is to be exceeded. A car that has a top speed less than that of the speed limit is unlikely to get stopped for speeding.

FR reserves the right to deal with any unexpected network congestion by discarding frames irrespective of the CIRs of the PVCs that may be currently using the network. In reality this is no different from the way that bridges or bridge/routers drop packets when congested. In general, however, the chances of frames that are within CIR getting discarded due to congestion are slim. Based on this, most major FRAD vendors now provide optional mechanisms whereby mission-critical SNA/APPN traffic is only transmitted within frames that do not exceed the CIR of their PVCs. This minimizes the possibility of SNA/APPN traffic-bearing frames from getting discarded due to congestion and then having to be retransmitted—which invariably wreaks havoc to the response times of SNA/APPN applications. This technique is described in detail in Section 6.7.

6.2.2 The Structure of Frame Relay Frames

A FR Frame has a framing structure more or less identical to that of a SDLC or HDLC frame. Just as with a SDLC frame, an FR frame is delimited at either end with a *Flag* byte. These 1-byte start and stop flags have the same b '01111110' bit composition as the flags used in SDLC. Following the start flag, an SDLC frame has a 1-byte SDLC address feld (i.e., 8-bit address). This address field is followed by a 1-byte control field. This control field contains the SDLC commands such as RR, SNRM, or Information Frame as well as the send and receive sequence numbers used to guarantee the delivery of frames.

Rather than having two separate 1-byte address and control fields à la SDLC, a typical FR Frame just has a 2-byte address field, which contains a 10-bit address. The FR address used to uniquely identify a VC, across a given DTE-to-DCE interface, is referred to in FR as the *Data Link Connection Identifier* (DLCI). The other 6 bits of the 2-byte address field are used as flags. If 16-bit addressing is to be used, a 3-byte address field is required, with 23-bit addressing necessitating a 4-byte address field. The same FR flags appear in all these Address fields irrespective of their length.

The flag bits that appear within the address field have the following designations: the Extended Addressing (EA) bit, the Forward Explicit Congestion Notification (FECN) bit, the Backward Explicit Congestion Notification (BECN) bit, the Discard Eligibility (DE) bit, and the Command/Response (CR) bit. An EA bit occurs as the last bit of each and every byte in an FR address field. Hence, there will be 2 EA bits within a 2-byte address field, 3 EA bits in 3-Byte address field, and so on. Setting the EA bit on indicates that further addressing bits are located within the next byte of the address field—hence its presence at the end of each byte.

The FECN and BECN bits can be optionally utilized by the network. If used, the network will set these 2 bits to indicate possible congestion within the WAN. The FECN is set by the network if congestion is being experienced in the direction the frame is traveling. The BECN is set by the network to notify the recipient that congestion is being experienced in the opposite direction to the one just traversed by the subject frame. Note that FECN and BECN are both, optionally, set by the network as opposed to one of the DTEs. FECN and BECN notifications by the network are purely for informational purposes.

DTEs are not obliged to react to FECN and BECN in anyway. FR, in marked contrast to SNA's or HPR's congestion control mechanisms, does not dictate that DTEs must reduce their rate of transmission to help dissipate the network congestion. However, recognizing FECN and BECN is a good way to determine whether it is prudent to exceed CIR at a given time. Exceeding CIR when there are FECN or BECN indications being sent by the network is not likely to be a productive exercise. Given that the network is already experiencing congestion it is unlikely to tolerate excess traffic beyond the CIR of a given PVC. Thus, there is a fairly high chance that frames exceeding CIR will be discarded. There is a possibility that, if the congestion is severe enough, even frames within the CIR may get discarded. In the context of the highway speed limit analogy used above, receiving FECNs or BECNs would be the equivalent of using a radar detector and having it signal that the police are checking the speed limit on this stretch of highway. Speeding is still an option, but unlikely to be a prudent one.

The DE bit also relates to congestion. Whereas the FECN and BECN bits can only be set by the network, DE can be set by either a DTE or the network. The DE bit, as implied by its name, provides a means of favoring one set of frames versus another in the event that frames have to be discarded due to congestion. Given a choice, FR networks will first try to alleviate congestion by only discarding frames that are marked as being "discard eligible" by having their DE bit set. Consequently, frames that are sent once the CIR of a PVC has been exceeded should have their DE bit set on. If the DTE does not automatically do this, the network DCE could set the DE bit on to notify other nodes that this frame should be discarded in preference to others since it is outside the CIR quota for that PVC. Once set by the DTE or the network the DE bit cannot be turned off.

The CR bit is currently not used by FR since it, unlike SDLC or HDLC, does not use commands or responses such as RR, Receive Not Ready (RNR), SNRM, and so on. Hence, FR in contrast to SDLC does not have a command field. The absence of commands and responses within each frame is not as incongruous as it may first appear. The main purpose of having commands and responses included within each frame is to control data flow (e.g., RNR flow control), and have a sequence-numbers-based mechanism for acknowledging the successful receipt of frames. FR, unlike SDLC or HDLC, does not worry about the guaranteed delivery of frames. Hence, FR per se dispenses with the overhead of using traditional SDLC or LLC:2-like Layer 2 commands, responses, and sequence numbers. *The RFC 1490 encapsulation scheme compensates for this by providing a framework that can be used to embed SDLC or LLC:2-like commands and responses, within each FR frame to create a traditional Layer 2 protocol flow within the context of FR.* The FRF.3 specification exploits this feature to ensure that SNA/APPN traffic transported across FR is protected via a full implementation of LLC:2 protocol around this data.

FR, however, obviously still needs a mechanism to check on the availability of VCs and ensure that end-to-end connections corresponding to a given VC are active. FR achieves these availability and activity status functions via its so-called *local management interface* (LMI). LMI information, for a given end user VC, is exchanged across a separate, reserved VC (e.g., VC number zero) that is established between each DTE and the FR network DCE. Thus LMI is referred to as an out-of-band signaling scheme that the management information is exchanged across a path other than the one that it refers to. (In a sense this is analogous to using SSCP-PU sessions to transport management information between an SNA node and the mainframe SSCP, rather than using a LU-LU or SSCP-LU session to convey this management information. See Appendix A.) These network management-related exchanges are typically done on a poll and acknowledgment basis for each databearing VC.

The address field in an FR frame will be immediately followed by a variable length I-field. The maximum permissible length of the I-field will be dictated by the network, with none likely to specify a length any less than 1,600 bytes. Some implementations already support I-fields as long as 8Kbytes or even more, though most applications tend not to exceed 2Kbytes. Just as with SDLC, the I-field is immediately followed by a two-byte FCS field that provides a bit corruption check for all the bits in the frame excluding the two flags at either end. Figure 6.5 shows the format of an FR frame and compares it to that of a SDLC frame.

As highlighted in Figure 6.5, FR and SDLC have nearly identical frame structures. Given this similarity IBM 3745s, 3746s, 3174s, and so on do not require any new hardware adapters in order to support FR. They can support FR using the same hardware (e.g., 3745 LIC Type 1 or 3) as used to support SDLC links. Thus, the migration from SDLC to FR in many cases only requires upgrades to the software (e.g., ACF/NCP 7.1 for 3745s or Configuration Support-C Release 5 micro-code for 3174s). New hardware adapters(e.g., 37xx HSS) will only be required if the existing adapters cannot support the higher link speeds possible with FR. Thus, if FR is being used as a means of increasing networking bandwidth there is a likelihood that new adapters may be necessary to accommodate the new link speeds.

The total framing overhead of both FR (with the typical 10-bit addressing) and SDLC (with the prevalent modulo-8 acknowledgment scheme) is 6 bytes—2 flag bytes, 2-byte address field, and 2-byte FCS. Unfortunately, SNA/APPN traffic cannot be successfully transported across FR just using the 6-byte framing given FR's lack of an inherent guaranteed delivery protocol. Thus, an additional header invariably has to be used with FR to realize guaranteed end-to-end delivery of SNA/APPN traffic. The format of this additional header is defined by RFC 1490 and FRF.3. Therefore, despite the initial cursory similarity between SDLC and FR, a 6-byte FR frame cannot be used by itself to transport SNA/APPN traffic as can be done with a 6-byte SDLC frame. Instead, in practice at least a 16-byte FR frame consisting of the standard 6 bytes augmented by a 10-byte RFC 1490/FRF.3 header is required to support SNA/APPN traffic.

6.3 COMING TO TERMS WITH RFC 1490, RFC 1294, FRF.3, AND BAN

RFC 1490 is a simple and straightforward encapsulation standard. It specifies a generic framework for transporting traffic based on other protocols (e.g., IP or SRB bridged) end-to-end across an FR network. The framework specified by RFC 1490 to realize this multiprotocol assimilation just consists of a variable length header structure with certain defined code points to identify the protocol of the traffic being transported. The actual length of the RFC 1490 header can vary from 2 to 10 bytes depending on the protocol type. RFC 1490 uses the Network Level Protocol ID (NLPID) values administered by ISO and

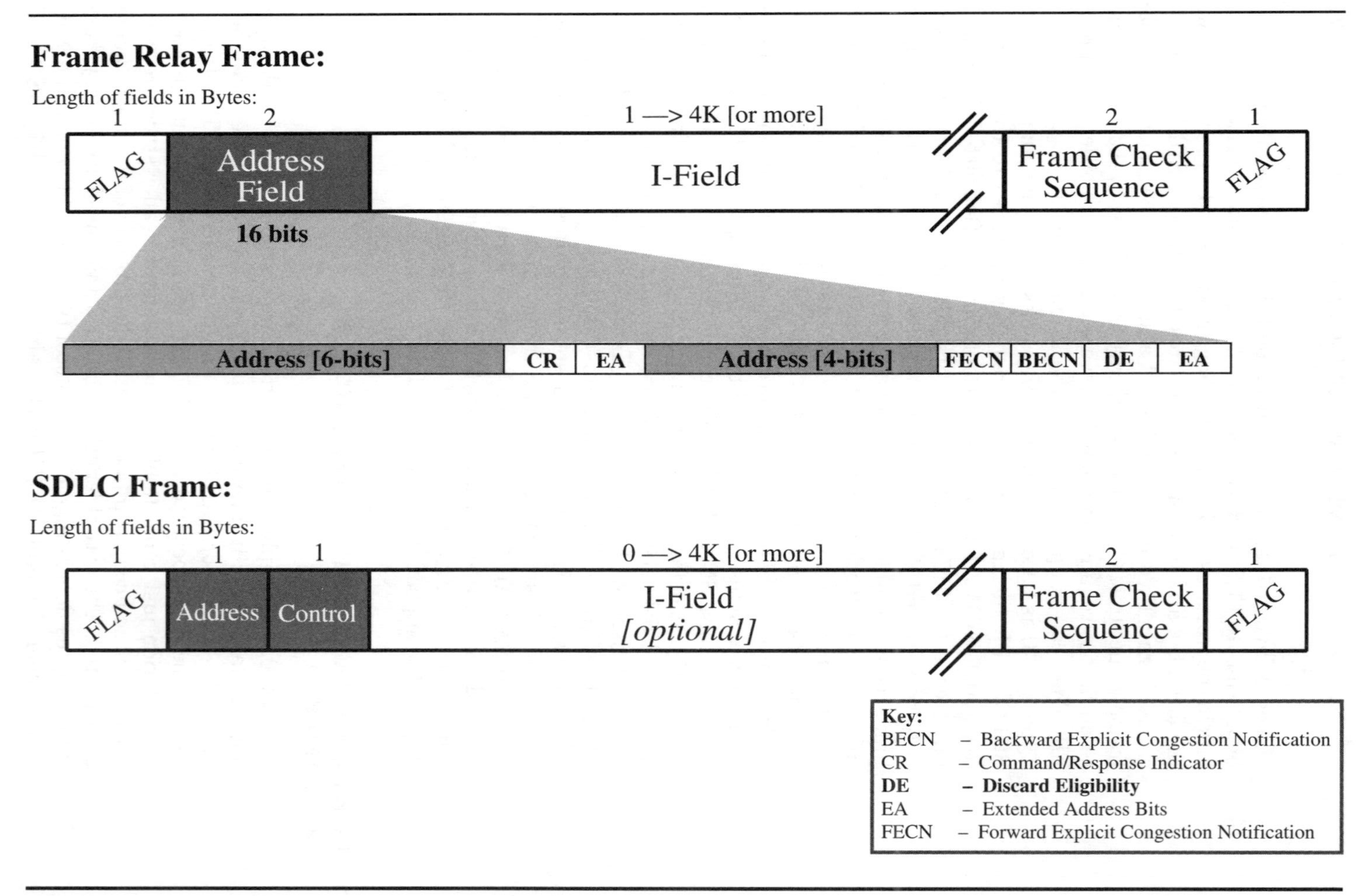

FIGURE 6.5 Format of a frame relay frame vis-à-vis SDLC frame.

CCITT to identify, at the highest level, the type of protocol being encapsulated within a given frame.

The 2- to 10-byte RFC 1490 header is inserted immediately after the FR address field at the start of the I-field. The actual data traffic is then inserted after the RFC 1490 header and will occupy the remainder of the I-field. A 2-byte-long RFC 1490 header is used when transporting IP datagrams over FR. This header just includes two fields: a 1-byte control field that indicates that the information is being transferred in connectionless unnumbered I-frame format, and a 1-byte NLPID that designates the protocol as being IP. Traffic using OSI's Connectionless Network Protocol (CLNP) is also transported using a similar 2-byte RFC 1490 header. A 9-byte header is typically used to transport bridged (e.g., SRB) traffic while a full, 10-byte header is used for SNA/APPN and NetBIOS traffic.

RFC 1490 per se does not specify the exact format, field contents, or the code points of the RFC 1490 header that is used to prefix SNA/APPN and NetBIOS traffic. Instead these are specified by the FRF.3 specification entitled *Multiprotocol Encapsulation Implementation Agreement*. The overall structure of the RFC 1490 header, per the FRF.3 specification for SNA/APPN and NetBIOS traffic is shown in Figure 6.6, which also illustrates how this header is used to prefix SNA/APPN PIU message units. Note that with standard RFC 1490/FRF.3 the SNA/APPN traffic is encapsulated in native form without recourse to any additional encapsulation.

A key aspect of this 10-byte RFC 1490/FRF.3 header is the presence of a 4-byte 802.2 LLC:2 header that is exactly the same as that used within MAC frames when SNA/APPN or NetBIOS traffic is transported across token-ring LANs. This LLC:2 subheader, in bytes 6 to 9 of the header, includes the customary Source and Destination SAP values (i.e., SSAP and DSAP) as well as the SDLC-like 2-byte control field that conveys commands such as RR and RNR and includes send and receive sequence numbers. Thanks to this LLC:2 subheader SNA/APPN and NetBIOS traffic can now continue to enjoy the comfort and protection of their preferred SDLC/LLC:2 link-layer protocol while traversing an FR WAN. With this subheader, an FR WAN, from a Layer-2 protocol perspective, essentially becomes just an extension of a token-ring LAN or even an SDLC link.

6.3.1 Correlating Destination Addresses with VCs and SSAPs

The SSAP field in the LLC:2 subheader provides a standard means of enabling traffic from different SNA/APPN or NetBIOS devices to be multiplexed across the same VC. All that an RFC 1490-based FRAD or

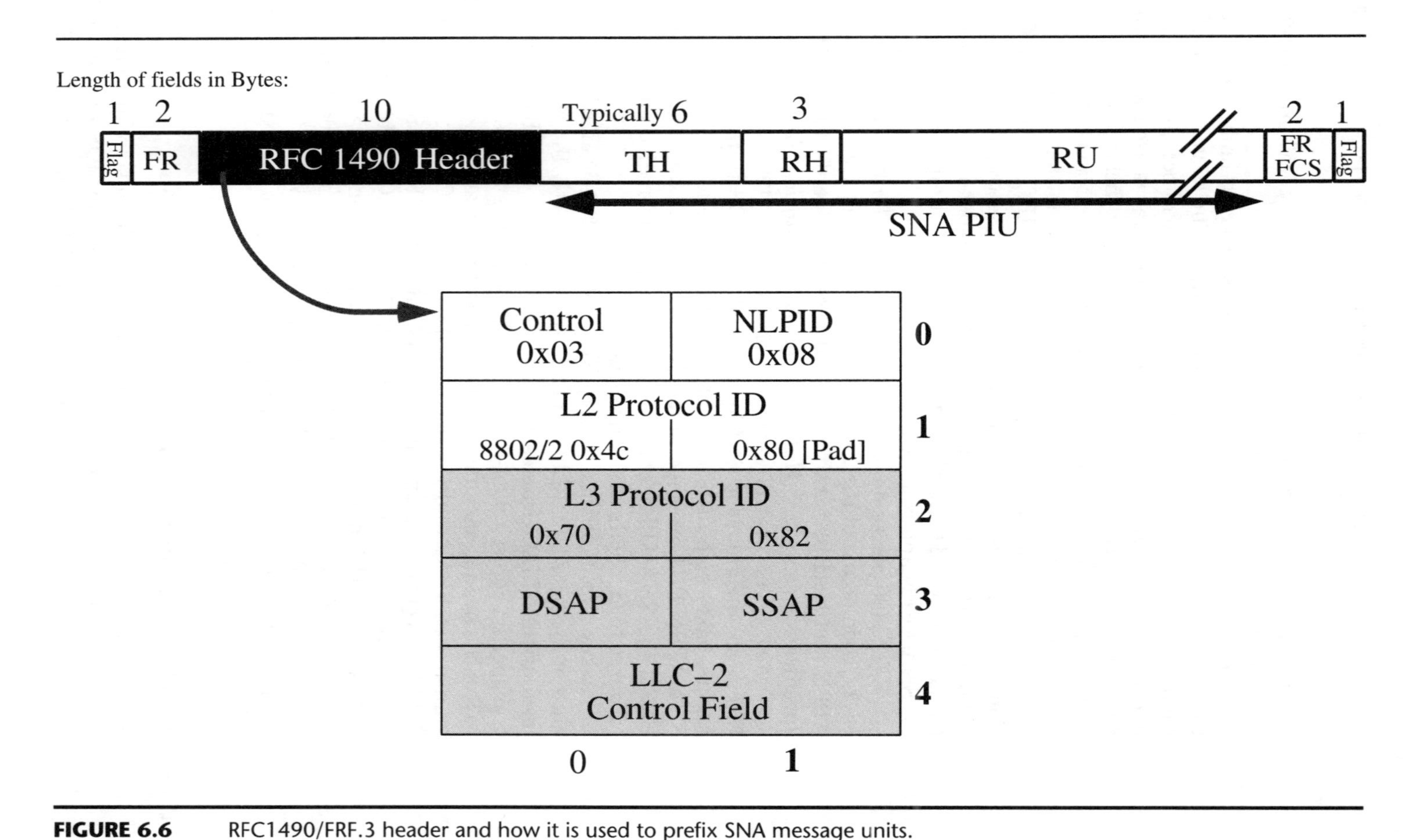

FIGURE 6.6 RFC1490/FRF.3 header and how it is used to prefix SNA message units.

bridge/router has to do is assign a unique SSAP value to each device that will be using the same VC. Today, it is not unusual to be able to multiplex up to 128 separate devices (e.g., SNA devices) per FR VC using this SSAP field. Such SSAP multiplexing is a standard feature of RFC 190-based solutions for supporting SNA/APPN traffic.

The only field that could initially be considered to be missing in the RFC 1490/FRF.3 header is that for a destination address—for example, a MAC address of the destination. While the presence of such a destination field may have simplified the implementation of certain RFC 1490 solutions, its absence is not, obviously, a show stopper. FR VCs can only be established on an end-to-end, point-to-point basis—in much the same way as leased lines. Thus, there are always one or more point-to-point VCs between a given source and destination DTE pairing. Hence, a DTE destination address (e.g., MAC address of a 3745) is only required at the source end to determine which VC should be used to reach that destination. Once traffic is placed on that VC there is no explicit need for a DTE destination address designation since the VC is only going to terminate at one destination—which hopefully is the intended one. (Note that having a DTE destination address is different from having a DLCI address to identify a specific VC between a given pair of DTEs.)

Standard RFC 1490 solutions for supporting SNA/APPN or NetBIOS traffic correlate destination MAC addresses with the appropriate VC to reach that destination using a manual definition process. Take, for example, a LAN at a remote location that has 30-odd devices —PCs, 3174s, and AS/400s—connected to a token-ring LAN that needs to access SNA applications on mainframe. Per the standard SNA software conventions described in the previous chapter for SRB and TCP/IP encapsulation, these SNA devices will identify the SNA LAN gateway (e.g., 3745) through which they hope to reach the mainframe via its MAC address. Given that they are all likely to be using the same gateway there is typically only going to be one destination MAC address involved. Since this destination gateway address has to usually be explicitly defined at each SNA device—typically via a gateway operand in the form of "GW="—this address will be readily known or easily accessible (from within a configuration file). All that now needs to be done is to associate this destination MAC address with the DLCI of the VC to be used to reach that destination.

Once this DLCI-to-destination MAC address mapping is in place, forwarding SNA/APPN traffic to its intended destination over the appropriate VC is not a problem. *Just as with DLSw, the RFC 190-based FRADs or bridge/routers will intercept SNA/APPN LAN traffic that is destined to another distant LAN on the other side of the FR*

WAN. The destination MAC address will determine the VC that will be used to forward the traffic. Now if traffic from multiple devices is going to be sharing that VC, a subaddressing scheme is required to identify the traffic from a particular device. This is realized by using separate SSAP values. (Incidentally, this approach of using separate SSAP values to uniquely identify multiple devices sharing a common VC is no different from the scheme used in many SDLC-to-LLC:2 conversion schemes. If the conversion scheme is only able to support one virtual MAC address to represent all of the SDLC devices, it will resort to using a different SSAP per SDLC device to uniquely identify them relative to the common MAC address.)

Having to manually define the destination MAC address to VC mapping is not as tedious as it appears. Typical mainframe-centric SNA networks or AS/400-based APPN networks do not have many destinations that have to be *initially* located. SNA and APPN networks are set up such that each peripheral (or end node) is interfaced to the network via one SNA LAN gateway (i.e., channel-attached or remote 37xx, 3172, 3174, or a PC-based gateway such as Novell's NetWare for SAA), or in the case of APPN, a Network Node. Routing between the LAN gateway (e.g., a 3745) and one or more mainframes is realized using SNA subarea-to-subarea routing. In the case of APPN, routing between the initial (or entry) NN and all other nodes in the network is achieved using APPN NN routing. Hence, the number of SNA/APPN related destination addresses that will have to be manually assigned to the relevant VCs may not be much more than one or two in most cases.

If more than four destinations are being actively used, it may be worthwhile to consider APPN NN routing or even DLSw, over FR, as opposed to a straight RFC 1490 solution. Refer to Table 1.1 in Chapter 1. IBM's BAN is an attempt to address this destination address location issue that ends up realizing this goal by essentially implementing SRB across FR!

In addition to specifying the generic variable length header scheme for encapsulating multiple protocols within FR, the RFC 1490 specification also deals with a scheme for segmenting such encapsulated frames if they are too big to be sent across an FR WAN as a single frame. This segmentation, when required, is realized by use of another header—the RFC 1490 Segmentation (or Fragmentation) Header.

RFC 1490 does not address any issues apart from the generic encapsulation header structure and the frame segmentation scheme. Thus, unlike DLSw, RFC 1490 per se does not deal with Local LLC:2 acknowledgment mechanisms, SDLC-to-LLC:2 conversion, or a search mechanism to dynamically locate MAC addresses or NetBIOS names.

This does not mean that RFC 1490-based solutions for SNA/APPN traffic is devoid of such feature. Far from it. What happens in practice is that such features are added on to the basic RFC 1490 encapsulation scheme typically by FRAD vendors but to an extent also by bridge/router vendors. The key value-added features available with RFC 1490 solutions include: local LLC:2 acknowledgment and comprehensive support for various esoteric link protocols in addition to SDLC.

Thus DLSw and RFC 1490 solutions, in terms of their feature repertoire, look very similar. The key difference is that certain value-added features on FRADs, in particular support for link traffic, could be based on proprietary vendor-specific schemes. This may preclude unrestricted multivendor interoperability between FRADs, whereas such interoperability, in theory, is the forte of DLSw. However, in networks that involve only SNA/APPN LAN traffic and SDLC-to-LLC:2 converted SDLC link traffic, multivendor interoperability, per RFC 1490/FRF.3, should be possible. RFC 1490 advocates are also likely to point out at this juncture that RFC 1490 is a true ratified IETF industry standard, whereas DLSw is only an Information RFC, which by definition is not a true industry standard.

6.3.2 RFC 1294 and IBM's Boundary Access Node Scheme

RFC 1294, which was developed circa 1992, was the original multiprotocol encapsulation scheme for FR. Like RFC 1490, it was based on the use of an encapsulation header, and restricted itself to essentially supporting IP, OSI CLNP, and bridged traffic. RFC 1490 which was developed around 1993, was an attempt to expand RFC 1294 to provide an open-ended and generic scheme that could be easily adopted to accommodate traffic using any type of protocol—for example, SNA/APPN or NetBIOS. RFC 1490 made RFC 1294 obsolete. However, the RFC 1490 header structure is such that it is transparently backward-compatible with the header structure used by RFC 1294. Thus in the case of IP, OSI CLNP, and bridged traffic in particular there is no difference between the RFC 1294 and RFC 1490 headers used to prefix this traffic.

Therefore, when talking about IP or bridged traffic transport over FR, the terms RFC 1294 and 1490 can essentially be used interchangeably. Old 1294 implementations will still work transparently and seamlessly with RFC 1490 solutions.

BAN is IBM's name for an RFC 1490-based scheme for transporting SNA/APPN traffic that is now offered on IBM 2210 and 6611 multiprotocol bridge/routers and is supported as an option by ACF/NCP Version 7 Release 3 (and greater) for 3745 serial port I/O. *BAN is essentially token-ring SRB across FR.* In the case of Ethernet or SDLC-based SNA/APPN traffic, the 2210 and 6611 bridge/routers extract the SNA PIU from its original frame and insert it into a new 802.5 token-ring MAC frame in order to transport that traffic over FR. BAN can function in two modes: straight bridging mode and the so-called DLSw terminated mode. In the latter mode BAN provides automatic Local LLC:2 acknowledgment, à la DLSw, on either side of the WAN. BAN was announced by IBM in March 1995, roughly a year after standard RFC 1490 support for SNA/APPN over FR was offered across 3745 serial ports with ACF/NCP 7.1.

The key difference between BAN and standard RFC 1490/FRF.3 is the amount of header overhead included with a given SNA PIU message unit. With RFC 1490 the only additional header used is the 10-byte RFC 1490/FRF.3 header as shown in Figures 6.6 and 6.7. With BAN, given its SRB orientation, the SNA PIUs are first included within full MAC/LLC frames. These are in turn then encapsulated within an FR frame using the RFC 1490 header for bridged traffic—which happens to be 9 bytes long. Figure 6.7 compares standard RFC 1490 encapsulation of an SNA PIU to that used by BAN. Ironically, the difference between the RFC 1490/FRF.3 overhead and that of BAN, is exactly the same as that of the difference between the IBM-inspired DLSw and Cisco's RSRB. Except in the case of DLSw IBM did the opposite of what it did with BAN. DLSw does not include the MAC/LLC framing, while RSRB does.

BAN's use of a MAC/LLC frame around the SNA PIU adds a minimum of 17 bytes of additional header. These 17 bytes consist of 12 bytes for the source and destination MAC address, the 4-byte LLC sub-header and the 1-byte frame control field at the start of the MAC/LLC frame. With the 9-byte RFC 1490 bridging header, BAN has a total of 26 bytes of encapsulation header compared to standard RFC 1490's 10 bytes. This adds 16 bytes of header per SNA PIU—which is in effect a 160% increase in the header overhead.

Given that FR bandwidth is relatively cost-effective, these additional 16 bytes of overhead per SNA PIU are usually dismissed as being irrelevant. It is not uncommon to have a million or more SNA message units a day flow across some large IBM-centric networks. In such large networks, these 16 bytes of additional header will translate to 16 million additional bytes per day being sent across the network—equivalent to the volume of traffic of a few hefty file transfers. Another way to look at this spurious header traffic is to note that if 56Kbps VCs and access links are being used it would take around 38 minutes a day to just to transport these headers, that is, half an hour per day worth of bandwidth consumed by the additional bytes.

BAN permits traffic from multiple devices to be multiplexed across

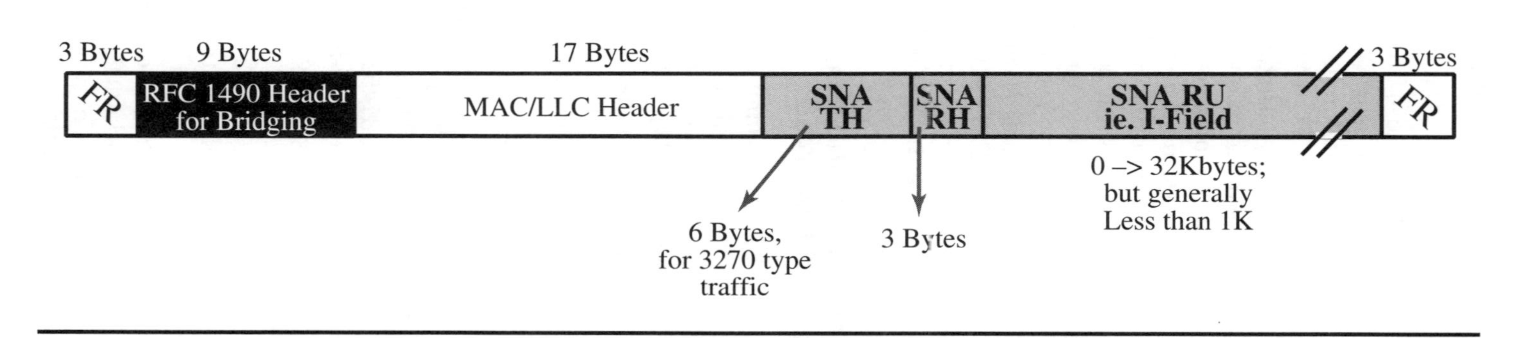

FIGURE 6.7 Comparison, to scale, of standard RFC 1490/FRF.3 encapsulation of SNA/APPN versus IBM's BAN.

one VC. It also allows SNA/APPN and TCP/IP traffic destined to, or originating from, a mainframe to share a single VC. Both these features as well as Local LLC:2 acknowledgment are supported by most standard RFC 1490 solutions. Thus the only supposed advantage of BAN over a standard RFC 1490 solution is its ability to dynamically locate destinations. As described above, this in practice is not a major requirement since most SNA/APPN networks do not have that many destinations. Consequently in many cases there is no real compelling justification for BAN over a standard RFC 1490 solution. Moreover, BAN, though it could be used purely between IBM bridge/routers, is geared for use across a 3745 serial port interface—hence the support in ACF/NCP 7.3. If a serial port interface is not being used there is even less justification for BAN. In general, enterprises should consider standard RFC 1490 solutions in preference to BAN as the means of realizing an optimum FR-based solution for SNA/APPN-capable multiprotocol LAN/WAN networking.

6.4 FRAME RELAY AND SNA/APPN-OVER-FRAME RELAY VIS-A-VIS IBM 3745S

IBM 3745 communications controllers have had FR support in some form or another since ACF/NCP Version 6 Release 1, which became available at the end of 1992. ACF/NCP 6.1 included FR DTE mode operation. With FR DTE, physically dispersed 3745s forming a distributed SNA subarea network could be interconnected to each other across a public (or in-house) FR network. In this respect, FR became an alternative to SDLC and X.25 as a means of interconnecting dispersed 3745s.

ACF/NCP 6.2, which became available in mid-1993, enabled a 3745 to act as an FR DCE—that is, a bona fide FR network switch—in addition to being able to operate as an FR DTE. With ACF/NCP 6.2, one could build a private, in-house FR network, as discussed in Chapter 3, using 3745s as the FR switches to create the FR network fabric. With such a 3745-based FR network one could freely interconnect any FR DTE devices (e.g., multiprotocol bridge/routers) to each other across the 3745 network.

This ACF/NCP 6.2 FR DCE capability is somewhat analogous to the X.25 SNA Interconnect (XI) facility that IBM has had on 37xx's since 1986, which permits X.25 DTEs to communicate with each other across an SNA network. The notion of companies implementing private networks using IBM equipment and leased lines, as opposed to using public networking services, has always been dear to IBM's heart given that SNA networks—cut to the chase—are but large, far-flung private networks. With a private FR network based on 3745s or IBM's 2220s it would be possible to have hybrid networks where the IBM equipment-based private network could be used in conjunction with a public FR or even ATM service.

The ACF/NCP 6.2 FR DCE facility is formally referred to by IBM as an FR Frame Handler (FRFH). This FRFH feature enables multiple dispersed 3745s in a SNA subarea network to be defined to each other as if they are all logically adjacent to each other even though in practice there may be intermediate 3745 nodes between some of the supposedly adjacent 3745s. With FRFH, in line with FR's Layer 2 orientation, all the frame processing at any intermediate 3745 is done at Layer 2, obviating the need for any higher-level SNA Path Control Network-based intermediate node processing. This FRFH Layer 2-only processing can significantly speed up data transfer within an SNA subarea network that has many intermediate node 3745s. This concept of speeding up intermediate node rerouting by doing all the relevant processing in Layer 2 has now been included within IBM's HPR, as will be discussed in the next chapter.

ACF/NCP 6.1 and 6.2 did not support serial port-based SNA over FR between a 3745 and an SNA peripheral node such as a 3174. FR was only possible either between 3745s, or between two or more FR DTEs (e.g., bridge/routers) across a 3745 network. (SNA traffic transported across an FR WAN by a bridge or bridge/router could obviously be forwarded to a 37xx via a token-ring interface.) To circumvent this SNA peripheral node limitation, vis-à-vis 3745 serial ports some vendors developed SDLC-to-FR mono-FRADs—an FR equivalent of a X.25 PAD for SDLC. With these FRADs, an SDLC device (e.g., 3174) could be connected to a 37xx SDLC port across an FR WAN by deploying a FRAD at each end. The true value of such FRADs was somewhat limited in that SDLC ports, as opposed to a single FR port that could support multiple VCs, still had to be used on the 37xx side.

The ability to directly connect SNA peripheral nodes (e.g., 3174) to a 3745 *serial port* using SNA-over-FR was only possible with ACF/NCP 7.1, which became available in early 1994. *Interestingly, this original 3745 serial port support for SNA/APPN traffic over frame relay was based on standard RFC 1490/FRF.3 encapsulation—that is, native mode encapsulation of SNA PIUs without a MAC/LLC wrapping around it.* IBM is committed to continuing this support for standard RFC 1490/FRF.3 on 3745s and 3746s for serial port-based I/O. In a sense it does not have a choice. IBM's SNA-over-FR support for 3174s, which became available with Configuration Support-C Release 5 micro-code roughly around the same time that ACF/NCP 7.1 came on-line, is based on RFC 1490/FRF.3. BAN support is now available on

ACF/NCP 7.1—but, ironically, only in the form of a *program temporary fix* (PTF).

ACF/NCP 7.3, which became available in mid-1995, supports both BAN and standard RFC 1490/FRF.3-based encapsulation for exchanging SNA/APPN or IP traffic across a 3745/46 serial FR port. This support for both encapsulation schemes permits backward compatibility with ACF/NCP 7.1 and 7.2. It also means that enterprises determined to use 3745/46 serial ports for FR interactions can do so with impunity using standard RFC 1490 solutions from multiple bridge/router and FRAD vendors, as opposed to the "header-happy" BAN approach.

6.4.1 Reasons for Bypassing 3745 Serial Port-Based FR Networking

Enterprises have multiple viable and attractive options when it comes to selecting a gateway for interfacing a mainframe to an FR WAN. Potential gateway options include: IBM 3745s, IBM 3746 Model 900 Expansion Frames attached to a 3745 (i.e., a 3745/46 combo), IBM 3172s, channel-attached bridge/routers à la Cisco, channel-attached intelligent hubs à la Cabletron, and so on.

When selecting such an FR-to-mainframe gateway, the fundamental choice that needs to be made is whether a 3745/46 serial port is going to be used as the interface to the FR WAN, or whether this interface to the FR WAN is going to be realized using a bridge/router or FRAD that is LAN-attached to the gateway. Given that IBM has supported FR across a serial port on 3745s since mid-1994 there has been a tacit belief—that IBM has duly encouraged and promoted—that such a serial port interface is the optimum means for interfacing an FR WAN to a mainframe. (The latest 3172 Model 3's, as of around mid-1994, also theoretically support FR serial ports that could be used to realize this FR interface. Most enterprises have not seriously considered this as a strategic option for realizing mainframe-to-FR connectivity.)

The 3745/46 serial port-based approach has some tangible disadvantages. These can be summarized as follows:

1. Only 3745/46s support FR over serial ports. This precludes the still fairly large installed base of the older 3720s and 3745s from being able to use this serial port approach for interfacing to an FR WAN. IBM recommends that these older 37xx's are upgraded to 3745s or 3746 Model 950s. (Ironically, FR serial port support will not be available on the new 3746 Model 950 until mid-1996.)

   A token-ring-based solution, where a low-cost token-ring-attached bridge/router or FRAD serves as the interface between the

37xx and the FR WAN will work with any 37xx—and will even do so with versions of ACF/NCP prior to 7.1. With such a token-ring-based solution, as shown in Figure 6.1, the 37xx does not have to be aware of FR in any form—hence the possibility of using pre-7.1 versions of ACF/NCP. With this type of solution, all the SNA/APPN and IP traffic transported across the FR WAN using RFC 1490(/FRF.3) or even BAN appears to the 37xx as if it emanated from devices directly attached to its token-ring interface—that is, virtual bridging across FR.

This LAN-based approach will also work with 3172s and compatibles. There is now the added possibility that an Ethernet interface could be used on the 3172, rather than a token-ring, since 3172s, unlike 37xxs, support SNA over Ethernet. Using a channel-attached bridge/router or channel-attached intelligent hub provides the possibility of eliminating the need for an intermediary bridge/router to mainframe gateway; see Figure 6.1.

2. Any existing 3745/46 serial port can theoretically be used as an FR port—without any modifications to the hardware. This is indeed an attractive feature. However, most 3745 ports are of LIC Type 1 or 3. A LIC Type 1 port can support a maximum data transfer rate of 19.2Kbps while that for a LIC Type 3 port is a slightly more respectable 256Kbps. A key advantage of using FR, as opposed to SDLC, is the ability to multiplex multiple VCs over a single access line or serial port. A port that can run at only 19.2Kbps does not provide much scope for multiple VC multiplexing.

   Even a 256Kbps port at best is likely to provide only adequate performance if restricted to 15 or so 56Kbps VCs. Fifteen VCs per serial port is not that many given that the 3745 theoretically supports up to 200 VCs per port. Not being able to support adequate VCs per port means that mutiple ports—and even more importantly multiple access links between the 3745 and the FR WAN—have to be used. While it is desirable to have two or more parallel access links to the FR WAN to realize a level of link-failure protection, having to deploy 10 or more such links will significantly dilute the fiscal appeal of an FR-based solution.

   IBM 3745s do have higher speed ports that could be used to achieve a higher VC multiplexing factor. These higher capacity ports are known as *high-speed scanners* (HSSs). An HSS can have one active port that can support link speeds up to 2Mbps. If a T1 (i.e., 1.54Mbps) access link is used from such an HSS port to an FR WAN it would be possible to multiplex around 30 to 40 or even more 56Kbps VCs across that link. At face value this type of HSS-based approach does appear to be an attractive proposition. There is, unfortunately, a catch. The U.S. list price for a 3745 HSS is around $23,000. By today's standards that is somewhat expensive for a single 2Mbps port, albeit even on a rock-solid and proven platform. Compare this with complete bridge/routers or FRADs with multiple 2Mbps ports for around $2,000—with IBM's 2210 ironically providing a good example. The bottom line here is that an HSS-based solution is not defensible from a cost standpoint.

3. Even if an enterprise decides to pursue an HSS-based serial port solution for FR, it is highly likely that the 3745 will also have a token-ring interface (i.e., the so-called 3745 Token-Ring Interface Coupler [TIC]) to support LAN-based traffic. A 3745 token-ring interface is roughly the same price as an HSS. Today it is difficult to find a 3745 that does not have a TIC. Given this nearly ubiquitous presence of TICs it will always be possible to opt for a LAN-based FR solution as opposed to using an HSS. In the unlikely even that a 3745 does not have a TIC, it would be more strategic to get a TIC as opposed to an HSS.

   At this juncture it is worth noting that an argument could be made that an HSS-based connection to FR is likely to be more reliable and robust than that achieved via a relatively inexpensive bridge/router or FRAD. There is merit to this argument. Hence, in mission-critical scenarios the HSS might be a valid and preferred option—irrespective of its cost—given that the lost opportunity cost of a few additional minutes of network downtime could easily exceed $30,000.

6.5 VALUE-ADDED RFC 1490/FRF.3 SOLUTIONS

Most leading bridge/routers now offer RFC 1490/FRF.3-based support for transporting SNA/APPN traffic over FR, right alongside bridging, TCP/IP encapsulation (i.e., DLSw(+)) and APPN NN routing. In addition, there is a cadre of FRAD vendors, which include Hypercom, Motorola, Netlink, and Sync Research, that offer compelling RFC 1490-based multiprotocol solutions with comprehensive support for SNA/APPN traffic as well as BSC.

Consequently, RFC 1490/FRF.3 is a direct alternative to DLSw(+), APPN NN, and even the IBM 2217 AnyNet Gateway solution—provided, of course, FR is being used as the central WAN. This total dependence on FR is the one major flaw of RFC 1490 solutions. Though the RFC 1490/FRF.3 scheme can be extended to work over other WAN schemes, in particular X.25, such non-FR-based implementations are not readily available. Thus, RFC 1490 solutions should, unfortunately, only be contemplated if FR WANs are an option. (RFC 1490 per se is also not used across ATM. Instead, there is a comparable ATM specific standard for such encapsulations.)

When FR is not an option, DLSw(+) and APPN NN are likely to be the most viable contenders for parallel network consolidation. APPN NN, however, is probably only optimum if the network has four or more mainframes, AS/400, or other SNA/APPN destinations (e.g., PC-based servers).

The RFC 1490/FRF.3 specification focusing just on an encapsulation scheme has not in any way precluded RFC 1490-based FRADs from being feature rich, value-added, "soup-to-nuts" solutions. Some of these extensions, albeit, are likely to be proprietary since there are no de facto industry standards that cover them (e.g., support for BSC, Async, voice over FR). The stigma attached to such proprietary value-added solutions should, however, be clearly analyzed. For a start, in most cases, the proprietary aspects apply only to the value-added functions.

Interoperability when it comes to exchanging SNA/APPN traffic is likely to be feasible, especially since most of these FRADs will talk directly to a 3745, sometimes even using BAN, through a serial port. If multiple different FRADs can effortlessly exchange SNA/APPN or IP data with a 3745 the chances are high that these FRADs could also talk to each other without undue difficulty. IP, IPX, and NetBIOS interoperability should also be possible. The problems are likely to be restricted to facilities such as BSC and voice. At this point enterprises need to make a judgment call. There is no doubt that multivendor interoperability is an attractive and enviable objective. However, if certain value-added facilities are only available in proprietary flavors, due to the absence of appropriate standards, it might be better to compromise the multivendor goal and gain the advantages of these value-added facilities. The other less savory option is to settle for a less feature-rich, least common denominator solution whose primary virtue is its standards compliance.

The typical repertoire of value-added facilities that can be found on today's RFC 1490-based FRADs include the following:

- Segregation of SNA/APPN and non-SNA traffic onto separate sets of VCs—This separate VC approach has its roots in the 3745 serial port approach for interfacing to FR. ACF/NCP 7.1 only supported SNA/APPN traffic. Hence, only SNA/APPN frames could be sent to a 3745 via a serial port. This precluded the possibility of using the same set of VCs for transporting both SNA/APPN and non-SNA traffic across the WAN. If a LAN-based solution is being used to interface a mainframe gateway there is really no need for using a separate set of VCs for SNA/APPN. The LAN-attached bridge/

router or FRAD, as shown in Figure 6.1, will take all the traffic off a single VC and put it on the LAN. The mainframe gateway can then just pick out the SNA/APPN or TCP/IP traffic destined for the mainframe of the LAN. Using the same set of VCs for all the traffic will obviously reduce costs. FR providers typically charge at least $5/month for each additional VC assigned to a DTE. In large networks this charge for additional VCs can start to add up.

- SDLC-to-LLC:2 conversion
- Comprehensive support for non-SDLC link protocols including 3270 BSC, 2780/3780 RJE BSC, Async, Burroughs Poll/Select, and so on
- Local LLC: 2 acknowledgment, or more likely *LLC:2 acknowledgment optimization—Optimization* does not eliminate sending the end-to-end acknowledgments given that these acknowledgments do provide assurance of frame delivery. Instead the FRAD will use a relatively large acknowledgment window size of around 16 or 32, irrespective of the window size being used by the end SNA/APPN applications—given that the LLC:2 protocol used within RFC 1490/FRF.3 supports window sizes as large as 127 (i.e., modulo-128). Such acknowledgment optimization ensures guaranteed delivery while at the same time minimizing the number of acknowledgments sent across the WAN.
- Traffic prioritization and bandwidth allocation, albeit just on the physical access link to the FR WAN—Once frames are within the FR WAN, the FRADs, obviously, cannot control how they are transported within the WAN.
- Ability to dynamically move traffic from a primary VC to a predesignated backup VC if there is a failure on the primary VC. This VC-based dynamic alternate routing function augments the WAN dynamic alternate routing that is supposed to be provided by the FR network.
- Ability to use multiple parallel physical access links between the FRAD and the FR WAN with load-balancing across these links and the automatic transfer of traffic from one failed link to another
- Support for voice over FR with either separate dedicated VC for the voice traffic or voice traffic being sent on the same VC as multiprotocol data traffic, with an automatic prioritization scheme that always gives voice traffic precedence over data traffic. (Data traffic is also typically segmented into 1,500-byte or smaller frames to ensure that there is no undue delay waiting for a data frame to be sent.) Some of these FRADs can now provide acceptable quality voice traffic using 16Kbps or less. Thus, it is possible, at least in theory, to have voice coexist with data over a 56Kbps VC.

6.6 COMPARING RFC 1490 WITH DLSW AS A TRANSPORT SCHEME FOR SNA/APPN

The key differences between RFC 1490/FRF.3 and DLSw when it comes to transporting SNA/APPN traffic are restricted to three specific areas. These three areas are: encapsulation overhead, dynamic alternate routing, and the ability to support large networks (i.e., scalability).

The 10-byte header-based RFC 1490/FRF.3 native mode encapsulation scheme for SNA/APPN traffic is as lean and mean as it gets. (IBM's BAN variant with its additional 16 bytes of MAC/LLC framing is, unfortunately, not as lean.) The RFC 1490/FRF.3-based native mode encapsulation overhead is roughly one-fourth that of DLSw, and nearly one-sixth that of RSRB. This low overhead, in the context of the hundreds of thousands, or even hundreds of millions, of SNA/APPN message units that typically flow across a WAN on a given day, can make a significant difference to network data usage charges relative to bandwidth, application response times, and overall network congestion.

When it comes to SNA/APPN, the predominant mission of both DLSw and RFC 1490 is to ensure that SNA/APPN traffic is safely and successfully transported, end-to-end, across a multiprotocol backbone WAN alongside non-SNA traffic. Both techniques rely on an encapsulation technique to realize this goal. RFC 1490/FRF.3 encapsulates SNA/APPN PIUs directly within FR frames with just a 10-byte FRF.3 header prefixing each SNA PIU, as shown in Figure 6.6. With RFC 1490 there are no other additional protocols or headers involved. RFC 1490 is thus a Layer 2-only, "ultra-light" encapsulation technique.

In marked contrast to the 16-byte overhead of RFC 1490 (i.e., 10-byte RFC 1490 header plus 6 bytes of FR framing), the Layer 3 overhead of DLSw, once an end-to-end TCP connection has been established, is 56 bytes as shown in the previous chapter. If DLSw is being used over FR, this overhead grows to 64 bytes, that is, 6 bytes of standard FR framing plus 2 bytes RFC 1490 header for IP traffic. This fourfold additional overhead of DLSw per SNA/APPN PIU, in practice, provides only a few tangible advantages over RFC 1490. These advantages are the dynamic location of remote destinations and built-in local LLC:2 acknowledgment. The former as previously discussed is of limited appeal given that most SNA/APPN networks do not contain that many destinations. The latter features is invariably available with most RFC 1490 solutions in some form or another. Thus in general it is hard to defend the additional overhead of DLSw by claiming that this overhead is compensated for by additional functionality.

The RSRB Layer 3 overhead is even greater, since Cisco rather than encapsulating just the SNA PIU encapsulates a complete MAC/

LLC frame à la BAN. Figure 6.8 compares, to scale, the framing structure of RFC 1490/FRF.3, IBM's BAN, DLSw, and RSRB. The minimum Layer 3 overhead of RSRB when used with TCP/IP is likely to be around 81 bytes.

The view that networking overhead is of little concern, given that networks are becoming increasingly faster and that bandwidth is continuing to become more cost-effective, has to be tempered with some common sense. Unnecessary overhead just saps bandwidth, increases overall network congestion, undermines network performance, and impacts application response times. Minimizing overhead and thus increasing the total useful bandwidth available for data transfer can help enterprises, particularly those involved in transaction processing, in two distinct manners. For a start, more work (e.g., processing more transactions or transfering more data) can be done with a given amount of bandwidth. In addition, cutting down overhead will also reduce the frequency at which additional bandwidth may be needed to cope with network growth. The other advantage that customers garner by reducing overhead and thus increasing network efficiency is a marked reduction in lost opportunity costs—that is, business or revenue lost by not being able to service clients fast enough. Reservation systems (e.g., airline, hotel, car rental) are classic examples of applications that have a tangible lost opportunity cost directly tied to how quickly transactions can be processed—that is, the number of clients that can be successfully serviced in a given period of time.

6.6.1 Dynamic Alternate Routing and Scalability

Apart from the disparity when it comes to encapsulation overhead, the other difference between DLSw and RFC 1490 relates to the sensitive issue of dynamic alternate routing. There has been a school of thought that advocated that DLSw should always be deployed on top of RFC 1490 to add value (e.g., Local LLC:2 acknowledgments) and resilience (i.e., dynamic alternate routing) to the FR solution. This is not a fair assessment. It is true that the potential for dynamic alternate routing is a key feature of DLSw given its IP roots. However, as discussed in the previous chapter, DLSw's rerouting capability cannot handle any failures to destination (i.e., central site) bridge/routers. This inability to reroute around failed destination bridge/routers is caused by DLSw's inherent dependence on end-to-end TCP connections.

RFC 1490 per se does not address any issues related to dynamic alternate routing. However, nondisruptive, transparent dynamic alternate routing is typically a standard feature of most FR WANs. In addition, most public FR services are built around contemporary digi-

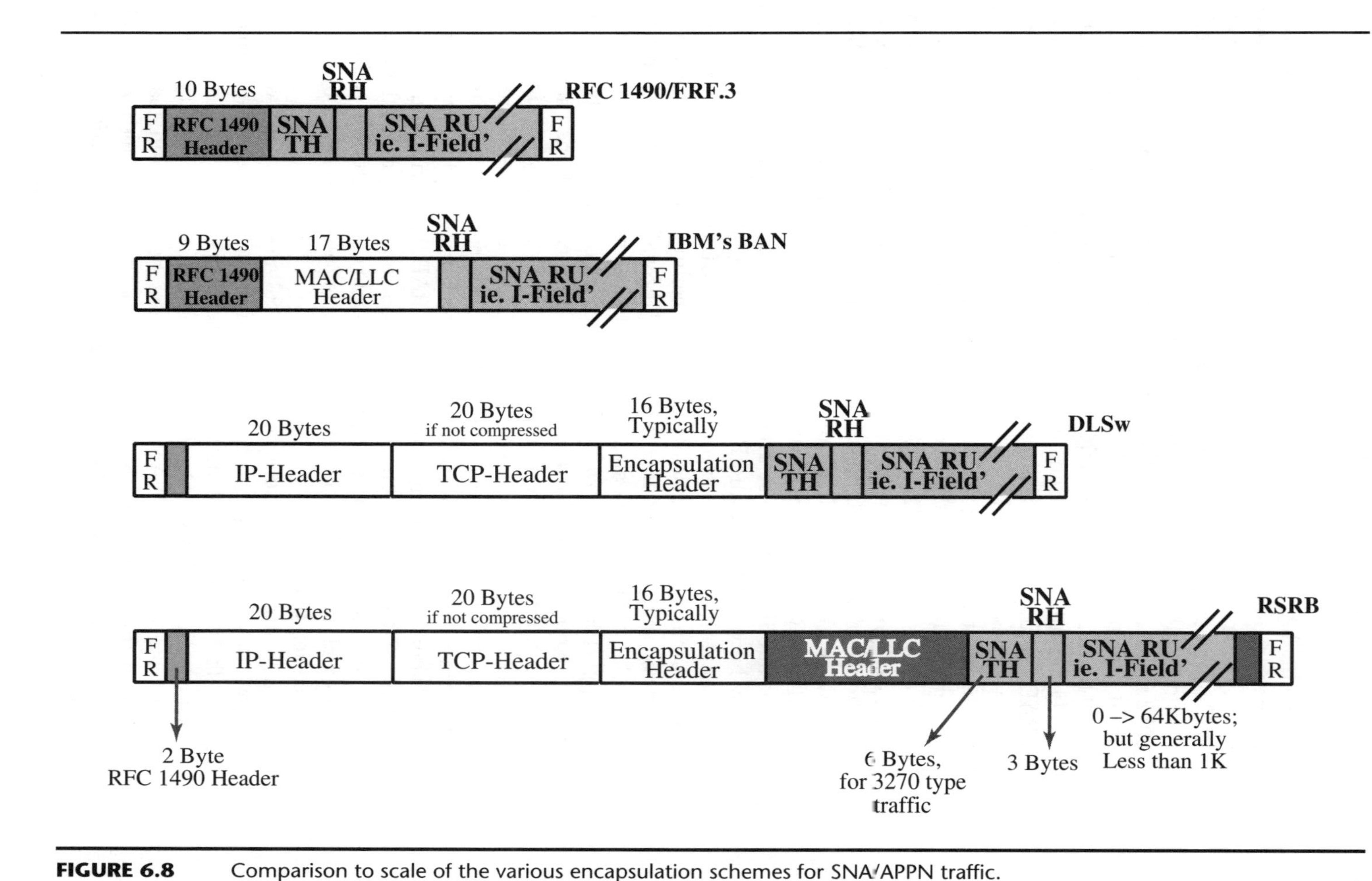

FIGURE 6.8 Comparison to scale of the various encapsulation schemes for SNA/APPN traffic.

tal technology and fiber-based links. Hence, they tend, in general, to be more robust and less prone to errors, either at the switching nodes or on the transmission links, than previous networks. If this level of WAN-provided fault-masking is not enough, most FRADs offer other network independent schemes for bypassing failed routes. Many now offer dial backup (over ISDN in some cases) while others also offer the ability to automatically switch traffic from a failed VC to another.

This VC switching capability could in theory be used to nondisruptively reroute traffic from a failed destination FRAD to a backup FRAD—a major coup vis-à-vis DLSw. Moreover, a few FRADs act as FR switches (i.e., FR DCEs) and thereby further augment the WAN transparent rerouting facilities by offering even more options for VC switching independent of the WAN. Most also support the use of parallel access links, à la SNA TGs, to the FR WAN, or provide dial backup. Thus, in general, it is fair to say that the robustness and resilience of RFC 1490-based solutions is at worst similar to that of DLSw—with a possibility that in some scenarios that it may even be significantly more sophisticated.

Scalability is the last area where there could be a significant disparity between DLSw and RFC 1490 implementations. As seen in the last chapter, DLSw's prerequisite of two TCP connections per bridge/router tends to automatically impose a ceiling as to the size of large networks that can be supported using DLSw. Today this ceiling is in the region of 500 remote sites. RFC 1490 does not use TCP connections to support SNA/APPN traffic. Hence, it does not, fortunately, run into any problems related to the number of TCP connections that have to be maintained. However, processing and memory resources are required in FRADs to support FR VCs. Thus, it is possible to envisage scalability issues at central site FRADs, analogous to that imposed by TCP connections, as the size of the network grows and many hundreds of VCs home in on a single destination FRAD. Deploying multiple central-site FRADs is obviously one solution to getting around VC overload. The same could be done with DLSw bridge/routers to reduce the number of TCP connections that will have to be sustained by a given destination bridge/router.

The bottom line here is that RFC 1490 does not in any way usurp DLSw. Both these network consolidation techniques have a role to play albeit in different networking scenarios. RFC 1490 is only an option with FR WANs. On the other hand DLSw can be very successfully used over X.25 WANs or networks built around leased lines running the internet point-to-point protocol (PPP). Another straightforward rule of thumb for deciding between DLSw and RFC 1490 relates to the protocol mix to be used over the WAN. DLSw should be seriously considered only if the network has some native TCP/IP traffic. Otherwise, TCP/IP and all of its administrative and bandwidth usage overhead will be introduced into the network just to support DLSw. This would be a case of the tail wagging the dog.

6.7 MINIMIZING THE POSSIBILITY OF SNA/APPN FRAMES BEING DISCARDED

FR reserves the right to deal with any unexpected network congestion by arbitrarily discarding packets—though it does endeavor not to discard frames that are within a VC's CIR quota. The discarding of any SNA/APPN traffic-bearing frames by an FR WAN can result in major fluctuations of SNA/APPN application program response times.

Discarding SNA/APPN message units, due to congestion or any other reason, is an alien and anathemic concept to the SNA/APPN community. All SNA/APPN interactions are session-based (i.e., connection-oriented). Thus, with a given session all message units are delivered end-to-end, error-free, and in sequence. The loss of a message unit always results in the disruption of end-to-end data flow while retransmission processes are invoked to recover the lost message unit and resequence the data flow. This disruption of data flow during a retransmission process will invariably be reflected in terms of altered response times.

The discarding of SNA/APPN frames to alleviate congestion will not typically result in data loss or transaction errors. Higher-level SNA sequence numbers at Layer 3 and 4, as well as application level transaction numbers, ensure that any SNA/APPN message units discarded by the network are successfully retransmitted and recovered. Such retransmissions, however, though ensuring data integrity, are not conducive to efficient networking.

SNA and APPN, in common with TCP/IP, does not support the notion of selective rejects—that is, the ability to request the retransmission of a single discarded or damaged packet/frame. Instead, SNA, APPN, and TCP/IP only have a mechanism to acknowledge the packets up to the first packet discarded or damaged. Consequently all the packets starting from the discarded or damaged packet all the way up to the end of the acknowledgment window cycle have to be retransmitted even if all the other packets following the one that was discarded (or damaged) had been successfully received at the other end in the first place.

Take an example where an IBM 3745 communications controller transmits seven SNA message units, numbered 1 to 7, each 512 bytes long, to a 3174 across a network. Now assume that due to congestion, the network discards message unit 3. Message units 1 and 2, as well as 4 to 7, are successfully received by the 3174. However, it cannot request that just message unit 3 be retransmitted. Instead, all it can do is to acknowledge the successful receipt of message units 1 and 2. The 3745 now has no option but to resend message units 3 to 7. Now recall that message unit 3 was discarded to alleviate congestion.

Retransmitting previously delivered message units would add to the congestion. The amount of retransmissions required will depend on two factors: the size of the original acknowledgment window (e.g., 8, 14, or even 128), and the position within this window of the first discarded packet. A larger number of packets will have to be retransmitted the larger the window size is (e.g., window size of 128 versus 8) and the earlier within the window that the first discarded packet appears (e.g., first packet in a window of 8 as opposed to the seventh packet in a window of 8). Such additional retransmissions could cause other message units within the block that is being retransmitted to be discarded. This, obviously, would result in more retransmissions, which could exacerbate the congestion further. Moreover, all these retransmissions will adversely impact response times, overall network performance, and efficiency, while all the time chewing up expensive bandwidth that could have been used for productive data transfer.

Discarding packets, in particular SNA packets, must no longer be an inevitable fact of life of multiprotocol LAN/WAN networking. Fortunately, overengineering the network with costly excess bandwidth to minimize the potential for congestion is also not the only, or even the most optimum, way to avoid packets being arbitrarily discarded. Conventional bridge/routers and FRADs can virtually eliminate the danger of packets being discarded—in the context of FR WANs. To do so, they have to first determine the various CIR thresholds of the FR network either by a dynamic network management query process, or via certain installation parameters defined when implementing the network. Once these CIR thresholds are known the bridge/routers or FRADs can then diligently meter the traffic, and in particular SNA/APPN traffic, being sent to the WAN to make sure that none of the CIR-based excess bandwidth thresholds are exceeded.

By thus keeping the traffic volumes below CIR the risk of having SNA/APPN traffic-bearing frames discarded by the network can be greatly minimized. Minimizing the number of SNA/APPN frames that are discarded ensures consistent and predictable application program response times.

Frame relay, in the form of its built-in and mandatory implemented bandwidth provisioning facilities such as CIR, Committed Burst Size (B_c), and Excess Burst Size (B_e), provides all the necessary control mechanisms to avoid congestion—and consequently the dropping of packets. Moreover, FR with its Local Management Interface (LMI) and ANNEX D-based network management schemes furnishes the relevant protocols whereby a central site bridge/router or FRAD can dynamically obtain the CIR, B_c, and B_e of *all* the end-to-end VC.

Governing (or metering) the traffic transmitted to the FR network so that it does not exceed CIR is also not contingent on the bridge/router or FRAD having large amounts of RAM for data buffering. Instead, the bridge/router or FRAD could use flow control techniques to govern the rate at which it accepts *input* data from its LAN and serial (e.g., SDLC or Async) ports. Now it can balance its data input rate against the nondiscard output rate to the FR network, thus minimizing the amount of data it has to buffer while awaiting bandwidth availability on the output side.

Controlling the rate at which a bridge/router or FRAD has to accept input data is particularly easy when dealing with SNA/APPN, Asynchronous, or BSC data. LLC:2 and SDLC, the data link control protocol schemes used by SNA/APPN, both include a built-in, Layer 2-specific flow control protocol that is administered using the RR and RNR commands. This RR/RNR protocol, though simple, is very powerful and functional. RNR, issued from a bridge/router or FRAD to a specific SNA device (e.g., PC on a LAN or a 3174 on a SDLC link), will temporarily stop it from transmitting SNA packets—until it receives a RR. Thus, by using RNR and RR, a bridge/router or FRAD can easily control the rate at which it receives input SNA packets. Async and BSC have comparable flow control mechanisms based respectively on XON/OFF and Temporary Text Delay (TTD) protocols.

With an FR solution it will be quite common for the CIRs of the VCs going to the individual remote sites (e.g., branch offices) to be less than the speed of the physical access links used between the remote sites and the frame relay network. For example, all the access links may be 56Kbps links. The VCs being used, however, may only have a CIR of 16Kbps. In many cases these 16Kbps CIR VCs would be replacing 9.6Kbps leased lines. There may not be a need for more bandwidth. Moreover, a VC with a CIR of 56/64Kbps will cost three times more than a VC with a CIR of 16Kbps. If 16Kbps will suffice, there is little merit in paying more for a larger CIR. When a CIR of a VC is smaller than that of the access link it uses—for example, a VC with a 16Kbps CIR using a 56Kbps access link—there is always a danger that the CIR of the VC will be exceeded given that data is transmitted over the link at the speed of the link (e.g., 56Kbps)—rather than at the CIR rate (i.e., 16Kbps).

This is where the output bandwidth limiting steps in. A bridge/

router or FRAD that is using frame relay can obtain the CIR, B_c, and B_e for each of the VCs it is supporting in a variety of different ways. It could use the network management LMI "STATUS ENQUIRY/STATUS" protocol to dynamically obtain these parameters from the frame network on a per VC basis. It could also use the frame relay ANNEX D network management scheme—if appropriate with vendor-specific extensions—to solicit this information directly from the remote bridge/routers or FRADs. The third option would be to have these parameters configured during the installation process. Irrespective of the method used, the central site bridge/router or FRAD, as well as all the remote site bridge/routers and FRADs, can gain access to the relevant CIR and B_c pertaining to the VCs that they have to support.

If there is a significant increase in the volume of traffic in this type of no-SNA/APPN-frame-discard network there will be a slight corresponding increase in response times—reflecting the extra time taken to deal with the additional traffic without violating the bandwidth limits. Thus, if response times were previously .5 second they may now rise to .6 or even .8 second—not the 3.5 seconds or greater that would have occurred if packets were being discarded.

6.8 REFLECTIONS

FR is probably the best thing to happen to IBM-centric networking since token-rings appeared in 1986. FR provides a plethora of ways to dramatically slash costs in IBM-centric networks. In many cases FR will provide the crucial, central core WAN for consolidated mutiprotocol LAN/WAN networks. Though based on an FR fabric this parallel network consolidation may be realized using multiple disparate technologies—with TCP/IP encapsulation à la DLSw(+), RFC 1490/FRF.3 encapsulation, and APPN NN routing being the three leading contenders. RFC 1490/FRF.3 encapsulation, with its 10-byte header is a low-overhead, highly efficient means for transporting SNA/APPN and NetBIOS traffic across an FR WAN. In the context of FR WANs, RFC 1490-based solutions can invariably be as good as, if not better than, DLSw-based solutions. Given that RFC 1490's encapsulation overhead is one-fourth that of DLSw, RFC 1490 solutions will invariably be more efficient when measured in terms of the ratio of data transported to the amount of overhead used to realize that volume of data transport. FR-based FRADs also excel in supporting non-SDLC link protocols. Some already offer voice-over-FR. FR, thus, can be viewed and exploited as the springboard to full-blown ATM.

CHAPTER 7

High Performance Routing: The Latest Incarnation of SNA

A QUICK GUIDE TO CHAPTER 7

High Performance Routing is the latest incarnation in the SNA family of architectures. It, as opposed to APPN, is destined to inherit SNA's mantle and propagate SNA's heritage of value-added, mission-critical networking into the next millennium. It boasts an impressive list of credentials to justify its right to be SNA's true successor.

Its indubitable strengths include: very low-fat, Layer 2 routing; dynamic alternate routing; state-of-the-art anticipatory congestion control; and elimination of intermediate node routing in addition to all of APPN's plug-and-play features. It is the first networking scheme from IBM to support dynamic alternate routing. Some of the protocols introduced with HPR are used by IBM's Networking BroadBand Services architecture, which can be thought of as an SNA for ATM-centric networking.

This chapter is an all-inclusive, total-immersion, no-holds-barred guide to HPR. After a brief overview of HPR, including the intriguing notion of multiprotocol HPR, the chapter starts off in earnest with a complete list of all of the networking facilities offered by HPR. HPR is unashamedly eclectic. It borrows concepts and methodologies from SNA/APPN, IP, and frame relay. The derivations of the quintessential features of HPR are also noted within this section.

Section 7.2 examines the anatomy of HPR and describes its three constituent functional components: Automatic Network Routing, Rapid Transport Protocol, and Adaptive Rate-Based Congestion Control. This section also deals with coexistence between HPR and SNA/APPN given that HPR per se only comes in the form of Network Nodes. Consequently the end nodes supported by HPR will end up being APPN or SNA nodes.

In keeping with its eclectic nature, HPR uses APPN's broadcast search mechanism to dynamically locate unknown LUs. Section 7.3 outlines the dynamics of this search process. It also describes how APPN ENs dynamically register their LUs with an HPR/APPN NN.

Section 7.4 builds on the network consolidation theme that has spanned all six of the previous chapters. DLSw(+), RFC 1490, and HPR/APPN NN routing are the three most viable options for such consolidation. Now that the characteristics of HPR have been enumerated, the pros and cons of HPR NN routing are compared to those of DLSw(+) and RFC 1490.

Ironically, many of HPR's strengths, such as dynamic alternate routing and congestion control, are nullified if HPR NNs are deployed around a FR WAN. This crucial issue of the merits of running HPR across a FR WAN is hammered out in Section 7.4.2.

In general, HPR, again attesting to its eclectic origins, has but a few weaknesses—and most of these are esoteric. Nonetheless, this chapter concludes by describing all of the idiosyncrasies that can be deemed to be weaknesses in HPR. This chapter also includes two sidebars: one on the Dependent LU technology required to ensure unrestricted support for SNA traffic vis-à-vis APPN/HPR, and the other on the General Data Stream used by APPN and HPR.

High Performance Routing, despite a name evocative of bridge/router technology, is the latest blessed descendant in the once stately family of SNA and APPN network architectures. HPR is thus destined in time to become SNA's rightful heir and champion SNA's heritage of value-added, mission-critical networking into the next millennium—albeit in the context of multiprotocol networks, as opposed to HPR/SNA-only networks.

HPR's primary raison d'être is to serve as a low-overhead, nimble, dynamic, and secure transport for SNA traffic within high-speed, broadband networks. HPR was initially positioned by IBM in the early

1990s, in the guise of APPN+, as an enhancement to APPN to improve its data routing dynamics and resilience. In reality, it is much more than that. HPR represents one of the most significant and dramatic metamorphoses undergone by SNA in its eventful and evolutive history. HPR's role in reshaping SNA is on a par with some of the most noteworthy landmark extensions to SNA, for example, cross-domain routing between multiple mainframes (1976); virtual routes and transmission groups (1979); LU 6.2 (1982); and Type 2.1 peer-to-peer nodes (1983).

HPR represents a refreshing new paradigm for SNA-centric peer-to-peer networking. *It is a Layer 2-based routing scheme that in the vein of frame relay and ATM eschews processing and error-checking at intermediate nodes.* This is a major departure from previous SNA and APPN routing schemes, which, in addition to being done in Layers 3 and 4, required considerable table lookups and processing at each intermediate node.

HPR is also the first major IBM networking scheme to include nondisruptive dynamic alternate rerouting in the event of path failure—hitherto a frustrating, inexcusable, and glaring omission in APPN, source-route bridging, and SNA. HPR also boasts a congestion control mechanism that works by trying to anticipate and avoid congestion from setting in, rather than trying to detect and dissipate it after the congestion has set in. All of these new value-added features are built on top of and totally integrated with the user-friendly plug-and-play networking infrastructure (e.g., automatic end-node registration and dynamic location of remote LUs) provided by APPN.

The first implementations of HPR were available in mid-1995 on ACF/VTAM Version 4 Release 3 and ACF/NCP Version 7 Release 3. The initial potential of HPR, with these two software products, is, unfortunately, restricted to VTAM-to-VTAM, mainframe-to-mainframe interactions such as LU 6.2-based file-transfer or distributed transaction processing between two CICS applications. HPR implementations on IBM AS/400s, OS/2, 3172s, and 2217 Multiprotocol Concentrators are likely to be readily available in 1996—as enhanced versions of the previously available APPN NN software. In addition, all vendors that now offer APPN NN routing on their multiprotocol bridge/routers (e.g., Cisco, Bay Networks, 3Com, IBM) have already stated that they are committed to upgrading their APPN software to encompass HPR during 1996.

The availability of HPR on such multiprotocol bridge/routers will presage HPR's predominant role in the future. This role is to provide a feature-rich, highly efficient Layer 2-based routing scheme that enables SNA traffic to be easily and effectively integrated alongside non-SNA traffic within bridge/router or ATM switch-based multiprotocol networks.

Figure 7.1 depicts a multiprotocol LAN/WAN network built around

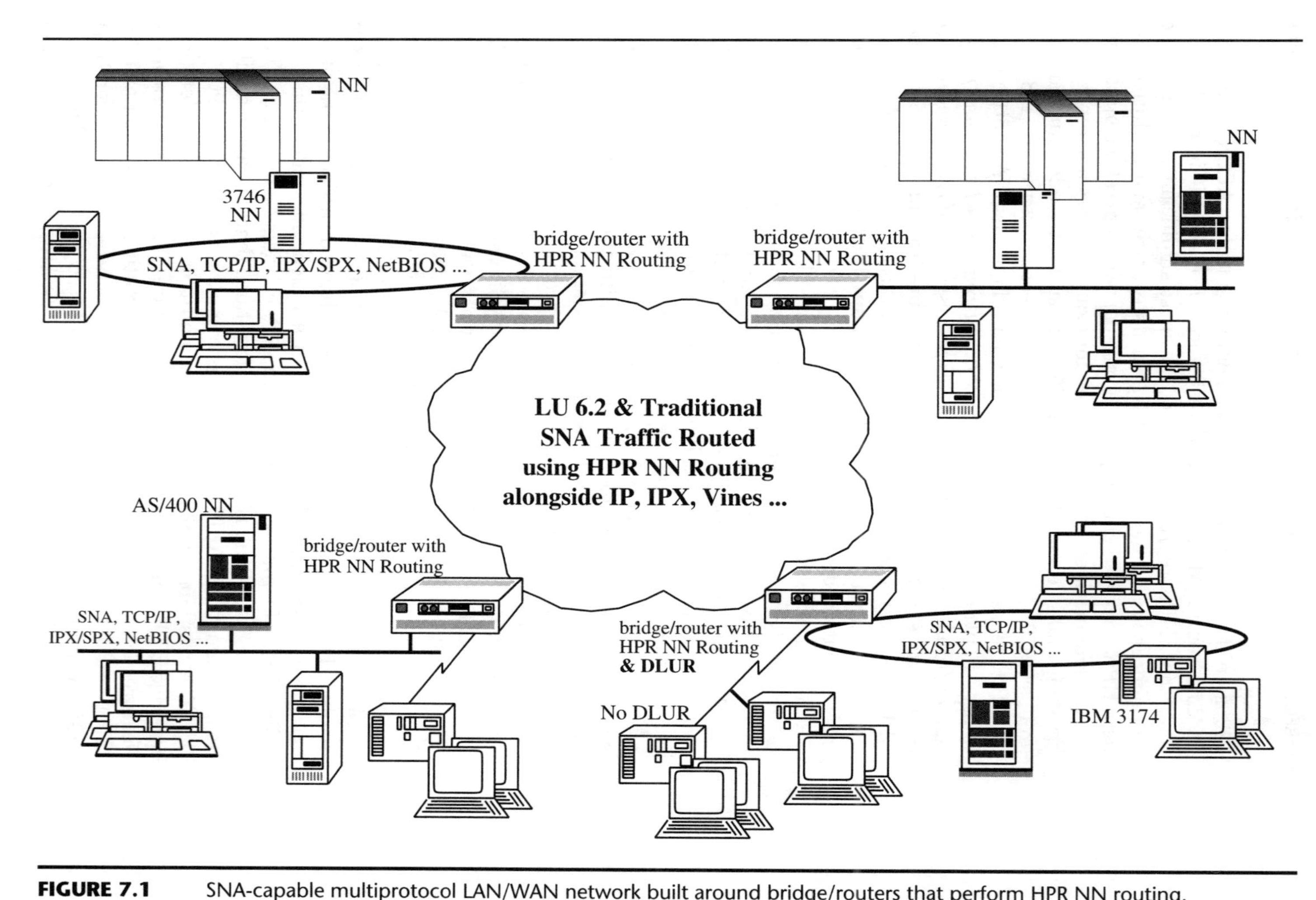

FIGURE 7.1 SNA-capable multiprotocol LAN/WAN network built around bridge/routers that perform HPR NN routing.

bridge/routers that support HPR NN routing. Note that the configuration shown in Figure 7.1 would be the same if the bridge/routers were using APPN NN routing instead of HPR NN routing. Facilitating bridge/router or ATM switch-centric SNA-capable multiprotocol networking, as shown in Figure 7.1, can be thought of as the somewhat conventional, APPN NN replacement role for HPR vis-à-vis high-speed, multiprotocol networking.

IBM, however, surreptitiously envisions another beguiling and unconventional role for HPR in the context of multiprotocol networking. In this other role, HPR will become the basis for a new generation of value-added, highly reliable enterprise WANs. HPR will be the controlling protocol for these WANs, and SNA will be the only type of traffic routed across them. Non-SNA traffic will either be converted to, or encapsulated within, SNA LU 6.2 message units by an AnyNet Gateway (e.g., IBM 2217) as it enters the HPR WAN. Another AnyNet Gateway at the other end of the WAN will either reconvert the traffic to its original form, or extract the non-SNA traffic from the SNA LU 6.2 message units.

This is the AnyNet Gateway approach to multiprotocol networking across a single protocol WAN—where in this instance that WAN protocol is HPR. Figure 7.2 illustrates the concept of a AnyNet Gateway-based multiprotocol network built around a single-protocol HPR WAN, while Figure 7.3 highlights how the single-protocol HPR WAN approach of multiprotocol networking using AnyNet Gateways is markedly different from the traditional bridge/router HPR NN routing approach. *Note that the AnyNet Gateway-based approach of transforming other protocols to HPR in order to transport them across a WAN is essentially the antithesis of DLSw(+).* DLSw(+) encapsulates SNA/APPN traffic with TCP/IP packets in order to transport them across a WAN, while the AnyNet/HPR converts TCP/IP traffic to HPR to route it across a WAN.

HPR is a more compelling single-protocol WAN for this AnyNet approach to multiprotocol networking than APPN. HPR excels over APPN in providing a Layer 2-based rapid transport routing scheme, nondisruptive alternate routing, and SNA-like parallel-link TGs (albeit in this case the availability is implementation-dependent). The AnyNet and HPR-centric approach to multiprotocol networking is discussed at length in the next chapter. IBM has been known to refer to this AnyNet cum HPR solution to multiprotocol networking as *multiprotocol HPR*.

HPR could be used across either LANs, frame relay, X.25, SDLC leased lines, or ATM. Bypassing error-checking and error-recovery at intermediate nodes is a key precept of HPR, a trait it has in common

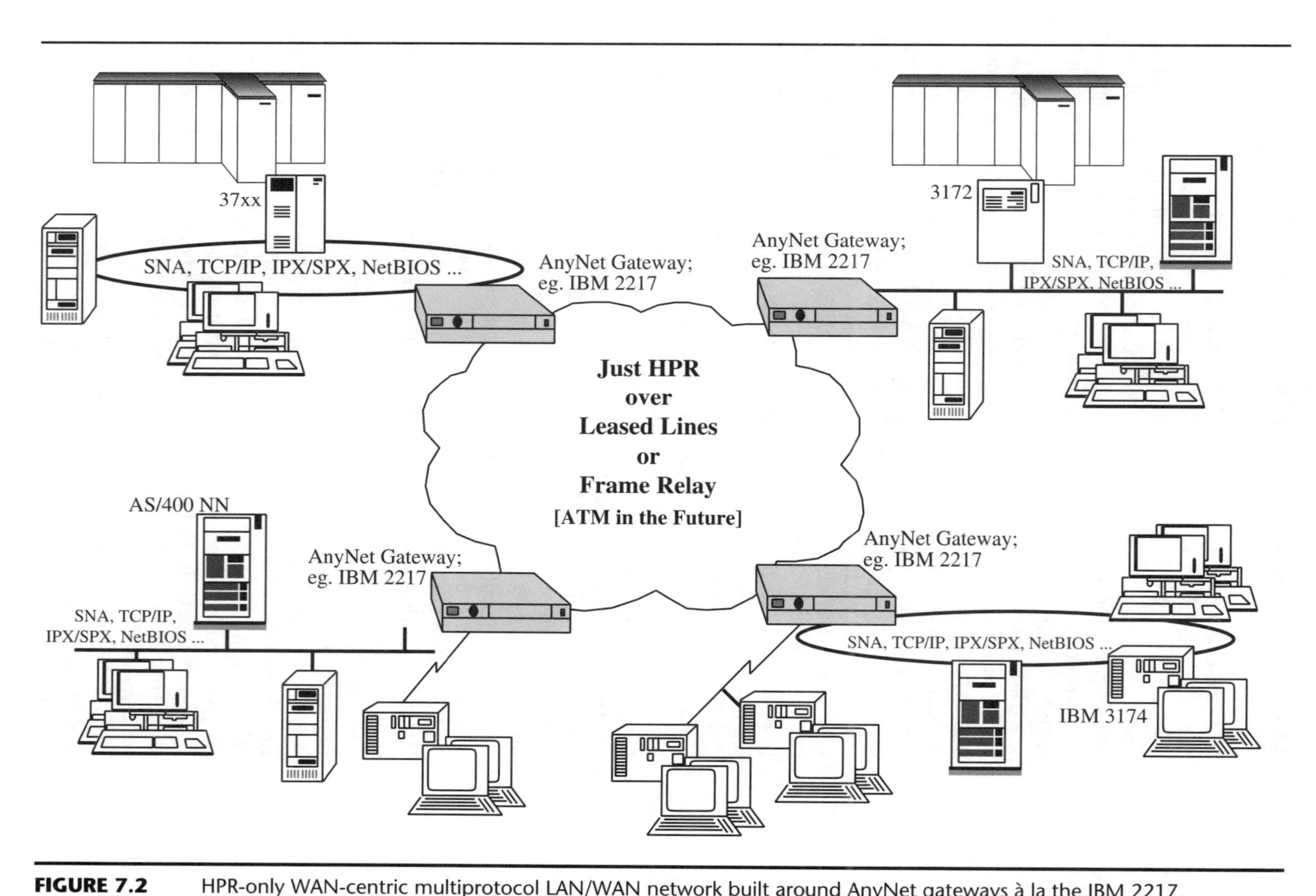

FIGURE 7.2 HPR-only WAN-centric multiprotocol LAN/WAN network built around AnyNet gateways à la the IBM 2217.

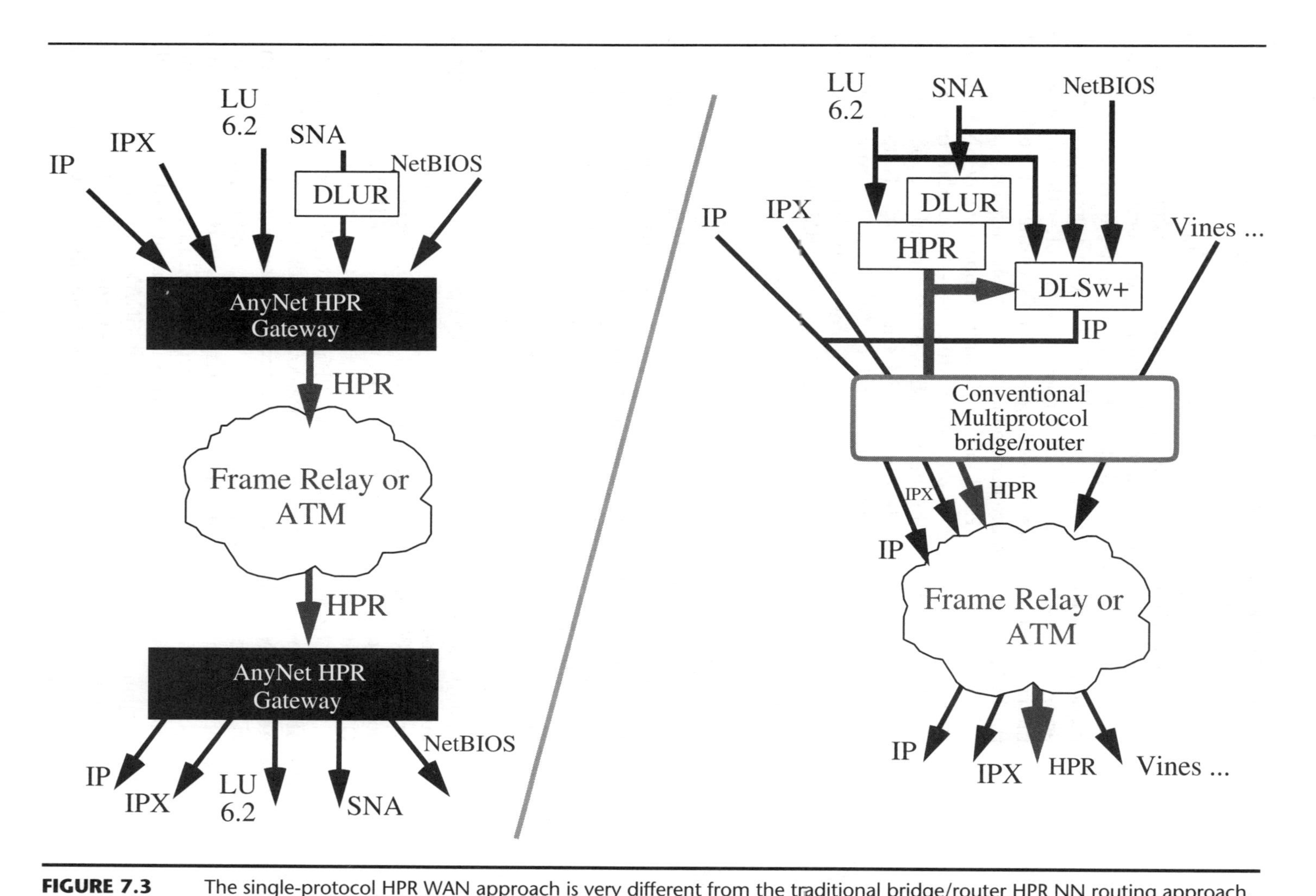

FIGURE 7.3 The single-protocol HPR WAN approach is very different from the traditional bridge/router HPR NN routing approach.

with frame relay. Consequently, wherever possible, HPR attempts to use a connectionless protocol across the links, for example, resorting to using the "no sequence number" Unnumbered Information (UI) frame capability in SDLC as opposed to the sequence-numbered I-frames mandatorily used by SNA and APPN. Though it prefers just doing error-checking at each end, HPR does, however, offer the option where intermediate-node error-checking and recovery may be performed if the links being used are deemed to be "noisy" or unreliable.

Just like APPN, HPR per se directly supports only SNA LU 6.2 peer-to-peer, program-to-program traffic. Non-LU 6.2 SNA traffic (e.g., SNA/3270 traffic), as in the case with APPN, is only supported using Dependent LU (DLU) technology. See the "DLUR in a Nutshell" sidebar. Like APPN, HPR includes architectural elements covered by IBM patents. Thus, as with APPN NNs, a license in the form of a Technology Package will be required from IBM for non-IBM implementations of HPR.

7.1 QUINTESSENTIAL HPR

With TCP/IP and SNA so well entrenched, and with APPN having masqueraded as SNA's eventual successor for the last decade, some have questioned the need for and the prospects of HPR. To be fair, SNA and TCP/IP, despite their ardent followings, are both now long in the tooth. They were both conceived in the 1970s when 19.2Kbps, as opposed to gigabits/sec, was considered broadband. While they have both gone through many function-enhancing evolutions they still have characteristics that hark back to their origins. APPN on the other hand was never diligently architected à la SNA and essentially grew out of product implementations—most notably the implementations on the S/36 and AS/400 around 1986 to 1989. Table 7.1 attempts to summarize the strengths and weaknesses of SNA, APPN, TCP/IP, and OSI in the context of HPR.

What IBM set out to do with HPR is very simple and obvious. HPR attempts to eradicate the major weaknesses inherent in SNA, TCP/IP, and APPN, while at the same time trying to leverage their strengths. HPR is unashamedly eclectic. HPR in essence is an inspired synthesis of the best features found in IP, SNA, APPN, and frame relay. HPR borrows from SNA the notion of session-based interactions that provide reliable communications with predictable and deterministic response characteristics. It also shares SNA's concern for congestion avoidance and control, though HPR uses its own state-of-the-art anticipatory scheme, which is very different from SNA's and APPN's window-size-based pacing schemes.

DLUR IN A NUTSHELL

HPR and APPN are capable of natively supporting only SNA LU 6.2 program-to-program traffic. To support traditional SNA traffic—such as SNA 3270 traffic, still the most predominant type of SNA traffic—without any restrictions being placed on the network configuration, APPN requires an extension referred to as Dependent Logical Unit Server & Requester (DLUS/DLUR).

The term *dependent* in this case alludes to the fact that these end user-serving LUs are totally dependent on ACF/VTAM on a mainframe for their directory (e.g., session setup) and Configuration (e.g., node activation/deactivation) services. In contrast, LU 6.2s can be configured to operate independently without being beholden to an ACF/VTAM. Such LUs are referred to as Independent LUs (ILUs). HPR and APPN were developed just to cater to ILUs.

DLUS/DLUR is essentially a client/server mechanism. DLUS resides within ACF/VTAM on a mainframe. DLUR is either embedded within an SNA device (e.g., 3174), or is implemented as a Surrogate DLUR Server in an APPN Node (e.g., bridge/router from Cisco, Bay Networks, or 3Com) and provides DLUR functionality to SNA devices attached to that NN.

ACF/VTAM 4.2 has had DLUS capability since mid-1994. The problem is with the paucity of SNA devices that have DLUR implemented within them. At present there are only two SNA devices with DLUR—IBM 3174 cluster controllers with Configuration Support C Release 5 micro-code and PCs running OS/2 with the latest CM/2.

If an SNA device (e.g., a PC LAN Gateway) does not have built-in DLUR there are essentially two ways in which it can be deployed within an APPN network: It can either be directly attached to a 37xx, or be front-ended by an APPN NN containing a Surrogate DLUR Server. The former option is referred to as the one-hop away configuration since the SNA device has to be logically adjacent to a 37xx. This type of "one-hop-away" configuration is too restrictive for normal HPR/APPN networking in that there can be no HPR/APPN NN between an SNA node and the 37xx.

In limited instances the role of the Surrogate DLUR Server could be provided by SNA LAN gateway PU Concentrator functionality. Cisco bridge/routers, Hypercom IEN FRADs, IBM's OS/2 Communications Manager/2, NetWare for SAA, and Microsoft SNA Server offer this type of PU Concentrator functionality.

The PU Concentrator option, however, is not as flexible as a Surrogate DLUR Server. (Note that Cisco bridge/routers offer both a PU Concentrator and a Surrogate DLUR Server.) The next release of Communications Manager/2, due by 1996, is also slated to offer a Surrogate Server capability.

TABLE 7.1 Major Strengths and Weaknesses of SNA, APPN, TCP/IP, & OSI

Strengths	Weaknesses
SNA	
• Highly reliable connection-oriented (i.e., session-based) communications • Predictable performance based on the use of predefined paths • Multilink Transmission Groups • Huge, 50,000 network strong installed base	• No dynamic alternate rerouting in the event of path failure • Hierarchical control scheme
APPN	
• Compelling plug-and-play networking • Sophisticated path selection scheme based on class of service	• No dynamic alternate rerouting in the event of path failure • No SNA-like Multilink Transmission Groups • Support for non-LU 6.2 sessions (e.g., 3270) just emerging
TCP/IP	
• Dynamic, connectionless networking • Large, ardent, and fast-growing following • Now the de facto open standard for intersystems interoperability • Basis of today's multiprotocol bridge/routers	• Following restricted to AS/400 customers • Not plug-and-play by any stretch of the imagination • Addressing scheme now running out of space and attempts to extend it ensnared in political intrigue • Low-octane congestion control mechanisms
OSI	
• Widely known, and extensively referenced, seven-layer model • X.400	• Little, or no, following! TCP/IP is now the new standard for interoperability.

The overall framework and infrastructure of HPR is based entirely on APPN. From an architectural perspective HPR could be viewed as just usurping APPN's Intermediate Session Routing (ISR) component without impinging on the rest of the architecture. Thus, HPR inherits APPN's peerless mechanisms for distributed directory-inspired plug-and-play networking, as well as COS-dictated traffic prioritization and path selection based on multiple criteria.

From frame relay, HPR copies the notion of doing all of its protocol processing at Layer 2, and doing so only at the start and end nodes, rather than at all intermediary nodes. At the same time HPR learns what true dynamic networking is all about from IP. HPR still relies on sessions à la SNA. However, these sessions are built on top of a routing scheme that is connectionless à la IP. Until HPR connectionless networking, much like dynamic alternate routing, was another networking notion that was totally alien to IBM. HPR's scheme of using sessions (i.e., connections) as a guaranteed delivery mechanism on top of a connectionless fabric is not uncommon. This what TCP is on top of IP.

The quintessential characteristics of HPR can thus be summarized as follows:

1. *All processing performed at Layer 2* à la frame relay

2. Nondisruptive, *dynamic alternate rerouting* in the event of path failure

3. Low-overhead, *connectionless internode routing* mechanism based on an SRB-like RIF that is prefixed to every HPR message unit, and specifies the path to be traversed by that message unit in terms of a hop-by-hop list of link addresses (or labels) between the various intermediary nodes

4. Link-level error-checking and recovery, as well as message unit segmentation or reassembly being performed only at the end nodes, that is, no intermediate node processing

5. *Selective retransmission* (i.e., selective reject) of failed or missing message units, as opposed to retransmitting all the message units between the one that failed or is missing, to the last message unit in the acknowledgment cycle as is the case with SDLC, LLC:2, or even TCP. (Assume that a sender transmits seven message units numbered 1 to 7 and that all but number 3 is received correctly at the other end. With selective retransmission only number 3 needs to be resent. The "go back to error" technique in use today requires that 3, 4, 5, 6 and 7 be resent even though 4 to 7 have already been received once without error.)

6. *Plug-and-play networking* of APPN, which is achieved through the synergistic combination of the following independent facilities:

- *The dynamic location of remote LUs* via a broadcast search mechanism that is augmented with cache directories à la DLSw(+) to preclude repeated searches for previously found LUs. This dynamic search mechanism is described in Section 7.3. In addition, optional central directory servers can be used to further minimize the amount of broadcast searches that need to be performed. When central directories are being used, source APPN/HPR nodes trying to locate an unknown LU would first interrogate a central directory server. If the central server does not have an entry for that LU it will conduct a broadcast search on behalf of the source node.
- *The dynamic registration of end node LUs* as described in Section 7.3.1. Such dynamic registration obviates the need to manually predefine the names of all the LUs resident in the various end nodes attached to a given APPN NN, as is the case when attaching a SNA peripheral node to an SNA subarea node. (Since mid-1992, with ACF/VTAM 3.4, IBM has offered a feature referred to as the Dynamic Definition of Dependent LUs [DDDLU], which belatedly provides dynamic peripheral node LU registration support along the lines of what is available in APPN. This DDDLU feature, though now supported by 3174s, is not widely used since most ACF/NCP or ACF/VTAM gens, with their reams of 3x74 LU definitions, were well in place by 1992.)
- *The connection network* concept that permits direct end node-to-end node data interchange on LANs, without all the traffic having to be relayed across one or more network nodes—but without the burden of each end node having to know the LAN MAC address of the other end node(s). See Appendix B. This concept of connection networks could also be very useful in certain frame relay applications. It would permit dispersed end nodes to talk to each other directly across a frame relay WAN without having to relay all message units through an NN (e.g., central 3745), and without having to know each other's Virtual Circuit DLCI addresses. Having unique addresses for each end node is a prerequisite for connection networks. In frame relay such unique VC DLCIs can be optionally obtained, albeit only in certain networks, via a feature known as Global Addressing. IBM has been repeatedly urged to extend APPN and HPR connection networks to embrace frame relay networks that offer *global addressing*. At a minimum this would require modifications to the existing APPN/HPR code in ACF/VTAM and ACF/NCP. Today, IBM has been reluctant to commit to frame relay-based connection networks.

  The optional implementation dependent facility whereby LAN resident NNs could automatically and periodically advertise their presence via a broadcast message over the LAN. Such an NN service advertising scheme (which is similar to the service advertising scheme used by Novell NetWare Servers) precludes the need to predefine at each end node the address of its NN server—even though this scheme is restricted to LAN configurations. (Broadcasting NN availability periodically across a WAN may not curry favor with many network administrators.)

7. A state-of-the-art *congestion-anticipating and avoidance scheme* based on a closed-loop feedback mechanism. This scheme continually measures the rate at which data is being delivered to a given receiver across the network and the rate at which the receiver is able to forward that data to the ultimate end user. If the rate at which data is being received and forwarded to the end user is roughly the same as that at which it is being transmitted, the receiver assumes that there is no congestion within the network or at the receiving node. On the other hand, if the receive rate or the forward rate is less than the transmit rate there is a likelihood that there is some congestion within the network or receiving node. If it detects such congestion it dynamically lowers the transmission rate until its feedback mechanism indicates that there is no longer any perceptible congestion.

8. *Multiple criteria-based, COS-oriented, path selection scheme* as found in APPN. Just as with APPN, up to nine physical link-related criteria can be specified for a given path. These nine criteria are: preferred cost to transmit a byte, acceptable connection cost per minute for a switched connection, desired amount of bandwidth, desirable propagation delay characteristics, physical link security requirements, preferred modem types, and three customer- or network-specifiable criteria. When one or more of these criteria are specified for a path that is to be established, APPN/HPR endeavors to ensure that all physical links that are to be traversed by this path conform to the stated criteria.

9. Dynamic promulgation of network configuration change information between all Network Nodes via APPN's *Topology Database Update* (TDU) scheme. This scheme for automatically adapting to any and all network topology changes (such as availability of network nodes and the links between such nodes) obviates the need for any human intervention each time an NN or link between NNs fails or is reactivated.

10. COS-based *transmission priority* (i.e., high, medium, or low) assigned to the traffic associated with each session à la APPN

11. Optional capability to automatically disconnect switched connections if they have not been used (i.e., no sessions or traffic across) for a predesignated amount of time. This useful feature, which is also available on SNA and APPN, saves switched connection costs by eliminating the possibility of nonproductive switched connections. This feature is referred to by IBM as *Limited Resource Link Deactivation*.

12. *Supporting SNA-like parallel-link TGs to facilitate load balancing and safeguard against single link failures* was an original goal of HPR. As with APPN it is now inexplicably absent from the official architecture. Certain HPR implementations are, however, committed to supporting TGs. It is also likely that the architecture will be updated at some point to embrace this valuable and quintessential SNA facility.

7.2 THE ANATOMY OF HPR

HPR adds three new functional components, all at Layer 2, to the APPN Network Node architecture. In addition, it replaces the APPN NN Control Point (CP) and the ISR component with an HPR Control Point. ISR, in APPN, is responsible for handling intermediate routing between APPN NNs. This function is totally superseded by HPR.

Note that HPR-only impacts APPN NNs. HPR NNs replace for APPN NNs. HPR functionality is not included within end nodes—at least not as yet. Instead, HPR NNs work seamlessly with APPN end nodes, Type 2.1 peer-to-peer nodes, and use DLU technology with SNA nodes. In each case, HPR NNs appear to end nodes as either APPN NN or Type 2.1 nodes. HPR node awareness is thus restricted just to other HPR nodes.

The three new Layer 2 functional components that make up HPR are as follows:

1. Automatic Network Routing (ANR)
2. Rapid Transport Protocol (RTP)
3. Adaptive Rate-Based (ARB) Congestion Control

Figure 7.4 shows how the HPR components relate to the APPN NN architecture, while Figure 7.5 shows how replacing the Layer 3 ISR with HPR's Layer 2-based routing scheme reduces the amount of data forwarding related processing that has to be done at each intermediate node. In Figure 7.4 note that traditional Layer 2 protocols such as frame relay, LLC:2 and SDLC occur underneath ANR from HPR's perspective.

7.2.1 Automatic Network Routing

ANR is a very low-overhead, connectionless routing mechanism for rapidly forwarding message units along a predetermined path. ANR uses a RIF, reminiscent of SRB, and now referred to as an ANR Routing Field (ANRF), to denote the path along which a given message unit is to be forwarded. Every HPR packet is prefixed by an ANRF that occurs within the Network Layer Header (NHDR) of an HPR Packet.

Figure 7.6 shows the structure of an HPR Network Layer Packet

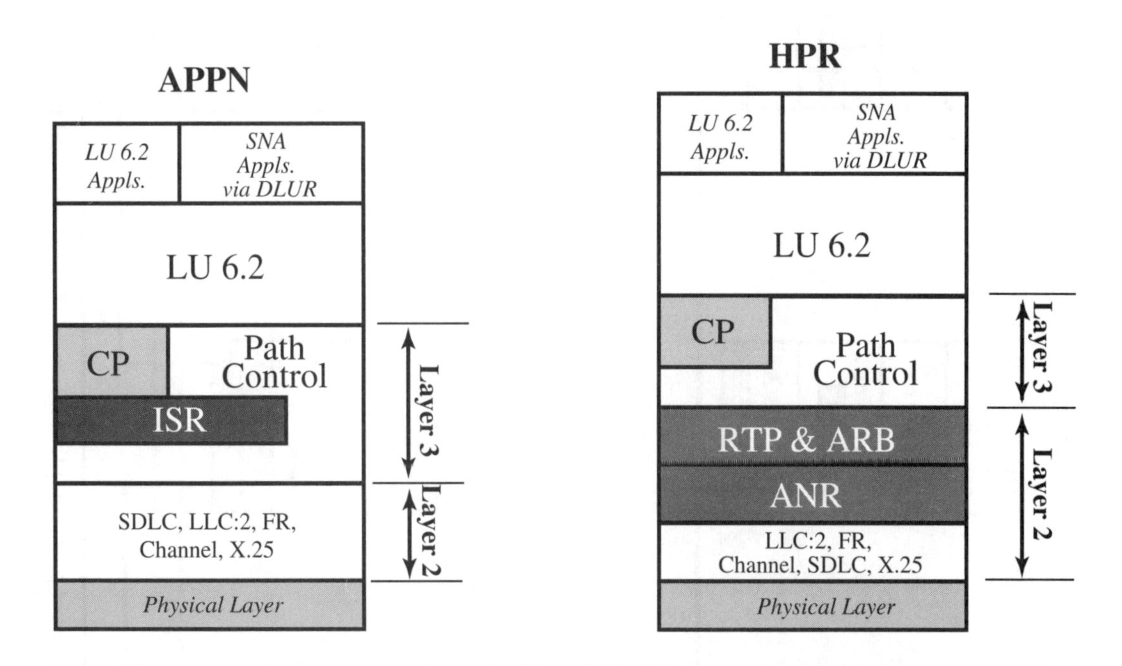

FIGURE 7.4 Comparison of the HPR NN protocol stack with that of APPNs.

(NLP) which starts with the ANR NHDR header. NLPs are only used between HPR nodes. Conventional SNA/APPN PIUs, prefixed with FID Type 2 (FID-2) Transmission Headers (THs) are used between end nodes and HPR NNs. Between HPR nodes that are using RTP NLPs are used to convey LU-LU session traffic, CP-CP session traffic and route setup messages. NLPs and FID-2 TH-prefixed PIUs may be transmitted across the same link. PIUs with FID-2 and NLPs can be differentiated by the settings they use on the very first 4 bits of their frames. An FID-2 TH will always have a bit setting of b'0010' (i.e., decimal 2) while the first 4 bits of NLSPs will never be set to b'0010'.

The NHDR and RTP Transport Header (THDR) that both occur at the start of an NLP, as shown in Figure 7.6, now occur in front of the SNA TH—which hitherto, both with SNA and APPN, was the first header in all SNA message units. In addition, HPR uses a new TH—the so-called FID-5. Figure 7.7 shows the format of an FID-5 TH and compares it to the FID-2 TH used by APPN and SNA Type 2 peripheral nodes. The FID-5 TH is the first new TH in this arena since the FID-4 and FID-F THs were introduced way back in 1978 to support SNA Virtual Routes and Transmission Groups. The FID-5 TH that

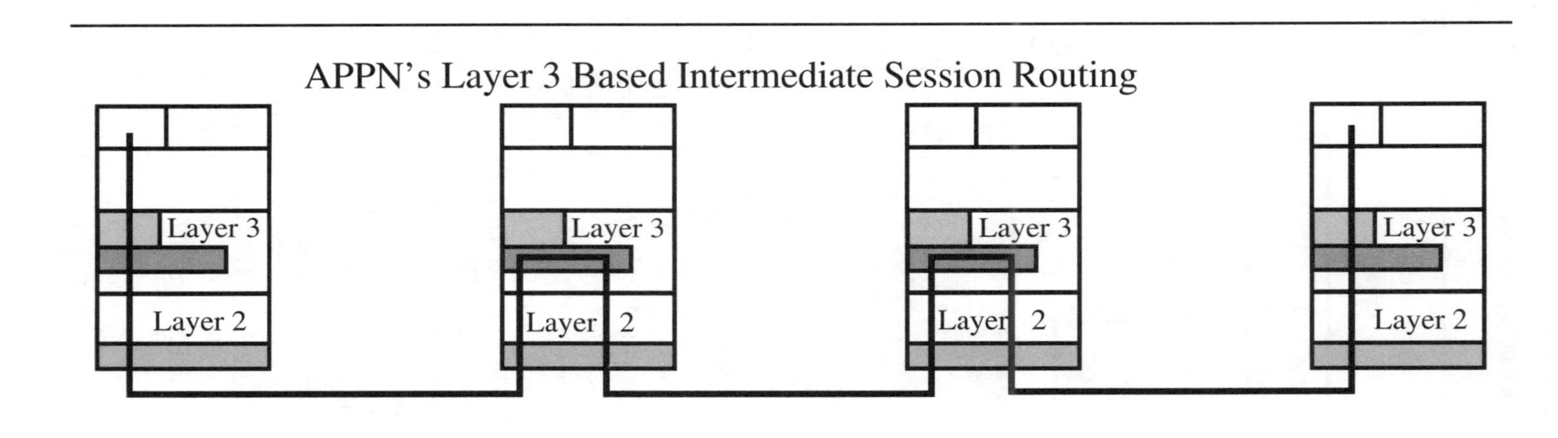

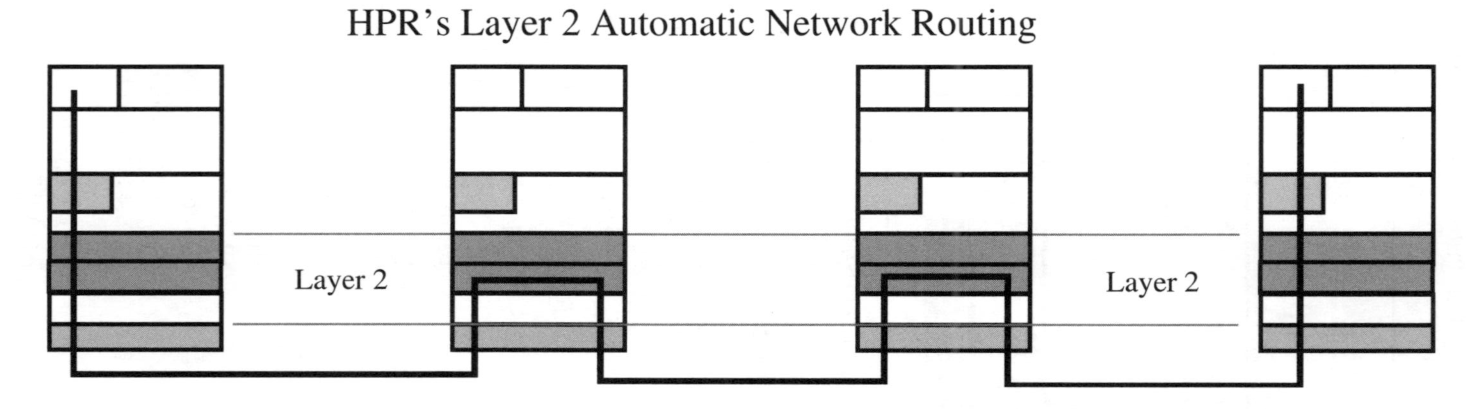

FIGURE 7.5 HPR's Layer 2-based routing minimizes the amount of processing that has to be done at intermediate nodes.

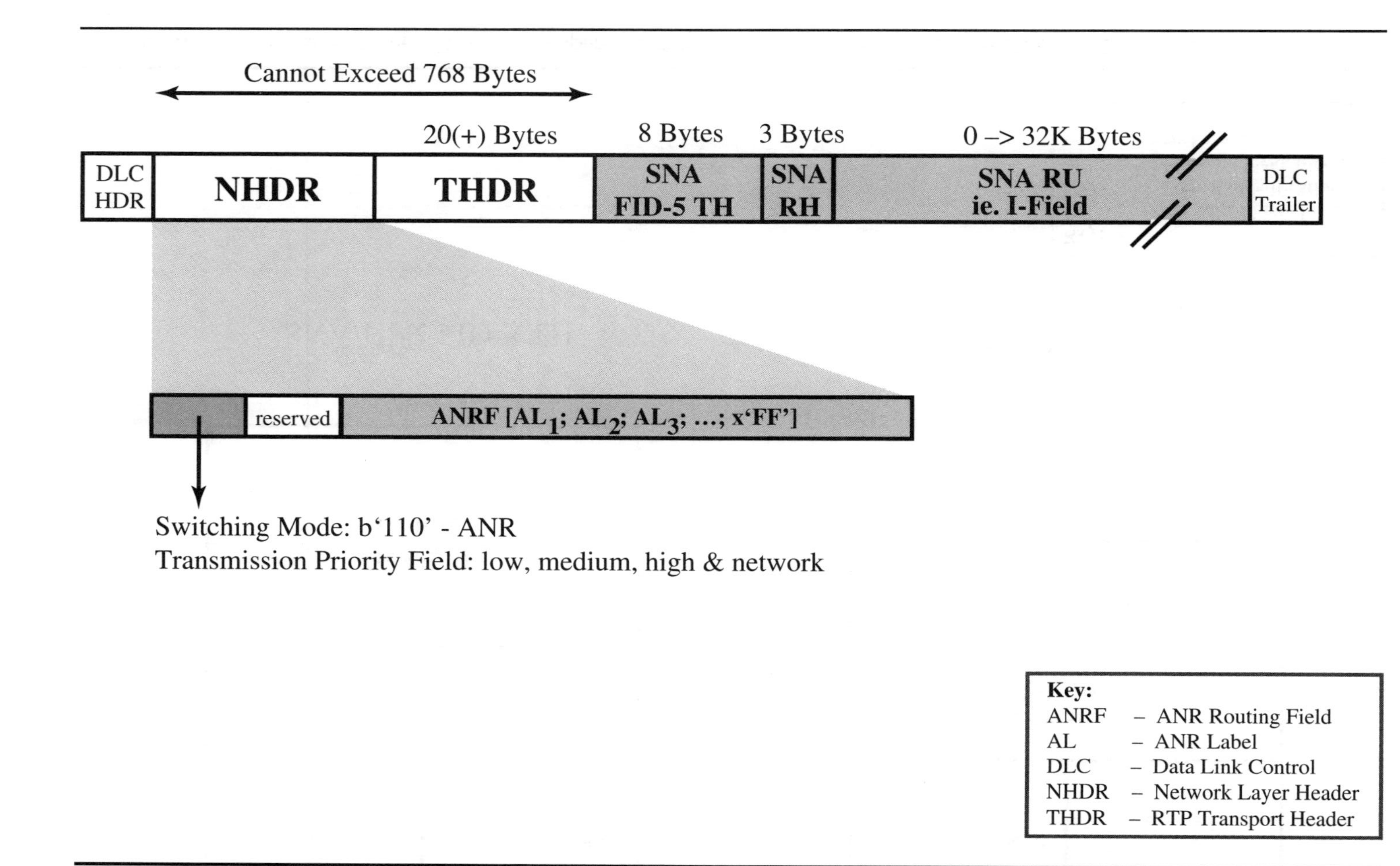

FIGURE 7.6 Format of an HPR network layer packet.

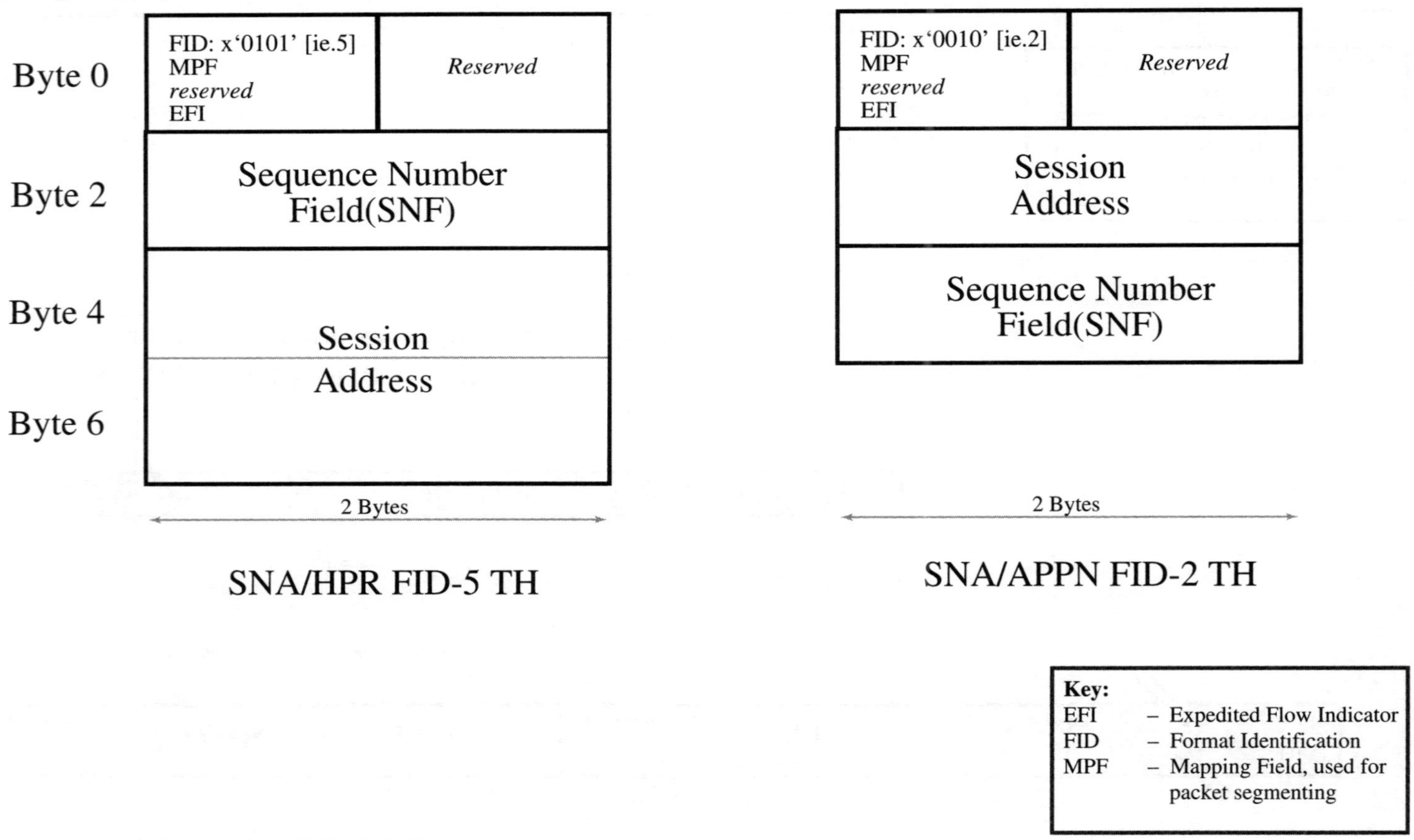

FIGURE 7.7 Format of the new FID-5 TH compared to the FID-2 used by APPN nodes and SNA type 2/2.1 nodes.

occurs following the THDR is used with LU-LU and CP-CP session traffic flowing between HPR NNs that use RTP.

The Route Selection Control Vector (RSCV) of APPN, described in Section 7.3, is analogous to an ANRF used by ANR. The RSCV, however, is only included in the BIND request used to set up a session (and in the response to that BIND). Each intermediate NN when forwarding the BIND request to its intended destination, using the hop-by-hop information embedded in the RSCV, establishes a session connector control block for that session that specifies the next hop relative to that node. Message units flowing on a session, instead of having RSCVs, have a session identifier that NNs use to refer to the session connector control block that then indicates what the next hop should be.

APPN's standard remote LU broadcast-search and the COS-based multiple criteria route calculation mechanisms are used by HPR to determine the initial end-to-end path to be used for an HPR-based session. This route is then specified in the ANRF in terms of a series of ANR labels that sequentially identify each intermediate link making up that route. The length of an ANR Label can vary from 1 to 8 bytes. It is expected that early implementations will typically use 2-byte labels.

HPR does not use a pointer (e.g., a counter) at the start of the ANRF to designate the next hop that needs to be traversed. Instead, HPR nodes discard the ANR label that designated the next hop as they forward a message unit to that hop. Thus, on receipt of a message unit, an HPR node only has to look at the first ANR label in the ANRF to determine where it should be forwarded. That label is then removed. This means that the length of an ANRF decreases as the message unit flows through the network.

IBM estimates that ANRF-based routing, which is all accomplished at Layer 2, will be at least three times faster than the APPN routing scheme. The elimination of session connector control blocks is also supposed to free up 500 bytes of memory, per session, at each intermediate node.

ANR, being the lowest layer of HPR, is also responsible for traffic prioritization per the high, medium, and low priority assigned to each session when it was established.

7.2.2 Rapid Transport Protocol

RTP, which sits above ANR, is a connection-oriented, end-to-end, full-duplex protocol. HPR uses RTP connections to provide a reliable, in-sequence (i.e., FIFO) transport mechanism for both LU-LU and CP-CP traffic. RTP is also responsible for performing ARB-based congestion avoidance data flow control.

RTP functions, including error recovery (via selective retransmission) and message unit segmenting/reassembly are only performed at the end nodes of an RTP connection. There is no RTP processing at intermediate nodes. See Figure 7.5. The elimination of RTP processing at intermediate nodes is again geared to improve the performance of HPR.

RTP, rather than ANR, is responsible for the end-user-transparent, nondisruptive dynamic path switching capability of HPR. RTP, however, realizes such dynamic rerouting by substituting a new ANRF in place of the ANRF corresponding to the path that failed. There are four conditions (or *triggers*) that will cause RTP to perform a dynamic path switch. These are:

1. The failure of a local link attached to one of the two end-to-end HPR nodes that have an RTP connection between them. (Note that there is no RTP processing at intermediate nodes for a given connection.)
2. HPR NN receiving a TDU indicating that a remote HPR transmission group has failed
3. Detection of an RTP connection failure, as indicated by repeated timeouts, during an attempted status exchange sequence, that would itself have been triggered on the expiration of a "keep-alive" timer. RTP utilizes a total of five timers at each end of an RTP connection, with two of them used to monitor the on-going availability of that connection.
4. Request from a human or automated network manager

Given that RTP is an end-to-end protocol, a path switch can only be detected and performed by one or both of the RTP end points. Unlike with other dynamic alternate rerouting schemes, a path switch cannot be performed by the intermediate node that detects the path failure. The inability for an intermediate node to perform a path switch will also delay the amount of time taken before the switch occurs.

When a dynamic path switch is required, HPR solicits the help of its coresident HPR CP to recalculate a new end-to-end path based on the original COS requirements. The HPR CP will attempt to do so using the latest network configuration status as reflected in its Network Topology Database. If an alternate route is located, a new ANRF corresponding to that route is generated. This new ANRF will then prefix all subsequent message units belonging to the RTP connections that were rerouted.

7.2.3 HPR and APPN Coexistence

With HPR positioned as an enhancement to APPN NNs, much attention is paid within the HPR specification to ensure smooth interoperability between APPN and HPR subnetworks or even nodes. This interoperability between APPN and HPR is realized via an *APPN/HPR boundary function.* This boundary function will be a service provided by selected HPR nodes. *The protocols used between an APPN node or subnetwork and an APPN/HPR boundary function will be exactly the same as that used between APPN nodes today.* Thus, APPN nodes, whether NNs or ENs, are oblivious to the fact that they are interacting with an HPR NN node, or communicating across an HPR subnetwork. Figure 7.8 depicts a network configuration that involves various levels of APPN and HPR interworking.

The APPN/HPR boundary function ensures that ANR, RTP, and ARB protocols are transparently used by APPN sessions being routed across an HPR network. APPN sessions, thus, still enjoy the dynamic rerouting, rapid routing, and congestion-avoidance facilities of HPR, while flowing through the HPR portion of the network.

The CP-to-CP protocols (i.e., remote LU search, end node LU registration, TDU exchange, XID/CP-capabilities, etc.) used by HPR CPs are unchanged from those used by APPN CPs. In addition, HPR also uses APPN's multiple COS criteria-based (e.g., cost per byte transmitted, cost per connect time, bandwidth, propagation delay, security, etc.) path selection mechanism. This ensures that the implementation effort for HPR can be kept to a minimum and that much of the existing APPN code (e.g., in OS/400 and OS/2) can be reused. It should also be noted that vendors who licensed APPN NN code from IBM (e.g., Cisco), or Data Connection Ltd. in the UK (e.g., Wellfleet), are expected to get the HPR code when available in mid-1996 as a standard no-cost update to their existing APPN license.

7.3 AUTOMATIC LOCATION OF REMOTE LUs AND REGISTRATION OF END NODE LUs

APPN has a broadcast search capability for locating remote LUs similar to that found in DLSw for locating unknown destination MAC addresses or NetBIOS names. In a sense it would even be fair to say that DLSw's search mechanism, replete with address/name caching, was inspired by APPN's scheme, which was developed in the mid-1980s. APPN's search scheme in turn appears to have been inspired by the multiple mainframe, cross-domain LU location facility that became available within SNA subarea networks around 1985, with ACF/VTAM 2.2. APPN's ability to dynamically locate remote LUs obviates the need to define destination LUs anywhere other than at their logical home NN server (i.e., NNs where the LUs are actually resident, or NNs to which the end nodes containing the LUs are attached).

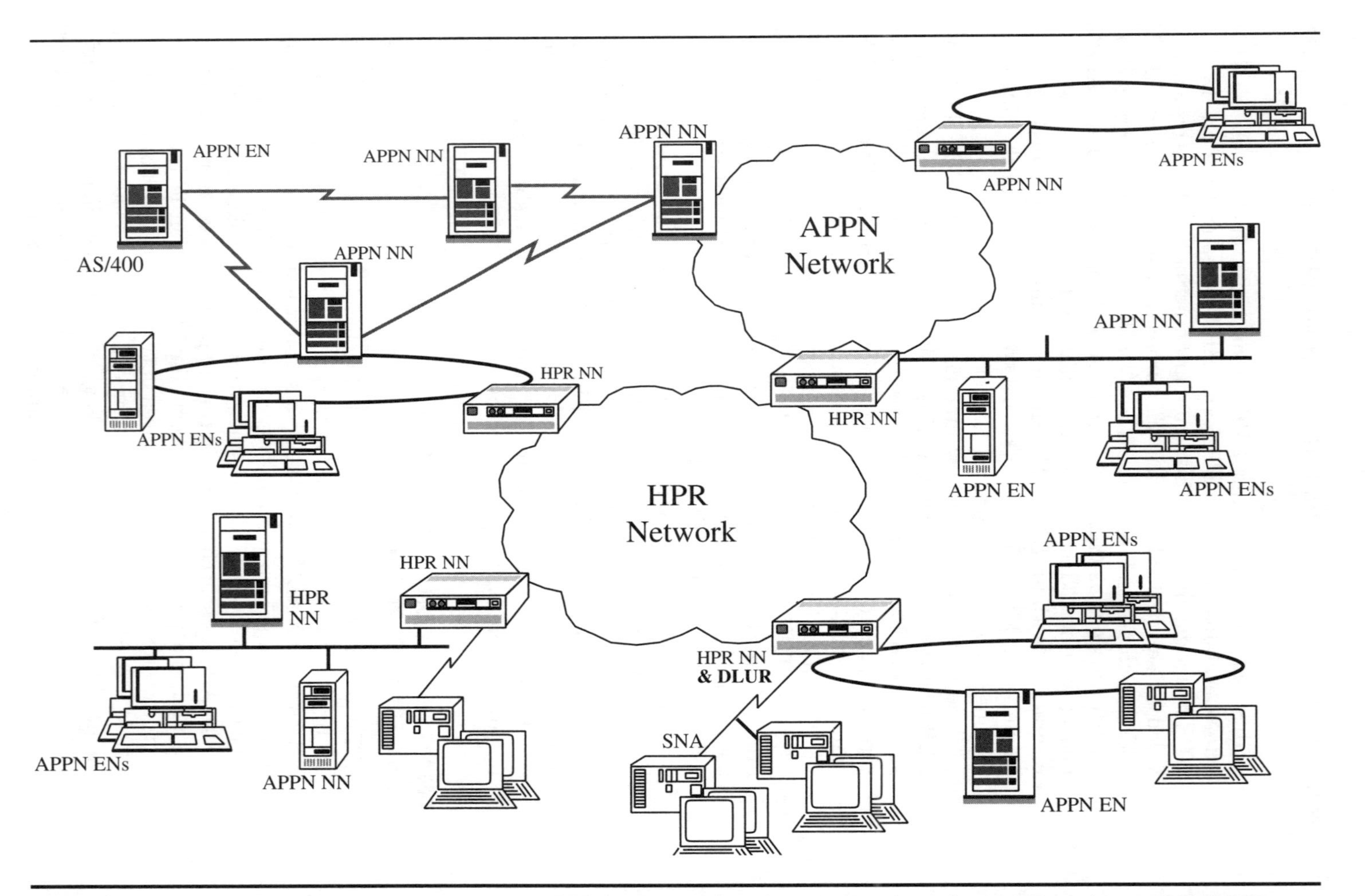

FIGURE 7.8 Seamless and transparent interoperability between HPR and SNA/APPN through HPR NNs acting as APPN/HPRR border nodes.

This remote LU location function is realized via the use of the Find Resource, Locate, Cross-Domain Initiate, and Found Resource APPN commands. APPN commands such as these are formally referred to as GDS Variables. Refer to "LU 6.2 General Datastream" sidebar.

LU 6.2 GENERAL DATASTREAM

All LU 6.2 program-to-program interactions are conducted using a native LU 6.2 datastream. This datastream is referred to as the *general datastream* (GDS). GDS is a very simple, unobtrusive, low-overhead datastream. It sets out to enforce a bare modicum of LU 6.2 specific uniformity, without in any way restricting the type or the length of the data structures that may be embedded within it. GDS expects all interactions to be formatted as self-delimiting rogical records. Each logical record is prefixed by a mandatory, 2-byte, logical-record-length (LL) header.

An LU 6.2 logical record thus consists of a LL header followed by a data field. The LL prefix specifies the entire length of the record, including the 2-byte LL prefix. Only 15 bits of the LL header are used to indicate the length of a record. Thus an LU 6.2 logical record, including the LL prefix, cannot exceed 32,767 bytes in length, that is, $2^{15} - 1$. This does not in any way limit the maximum length of data that may be exchanged in the form of a single end user record. End user data records that exceed the maximum length limits of a single logical record are segmented into multiple logical records, each logical record with its own LL prefix. Bit 0 of the LL field is used as a concatenation bit for this purpose.

A single, stand-alone LL-prefixed logical record, or two or more segmented logical records forming a single end user data record, is known in GDS as a *GDS variable*. In this context, all GDS-based interactions occur as a result of GDS variables being exchanged between LU 6.2s.

The only, or first, logical record included in a GDS variable may also optionally contain a 2-byte format identifier, referred to as the *GDS identifier* (GDS ID). The GDS ID is included between the LL prefix and the actual data field. The 2 bytes of the GDS ID, when present, also have to be reflected in the total record length specified in the LL prefix. The LU 6.2 architecture allocates a set of GDS IDs for use by end user applications, as well as by service applications such as APPN, HPR, SNA/DS, and so on.

LU 6.2s in APPN or Type 2.1 nodes attempt to establish sessions with partner LUs by issuing unsolicited negotiable (i.e., session parameters can be altered by either side) SNA Extended-BIND requests. BIND requests are explained in Appendix A. Unlike with BIND requests in subarea networks, these APPN BINDs will not identify either the destination LU or the origin LU by unique network addresses (i.e., the DAF and OAF addresses in SNA THs). Instead, the destination LU will be designated via its LU name (qualified by a network identifier when necessary) within the body of the BIND, while the session itself will be identified by a temporary *Session Index*.

To construct and issue such a BIND, the origin LU needs to obtain a variety of information from the CP located within its node. This prerequisite information includes: the link across which the BIND should be issued to reach its intended target, an unused session index for the proposed session, as well as a suitable BIND image consistent with the capabilities of the target LU. The target LU may be resident in a distant EN-only accessible via a route that traverses through multiple NNs. Hence, just the address of a link across which to issue a BIND is no longer adequate. With APPN or HPR, a complete, routing field indicating how to reach the destination LU needs to be established and appended to the BIND request.

An LU in the process of constructing a BIND will seek the various items of information it requires from its CP by conversing with it via implemenation-specific, strictly intranode, message units. For example, to obtain the routing information to reach the destination LU, as well as a suitable BIND Image, the origin LU may use a local Session Initiate request, analogous to the INIT-SELF (i.e., Logon) request used in subarea networks. This local request will specify the network identifier qualified name (e.g., <network name:LU name) of the target destination LU.

An EN CP on receiving such a request specifying an unknown LU name will issue a Find Resource command to its NN Server. The purpose of this Find Resource command is to establish the location of, and the optimum route to, the node containing the destination LU.

The network qualified name of the destination LU whose location needs to be found is specified within the Find Resource variable by means of a Search Argument Directory Entry Control Vector (CV). This is a straightforward structure that holds a 1- to 17-byte LU Name and has a type field that designates the name being carried by the CV as that of an LU.

The Find Resource Variable will also contain an Associated Resource Entry CV that will specify the name of the EN CP name issuing the Find, as well as a Directory Entry CV for the name of the origin LU that is seeking the destination LU.

7.3.1 APPN's Broadcast Search Mechanism

On receipt of a Find Resource variable, an NN Server CP will check its main directory to see if there is an entry for the destination LU name. If the LU happens to be located in another end node served by that NN Server, its name would have either been automatically registered if the node was an APPN EN, or manually entered if it was a Type 2.1 node. If the name is not found in the main directory, the CP will then search the cache directory it maintains of the results of previous LU searches. If the name is also not found in the cache, NN CP will initiate a search for the target LU. If the network has a central directory server the NN will send the search request directly to that server.

An APPN broadcast search is conducted between NN Control Points using Locate commands. A locate command essentially duplicates the CVs included in the original Find Resource issued by the source LU. However, before the search-initiating NN CP dispatches the first Locate, it adds an Associated Resource Entry CV to its name. This CV is added to the Associated Resource Entry that was included in the Find Resource command and specified the name of the source LU and the name of the end node where it was resident. Thus, the Locate will contain a complete, Search Originator hierarchy of the form: NN name, EN name, and name of source LU. This unique search originator identifier ensures that the responses to the locate request can be delivered back to the correct NN.

A Cross-Domain Initiate command may also be concatenated to the Locate command. This command by means of CVs such as COS/TPF and TG Descriptor specify COS characteristics for the proposed session, and the links available between the original EN and its NN Server.

The search originating NN CP initiates the broadcast search by issuing Locate/Cross-Domain Initiate variables to each of its adjacent NNs. It then waits to receive replies to the Locate from each of these nodes. An NN node receiving a Locate propagates it to all of its adjacent nodes before it checks its directory to ascertain if the subject LU is within its domain. (The domain of an NN CP in this context is all the LUs in the NN itself, plus all the LUs in adjacent ENs and Type 2.1 nodes that appear in its directory.) The reason the Locate is propagated *prior* to the local directory being checked is to ensure that the search reaches all corners of the network as soon as possible without getting held up at intermediary nodes. This search process between NNs will continue until a Locate arrives at the NN within whose domain the destination LU is resident.

If the LU sought is located in an EN, as opposed to the NN itself, the NN attempts to confirm that the destination LU is still available at that EN, prior to replying to the Locate. This confirmation will be sought by means of a Directed Search command. A directed search is realized by the NN forwarding a Find Resource to the EN that is supposed to contain the target LU. If the LU is still resident and active within that node, the EN will respond positively to the Find Resource by returning a Found Resource response. This response essentially echoes back the CVs included in the Find Resource.

On receipt of a Found Resource, the NN begins to construct a positive Locate Reply as well as a Cross-Domain Initiate Reply. In the Cross-Domain Initiate reply, it includes a TG Descriptor and a TG Characteristics CV to specify the exact NN-to-EN link over which it conducted the Directed Search. The name of the node containing the destination LU sought will be returned in the Locate reply. If the LU is resident in an end node, the name of the NN to which that end node is attached will also be included.

The positive reply is then returned, in reverse, along the same NN chain on which the Locates were propagated. This is achieved by the simple process of each NN now forwarding this reply as its outstanding response to the previously received Locate. Eventually, these replies meander back to the NN that initiated the broadcast search.

Using the information in the replies, as well as the path availability information reflected in its Topology Database, the NN will now calculate the optimum route to the destination LU that satisfies the COS criteria for the session. Calculating an optimum path based on COS criteria can be a lengthy and processor-intensive process depending on the number of links involved. APPN provides a means of optionally bypassing this optimum path calculation process. Instead of doing COS calculation, the APPN NN could maintain a spanning tree algorithm-based NN-to-NN routing tree such as that used by transparent bridges. Given a particular destination node the source NN may elect to use the route to this destination as specified by this spanning tree rather than trying to calculate a different route.

The optimum route calculated by the NN CP will be specified, in the form of a directed list, in a Route Selection CV, which is appended to the Cross-Domain Initiate Reply variable. This variable, along with the Locate reply, is then returned to the EN CP that initially inquired as to the whereabouts of the target LU. The information returned in these replies, in particular the Route Selection CV (RSCV), then gets conveyed back to the LU attempting to establish the session. It will synthesize and process this information, and finally append the RSCV as received from the NN CP to the resulting Extended BIND. Figure 7.9 illustrates the overall dynamics of an APPN broadcast search.

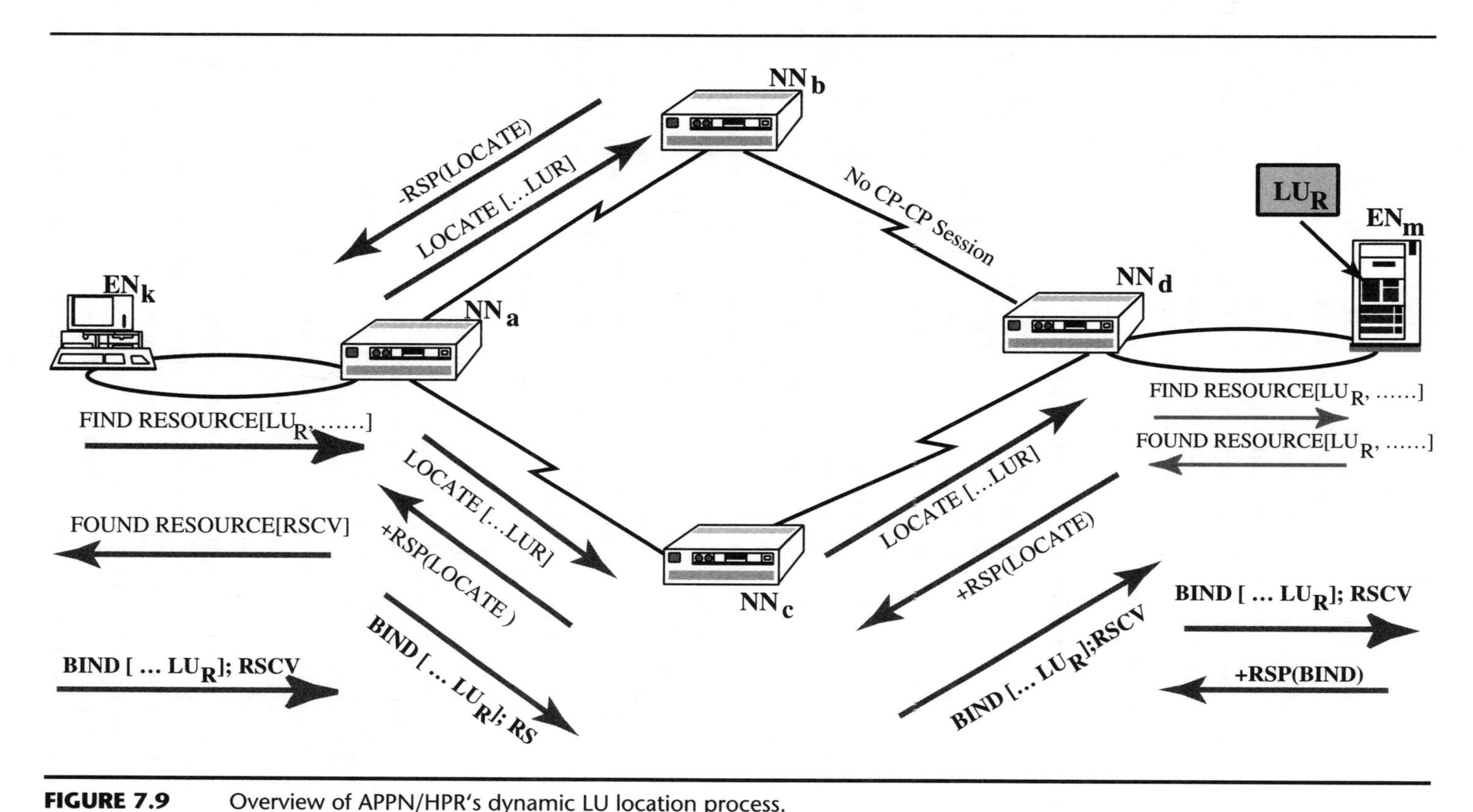

FIGURE 7.9 Overview of APPN/HPR's dynamic LU location process.

In the case of standard APPN, BIND will then be forwarded to the NN Server. The Path Control Network component in the NN will then route it to the next hop based on the route indicated in the RSCV. This hop-by-hop routing will continue until the BIND is eventually delivered to its target LU. In the case of HPR, the HPR NN receiving that BIND will set about to restate that route in terms of a series of ANR labels that sequentially identify each intermediate link and node making up that route.

7.3.2 Automatic Registration of End Node LUs

The ability to automatically register LUs resident within APPN ENs is a powerful and standard feature of APPN. This automatic registration process obviates the need to manually define EN LUs at NNs, as is the norm when dealing with most peripheral nodes in traditional SNA environments.

When an EN establishes a data link layer (e.g., LLC:2) connection with its NN server, they immediately exchange CP Capabilities commands that enable them to ascertain which CP facilities are supported by each CP (e.g., type of network management they support). Once the CP capabilities have been ascertained, the LU registration process can commence. The EN CP will specify the LUs that it wishes to have registered (i.e., entered in the NN Server's Directory) by means of a Register Resource command.

The LU names will be denoted within this variable by a series of Associated Resource CVs, which will indicate the name of the EN, and then the names of all the LUs to be registered. The command may also, optionally, contain a Directory Entry Correlator. This correlator, when present, holds a unique implementation-specific identifier. If a negative (i.e., error) reply to the Register request has to be returned by the NN, it will echo back this correlator.

Curiously, the NN server does not return an explicit positive reply to a Register Resource request to denote successful registration of the subject LUs. The only time a reply would be sent is if the NN Server wants to report an error. Hence, the only replies are negative replies.

An EN CP can also request its NN server to remove previously registered LU names from its Directory. It would do so if certain LUs are either removed from that node or are made temporarily inactive. The EN CP does so by sending a Delete Resource command. A Delete Resource variable uses the same repertoire of CVs as a Register Resource variable, namely, Command Parameter CV, Directory Entry Correlator CV (optional), Associated Resource Entry CV for the EN CP Name, and Directory Entry CV for each LU Name to be de-registered.

7.4 CHOOSING BETWEEN HPR/APPN NN ROUTING, DLSw, AND RFC 1490

Nearly all the leading bridge/router vendors now offer APPN NN routing as a strategic and SNA-oriented means of integrating SNA/APPN traffic into a multiprotocol LAN/WAN network. Interestingly, none of the FRAD vendors offers APPN NN. Instead they concentrate on RFC 1490 as the ultimate panacea for implementing SNA/APPN-capable multiprotocol networks—albeit only if FR is to be used as the WAN fabric. By 1997, all the bridge/router vendors now offering APPN NN will have upgraded their software to provide HPR NN routing. Moreover, all the APPN NN implementations now available, with the ironic exception of IBM's, include an integrated surrogate DLU/Requester server.

Thanks to this server, the APPN NNs in these bridge/routers can effortlessly and transparently route non-LU 6.2 traffic from any downstream device alongside LU 6.2 traffic. Thus, APPN NN routing at long last is no longer restricted to LU 6.2 traffic and DLU traffic from devices that include their own built-in DLU Requester soft-ware.

HPR/APPN routing is thus a direct alternative to DLSw(+) and RFC 1490 as a means of effectively consolidating parallel networks. The challenge of making the right choice between DLSw(+)- and RFC 1490-based encapsulation was discussed in the previous chapter. Now the pros and cons of APPN/HPR NN routing need to be evaluated against DLSw(+) and RFC 1490.

A first-cut choice between these three very different options would be to examine the projected WAN traffic protocol mix for the 1997–1998 time period. If SNA/APPN traffic is still expected to account for 75% or more of this traffic mix, HPR/APPN NN routing or even an IBM 2217-type AnyNet Gateway should be carefully evaluated as a potentially attractive and applicable solution. The other key criterion to consider is the number of IBM mainframes or AS/400s likely to be in the consolidated network.

HPR/APPN NN routing or a 2217 solution is likely to be particularly appealing to enterprises that have four or more IBM mainframes, with different sets of applications on each, and with remote users who need regular access to most of these applications. Such SNA-centric solutions will obviously also have an intrinsic appeal to enterprises currently running AS/400-based APPN networks.

At this juncture it is, however, worth noting that some enterprises with AS/400 APPN networks are now seriously thinking of moving to a totally IP-based solution using the top-notch TCP/IP support now available on AS/400s. Rather than using full-stack SNA at the end nodes (e.g., PCs) these enterprises are electing to use tn5250—the 5250-

terminal-family-oriented version of Telnet-based tn3270. The TCP/IP support on the AS/400 includes a full-function tn3270 server that enables tn3270 clients to readily access SNA applications running on the AS/400s.

Now IP rather than APPN will be used to provide access to the various SNA applications running on the different AS/400s. With this type of tn5250 solution the enterprise does not even have to worry about DLSw given that there is no SNA/APPN traffic flowing through the network. This type of solution is most likely to appeal to enterprises that are making a concerted effort to cut down their SNA usage in favor of new, typically Unix-based TCP/IP applications. Enterprises whose traffic mix is still predominantly SNA/APPN will want to pursue solutions that at least accommodate SNA/APPN as opposed to banishing SNA/APPN traffic altogether from the network.

7.4.1 HPR/APPN NN-Based Routing Between Mainframes

In networks with multiple mainframes or AS/400s, HPR/APPN NN routing, in marked contrast to both DLSw(+) and RFC 1490, will provide a native SNA-session-level routing mechanism. This APPN NN routing will enable remote users to gain direct access to the SNA application they wish to reach irrespective of what mainframe or AS/400 it is executing on. If DLSw(+) or RFC 1490 were being used in such a multiple-host environment, the remote users would first have to access a SNA/APPN gateway (e.g., IBM 3745 or AS/400 NN) and then get routed from that gateway to their ultimate destination host.

This two-hop routing to the destination host will not be restricted to the initial application locating logon message unit. All the data traffic will also end up traversing a two-hop path as shown in Figure 2.10 in Chapter 2. This two-hop routing is necessitated by the fact that both DLSw(+) and RFC 1490 are essentially Layer 2 bridging schemes. They both rely on the MAC address of a destination gateway (e.g., 3745 communications controller) as their only means of identifying a required SNA destination.

Most remote SNA devices, in addition, do not provide a mechanism whereby the MAC address of their SNA gateway can be dynamically modified depending on which mainframe they wish to reach. Thus with any MAC address-based bridging scheme, whether it be source-route bridging, DLSW(+), or RFC 1490, any given SNA device can only be initially routed to the gateway corresponding to the predefined MAC address.

HPR/APPN routing does not use MAC addresses to locate destinations, which in the case of APPN are always SNA application programs (or to be more precise the LUs serving such applications). Instead, HPR/APPN routing, like SNA, uses the actual SNA application name (i.e., LU name) as its routing metric. If the name of the SNA/APPN node containing the desired application is not in the cache directory of the source node, it will conduct an APPN-based broadcast search as discussed above. When an application LU is thus located, its location, in terms of a node name, will be cached to preclude the need for repeated searches. Once the name of the node containing the desired application is known, the source node will use the data contained in its Network Topology Database along with the COS to determine the best path through the network to that application.

HPR/APPN NN routing by a bridge/router is ideal for multimainframe or multi-AS/400 environments where this type of application name-based routing would eliminate two-hop routing. It would also ensure that APPN's sophisticated, nine-criteria-based path selection scheme can be used to select the optimum path to a given application.

In a network with a single mainframe, APPN NN routing will definitely be superfluous given that there is only one possible destination. DLSw(+) or RFC 1490 will invariably be better options—with the latter likely to be the most efficient if frame relay is being used.

When planning to use HPR/APPN NN routing in a multimainframe network it is imperative to note that this routing will not exactly replicate the path that may have been used by traditional SNA subarea-to-subarea routing. Subarea-to-subarea routing takes place across fixed, predefined paths. HPR/APPN, in contrast, dynamically selects the optimum path based on the nine-criteria COS and the current state of the network topology. This means that a path selected between a given source and a destination using HPR/APPN routing might not be the same as the path that would have been used with SNA. Moreover the HPR/APPN path may vary each time a new session is set up depending on the network topology at that instance.

This incompatibility between the two routing schemes is not really a major issue provided that network designers are cognizant of it — especially when provisioning bandwidth for the various APPN routes across the network. If one designs the new multiprotocol network assuming that HPR/APPN will use the same routes that SNA would have used, the actual traffic patterns within the network may prove to be very different, with some routes experiencing overload while others are hardly used; see Figure 2.10. The only way to ensure that HPR/APPN routing follows the same paths that were being used by SNA is to manually customize all of the HPR/APPN NN COS tables. With a large network, this may not be a trivial task and should not be undertaken lightly.

7.4.2 The Revenge of Frame Relay vis-à-vis HPR

The potentially glittering prospects of an HPR-centric networking scenario could lose much of its luster if this network were to be run across a frame relay WAN, as opposed to long-haul leased lines between HPR NNs. If frame relay is to be the backbone WAN, it as opposed to HPR will dictate much of the attributes of that WAN. Hence, many of the features of HPR, such as dynamic alternate routing, anticipatory congestion control, Layer 2 routing, no intermediate node routing, and so on, become superfluous and redundant. These functions, or to be precise their equivalents (e.g., dynamic alternate routing, congestion control, Layer 2 routing, elimination of intermediate node processing, etc.), will be provided by the frame relay WAN rather than by HPR.

Thus many of the key features of both HPR and APPN become nullified if HPR/APPN nodes are deployed around the periphery of a frame relay WAN. In addition, APPN and HPR will also excel only if many of the nodes they support are APPN end nodes, and if traffic from these end nodes has to be continually routed to multiple destinations.

Most of today's mainframe-centric SNA networks will not meet these criteria. There will not be that many destinations (i.e., mainframes) and there will be a paucity of true APPN ENs with most of the remote nodes being SNA nodes—that require some level of DLUR support in order to communicate across the APPN network. Thus, if a public frame relay service is envisioned as the basis for transitioning such SNA networks, with only one or two mainframes, into multiprotocol networks, APPN/HPR NN routing or 2217 gateways may not be the best solution. If the number of destinations is few, and most of the nodes are SNA, RFC 1490 first and then DLSw(+) should take automatic precedence over APPN in the context of frame relay WANs.

APPN and HPR both being a set of services, as opposed to a type of traffic, results in there being some intriguing intertwined relationships between HPR/APPN, and DLSw(+) and RFC 1490. For a start, both DLSw(+) and RFC 1490 freely support LU 6.2 as well as all other SNA traffic types. Thus, the so-called APPN traffic, which is in reality just LU 6.2 traffic flows, can always be supported across a DLSw(+) or RFC 1490 network, without the need for the router providing DLSw(+) or RFC 1490 to also be an HPR or APPN NN.

Using DLSw(+) or RFC 1490 as opposed to HPR or APPN NN routing also obviates the need for Surrogate DLUR capabilities within the bridge/router. DLUR is only an issue when using APPN. Without APPN, any non-LU 6.2 traffic can be supported without any problems with either DLSw(+) or RFC 1490 without recourse to DLUR.

7.4.3 Hybrid Solutions Involving DLSw(+) or RFC 1490 with HPR/APPN Routing

One can easily implement a DLSw(+)- or RFC 1490-based multiprotocol bridge/router or FRAD network that supports all the end-to-end APPN interactions between any set of APPN ENs and NNs—such as a collection of AS/400s. The bridge/router or FRAD-based backbone will not provide any APPN services (e.g., directory or LU registration) or support NN-based routing across the backbone. Instead, all the routes between the APPN nodes will be in the form of end-to-end point-to-point connections, across the backbone, that would have been established using a Destination MAC address-based bridging scheme.

In this type of DLSw(+) or RFC 1490 network, the bridge/router backbone is in essence providing a set of logical point-to-point links between the APPN nodes. There are existing bridge/router networks that support APPN nodes in this manner—typically using SRB or DLSw(+). Figure 2.2 shows a network where HPR/APPN services are being run end-to-end across a DLSw network.

This ability to freely support APPN/HPR across DLSw(+) or RFC 1490 also leads to the possibility of deploying hybrid bridge/router networks where some of the bridge/routers contain APPN NN routing while the others do not. The bridge/routers without NNs will use DLSw(+) or RFC 1490 to transport the APPN traffic to/from the bridge/routers acting as NNs.

There can be some limitations in trying to support a series of HPR/APPN nodes across a bridge/router network where the bridge/routers—or at least the bridge/routers at the outer edge of the network—are not providing HPR/APPN NN services. For a start, there will be no dynamic, COS-based HPR/APPN routing across the network. In addition, if all the NNs are upstream, and all the downstream nodes are ENs, all the NN services required by the ENs—such as LU registration or remote LU location—will involve interactions across the network. If the bridge/routers were NNs, all such services could be done locally, obviating the need for interactions across the backbone. If this type of EN to NN interaction is likely to be common, as might be the case with a large AS/400-centric APPN network, it would be advantageous to consider an HPR/APPN NN-based bridge/router solution. If the SNA traffic is very high then even a 2217-based solution should be contemplated.

HPR, APPN, and RFC 1490 will again come together when bridge/routers are using HPR/APPN NN routing across a frame relay WAN. The bridge/routers will use RFC 1490 to transport all the HPR/APPN flows—whether they be NN-to-NN control flows, actual LU 6.2 data traffic, or SNA traffic being supported via DLUR—across the FR network.

This relationship between HPR/APPN and RFC 1490 is essentially the same as that between DLSw(+) and RFC 1490. DLSw(+) traffic, which is all IP-based, is transported across frame relay using RFC 1490 encapsulation. Similarly, RFC 1490 could end up supporting HPR/APPN traffic in two distinct modes. For a start it could act as just the Layer 2 transport mechanism, below HPR's RTP and ANR layers, for data that is being exchanged between HPR NN routed data. If on the other hand, HPR NN-routing is not being used, RFC 1490 can still provide a native transport scheme for all SNA traffic types, independent of HPR, via its standard encapsulation scheme.

7.4.4 Standards, Scalability, and Efficiency with HPR

APPN/HPR NN routing shares many attributes in common with DLSw(+). They can both dynamically locate remote destinations, and thus tend to be better suited than RFC 1490 for environments that have multiple destinations. Like DLSw(+), APPN/HPR is also not a bona fide industry standard. The so-called APPN, HPR, and DLSw standards are all administered by the IBM-sponsored APPN Implementors' Workshop. APPN and HPR not being true standards should not in practice be a major stumbling block that in any way precludes multivendor interoperability. SNA never was and will now never be a true standard. It was, however, for the longest time the predominant basis for multivendor interoperability within the commercial sector.

In the case of APPN/HPR NN, multivendor interoperability in practice will be even less of a problem. At present all the APPN NN implementations are based on code obtained from one of two sources—IBM or Data Connection Ltd. (DCL). This is also likely to be the case with HPR. Given that IBM and DCL currently work hand in glove to ensure compatibility between their code bases, interoperability at present is not really an issue—at least until there are other source code vendors or a major rift between IBM and DCL.

The overhead associated with APPN's broadcast searches and the topology updates could reach unmanageable, and unbearable, levels in networks consisting of many hundreds or more NNs. (It is worth noting that APPN's toplogy updates only occur when there is an explicit change in the network as opposed to on a periodic basis as is the case with IP's RIP.) To circumvent this scalability problem APPN provides two separate solutions: *border nodes* and *central directory servers*.

Border nodes permit APPN networks to be partitioned into subnetworks so that topology update traffic can be restricted to a given partition. Border nodes, however, do not in any way restrict transparent interoperability (i.e., session establishment and broadcast search based destination location) across the partitions. The amount and scope of broadcast searches used to locate unknown destinations can be curtailed using the central directory servers, which augment the cache directories maintained by each NN. Though the specification for the APPN Border Node has been available for awhile, at present border node implementations are available only on AS/400s and mainframes.

In marked contrast to DLSw(+) and RFC 1490, HPR and APPN—at least outside the realm of the 2217—are not encapsulation schemes. HPR or APPN NN routing, in conjunction with DLUR where necessary, route SNA message units in native mode, that is, without encapsulating the traffic in any other protocol. In this respect HPR and APPN NN routing is equivalent to IP or IPX routing.

To transport SNA, IP, or IPX data between bridge/routers, one will have no choice but to use a Layer 2 protocol. If the bridge/routers in question are interconnected via frame relay, this Layer 2 protocol will be RFC 1490—with its lean and mean 16-byte overhead encapsulation. Thus, HPR and APPN per se do not have any encapsulation overhead related efficiency issues, as does DLSw(+).

The broadcast search and topology-update-related APPN traffic will without doubt introduce some level of APPN-specific overhead traffic into the network. The amount of this overhead traffic could be kept under control, in even large networks, by the use of border nodes and central directory servers. APPN TDUs are never exchanged on a periodic basis when the network topology is stable. All bridge/router routable protocols also have some level of overhead, for example, RIP- or OSPF-based routing table updates in IP, etc.

7.5 HPR'S WEAKNESSES

Despite being based on the lessons learned from other major networking schemes, HPR, as is inevitable, does have some weaknesses. In general, these weaknesses are in no way as severe as those of APPN. HPR's technical weaknesses on the whole are mainly esoteric and unlikely to in any way severely diminish its practical appeal and popularity.

HPR's major weaknesses can be summarized as follows:

1. Dynamic path swapping can only be done by the end nodes supporting an end-to-end RTP connection, rather than by an intermediate node that actually detects the path failure. This does mean that in the event of a path failure, message units already transmitted are likely to be lost, necessitating retransmissions, and there will be a delay before a new path is found.

2. Having to remove the next hop ANR label from the ANRF at each intermediate node necessitates processor cycle and time-consuming recreation of the ANRF field. It would have been much simpler to have a pointer that was incremented at each node to point to the relevant next hop in the ANRF. The argument that removing the ANR labels progressively reduces the length of the HPR message units as it flows across the network, though valid, does not totally justify this technique. Most ANR labels are expected to be 2 bytes in length. Shedding such 2-byte labels would result in a total saving of 20 bytes when the message reaches its final hop if the route involved 10 intermediate nodes. Most routes should have less than five intermediate nodes. Speed, and speed at all costs, is supposed to be the driving force behind HPR. The recreation of the ANRF is, however, unfortunately at odds with that goal. (Ironically, this ANRF recreation is even more time consuming than the transmission header recreation required by APPN as a result of a different session identifier being used between each intermediate hop.)
3. At present the total length of the HPR headers, including the ANRF, cannot exceed 768 bytes. Though nowhere as severe as the seven-hop limitation inherent in SRB due to its 18-byte maximum RIF, this does mean that HPR, like SRB—and for that matter APPN—has a theoretical maximum ceiling on the number of intermediate nodes that can occur on a given route. (The APPN limit, which can range between eight to 30 intermediate nodes, depending on the length of the names assigned to CPs and links, is due to the RSCV not being able to exceed 256 bytes.) The exact intermediate node ceiling for HPR, which will always be in excess of 20 intermediate nodes, will depend on the length of the ANR labels being used. (ANR labels can be from 1 to 8 bytes long.)
4. The variable length ANR labels do not have a length designation or a label-delimiting scheme. Consequently, manual intervention is required to avoid potential ambiguities that may be caused by a node using different length labels that have the same start sequence!
5. There is a potential danger that ARB's congestion avoidance mechanism might unnecessarily slow down data transmission by trying to react, after the fact, to transient, fast-clearing congestion situations. In high-speed networks, in particular ATM, congestion conditions are not expected to last long. Any congestion that might occur is likely to dissipate very quickly. However, with ARB's closed loop-back scheme there will be some lag-time before either side detects that the congestion it was reacting to is no longer there. During that time data transmission would have been throt-

tled back. The issue here is that ARB may end up throttling back for longer than it needs to, thus slowing down its overall data transmission rate.

6. Verbose protocols with lavish use of control vectors that in some cases perform redundant exchanges are an unfortunate holdover from the protocol profligacy that came to be with APPN in 1976, though to be fair HPR is nowhere as extravagant as APPN.

7.6 REFLECTIONS

HPR, the heir elect to the still sprawling SNA empire, posseses an impressive list of credentials. HPR's strengths include: agile and very low-fat, Layer 2 routing; dynamic alternate routing; state-of-the-art anticipatory congestion control; elimination of intermediate node routing in addition to all of APPN's strengths such as plug-and-play and multiple criteria-based COS path selection. Unfortunately, most of these HPR value-added features are available only if one implements the entire backbone using HPR NNs across long-haul leased lines. Many of HPR's key features, such as dynamic alternate routing, congestion avoidance, and Layer 2-based routing, become nullified if HPR NNs are deployed around the periphery of a frame relay WAN. In such scenarios frame relay as opposed to HPR will dictate the overall dynamics of the WAN.

HPR/APPN NN routing now joins DLSw(+) and RFC 1490 as yet another, supposedly strategic means for implementing viable SNA/APPN capable multiprotocol LAN/WAN networks. Most of the leading bridge/router vendors now offer all three techniques within the same bridge/router with IBM in addition also promoting the 2217 AnyNet Gateway approach.

In networks with multiple SNA/APPN destinations (e.g., mainframes or AS/400s) HPR/APPN is likely to offer a better solution than the other two techniques. With APPN, a bridge/router can perform SNA application name-based dynamic routing to multiple destinations. This is a definite improvement on not having any type of SNA-like routing in bridge/router networks—even though it is only fair to note that APPN routing does not exactly mimic SNA routing.

In environments with few SNA/APPN destinations, DLSw(+) or RFC 1490 is likely to have the edge over APPN. If in such cases, FR is to be used as the WAN backbone, RFC 1490 may turn out to be the more efficient and powerful means for integrating the SNA/APPN traffic with the other multiprotocol traffic. However if an IP-based backbone is going to be the WAN transport mechanism, DLSw is a good option.

CHAPTER 8

Multiprotocol LAN/WAN Networking over a Single-Protocol IP or HPR WAN

A QUICK GUIDE TO CHAPTER 8

Multiprotocol LAN/WAN networking can be conducted across a WAN that supports and runs only one protocol. This single protocol used by the WAN could be IP or HPR/SNA. This type of single WAN protocol-based multiprotocol networking can have considerable appeal if the bulk of the traffic flowing across a WAN consists of either IP or SNA message units. With this type of single-protocol approach the WAN can conform to and provide native mode support for its majority traffic component.

Proven technology is readily available to realize IP-centric or SNA/HPR-centric multiprotocol networks. IP-to-SNA gateways, tn3270(e), DLSw(+), Desktop DLSw, and even IBM's AnyNet can be used to implement IP-only WANs that are nonetheless capable of adroitly accommodating any and all types of SNA/APPN interactions. With the burgeoning popularity of IP many enterprises are seriously considering the option of standardizing their WAN on IP and using one or more of the these technologies to support SNA/APPN.

Section 8.1 covers all the different options for building SNA/APPN-capable IP-only WANs. This section also examines whether an enterprise should opt for a TCP/IP and SNA/APPN coexistence scheme such as DLSw(+), or whether it makes sense to banish SNA/APPN all together from the WAN à la tn3270(e).

Standardizing on an IP-only WAN does not necessarily mean that TCP/IP software is required on IBM mainframes.

FTP-based file transfers and tn3270(e)-based application access is now possible using channel-attached TCP/IP off-load gateways. Sections 8.1.3 and 8.1.4 look at all the various possibilities for profitably deploying the multitude of different IP-to-SNA gateway options now available including Reverse Protocol Conversion gateways that permit 3270s to access Unix applications.

In marked contrast to the multiple disparate options for realizing an IP-only network, IBM's AnyNet protocol conversion technology is the only viable solution, today, for implementing an HPR/SNA-centric multiprotocol network. This AnyNet technology has its roots in the so-called *Networking Blueprint* initiative unveiled by IBM in 1992.

This Blueprint postulated the possibility of network-neutral applications and the viability of running SNA/APPN applications across IP or vice versa. Section 8.2 looks at the Networking Blueprint, and its successor, the *Open Blueprint*. The Common Transport Semantic (CTS) component of these Blueprints is the architectural basis for AnyNet.

Sections 8.2.2 and 8.2.3 describe the protocol conversion, protocol compensation, and address mapping functions specified by CTS and implemented by AnyNet. The remainder of this chapter looks at the AnyNet product repertoire and how AnyNet can be used to realize both pragmatic as well as esoteric multiprotocol LAN/WAN solutions.

In environments where the predominant traffic type already is, or is soon expected to be, either TCP/IP or SNA/APPN, multiprotocol LAN/WAN networking can be realized over a single-protocol WAN backbone. The single protocol used on the WAN will be that of the prevalent traffic type—IP if TCP/IP traffic predominates and HPR/APPN if SNA/APPN prevails. The lure of this type of single-protocol WAN solution, as opposed to trying to juggle multiple disparate protocols across the WAN, is obvious. With the single-protocol approach, the WAN conforms to and provides unstinted native-mode support for its majority traffic component—without compromise and with little contention. This is in effect majority rule vis-à-vis multiprotocol networking.

Multiprotocol networking across a single-protocol WAN is particularly appealing when the repertoire of protocols being used is small and restricted to just TCP/IP, SNA/APPN, IPX/SPX, and NetBIOS. If the IPX/SPX, and NetBIOS traffic is mainly restricted to intracampus applications (e.g., file or print server) and does not account for much of the WAN traffic, the attractiveness of this approach skyrockets even further. The challenge in the end will essentially boil down to trying to support SNA/APPN traffic across an IP WAN or IP across an HPR/APPN WAN. Fortunately, as has been discussed in previous chapters, there is proven and effective technology available to realize, with impunity, either of these permutations.

SNA/APPN interactions can be successfully conducted across an IP WAN in a variety of very distinct ways. For a start there is DLSw(+). DLSw(+) permits bona fide, unadulterated SNA/APPN traffic, whether SNA- or SDLC-based, to be transparently and effortlessly transported across an IP-based WAN. There is even the notion of *Desktop DLSw* (DDLSw) where the SNA traffic is encapsulated within TCP/IP packets at its source (e.g., PC or workstation) even before it reaches a LAN, let alone a bridge/router. (Cisco, the original proponents of this desktop encapsulation scheme, refers to its scheme, which is now supported by some SNA/3270 emulation vendors such as Wall Data, as the *Native Client Interface Architecture* [NCIA].) Figure 8.1 illustrates the overall concept of DDLSw and how it precludes the need to even handle SNA/APPN traffic on LANs.

IBM's AnyNet protocol conversion technology is ironically another viable option for doing any kind of SNA interactions over an IP network. Whereas DLSw(+) encapsulates complete SNA PIU message units within TCP/IP packets, AnyNet will convert the SNA PIU into a TCP/IP packet. Both DLSw(+) and AnyNet, though using fundamentally different techniques, comprehensively and without any exceptions address the issue of transporting traditional SNA (e.g., SNA/3270) and LU 6.2 traffic end-to-end across an IP WAN.

Both these techniques are totally transparent to the SNA devices (e.g., 3174s), SNA applications (e.g., TSO) and SNA system software (e.g., ACF/VTAM and ACF/NCP) and do not require any changes to be made to the overall end-to-end infrastructure and configuration. They both make the IP WAN appear to the SNA devices and applications as if it is nothing but a leased-line network or one gigantic single LAN. Neither, however, is ideally suited for networks with multiple mainframes or AS/400s where remote users need access to more than one mainframe or AS/400. Remote SNA switching, which is akin to DLSw on steroids and was described in Chapter 5, would be a more preferable option. Transporting genuine SNA PIU message units end-to-end via conversion or encapsulation is, however, not the only way in which SNA interactions may be conducted across an IP network.

Another viable approach, which is becoming increasingly popular,

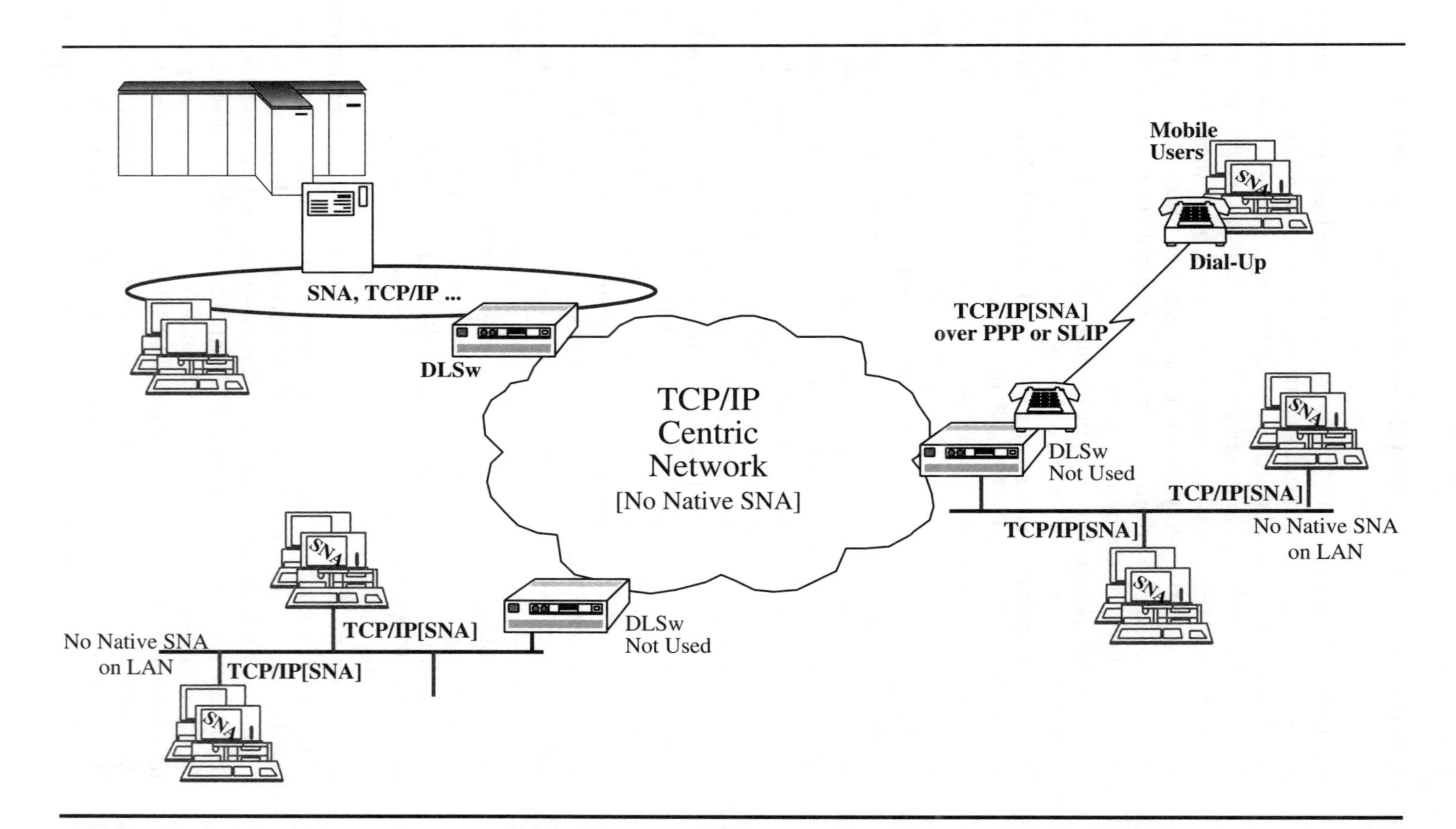

FIGURE 8.1 The notion of desktop DLSw (DDLSw) where SNA/APPN traffic is encapsulated in TCP/IP right within a PC.

is to use IP-to-SNA PU Controller Gateways, as described in Chapter 2, at, or adjacent to, the mainframes and just transport TCP/IP interactions across the WAN. TCP/IP Telnet protocol-based tn3270 (and now the extended version known as tn3270e) is the classic example of this type of no-end-to-end SNA approach to accessing mainframe or AS/400 SNA applications across an IP WAN. With tn3270, SNA, and 3270 datastream is restricted to the mainframe or to a tn3270(e) server attached to the mainframe. With tn3270(e) not even 3270 datastream is sent across the WAN. The screen painting and data capture at the PCs and workstations using tn3270(e) is done using Telnet.

Today, there is another way to realize tn3270. All the leading PU Controller or PU Concentrator SNA LAN gateways, in particular Novell's NetWare for SAA, Microsoft's SNA Server, and Eicon's SNA Gateway, now support TCP/IP between the gateway and the 3270/5250 emulators running on PCs and workstations. With such Gateways, PC/workstation users can enjoy full-function, feature-rich 3270/5250 emulation at the desktop, replete with local printing, while still restricting the data flows across the WAN to TCP/IP transactions—albeit with 3270 datastream blocks now being carried end-to-end, encapsulated within TCP/IP packets.

The options for transporting IP traffic over SNA/APPN/HPR WANs are more limited than those available for SNA-over-IP. In general, the only generic, readily available and actively supported technology for IP-over-SNA boils down to IBM's AnyNet products, Novell's SNA Links, and a few SNMP-over-SNA solutions offered by a few hub vendors (e.g., Cabletron). AnyNet protocol conversion, whether from SNA-to-IP or IP-to-SNA is available in two distinct forms: AnyNet Access Node software installed in end nodes (e.g., OS/2 PCs, RS/600s, AS/400s, or even MVS mainframes), and AnyNet Gateways that provide AnyNet services to multiple devices that do not contain any AnyNet software.

Today's AnyNet Gateways can handle IP traffic either by converting it to SNA message units or by encapsulating it within SNA LU 6.2 message units. The notion of IP-over-SNA via LU 6.2 encapsulation, which happens to be the exact opposite of DLSw(+), was pioneered by an OS/2-based IBM software product known as *LAN-to-LAN over WAN Program* (LTLW). LTLW, which supports the transport of IP, IPX/SPX, and NetBIOS traffic across SNA WANs through LU 6.2 encapsulation, is now marketed as an AnyNet Gateway.

LTLW is the epitome of a archetypal LAN-over-SNA solution in that it does not try to convert non-SNA traffic to SNA. Instead it, like DLSw(+), is a straightforward encapsulation (or *tunneling*) scheme. Novell's SNA Links is essentially Novell's version of LTLW. Around

1993, there were other LAN-over-SNA solutions, most trying to add value to LTLW, such as ANR's L2-SR, CCI's Eclipse 7020, and APT's APTNet. Today, AnyNet has the dubious honor of being the only real champion of LAN-over-SNA networking. Of the AnyNet offerings available, the IBM 2217 MultiProtocol Concentrator, which can be thought of as AnyNet-in-a-box, would appear to be the most compelling.

The IBM 2217 which is now promoted by IBM, albeit somewhat sotto voce, as a strategic means of implementing an HPR/APPN WAN-based solution for multiprotocol networking includes both AnyNet IP-to-SNA conversion and LTLW encapsulation software. Enterprises which expect that SNA/APPN traffic will still account for at least 75% of their 1997–1998 WAN traffic mix should seriously consider a 2217-based solution, particularly if there is also a need for routing between multiple mainframes. By mid-1996 the 2217, which will likely be in Release 2 livery, will be capable of offering an HPR-centric multiprotocol solution that while propagating SNA's heritage of value-added, mission-critical networking will do so not just for SNA/APPN traffic but also for non-SNA traffic.

8.1 THE TREND TOWARD IP-CENTRIC NETWORKING IN IBM ENVIRONMENTS

TCP/IP now flourishes within enterprises that just a few years ago were "true blue" and relied exclusively on SNA for accessing mission-critical applications. TCP/IP's burgeoning popularity within the IBM world can be directly attributed to three interrelated factors, namely:

1. Most new enterprise applications are Unix-based and use TCP/IP.
2. TCP/IP appears to be the underlying basis, even if it is just for network management, of most LAN, multiprotocol, and ATM products and solutions.
3. IBM is actively promoting Unix and TCP/IP.

Unix systems with their tightly integrated support for TCP/IP has now become the preferred platform on which new enterprise applications are being developed and run. Nearly half of all mainframe shops now use one or more Unix applications in parallel to their so-called legacy applications. This trend toward Unix-based applications will continue to gather momentum as RISC technology, the ideal hardware base for Unix, continues to relentlessly lower the cost of high-speed, high-capacity computing.

The growing popularity of TCP/IP-speaking Unix applications is obviously another significant contributing factor leading to the landslide move toward LAN-centric, multiprotocol networking. All the devices necessary to implement LANs (e.g., hubs and LAN switches) and multiprotocol networks (e.g., bridge/routers, ATM switches) invariably have a TCP/IP bias—if nothing else they all rely on TCP/IP-based SNMP as their preferred, and in most cases only, management scheme.

Thus, just the mere process of moving toward LANs and multiprotocol networking introduces more TCP/IP and TCP/IP-based network management into the network. Moreover, nearly all of today's brand-new networking solutions, even from IBM, invariably tend to first support TCP/IP. ATM is a good case in point given that all the *initial* standards for transporting data over ATM concentrated on TCP/IP and TCP/IP encapsulation schemes. Standardizing HPR-based data transport across ATM, in contrast, is still in its infancy, and has yet to receive any widespread multivendor attention.

Compounding all of the above TCP/IP catalysts is the fact that IBM itself is now a devout believer and promoter of TCP/IP. IBM now provides comprehensive support for TCP/IP across all of its strategic system offerings from mainframes to OS/2. Even the 3174, which used to be the quintessence of an SNA device, now permits terminals attached to it to use Telnet to access TCP/IP applications!

The bottom line of all of this is that TCP/IP is now as important as SNA, if not more so, when it comes to contemporary IBM-centric networking. Two surveys done five years apart by the eminent IBM watchers Xephon PLC (UK) highlights how TCP/IP's stature and stock has risen within the IBM world. A worldwide survey of over 900 mainframe customers in 1991 indicated that TCP/IP was not very important. A similar survey conducted in 1995, involving nearly 440 worldwide mainframe customers, shows that TCP/IP now only comes behind SNA and SQL when it comes to standards that are considered to be very important when considering new systems. Just to put this in perspective, APPN came in 12th, with SNMP coming in at fourth place just behind TCP/IP, pushing NetView/390 to fifth place. This phenomenal popularity of TCP/IP only has one other precedent—that of the so-called PC explosion.

With TCP/IP now deemed to be very strategic as the cornerstone of multiprotocol networking and its apparent omnipresence when it comes to current Unix applications and networking solutions, some hitherto SNA-oriented enterprises have already standardized on IP as their new networking fabric. Many others are planing to do so shortly. Many also have over the last few years installed TCP/IP on their mainframes to facilitate TCP/IP-based application access via tn3270, and file transfers via TCP applications using the FTP protocol. The desire to standardize on IP is understandable. IP has now become what SNA was in the early 1980s and what OSI was supposed to be—a strategic, widely supported framework for building multivendor distributed computing environments. Consequently, there is a strong prevailing sentiment among IBM networking professionals that they have to make a concerted effort to standardize on TCP/IP to ensure that their networks do not become technically obsolete and bankrupt.

With the possible exception of the 2217, which was much too little much too late, IBM has done very little to allay this fear and promote SNA/HPR as an alternative to IP-centricity. *Hence, SNA is now rapidly becoming relegated to a legacy protocol that has to be tolerated and accommodated until the mainframe SNA applications are ported over to Unix.* However, as discussed earlier, given their innate mission-criticality this transition to non-SNA application is not going to happen overnight. Plus there is the huge, 20-year plus investment that has been made in these applications.

SNA mission-critical applications will still be around 10 to 15 years from now. That is a long time. Too long in many cases for a solution in which SNA traffic is tolerated, somewhat under duress, within the context of an IP-based network. Hence, enterprises are urged to conscientiously evaluate all options and implications before electing to standardize on a totally IP-oriented backbone especially if the SNA/APPN traffic that needs to be transported across that network over the next few years still account for 60% of the total traffic mix. In instances where SNA interactions are going to be a dominant component of the traffic mix, an SNA coexistence scheme, such as DLSw(+), might end up providing more long-term flexibility than an option such as tn3270 that eliminates end-to-end SNA across the network. The bottom line here is the same as that which has been regularly advocated to this point: traffic mix should be the key criterion that dictates the composition and characteristics of a new, consolidated multiprotocol network. The remainder of this section elaborates on the various issues that should be considered.

8.1.1 Two Ways of Realizing an IP-Centric Backbone

There are essentially two distinct ways that an enterprise can go about standardizing on an IP-centric networking solution. The first is to opt for a bridge/router-based solution, to date the most common approach. The other approach, which has nothing to do with technology per se, is to dogmatically insist that irrespective of how the IP network has been implemented, the only traffic that may flow across it is IP datagrams.

The only-IP-across-the-backbone approach is really only justifiable in a few scenarios. The first is when 75% or more of the traffic flowing across the network is between native TCP/IP applications (e.g., Unix systems). The other instance is if concrete measures are already in place to migrate all the SNA applications to Unix systems within the next few years. If, on the other hand, SNA interactions are expected to be around, in significant volume, for the next five years or more, a bridge/router-based approach may provide the best of both worlds. This is an IP-centric solution that is also capable of adroitly supporting any and all types of end-to-end SNA/APPN traffic without any restrictions.

All of today's multiprotocol bridge/routers are inherently IP-oriented. They have evolved from IP-only routers developed in the 1980s for IP-only networks even then being heavily used in academic and manufacturing scenarios, as well as for sustaining the already active Internet. Consequently, bridge/routers can be readily used to implement IP-only networks. However, deploying bridge/routers in IBM environments purely as a means of implementing an IP-only network could be unnecessarily restrictive and short-sighted if other protocols such as IPX/SPX and SNA/APPN need to be supported across the WAN.

Today's bridge/routers, despite their innate IP orientation and SNMP-based management schemes, offer a plethora of multiprotocol capabilities particularly for accommodating end-to-end SNA/APPN traffic, such as DLSw(+), APPN/HPR NN routing, and even bridging. They also offer a tightly integrated and proven capability to route IPX/SPX (and other) traffic alongside IP-traffic. Figure 7.3 illustrated how bridge/routers route multiple protocols across the network—as opposed to encapsulating all protocols within TCP/IP and then just routing IP datagrams across the WAN.

This is a very important concept to remember when evaluating a so-called IP-only solution. While it is possible to have SNA/APPN and NetBIOS transported end-to-end encapsulated within TCP/IP using DLSw(+), this is not going to be the case for all other protocols. In many cases, particularly with IPX/SPX, DECnet, Banyan Vines, OSI, and so on, the bridge/router vendors are likely to recommend that such traffic is routed, in native mode, alongside IP—as opposed to trying to come up with a scheme whereby this traffic has to be encapsulated within IP. Routing a few protocols side-by-side, across a WAN, is not going to be a major issue with today's bridge/routers, especially in terms of management. The SNMP-based management scheme for the bridge/router network will encompass all the traffic being routed across the network rather than being restricted just to the TCP/IP traffic.

Routing multiple protocols, as opposed to just IP, does have some, albeit relatively insignificant, drawbacks. For a start, each protocol will use its own routing table update protocol; for example, traditional IPX uses a routing table update protocol analogous to IP's RIP. Thus, there will be a routing table maintenance-related overhead associated with each protocol rather than just IP. However, if other protocols like IPX are to be used end to end, the routing tables they are using must be updated in some form to ensure that these protocols still retain unimpaired visibility of their own subnetwork. Thus, while an encapsulation scheme could reduce this overhead—for example, by only transporting updates when there has been an explicit change to the tables—the routing table upkeep overhead cannot be totally eliminated.

Another limitation could be that of traffic volume accounting. In general bridge/routers do not excel in providing traffic accounting schemes. If available, they tend to be somewhat sketchy (e.g., without time-stamps or based on wraparound buffers without any indication if buffer wraparound has occurred) and limited to IP traffic. There will thus usually be no explicit accounting data, on a per protocol basis, for non-IP traffic that is being routed. However, given that accounting in general is still not deemed imperative in multiprotocol networks, this is not a major showstopper today.

If multiple protocols are to be routed across a bridge/router network, then the possibility of using APPN/HPR NN routing should not be arbitrarily overruled since APPN/HPR in this instance just becomes yet another protocol being routed across the WAN. In environments with multiple mainframes or AS/400s, APPN/HPR routing will provide direct one-hop routing, whereas using DLSw(+) may entail two-hop routing as was described in the last chapter. If in such multihost environments DLSw(+) is to be used in order to ensure an IP-only backbone, then remote SNA Switching, which is compatible with DLSw(+), should be considered as a way of overcoming the two-hop routing problem.

8.1.2 DLSw(+) and tn3270-like IP to SNA Gateway-Based Solutions

In an IP-only network either DLSw(+) or a tn3270-type IP-to-SNA gateway can be used to transport SNA interactions across the network. The underlying difference between these two approaches is that DLSw(+) transports actual SNA PIU message units end-to-end while the gateway approach terminates SNA at the gateway. Figure 8.2 illustrates the difference between using DLSw(+) to transport SNA PIUs end-to-end and that of using a gateway to translate IP packets to SNA. It also depicts how hybrid configurations of DLSw(+) and gateways can be used within the same network.

The advantage of DLSw(+) is flexibility and compatibility, while that of a gateway approach is primarily lower cost in that there are no downstream SNA devices that need to be supported. If traffic from

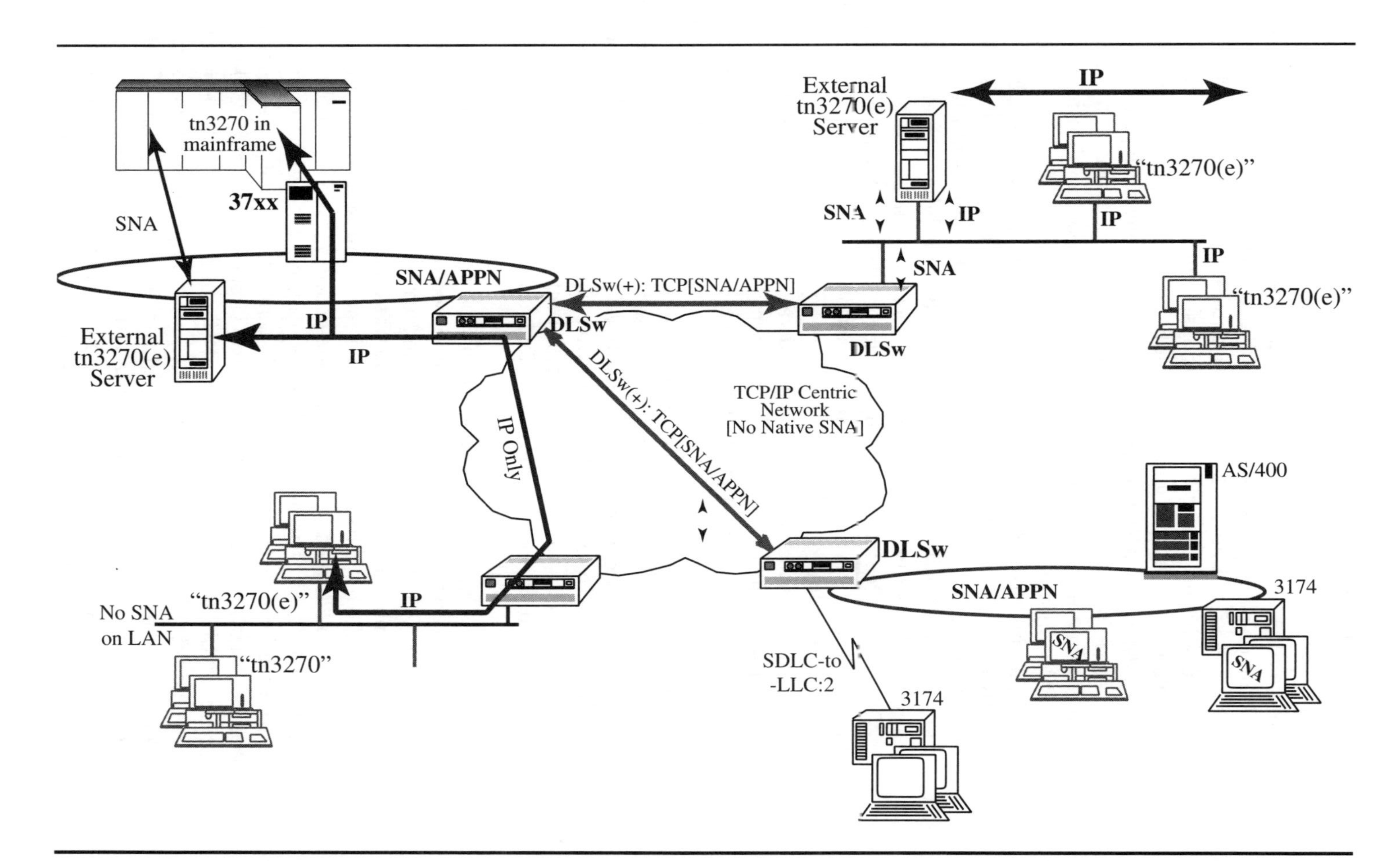

FIGURE 8.2 Various options for using IP-to-SNA gateways of DLSw(+) to realize a TCP/IP-centric SNA that has no native SNA/APPN traffic flowing across it.

genuine SNA devices (e.g., AS/400s, 3x74s without the Telnet feature, 4700 financial systems, automated teller machines) has to be transported across the network, DLSw(+) automatically becomes the only IP-based solution. The gateway solutions, which are in effect SNA PU Controller Gateways, in general are only targeted at PCs and workstations that require access to SNA applications. Obviously, hybrid solutions are a possibility—with DLSw(+) supporting traffic from SNA devices while a gateway scheme is used to support application access from PCs/workstations.

Using just TCP/IP to the PCs/workstations means that SNA does not obviously go end-to-end across the network. For a start, this means that there can be no SNA-based NetView/390 visibility of the activity status of the PCs/workstations. NetView/390 visibility will end at the gateway. It also means that SNA protocols are not used across the WAN for error-recovery and congestion control (e.g., pacing). Table 2.2 highlights the strengths and weaknesses of this gateway approach under the heading of PU Controller Gateway.

Obviously, such gateways do not always have to be deployed adjacent to the mainframes, that is, upstream from the WAN. Since most are PC or Unix workstation-based they can also be deployed remotely, downstream of the WAN. They will now, however, only talk standard SNA across the WAN. Since the goal is not to transport native SNA across the WAN, DLSw(+) can now be used to transport the SNA traffic to/from the remote gateways. This is yet another hybrid scenario in which DLSw(+) and IP-to-SNA gateways can be used in tandem. Figure 8.2 depicts the various hybrid configurations possible with DLSw(+) and IP-to-SNA gateways for supporting SNA interactions across an IP backbone.

IP-to-SNA gateways, whether deployed upstream or downstream, come in two distinct flavors: tn3270(e)-based or 3270 emulation-based. The difference between these two is the former uses the Telnet protocol for its terminal emulation on PCs/workstations, whereas the latter uses standard 3270 datastream to define and control its screen painting (e.g., displaying 3270 fields). The emulation-based gateway solutions (e.g., NetWare for SAA, Microsoft SNA Server) are likely to offer significantly more functionality than a plain tn3270 scheme. For a start, emulation-based gateways will provide extensive support for: local 3270-based printing; LU 6.2 program-to-program communications; 3270-oriented file-transfer schemes (e.g., IND$FILE), and comprehensive support for 3270 keyboard emulation replete with the "SYS REQ" (i.e., System Request key) type functions to facilitate session switching.

Though widely used, tn3270 was never considered by many to be the universal solution for providing TCP/IP-based access to SNA applications, given some of the limitations of tn3270. Most tn3270 implementations do not support printers. This is an unfortunate restriction given that most SNA/3270 applications rely on some form of remote printing. Compatibility with SNA/3270 applications, particularly when the tn3270 software is resident on the mainframe, is also an issue. Most tn3270 implementations also did not support all the 3270 key stroke permutations, thus imposing certain frustrating restrictions when trying to use certain applications or switch between applications.

Some of these limitations, most notably that of local printing, have been addressed by tn3270e. However, annoying limitations persist. Most tn3270e implementations do not support LU 6.2-based communications—instead they concentrate on traditional terminal-based access to terminals. Typically all the limitations of tn3270(e) can be overcome by using a full-function 3270 emulator. Obviously the issue here is one of cost. Given its Telnet roots, the PC/workstation component of tn3270, if not the mainframe-side gateway, has traditionally been inexpensive. Today, tn3270 for Windows or Unix workstations is available as no-charge free-ware from various bulletin boards. On the other hand, good 3270 emulation software is rarely available for under $100 per PC/workstation.

Yet again hybrid solutions are an option. Today, most, if not all, gateways that support 3270 emulation also support tn3270(e). Thus, cost compromises can be made by providing heavy-duty SNA users with 3270 emulation, while casual users who only require ad hoc access to SNA applications could be given tn3270(e).

The diversity of IP-to-SNA gateways does not, unfortunately, even end here. When it comes to such gateways the prevailing theme is that of variety and a multiplicity of disparate ways to achieve the same end result. The following section looks at the various possible IP-to-SNA gateway solutions that are available today.

8.1.3 Channel-Attached Gateways and Off-Loading TCP/IP from Mainframes

Today, tn3270(e) solutions are not contingent on tn3270 and TCP/IP software being resident on the mainframes, as was the case with the original tn3270 implementations. Instead it is possible, and now quite common, to have external tn3270(e) servers. These external servers can be interfaced to a mainframe through direct channel attachment, a LAN interface, or even an SDLC link.

Given that gateways initially developed to support 3270 emulation now also support tn3270(e), these external tn3270(e) servers do not have to be restricted to just dealing with tn3270 traffic. Instead, they can be full-function gateways that support both 3270 emulation and tn3270(e). In the casesf of NetWare for SAA and Microsoft SNA Server, this type of gateway can even be channel-attached using offerings from Bus-Tech or Cabletron. Therefore, *some* of the key permutations possible for realizing an IP-to-SNA Gateway solution for accessing SNA applications across an IP backbone include the following:

- tn3270 on PCs/workstations with tn3270 software on the mainframe
- tn3270 on PCs/workstations with an external, possibly channel-attached, tn3270 server
- tn3270 on PCs/workstations with an external, possibly channel-attached, NetWare for SAA or Microsoft SNA Server Gateway
- Full 3270 emulation on PCs/workstations with an external, possibly channel-attached, NetWare for SAA or Microsoft SNA Server Gateway
- Hybrid solutions with tn3270 on some PCs/workstations and full 3270 emulation on other PCs/workstations with an external, possibly channel-attached, NetWare for SAA or Microsoft SNA Server Gateway
- Full 3270 emulation on PCs/workstations with a *remote* PU Controller gateway (e.g., NetWare for SAA or Microsoft SNA Server Gateway) with DLSw(+) being used to transport the SNA traffic to/from Gateway across the IP WAN

In the early 1990s quite a few enterprises installed TCP/IP software from IBM or third-party vendors on their mainframes right alongside mission-critical SNA applications. Supporting tn3270 was one rationale for TCP/IP on mainframes. Another was to facilitate FTP-based file transfers between mainframes and Unix systems or workstations. If tn3270 and FTP are the only TCP/IP applications being used on the mainframe it is no longer imperative that TCP/IP software is installed on the mainframe.

There are value-added gateway solutions, some channel-attached, that can now be easily used to realize both tn3270(e) and FTP-based file transfer to/from a mainframe without the need for mainframe TCP/IP. Mainframe TCP/IP, especially if purchased from IBM, can be rather expensive with a purchase price for some of the larger mainframes in the $100,000 range and a $6,000-per-month maintenance cost. In addition, mainframe TCP/IP can consume a fair amount of processor cycles—hence the growing market in TCP/IP off-load solutions from both IBM and other vendors. Thus, any solution that obviates the need for TCP/IP on the mainframe while still providing comprehensive support for tn3270(e) and FTP can be extremely attractive.

These tn3270 and FTP gateways are totally compatible with downstream tn3270 and FTP clients running on PCs/workstations or Unix systems. To ensure this compatibility such gateways obviously use standard TCP/IP interactions with the downstream clients. However, they do not use TCP/IP to talk to the mainframe. For a start they cannot since there is no longer any TCP/IP software in the mainframe. Instead they use standard SNA–SNA/3270 for tn3270 applications and LU 6.2 for FTP applications. Hence, such gateways appear to the SNA mainframe software as bona fide SNA devices—for example, an AS/400. This type of tn3270 and FTP gateway solutions, in channel-attached form, is available from vendors such as Bus-Tech, CNT/Brixton, Cabletron, and so on. Figure 8.3 depicts the general dynamics of a tn3270 and FTP off-load gateway.

Even if tn3270 and FTP are off-loaded from mainframes there might be other reasons for running TCP/IP software or Unix applications on a mainframe. It is relatively common, particularly so within the manufacturing and scientific community, to have bona fide TCP/IP applications running on IBM mainframes. In some cases these applications would be running on top of a true Unix operating system, such as IBM's AIX/ESA operating system for mainframes. Now there is another away: *OpenEdition MVS*. OpenEdition MVS permits Unix applications that conform to the POSIX standard to be run directly on IBM's flagship MVS/ESA operating system, using IBM's implementation of the OSF Distributed Communications Environment (DCE) to realize their TCP/IP-oriented communications needs.

TCP/IP off-load solutions are targeted at enterprises that have no choice but to run TCP/IP on the mainframe—typically to support native TCP/IP applications. TCP/IP off-load is available on a range of different offerings including: IBM 3172 Model 003 Gateways, Cisco's channel-attached bridge/routers and CNT/Brixton Convergence offerings. Off-load eliminates mainframe resources, in particular processor cycles, from having to be used to perform TCP and IP protocol-related functions. The IBM 3172 facility only sets out to reduce mainframe cycles and does not claim to improve end-to-end TCP/IP performance. Cisco and CNT/Brixton off-loads reduce mainframe cycles as well as improving end-to-end performance.

8.1.4 Pros and Cons of Desktop DLSw

DDLSw or Cisco's NCIA is a way of restricting SNA to the outermost periphery of an IP-based network. With DDLSw or NCIA, SNA never leaves a PC/workstation or a mainframe data center. Unencapsulated SNA traffic does not even flow on LANs at remote sites, let alone the

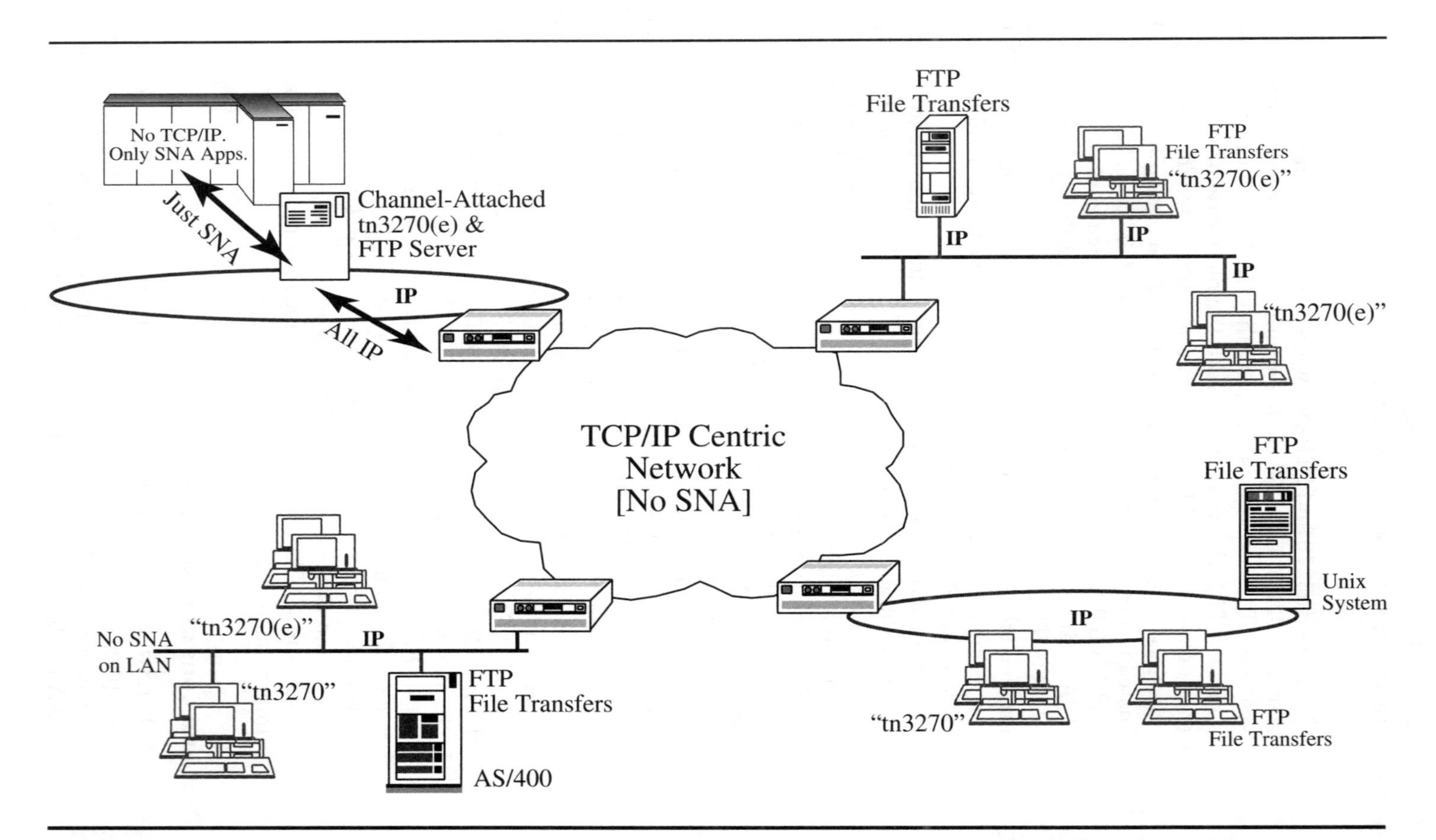

FIGURE 8.3 Using a channel-attached tn3270e and FTP Server to off-load TCP/IP processing and applications from the mainframe.

WAN. With DDLSw and NCIA, as shown in Figure 8.1, SNA traffic is encapsulated within TCP/IP right inside a PC/workstation by the SNA/3270 emulation software. With standard DLSw(+) such encapsulation only happens once the SNA traffic has reached a bridge/router across a LAN or SDLC link.

DDLSw per se, as supported by vendors such as Eicon Technology, uses DLSw encapsulation (i.e., no MAC framing inside TCP/IP packet) and protocols while NCIA at present is based on Cisco's RSRB framing (i.e., MAC frame within TCP/IP packet) and protocols. (In the remainder of this section the term *DDLSw* is used to embrace NCIA unless otherwise stated.)

DDLSw does not have a mainframe component—as yet. Thus SNA traffic to a mainframe has to be extracted from its TCP/IP packet by a bridge/router adjacent to the mainframe; see Figure 8.2. This bridge/router will also be responsible for encapsulating SNA traffic from the mainframe prior to it arriving at the LAN. In the case of NCIA this SNA encapsulation and extraction could be performed by a channel-attached bridge/router. Figure 8.4 depicts NCIA being used with a channel-attached Cisco bridge/router. Thus, at present, DDLSw and NCIA at the mainframe side is no different from using standard bridge/router-based DLSw(+). This, however, could change. It would be possible to develop a mainframe DLSw(+) that would do the TCP/IP encapsulation/de-encapsulation functions within the mainframe. In this case, TCP/IP software will be required on the mainframe to support this DLSw(+) application.

DDLSw should not in any way be confused with tn3270(e). They are very different schemes. With DDLSw, just as with DLSw, SNA is used end-to-end, mainframe-to-PC/workstation. It just happens to be encapsulated within TCP/IP for much of this end-to-end trip. With DDLSw, NetView/390 retains visibility of the actual physical PCs/workstations. With tn3270(e), SNA is not used end-to-end and NetView/390 does not actually see the physical PCs/workstations accessing the mainframe applications.

NCIA is very different to tn3270 in that true SNA is used end-to-end, from the end station to the mainframe—albeit encapsulated within TCP/IP. With tn3270, TCP/IP Telnet, rather than SNA, is used at the end station—with Telnet to SNA/3270 protocol conversion taking place either in the mainframe or at a server adjacent to the mainframe.

There is no real justification for using DDLSw with LAN-attached PCs/workstations. The argument that it relieves bridge/routers from having to do DLSw(+) does not really wash. DLSw(+)-capable bridge/routers are getting increasingly more powerful and less expensive. This, on the other hand, is not the case with the *installed base* of PCs.

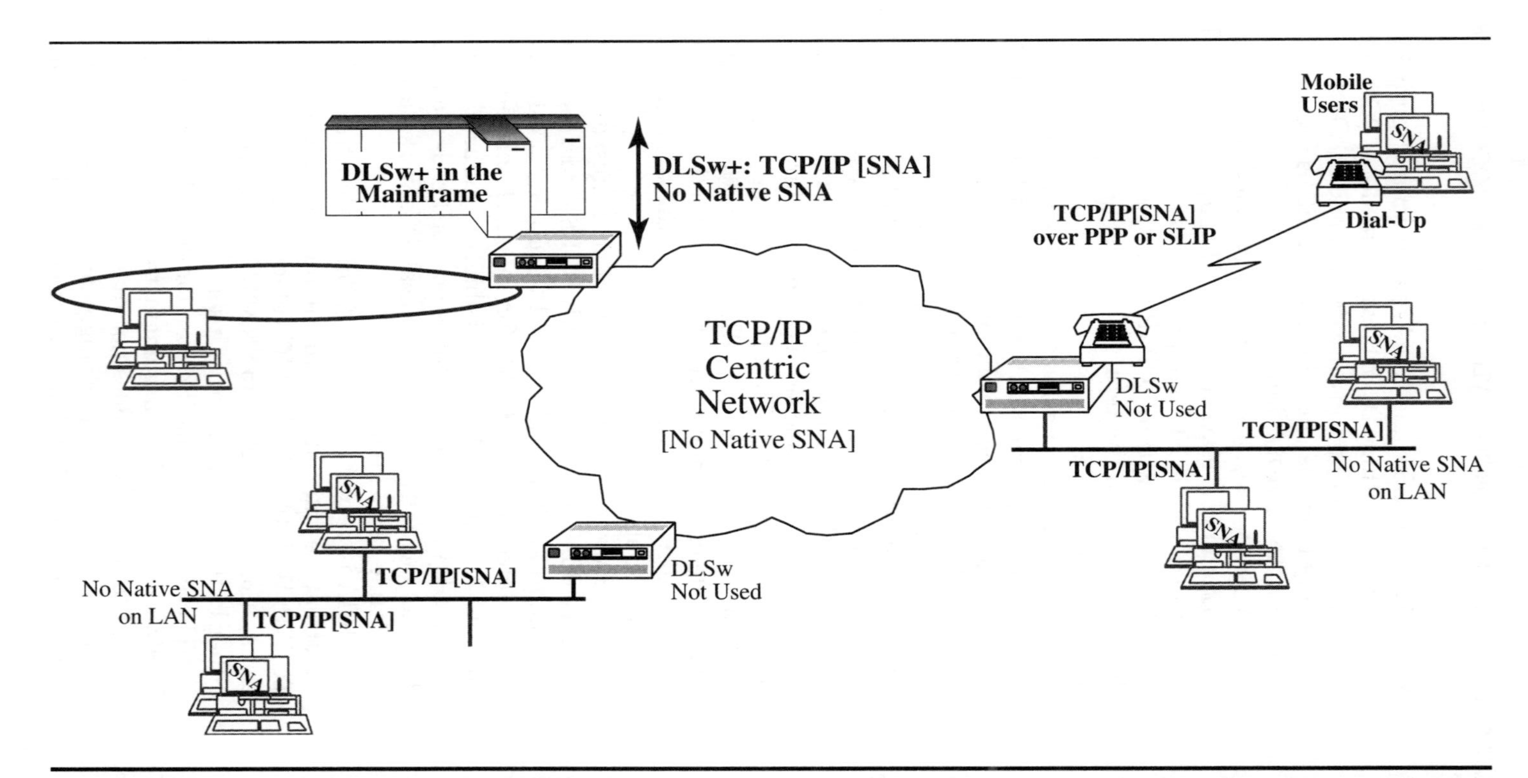

FIGURE 8.4 Taking the concept of desktop DLSw all the way to include DLSw in the mainframe and thus force SNA to the very periphery of the network.

With DDLSw, Intel 386DX-based PCs, already laboring under the strain of running Windows 95, are now expected to do more networking-oriented functions. To exacerbate this, Windows 95, OS/2 Warp, Windows NT, and Unix now support preemptive multitasking, that is, the ability to concurrently run multiple applications.

With preemptive multitasking, processor cycles not being used by one application can be gainfully exploited by others. DDLSw does not off-load networking functions from PCs, thereby freeing up processor cycles for end user applications (e.g., spreadsheets, graphics, etc.). Instead, it increases the amount of overhead processing that has to be done at the desktop. Ironically, DDLSw is in effect a DLSw off-load for bridge/routers! Remember that the issue here is not the viability or attractiveness of DLSw. The value of DLSw, in IP-centric networks, has already been clearly established. The issue here is whether LAN-attached PCs/workstations should do DDLSw as opposed to bridge/routers.

DDLSw does offer some advantages to mobile SNA users who on a regular basis have to dial into a bridge/router from a laptop to gain access to a corporate network. There are no commonly accepted, or notably efficient, standards for transporting SNA in asynchronous (i.e., start-stop) mode. With DDLSw, at least, dial-up users requiring SNA-based application access will be able to easily use Serial Line Internet Protocol (SLIP) or the Point-to-Point Protocol (PPP), the two de facto standards for remotely accessing bridge/routers.

LAN-attached PCs/workstations do not need TCP/IP-oriented protocols such as SLIP or PPP to talk to a bridge/router. They can do so with impunity using native MAC/LLC:2 protocols. It is also worth noting that DDLSw does not obviate the need for MAC frames being used across the LAN between the PCs/workstations and the bridge/router. The TCP/IP packet containing the SNA PIU still has to be transported across the LAN within a MAC frame. Thus, DDLSw just adds more headers, and consequently traffic, to the frames going over the LAN.

DDLSw (more so than NCIA) at present has one near insurmountable problem—it is not scalable. In Chapter 5 the scalability problem of DLSw brought about by DLSw's need for two active TCP connections per remote site was discussed in detail. DDLSw compounds this problem. With DLSw, each remote bridge/router requires two TCP connections. This, at present, precludes DLSw from being successfully used in networks with, say, over 500 remote sites. With DDLSw, each PC/workstation requires two TCP connections. This will preclude DDLSw from being used concurrently by more than around 500 PCs/workstations. From an SNA standpoint, a network with only 500 concurrent 3270 users is a relatively small network.

There are other issues such as the added network administrator cost of having to administer a DDLSw-related IP-address to each PC/workstation rather than being able to restrict such address administration just to bridge/routers. Everything considered, DDLSw rarely has any real justifiable value in dial-up applications. Any temptation to use DDLSw on LAN-attached PCs/workstations should be carefully analyzed and then quietly curbed.

Ironically, IBM's AnyNet, as opposed to DDLSw or NCIA, can in theory prove to be a better option if there really is a burning desire to restrict SNA to the very periphery of an IP network. AnyNet, in marked contrast to DDLSw and NCIA, already has a mainframe component—AnyNet/MVS. The Windows and AIX (i.e., Unix) versions of AnyNet, however, as yet only support LU 6.2 traffic. Only the OS/2 version, namely, AnyNet/2 Version 2.1, at present supports both SNA/3270 and LU 6.2. This paucity of AnyNet support on the client side is in reality not that different from the availability of DDLSw and NCIA offerings. DDLSw and NCIA support, at best, is only available on less than a handful of SNA/3270 emulators—yet another hurdle when trying to realize this type of solution to ban unencapsulated SNA from entering the network.

8.1.5 Reverse Protocol Conversion

The last piece of the puzzle, in quite a few instances, when deciding to standardize on an IP-centric backbone is trying to productively assimilate the installed base of 3270 terminals into the new network. Writing off and getting rid of this installed base, which could be sizable, though the most expedient and tempting option, may not be immediately viable due to either fiscal constraints or the desire to minimize the overall disruption associated with the network cut-over. In effect this is the same dilemma faced by enterprises moving to a multiprotocol LAN/WAN-centric network that are still saddled with large numbers of link-attached SNA devices. While they would like to standardize on LAN-attached devices, this is not always possible. Hence, they resort to SDLC-to-LLC:2 conversion. Fortunately, there is a corresponding technological solution to facilitate the integration of 3270 terminals into a non-SNA-based network. This technology is referred to as *Reverse Protocol Conversion*.

Reverse protocol conversion permits either bona fide 3270 terminals attached to a 3x74 control unit, or PCs/workstations running a full SNA/3270 software stack (e.g., IBM Personal Communications/3270), to access TCP/IP and other non-SNA (e.g., DEC, Tandem, Stratus) applications. It is in effect the converse of tn3270. There is at present considerable demand for reverse protocol conversion. There is still an installed base in excess of 2 million 3270 terminals. In addition,

PC/workstation users who spend the bulk of their time accessing SNA applications often prefer, both for cost and consistency reasons, to run SNA/3270 software, rather than just having TCP/IP software (tn3270) or swapping between SNA/3270 and TCP/IP software stacks depending on what application they have to access.

Reverse protocol conversion can be realized in a variety of ways. In the mid-1980s, mainframe-resident conversion programs, working in conjunction with modified SNA protocol converters that provided SNA ports on one side, and asynchronous ports on the other side, were the only viable solutions. A-Net, a product marketed by IBM for a while, was an example of such a mainframe-based solution. Mainframe-based solutions were never optimum. They consumed expensive mainframe resources, and tended to have sluggish performance since all of the 3270 traffic had to continually be relayed through a mainframe and a 37xx.

Today, IBM 3174 microcode control software includes built-in Telnet client support to permit direct access to TCP/IP applications from 3270 terminals. The TCP/IP access will be realized across a token-ring or Ethernet LAN. Thus the 3174 has to be LAN-attached and have access to the systems running the TCP/IP applications either directly across the LAN or through a bridge or bridge/router network. If such LAN-based access is not convenient, there is another alternative approach. The 3174 now has an optional Asynchronous Emulation Adapter that enables 3270 terminals to act as asynchronous terminals. These solutions are, however, unfortunately contingent on customers having 3174s, as well as up-to-date 3174s with fairly recent microcode.

A generic reverse protocol conversion product that will work with any 3270 controller or PC/workstation configuration is available from CNT/Brixton as a part of their Convergence family of solutions. This CNT/Brixton software can run any of the popular Unix workstations. This reverse protocol conversion approach, just like CNT/Brixton's companion remote SNA session switching solution that was described in Chapter 5, is based on deploying a mini-SNA-mainframe capability replete with SNA SSCP functionality. This mini-SSCP function is a subset of the SSCP functionality provided by a bona fide Type 5 node such as ACF/VTAM, and is the same software as that used to provide Remote SNA Session. The SSCP function permits this software to masquerade as an SNA mainframe.

When activated, it issues the SSCP session activation requests (i.e., the so-called ACTPUs and ACTLUs) to the physical 3270 devices. Thus, the 3270s end up establishing these SSCP control sessions with the CNT/Brixton SSCP as opposed to with a mainframe-resident ACF/

VTAM. Consequently, the CNT/Brixton SSCP ends up controlling and "owning" all the downstream 3270 devices. (In technical terminology, this means that the PUs and LUs in the various SNA devices end up establishing their SSCP-PU and SSCP-LU sessions with the CNT/Brixton SSCP.)

When performing reverse protocol conversion, the CNT/Brixton software intercepts and terminates all of the SNA sessions. This again is analogous to what happens with remote SNA session switching. The software then strips the 3270 datastream field from within the SNA PIU frame and inserts it as data within a TCP/IP packet. The TCP/IP packet is then routed to the relevant TCP/IP application across the TCP/IP backbone. To ensure that the downstream 3270 devices, front-ended and owned by the CNT/Brixton SSCP, can also easily access an SNA application running on mainframes, the software provides a no-intervention pass-through SNA gateway that forwards unaltered SNA PIUs to the mainframes. Figure 8.5 shows a reverse protocol conversion configuration based on the CNT/Brixton mini-SSCP approach. This reverse protocol conversion can also be used in conjunction with remote SNA session switching such that some 3270s are accessing TCP/IP applications while others are switching between various SNA applications running on different mainframes.

The bottom line here is that there is a plethora of technology such as tn3270(e), IP-to-SNA gateways, DLSw(+), and reverse protocol conversion to facilitate interworking between SNA and TCP/IP. Hence standardizing on an IP-centric backbone while still running mission-critical SNA applications is no longer a daunting proposition. Obviously, as has been repeatedly stressed, such an IP-centric solution should only be considered if SNA traffic volumes are already down and continuing to decline as a result of SNA applications being usurped by Unix applications. If SNA traffic still accounts for over 50% of the WAN traffic mix and is likely to be so for many years to come, then a bridge/router or FRAD-based multiprotocol solution might be more appropriate than a pure IP-only backbone.

8.2 THE NETWORKING BLUEPRINT AND THE OPEN BLUEPRINT

AnyNet is the direct and only progeny of IBM's "Networking Blueprint" vision that was unveiled in early 1992. The Blueprint was uncannily perspicacious, topical, intuitively simple, and above all iconoclastic. It set out to categorically dispel the time-honored belief that all multiaccess application programs had to be branded at birth as belonging to one and only one network type. This automatic application-to-network-type branding that has taken place ever since people started writing

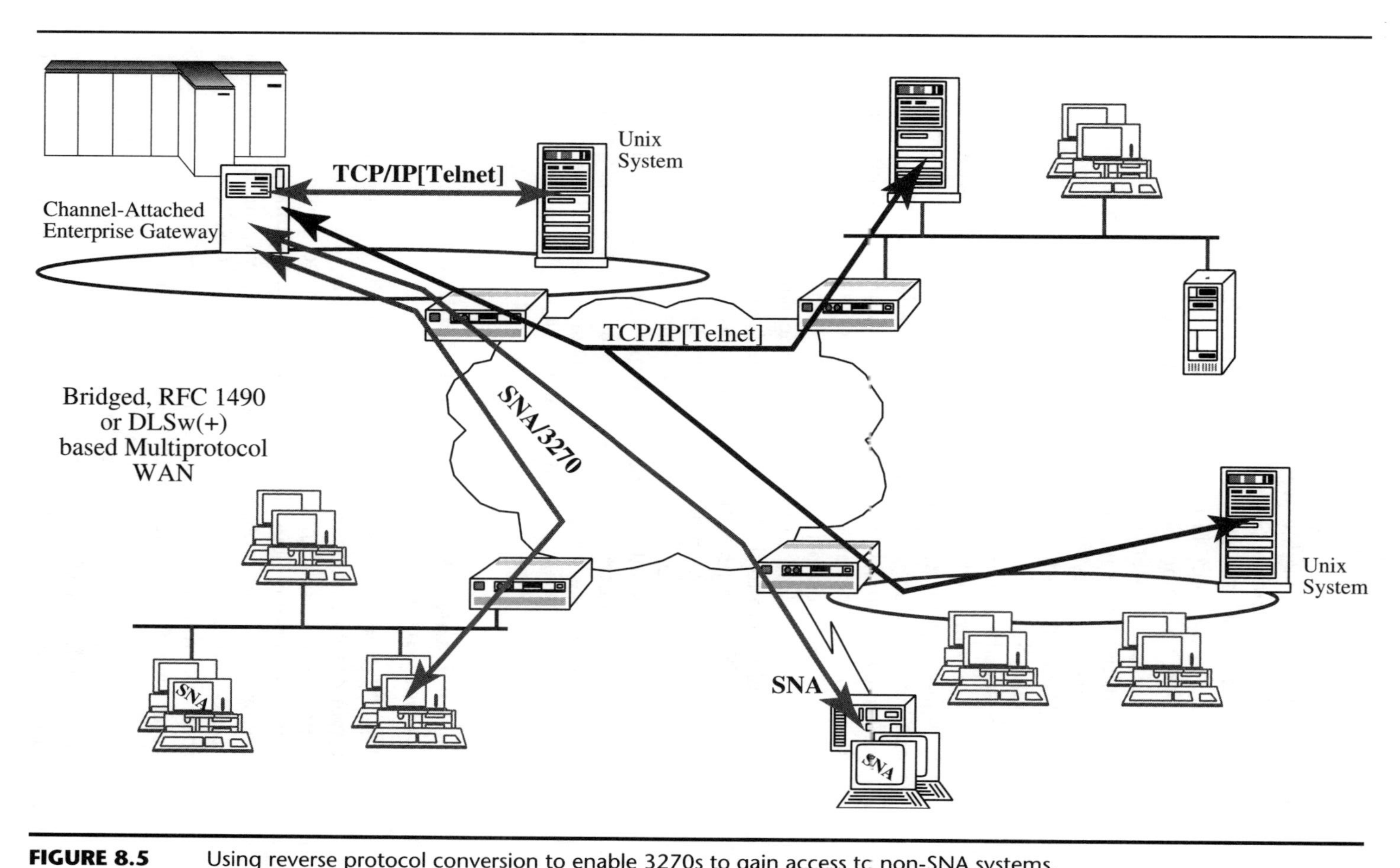

FIGURE 8.5 Using reverse protocol conversion to enable 3270s to gain access tc non-SNA systems.

applications accessible over a network has led to X.400 being classed as an X.25 application, SNMP and FTP being classed as TCP/IP applications, and TSO and DB2 being known as SNA applications.

In reality there is no reason why the higher-level problem-solving portion of an application (i.e., Layer 6 and above) has to be tightly glued onto a single networking stack. In the past, it was only a case of cultural and political one-upmanship that resulted in organizations, with IBM being as guilty as any, writing applications that were tightly and inescapably intertwined with their then favorite networking scheme, for example, SNA, X.25, or TCP/IP. Hence, applications were always network-specific when they should and could have been network-neutral.

A network-neutral application would just concentrate on its problem-solving charter. It would rely on a generic and network-type-independent interface (i.e., set of function calls à la RPCs or ACF/VTAM-like macros) for its networking services and functions. Once written, such an application, in theory, could be easily ported to run over various different network types. A layer of network specific "glue" software would be required to mesh the generic networking interface used by the application to that of a particular protocol stack.

The Networking Blueprint advocated the notion of network-neutral (or network-type-independent) applications. It presented a framework for developing such applications. The Networking Blueprint framework sets out to cleanly decouple the higher levels of an application from the network-type-specific lower layers in order to achieve its goal of such network-neutral applications. Figure 8.6 shows the overall architectural framework of the Networking Blueprint as presented by IBM in 1992.

8.2.1 The Common Transport Semantics and Multiprotocol Transport Networking

The CTS layer of the Blueprint sits above the transport layer (i.e., Layer 4) functions as envisaged by the OSI 7-Layer model. CTS represented somewhat innovative IBM-developed protocol mapping and conversion technology. The CTS conversion technology would take the Layer 5 output protocol from any of the various network services stacks that sit above the CTS (e.g., LU 6.2, OSF DEC, OSI T.P.) and then reshape that output to match the needs of the network type across which the application needs to be run.

This Layer 5 output conversion is not a permanent one-way-only conversion that, for example, changes LU 6.2 output to be totally compatible OSF DCE's TCP/IP-based protocols. Instead, the conversion is only to ensure that the Layer 5 output can be transported across a

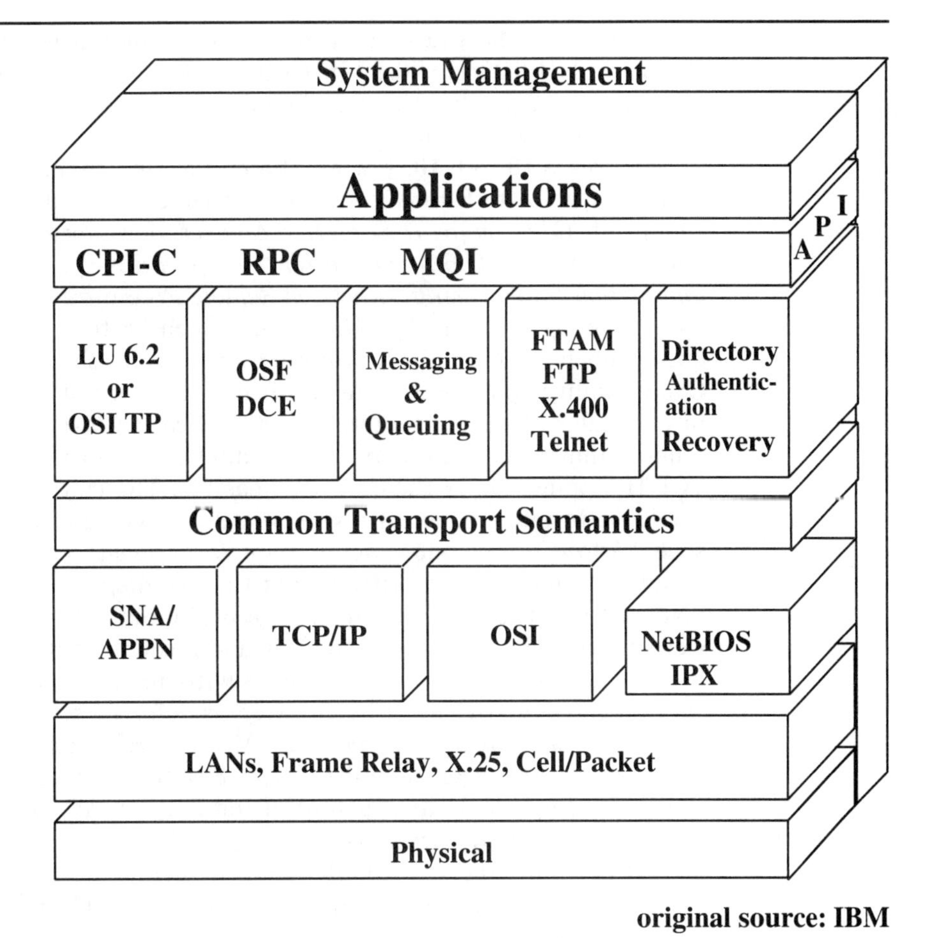

FIGURE 8.6 The architectural framework of IBM's Networking Blueprint.

nonnative network to its intended destination node. At the destination node, another CTS component will *convert the Layer 5 output back to its original form* so that it can be transparently passed up to the network services stack corresponding to the one that created the output.

This "A-to-B-to-A" conversion is the fundamental basis of CTS. Therefore, CTS technology will convert the SNA output from a LU 6.2 stack so that it can be transported across a TCP/IP network to its destination node. However, at the destination node, the output message unit that was conveyed using TCP/IP will be converted back into a bona fide SNA message unit and passed up to a LU 6.2 stack. The CTS conversions at the source and destination nodes together result in: "LU6.2 to TCP/IP to LU 6.2." It is very important to note that CTS, at least as yet, only talks about and deals with "same to same across an alien network" conversion.

Figures 8.10 to 8.13, toward the end of this chapter, graphically highlight the "A-to-B-to-A" conversion scheme used by the Networking Blueprint. *CTS, or for that matter AnyNet, does not even attempt to address how an LU 6.2 application may talk directly with a TCP/IP application that is using sockets or Remote Procedure Calls (RPCs) as its API for generating TCP/IP traffic.* Such "A-to-B" conversion requires much more than just protocol conversion at the transport layer. Compatibility and process synchronization is required at higher levels, in particular at the presentation and application layers. An AnyNet+ scheme that would allow an LU 6.2 application to talk to a sockets TCP/IP application is theoretically feasible. The challenge, however, would be to make such a conversion scheme generic. In most cases, such "A-to-B" conversion would require at least some application-specific customization to ensure that the two disparate ends (i.e., the SNA end and the TCP/IP end) were coordinated with each other and were in sync when it came to sending and receiving data.

The formal architecture that deals with all of the protocol conversion technology envisaged by the CTS is referred to by IBM as the *Multiprotocol Transport Networking* (MPTN) architecture. The MPTN architecture has been submitted by IBM to the Unix-oriented, X/Open standards body as a potential industry standard for realizing network-neutral applications. Consequently, the specifications for MPTN are now freely available from X/Open. The AnyNet protocol conversion products are an implementation of MPTN. (IBM now also uses the term AnyNet to include LAN-over-SNA encapsulation technology à la IBM's LTLW product. In general, though, most AnyNet products are still conversion- as opposed to encapsulation-oriented. In the remainder of this chapter, unless otherwise noted, AnyNet is used to denote MPTN-based solutions as opposed to LTLW-based encapsulation products.) In many cases the terms MPTN, CTS, and AnyNet can be used interchangeably to refer to the Networking Blueprint's protocol conversion technology for realizing network-neutral applications.

8.2.2 The MPTN Conversion Approach vis-à-vis DLSw(+)

When dealing with SNA across a TCP/IP network, the MPTN process can be thought of as being roughly analogous to that of DLSw(+) in that it just transports SNA data across a TCP/IP network. The fundamental difference between DLSw(+) and MPTN is that the former is a straight encapsulation scheme while MPTN is a conversion scheme. With DLSw(+) the entire SNA PIU including the SNA TH is encapsulated, untouched, and hence obviously unadulterated, within a TCP/IP packet. MPTN, in contrast, tries not to include the entire SNA PIU within a TCP/IP packet. Instead it maps the information carried in the SNA TH (e.g., destination address) to comparable fields in the IP header. It also uses a MPTN-specific header to carry information that cannot be included within the IP header. The SNA Layer 5 Sequence Number Field (SNF), which is a crucial 2-byte field in all SNA THs, is included within the MPTN header. Then, the remainder of the SNA PIU, that is, the RH and the RU, along with this MPTN Header are included with a TCP/IP packet for transportation across the IP network. Figure 8.7 depicts the difference between DLSw(+)'s encapsulation scheme versus MPTN's conversion approach.

Whether MPTN conversion is better than DLSw(+) when it comes to transporting SNA/APPN over TCP/IP is a highly debatable and contentious issue. What should be noted in this context is that MPTN, and consequently AnyNet, was not envisaged just to address the need to transport SNA/APPN or NetBIOS over TCP/IP. On the other hand, that is the express charter of DLSw(+). Thus comparing MPTN, or AnyNet, with DLSw(+) is somewhat inappropriate and unfair.

In terms of overall header lengths, the MPTN approach as shown in Figure 8.10 can be said to be slightly more efficient than DLSw, in that it does not include the TH and the typically 7-byte MPTN Header is shorter than the 16-byte DLSw Header. Given that the FID-2 THs used by SNA Type2/2.1 nodes (e.g., PCs, 3174s), and APPN nodes (e.g., AS/400s) are 6 bytes long, MPTN's TH elimination scheme coupled with its 7-byte header saves 15 bytes of header per SNA PIU compared to the encapsulation scheme used by DLSw. (The savings will be greater when compared to DLSw+, which also includes MAC/LLC framing around the PIU.)

This 15-byte saving in header lengths, however, has to be offset by the fact that MPTN's conversion approach is likely to require slightly more processing effort, at either end. The bottom line here is that when it comes to transporting SNA/APPN over TCP/IP, both the MPTN approach, as implemented by AnyNet and the IBM 2217, and DLSw(+) are viable techniques with neither having a tangible technical edge over the other. Thus, the choice between AnyNet/2217 and DLSw(+) will invariably boil down to other factors—for example, that DLSw(+) is available on full-function, multiprotocol bridge/routers whereas the AnyNet/2217 solutions typically only support a limited repertoire of protocols.

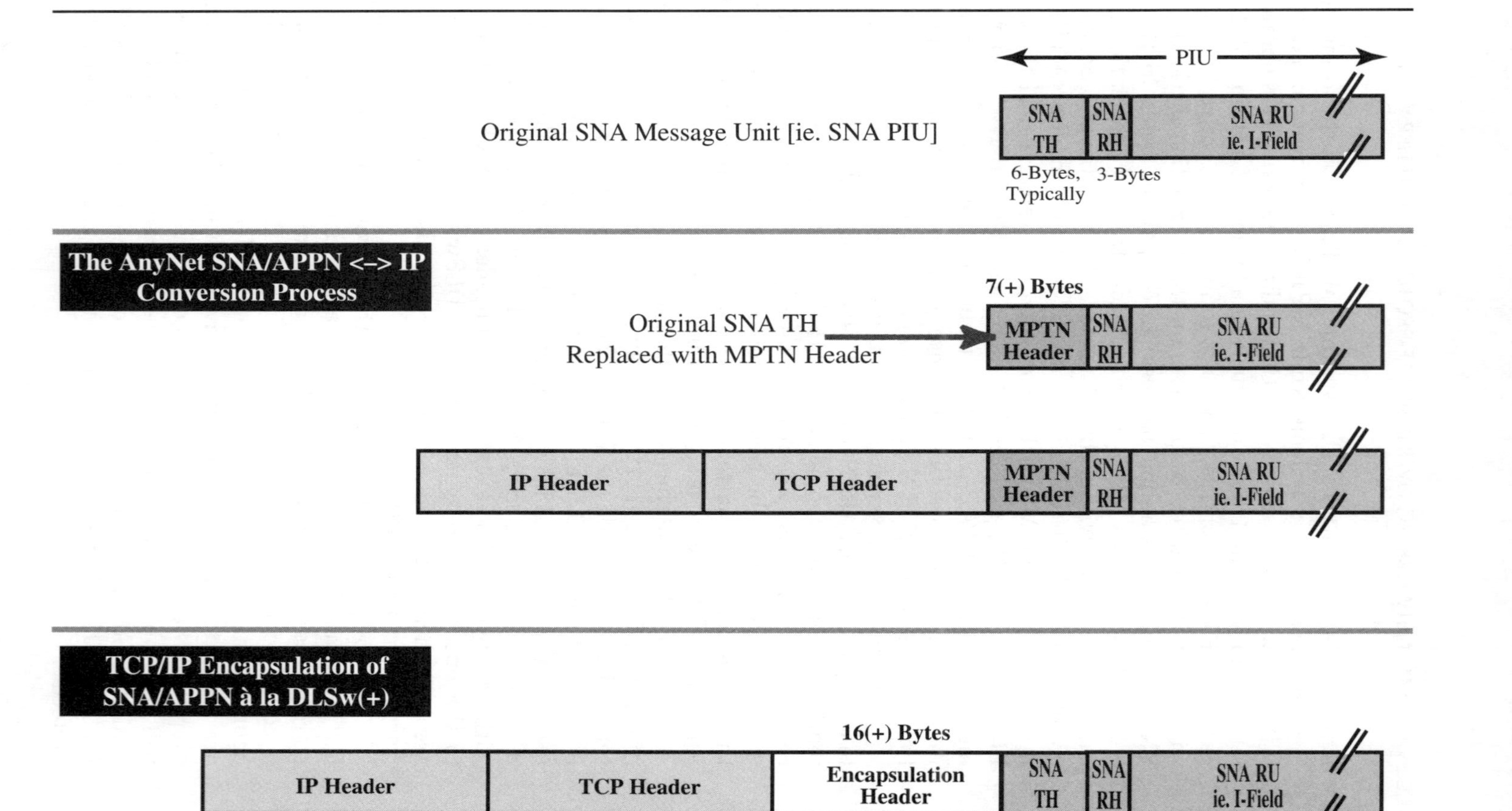

FIGURE 8.7 How the end result of the AnyNet conversion differs from TCP/IP encapsualtion à la DLSw(+).

8.2.2 MPTN's Functional Compensation and Address Mapping

The MPTN conversion process, in addition to performing format translations such as changing headers from one format to another, would also compensate for functions used by the source protocol but not supported by the destination (or target) protocol. It also takes care of mapping destination addresses from one protocol to the other in order to transport the source message units to their destination across a nonnative network using a different transport protocol, for example, mapping destination IP addresses in order to deliver SNA message units to their destination node across a TCP/IP network.

The compensation facility of MPTN is required because the transport protocols in popular use each have their own unique set of functional capabilities. Consequently each protocol will have some functions not supported by other protocols. For example, SNA supports the notion of expedited flows whereby certain requests within a given session can leap-frog normal-flow requests at all queuing points so that they can reach their destination prior to normal-flow requests that were sent ahead of them. SNA typically uses expedited flow to send network control requests that may be needed to maintain session integrity. In essence, expedited flow is a means for assigning an ultra-high priority to specific requests within a given session.

NetBIOS does not have a built-in facility to permit such expedited flows. TCP does have a so-called urgent mode. It is, however, in effect only a means of *notifying* the other end that urgent data is being sent across the connection, as opposed to an SNA-like mechanism to actually ensure that urgent data is guaranteed precedence over normal flow data. Thus, when converting SNA to TCP/IP, the MPTN conversion logic needs to provide a compensation mechanism whereby TCP's urgent mode can be suitably extrapolated to provide an acceptable emulation of SNA's expedited flows.

Expedited flow is but one example of an instance where functional compensation is required. There are others. TCP/IP and NetBIOS, as well as most of today's LAN-oriented protocols, support multicast addressing whereby a copy of a single message unit can be sent to a group of destinations who all share a common group address (or multicast address) in addition to their own unique address. SNA and, for that matter, SDLC both support this concept of multicasting. (SDLC has a broadcast address, but not a mechanism of subsetting such a broadcast so that it only goes to multiple, but not all, destinations.) The only way to do the equivalent of a multicast in SNA is to individually transmit a copy of the intended message unit to each recipient—addressing each by their own unique SNA Network

Address. Hence, when doing protocol conversion from TCP/IP or NetBIOS to SNA, MPTN has to compensate for SNA's lack of support for multicasting.

SNA and NetBIOS also transmit data in the form of records, that is, structured and delimited fields recognized by both ends. TCP/IP, on the other hand, sends data in the form of streams of bits with no explicit record boundaries. This is yet another function that has to be compensated for when trying to do meaningful protocol conversion between SNA or NetBIOS and TCP/IP.

In addition to providing functional compensation, MPTN also has to tackle the whole issue of converting between the disparate addressing schemes used by the protocols in order to deliver message units from the source node to the destination node. In some instances this address conversion may also have to embrace name (e.g., LU name) to address mapping if the source protocol uses names as opposed to addresses to initially identify destinations as is the case with SNA and APPN. Subsequently, when doing SNA/APPN to TCP/IP conversion, for example, MPTN requires a mechanism so that it can associate and convert SNA LU names to the appropriate IP address, at the transport level, for the purposes of having the message units delivered end-to-end across TCP/IP network to the destination node.

Note that the address mapping is only for the purposes of transporting the data at Layers 2–4 across the network to the destination node. At the destination node, the transport address will be again converted to identify the actual destination per the destination address specified by the transmitter. The need to convert the transport address back to correspond to the native destination address is again a facet of MPTN only doing "A-to-B-to-A" conversion.

This address reconversion at the destination node is another parallel between MPTN and DLSw(+) when it comes to transporting SNA/APPN over TCP/IP. DLSw(+) uses Layer 2 MAC addresses on an end-to-end basis to identify the SNA/APPN destination. It, however, uses an IP address to identify the bridge/router to which it has to deliver the encapsulated SNA/APPN message units so that they can be de-encapsulated and forwarded to the appropriate MAC address. The MPTN scheme is similar, except that the destination address, which is now a Layer 5 (or above) address, has to be recreated from the transport address used to deliver the messages to their destination node. The similarity is that in both cases the address used to transport the message units across the network is different from the address used to determine the eventual destination of the message units.

MPTN specifies three techniques for realizing address mapping. These are:

1. Algorithmic Mapping
2. Protocol Specific Directory
3. Dynamic Address Mapping

With Algorithmic Mapping, as implied by the name, an algorithm is used to convert between the two address schemes. This technique is typically used when transporting TCP/IP across an SNA/APPN network. The destination IP address has to be mapped to the name of the SNA LU that will accept the TCP/IP data traffic at the destination node and pass it on to the MPTN component so that it can be reconverted back to TCP/IP. IP addresses consist of a network ID component that identifies the IP subnetwork within the overall network and a host ID that identifies the destination node within that subnetwork. (This addressing structure in reality is no different from the SNA network address scheme that consists of the subarea address that identifies the SNA subnetwork and an element address that identifies the destination within that subarea.)

The MPTN algorithmic scheme for IP-to-SNA/APPN address mapping uses a two-step process. It will first map the network ID component of the IP address to an SNA/APPN network identifier (i.e., SNA/APPN Net ID) using a table lookup scheme in conjunction with a definition table previously set up by a Network Administrator. It will then convert the host ID to a destination LU name using a predefined algorithm. The algorithm used may convert the numeric host ID into a six-character text field. It will then prefix it with two alphabetic characters previously allocated to be the start characters of all destination LUs used by MPTN (i.e., AnyNet) within a given network.

The Protocol Specific Directory approach is typically used when transporting other protocols (e.g., SNA/APPN) across a TCP/IP network. TCP/IP networks have domain name servers that provide them with global directory services, that is, mapping destination names to the appropriate IP addresses. MPTN extends this DNS name resolution capability to embrace alien (e.g., SNA LU) names in addition to TCP/IP names.

MPTN's Dynamic Addressing Scheme is used when neither of the other two techniques can be gainfully utilized to achieve address mapping. It is typically used when transporting other protocols (e.g., IP) over NetBIOS or IPX. With this scheme a network administrator must define (or register) how a given destination address needs to be converted vis-à-vis MPTN. Once defined, MPTN will dynamically lookup the relevant entry for a given address and perform the necessary mapping.

8.2.3 The Networking Blueprint: Theory and Practice

The Blueprint per se is targeted at making new applications that are going to be developed network-neutral as opposed to making existing network-specific applications (e.g., SNMP) network-neutral. Consequently the Blueprint, as shown in Figure 8.8, concentrates heavily on APIs, Network Services, and Systems Management (à la SystemView). The Blueprint endorses three strategic APIs, which in turn serve as the exclusive conduits to three separate sets of cohesive networking services. In addition a repertoire of global, commonly needed, backbone utility functions are also identified. These backbone utility functions, which are meant to augment the functions provided by the networking services stacks include: directory services (à la X.500), mail and data distribution services (e.g., X.400), file access services (e.g., FTAM, FTP), security and authentication services, database recovery services, and so on.

The vision of the Blueprint is that the new network-neutral applications would be developed using one of the appropriate APIs and the network services accessible via that API. Exactly how an application would gain access to the backbone functions through the API it was using was never fully articulated. Once an application was developed in this manner CTS would step in and do its protocol conversion magic to ensure that the application could run across a nonnative network.

The three APIs endorsed by the Blueprint are: the SAA Common Programming Interface for Communications (CPI-C) first introduced by IBM in 1987, RPCs à la Unix, and the relatively new Message Queuing Interface (MQI). CPI-C is positioned as the preferred API for accessing either SNA LU 6.2 services, or OSI Transaction Processing (TP) services. The RPC mechanism, in its turn, provides access to the services repertoire offered by the OSF DCE, while MQI is the interface to the Message Queuing services. Right up front the Blueprint thus offers application developers the chance of being able to pick an API of their choice, and then a service set appropriate to that API.

LU 6.2 and OSI TP are targeted at transaction-processing applications, most likely interactive, that require extensive database access and update facilities, that are likely to span multiple, dispersed databases. The OSF DCE is aimed at Unix-oriented developers and particularly for batch-mode applications that do not require much interaction with distributed, relational databases. (OSF DCE support is now available on IBM mainframes and AS/400s.)

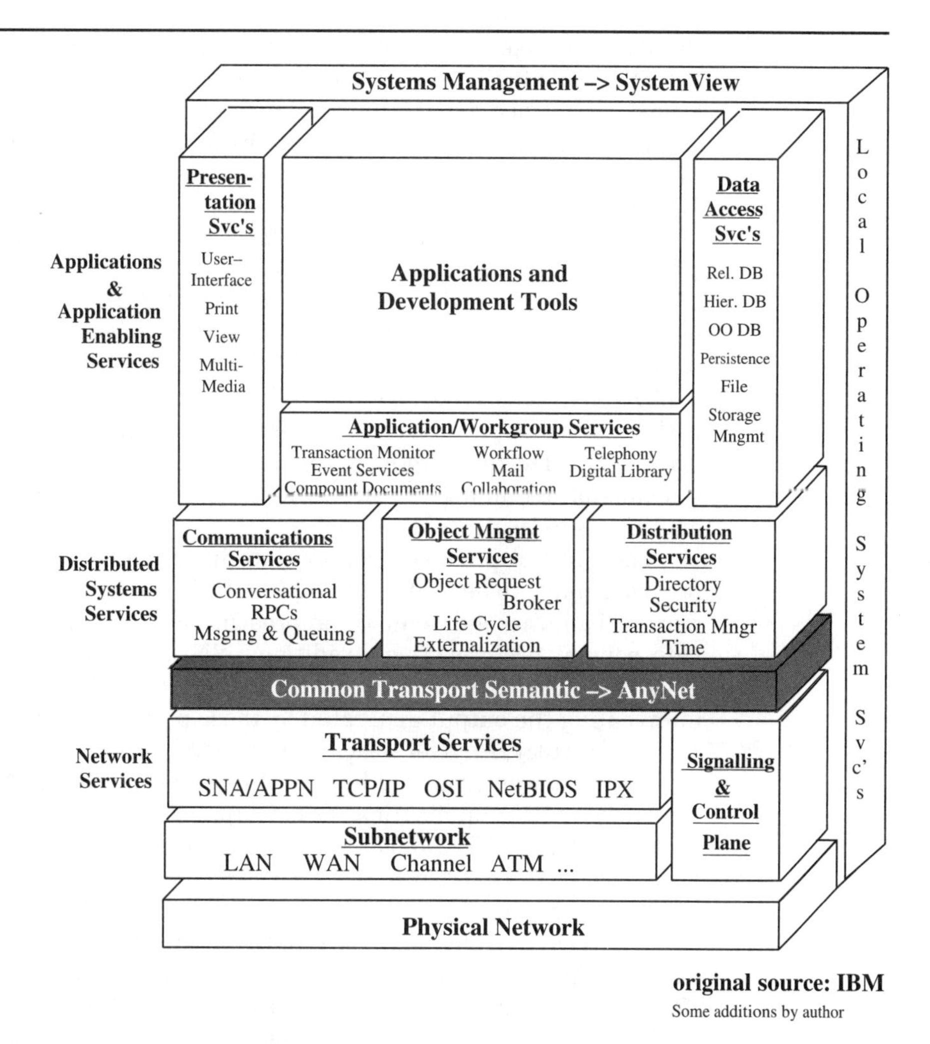

FIGURE 8.8 The architectural framework of IBM's Open Blueprint.

The message-queuing service is in effect a cross between RPCs and LU 6.2. It strives to offer the intuitive ease of use of RPCs, but to do so within a value-added framework that incorporates built-in facilities such as error-recovery and database resource resynchronization. In the case of RPCs such value-added functions had to be taken care of by the programmer, as opposed to the system software. Message queuing is based on the concept of interprogram interactions—on an asynchronous (i.e., not real-time) basis—via requests and responses exchanged through message queues maintained by the individual programs. It is a technique ideally suited for interactions between disparate platforms. It is aimed at emerging applications such as EDI, EFT, and so on. IBM, in conjunction with System Strategies Inc., now provides message-queuing services on MVS, VSE, OS/400, OS/2, RS/6000, DEC Vax, Tandem, and Stratus platforms.

In practice IBM could not wait for a whole new wave of new Blueprint-inspired applications to be developed that diligently adhered to CPI-C or MQI in order to exploit its MPTN (i.e., CTS and AnyNet) based vision of running TCP/IP applications across SNA/APPN networks, SNA applications across TCP/IP, or NetBIOS applications across SNA/APPN. What it needed was a scheme whereby MPTN could be used with existing applications as opposed to new applications. With the AnyNet repertoire of products this is what IBM now essentially has—a means of extending the network-neutral vision of the Blueprint to encompass existing applications.

The AnyNet products, rather than insisting that network-neutral programs have to be written using CPI-C, RPCs, or MQIs, take a much more pragmatic and realistic approach. AnyNet focuses on working with popular protocols (e.g., traditional SNA and LU 6.2) and common APIs (e.g., TCP/IP's sockets interface and NetBIOS's NetBEUI)—and intercepting the output generated by these protocols and APIs at Layer 5. Consequently, AnyNet in some sense can be thought of as going well beyond what was promised by the Blueprint. Hence, AnyNet will support, for any SNA application, sockets applications independent of OSF DCE, and NetBEUI applications even though Blueprint does not even acknowledge the existence of NetBIOS applications.

However, there are other situations where AnyNet can be accused of not covering all the bases originally promised by the Blueprint. For example, AnyNet, as yet, does not support any kind of application over IPX even though the Blueprint shows the possibility of running any application over IPX. AnyNet, somewhat understandably, also does not provide any support for OSI, though the Blueprint, reflecting its 1991–1992 vintage, gave prominent billing to OSI alongside SNA/APPN and TCP/IP. The bottom line here is that the Blueprint was the theory and that AnyNet is a pragmatic implementation of it in practice. The AnyNet products are described in more detail in the next section.

8.2.4 The New Open Blueprint

Undeterred in any way that the Blueprint vision essentially boiled down to AnyNet's more down-to-earth approach, IBM has significantly revamped the original Blueprint. The new Blueprint, that was unveiled in mid-1995 atypically with hardly any fanfare, is referred to as the *Open Blueprint*. Figure 8.8 shows the overall architectural framework of the new Open Blueprint as unveiled by IBM. A comparison of Figures 8.8 and 8.6 will immediately show that the overall structure and the bottom four layers of this new framework are nearly the same as that of the original Blueprint. The only change to the bottom four layers is the addition of the Signaling and Control Plane enhancement to cater to IDSN-based connection establishment and sustenance in ATM networks.

The upper layers of the Open Blueprint essentially expand upon and augment what were the Application Services, the APIs, and the Applications and Application Enabler components of the Blueprint. The Communications Services and Distribution Services components of the Open Blueprint in effect encompass all the communications services originally postulated by the Blueprint. What is new, but not unexpected, is the Object Management Services component.

The upper layers of this new architecture are uncannily reminiscent of the now unfortunately deceased Systems Applications Architecture (SAA). SAA was an inspired though hugely ambitious IBM initiative. Its original goal, when it was launched with much hoopla and hyperbole in early 1987, was to define a hardware platform-independent application execution environment—that would sit on top of mainframes, AS/400s, PCs, and Unix workstations. This hardware-independent application execution environment was complemented by a standard repertoire of communications services, and a standard style guide for developing the graphical user interfaces for this new breed of *platform-neutral* applications. Though not explicitly spelled out, platform-neutral applications would in effect also be network-neutral à la the Blueprint, since different platforms would each have their own native and preferred networking scheme.

The platform-independent program execution environment was defined in terms of seven ANSI standard programming languages (e.g., C, COBOL, FORTRAN, PL/I, RPG, etc.) and six APIs to key services. These services included: database access, database query services, SNA LU 6.2 (accessed via CPI-C per the Blueprint), and OSI TP. SAA just like the Blueprint, focused on new applications rather than existing applications. Thus, the theory with SAA was that if applications were written in one of the seven prescribed languages and accessed all the necessary services using one of the prescribed APIs, then these applications would be portable and could be easily moved from one platform to another. Note the similarity to the Blueprint vision, which came five years to the month after the unveiling of SAA and around the time that SAA was in its final death throes.

IBM's problem with SAA was that it was unable to provide the common program execution environment across all of its key platforms, such as mainframes, AS/400s, RS/6000s, and PCs. Without this there was no possibility of application portability, which in the end really was the founding precept of SAA. With application portability now but an impossible dream, SAA languished, inappropriately, for awhile as a kind of SNA+ (hence, for example, why Novell's SNA/3270 gateway is referred to as NetWare for SAA as opposed to NetWare for SNA/APPN). By early 1990, SAA was all but dead and all that really remained were communications and system management parts of it, with CPI-C, APPN, and SystemView being the most noteworthy.

Around 1992, IBM started talking about WorkPlace. This was in many ways another attempt to have a stab at SAA's Common Programming Execution environment, but now exploiting the common hardware and operating system possibilities brought about by the advent of the PowerPC and the PowerOpen/AIX standard Operating System for PowerPCs. (Note that the PowerPC RISC hardware is found on SP/2 scalable parallel mainframes, RISC AS/400s, PCs, and PowerMacs!)

The WorkPlace architecture consisted of four distinct components. These were: a micro-kernel, a set of common services, an object layer, and the GUI desktop. There were to be two versions of the micro-kernel: one for OS/2 and the other for OS/400. The purpose of these two micro-kernels was to provide a processor-independent execution environment for OS/2 and OS/400—that is, for PCs, PowerMacs, and AS/400s. Thanks to these micro-kernels, the upper-layer WorkPlace software would have been able to run on both PowerPCs as well as its native, original hardware (i.e., Intel x86 for OS/2 and AS/400 for OS/400).

The micro-kernel approach was a tangible means of trying to realize SAA's Common Programming Execution environment. Its goal was to shield the operating system from the hardware. To maintain consistency, and to retain the option for future enhancement, IBM also talked about a micro-kernel for the RS/6000. The need for this micro-kernel is debatable, given that IBM's compilers, as well as Advanced Interactive Executive (AIX), mask the few instruction set differences between the PowerPC and the RS/600 POWER processors.

The object of the common services component of WorkPlace was to provide the same utility functions—file server, transaction processor, DBMS—across all three execution environments. This component would have consisted of software that ran directly on top of the micro-kernel.

IBM intended to deliver three WorkPlace products—referred to as WorkPlace for OS/2, WorkPlace for OS/400, and WorkPlace for AIX—by 1995. A WorkPlace environment per se was not planned for Multiple Virtual Storage (MVS), which was always positioned as a server for WorkPlace clients. Though IBM is still supposed to be working on WorkPlace, there are indications that this initiative too is following in the unfortunate footsteps of SAA. The new Open Blueprint is in effect the latest chapter in the SAA and WorkPlace saga.

The upper layers of the Open Blueprint are nothing other than yet another crack at SAA's holy grail of developing platform- and network-neutral applications. There is no doubt that the world does need a scheme to realize such hardware platform independence and network-centric computing. Ironically, MicroSoft is making some major strides in making its applications (e.g., Word, PowerPoint, Excel, etc.) platform-independent. The latest versions of these market-leading applications use a common code base that is then mapped to Windows and the Mac OSs.

The Open Blueprint, like the Networking Blueprint, should serve as a basic template for products that try to implement at least a pragmatic subset of the overall vision. AnyNet was the outcome of the original Blueprint. It will, as an implementation of CTS, continue to play a role within the context of the Open Blueprint. AnyNet, though by no means as comprehensive as it could be, is at least a partial solution to network-neutrality. A similar practical initiative at the Application/Workgroup level to provide effortless mapping between applications and new services would be a very useful outcome of the Open Blueprint and be warmly welcomed. However, given the grandiose and overarching vision of this Blueprint it will be more difficult to come up with an AnyNet-like subset as was possible with the Networking Blueprint.

8.3 IBM'S ANYNET PRODUCTS

The initial AnyNet products were all pragmatic implementations of the MPTN architecture. The first AnyNet product, AnyNet/MVS, was available in mid-1993 just over a year after the Networking Blueprint had been unveiled. AnyNet/MVS, which was initially referred to as the Multiprotocol Transport Feature for MVS/ESA, supported both LU 6.2 across TCP/IP networks, as well as sockets applications across SNA/APPN applications. It included AnyNet code for OS/2 that could be downloaded to PCs running OS/2 from the mainframe. Consequently, AnyNet could be used between a mainframe and OS/2 workstation, two mainframes, or two OS/2 workstations. AnyNet was conclusive proof as to the viability of Layer 4 protocol conversion, as well as the possibility of network-neutral applications, as promised by the Blueprint.

Today, there is a relatively large repertoire of AnyNet offerings across all of IBM's so-called strategic platforms. AnyNet solutions are now available on mainframes (albeit just with MVS), AS/400s, systems running IBM's AIX version of Unix (e.g., RS/6000s and SP/2s), and PCs running either OS/2 or Windows. The protocol conversion permutations supported by AnyNet include: SNA (both traditional and LU 6.2) over TCP/IP, sockets (i.e., TCP/IP) over SNA/APPN, NetBEUI (i.e., NetBIOS) over SNA, sockets over NetBIOS, sockets over IPX. Figure 8.9 depicts the protocol conversions currently supported by AnyNet using the Blueprint framework to emphasize the relationship between AnyNet, MPTN, CTS, and eventually the Networking Blueprint.

The AnyNet protocol conversion products also come in two distinct forms: AnyNet access products and AnyNet Gateways. Access node software is installed in end nodes (e.g., OS/2 PCs, RS/600s, AS/400s, or even MVS mainframes) and provide API-oriented (e.g., sockets, CPI-C NetBEUI) protocol conversion for applications running in that node. With AnyNet access products, the protocol leaving a device (e.g., mainframe) would already have been converted to the transport protocol to be used across the network.

With AnyNet access products the protocol conversion is done within the device using the device's processor. Thus, in the case of SNA over TCP/IP, AnyNet in effect is doing what is advocated by Cisco's NCIA and Desktop DLSw—that is, ensuring that SNA is restricted to being within the end nodes. (Note, however, that NCIA and DDLSw rely on encapsulating SNA within TCP/IP packets while AnyNet will at least eliminate the SNA TH by converting the SNA PIU into a TCP/IP packet containing a MPTN header, an SNA RH, and the databearing SNA RU.) Figures 8.10 and 8.11 depict the types of A-to-B-to-A conversions possible with AnyNet access products. Note in Figure 8.10 that AnyNet/MVS requires TCP/IP software on the mainframe in order to support SNA over TCP/IP.

SNA over TCP/IP and sockets over SNA/APPN are the forte of the AnyNet access products. Both of these two conversions are supported on all platforms from mainframes down to PCs running Windows. All the other conversions mentioned above, such as NetBEUI over SNA and sockets over IPX are only available under OS/2. Ironically, the bulk of the AnyNet products deployed to date, and in particular the access products, have been used to transport SNA across TCP/IP, as opposed to the other way around. This is but another reflection of the gathering momentum toward IP-centric backbones and AnyNet's considerable prowess in helping to realize such IP-centric solutions.

8.3.1 AnyNet Gateways and LAN-over-SNA

An AnyNet Gateway typically sits in a stand-alone box (e.g., OS/2 workstation or IBM 2217) and provides AnyNet protocol conversion

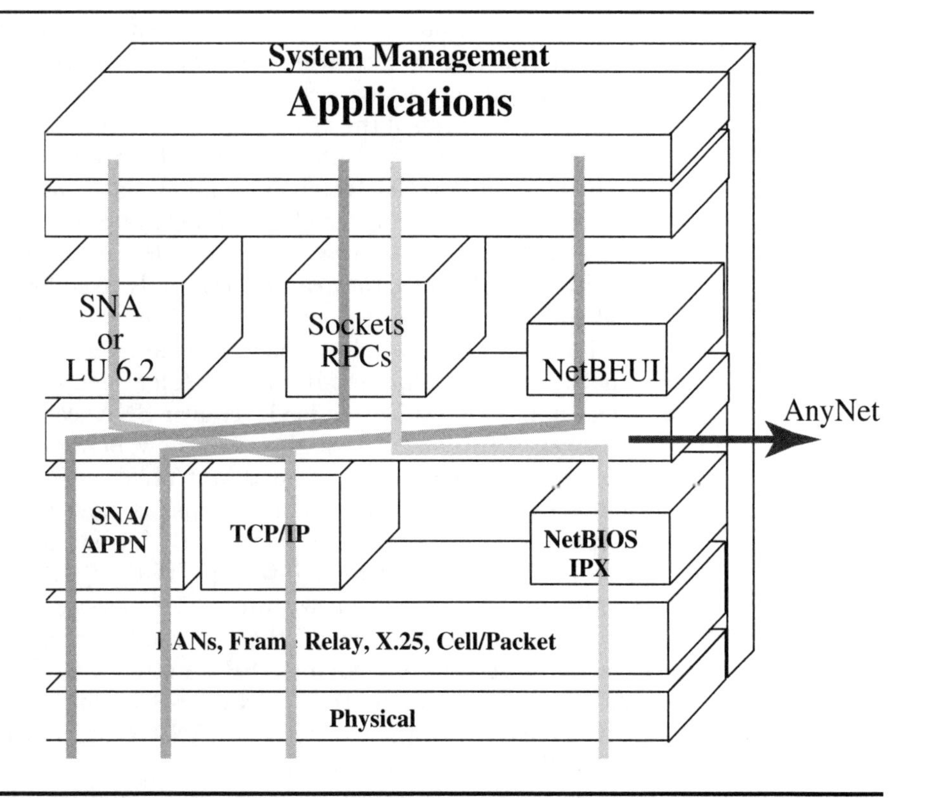

FIGURE 8.9 Some of the key protocol conversions now supported by the AnyNet suite of products.

services to multiple devices that do not contain any AnyNet software. The only exception to this is that AnyNet/MVS, though also an access offering, can double up as a gateway to enable SNA interactions to take place across a TCP/IP network. With AnyNet Gateways all the protocol conversion is done within the gateway. The devices using the gateway only transmit and receive their own native protocols.

AnyNet Gateways can be used to provide interconnection between systems using different transport systems though still supporting the same class of applications. For example, an AnyNet Gateway could be used to interconnect a OS/2 workstation running a sockets application on top of SNA using the AnyNet/2 access product with its counterpart running in native mode on a Unix workstation that has no AnyNet. Figure 8.12 shows the possibility of talking sockets-to-sockets between one workstation using SNA and the other using native TCP/IP via an

AnyNet Gateway. Note, once again that even though the transport protocols in the two workstations are different, the final AnyNet conversion is still sockets-to-sockets, as opposed to CPI-C to sockets. Figure 8.13 shows some of the other protocol conversion possibilities of AnyNet Gateways. AnyNet Gateways are software-only products. The 2217, which is an AnyNet-Gateway-in-a-box, is at present, the only gateway that is available as a self-contained system.

Given that AnyNet Gateways eliminate the need to have AnyNet software in the end nodes, they can be profitably used for SNMP across SNA network management applications where SNMP-based devices such as hubs or LAN switches need to be managed across an SNA/APPN network. Ironically, SNMP over SNA for such management scenarios was a major justification for LAN-over-SNA technology.

LAN-over-SNA can be thought of as now being a subset of the AnyNet initiative, especially given that LTLW, which was the only marginally successful product of this type, is now marketed as an AnyNet product. LAN-over-SNA was strictly based on encapsulation. A LAN-over-SNA router would intercept TCP/IP, IPX/SPX, or NetBIOS traffic and encapsulate it within LU 6.2 message units. Type 2.1 node, peer-to-peer protocol would then be used to route these LU 6.2 message units across an SNA or APPN network to the destination LAN. At the destination, another LAN-over-SNA router would receive the LU 6.2 message units, extract the LAN traffic, and put it on the LAN. It was bridging across SNA/APPN.

LAN-over-SNA when used across a traditional mainframe-centric SNA network did not result in the LAN traffic bearing LU 6.2 message units having to be continually relayed through the mainframe. Instead, these solutions relied on the Type 2.1 Integration feature, available in ACF/VTAM and ACF/NCP since mid-1987, that enabled Type 2.1 peer-to-peer traffic to be routed directly through a 37xx, without any intervention from ACF/VTAM, once the original peer-to-peer LU-LU session(s) had been established between the end-to-end LAn-over-SNA routers. Figure 7.2 shows a representative 2217-based AnyNet (LTLW) configuration as an example of what LAN-over-SNA was all about.

The rationale for LAN-over-SNA is exactly the same as that for AnyNet's IP over SNA—namely the possibility of implementing an SNA-only backbone that is capable of supporting a limited set of non-SNA protocols. This is also the raison d'être for the 2217. However, in part due to IBM's inexcusable apathy in aggressively promoting the advantages of SNA-HPR only backbones, most enterprises when deciding on standardizing on a single-protocol backbone opted for IP as opposed to SNA. Thus, LAN-over-SNA had limited market appeal akin to the market acceptance of IP over SNA compared to that of SNA over IP.

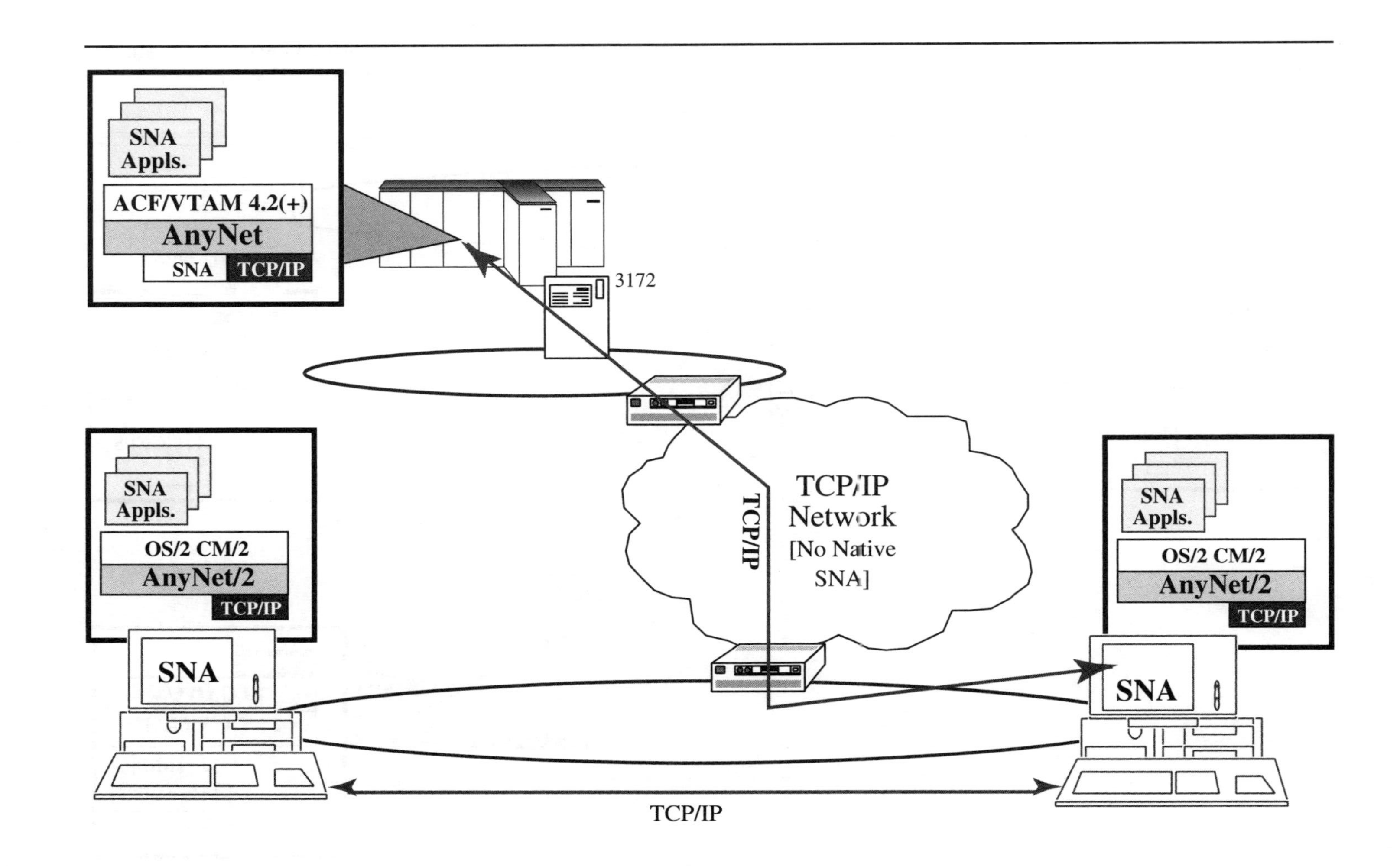

FIGURE 8.10 Using AnyNet to do SNA SNA across TCP/IP.

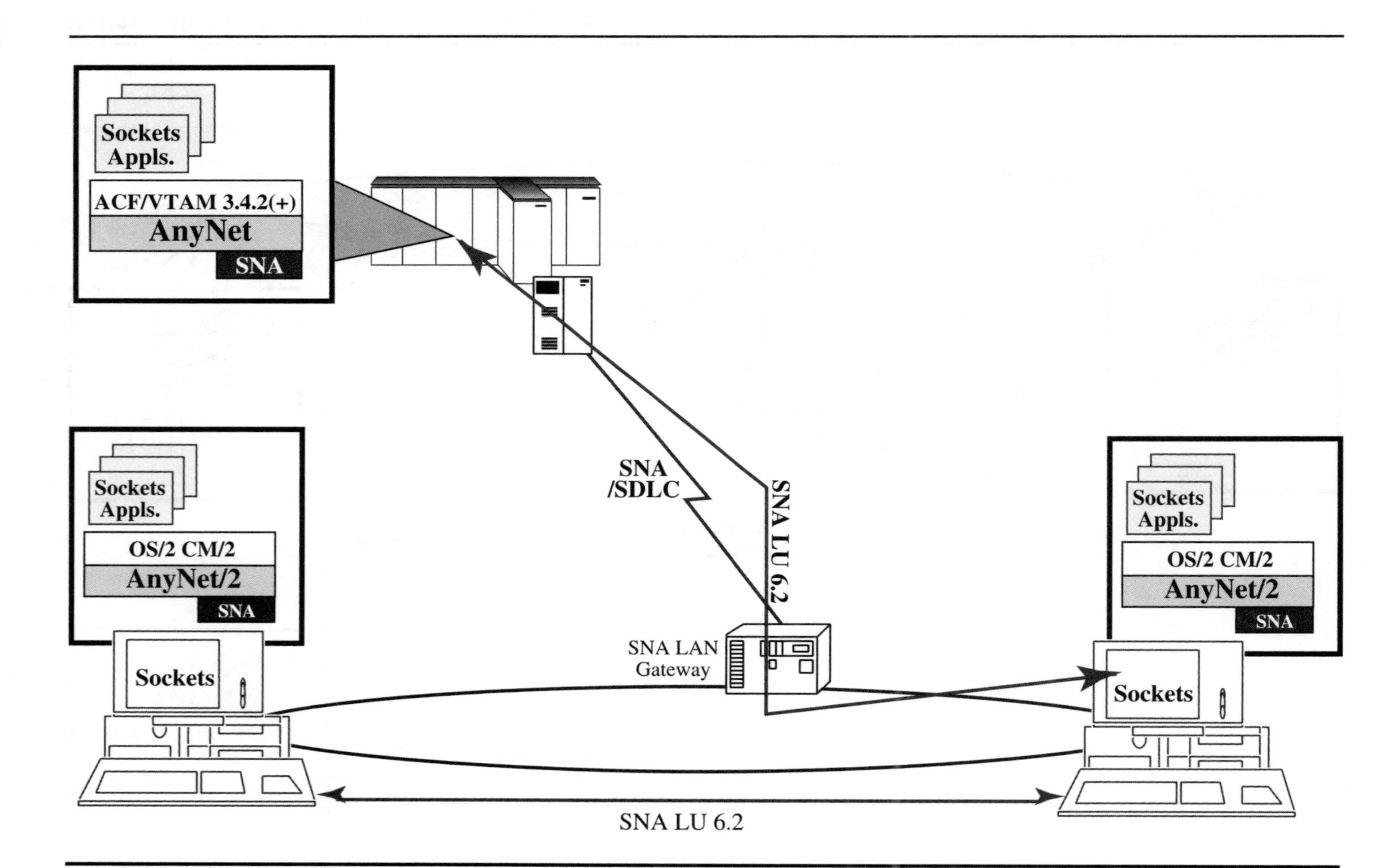

FIGURE 8.11 Using AnyNet to do sockets <—> sockets across an SNA.

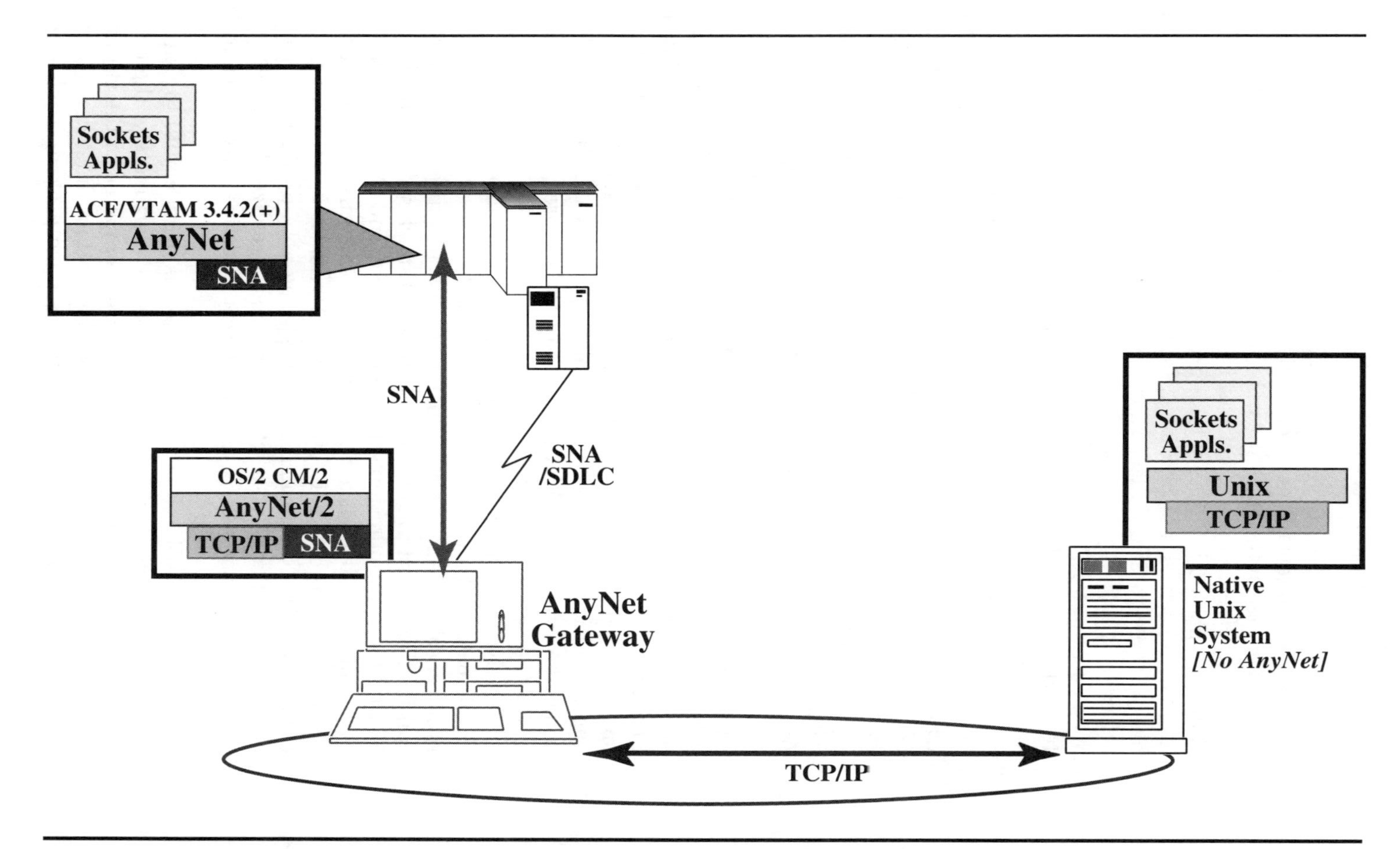

FIGURE 8.12 Doing sockets <—> sockets across both SNA and TCP/IP using an AnyNet gateway.

AnyNet Gateways can generally be used to realize any protocol conversion permutation that can be achieved using the AnyNet access products. Given a choice, networking solutions based around AnyNet Gateways should invariably take precedence over solutions concocted around AnyNet access products. For a start, the gateway solution is likely to be significantly less expensive. With access solutions, AnyNet software, which as yet is not free, is required on each and every device. When dealing with thousands of PCs, this software cost could be hefty. In marked contrast, a single AnyNet Gateway could easily perform the same protocol conversion task (e.g., SNA to IP) for 30 or 40 LAN-attached devices. The performance of the gateway should also no longer be a major concern. A good 486-based PC running AnyNet Gateway software should be able to perform protocol conversion for all the devices in a mid-size LAN (e.g., 40 users). Thus there is an immediate cost saving just in terms of the AnyNet software.

The software cost of an access solution, however, may not be restricted to just the AnyNet software. When doing SNA over IP, AnyNet access solutions require TCP/IP software to be colocated with the AnyNet software at each node—including the mainframe. TCP/IP software for mainframes, as has been discussed before, is hugely expensive. The same SNA over IP solution can be realized with gateways with no TCP/IP software required at any of the SNA end nodes. Therefore, as a rule of thumb, if considering AnyNet always try and opt for a gateway-based solution as opposed to an access software-based solution. Moreover, if considering an SNA-HPR-centric backbone, the AnyNet/LTLW Gateway supports more protocols than do any of the access products.

AnyNet, despite its relationship to the Blueprint and its considerable potential for delivering either SNA/APPN or IP-centric multiprotocol solutions, has not been a major market success. Much of this can be directly attributed to lackadaisical marketing on IBM's part. With the multivendor impetus around DLSw and its near ubiquitous availability on bridge/routers, including software routers such as Novell's Multiprotocol Router that can run on Novell servers, the demand for AnyNet as a means of realizing SNA-over-IP has been on the decline, especially so when enterprises realize that AnyNet access solutions are rarely the most cost effective way of achieving SNA-over-IP.

AnyNet(/LTLW) is now unique in terms of its ability to deliver a viable SNA-centric multiprotocol solution. Unfortunately, the clamor for SNA-centric solutions is but a pin-drop compared to the stampede toward IP-centric solutions. Thus the bottom line here is that AnyNet's value and appeal may already have peaked. AnyNet, over

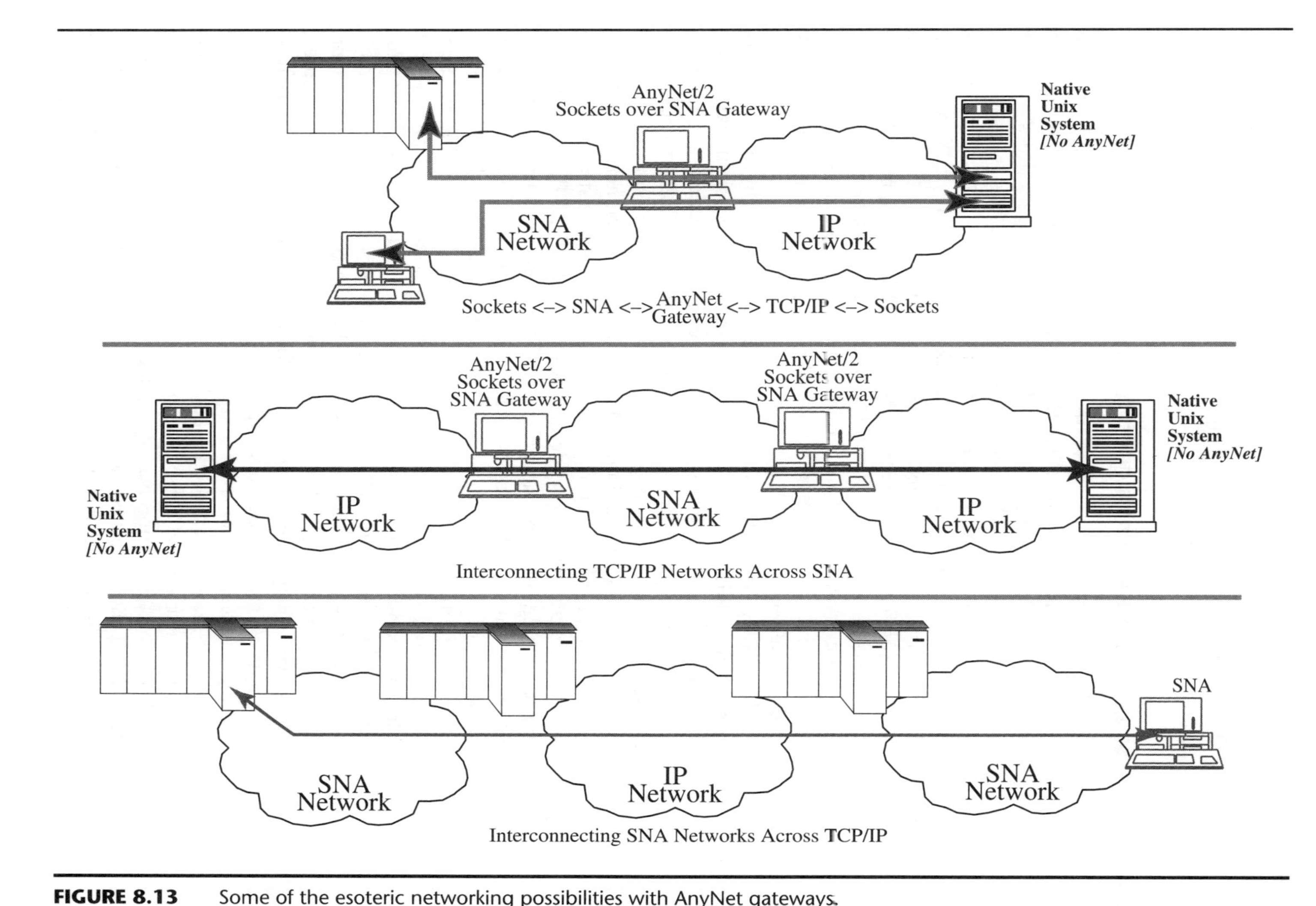

FIGURE 8.13 Some of the esoteric networking possibilities with AnyNet gateways.

the next few years, could end up quietly following the LAN-over-SNA products into obscurity.

8.4 THE IBM 2217 AND HPR/APPN-CENTRIC NETWORKING

The 2217 MultiProtocol Converter was a last ditch attempt by IBM to present a semicompelling SNA/HPR-centric multiprotocol solution to counter the rampaging success of IP-centric bridge/routers in what were until recently SNA-only networks. The 2217 is essentially AnyNet IP-over-SNA protocol conversion and LTLW in a box. It was much too little, much too late. The 2217 supports IP, IPX, and NetBIOS across traditional SNA or APPN networks. IP is transported via AnyNet conversion while IPX and NetBIOS are supported through LTLW-based conversion. If the SNA/APPN network being used by the 2217 is built on top of frame relay, other protocols such as DECnet, Banyan Vines, and AppleTalk can be supported using source-route bridging between the 2217s.

In reality, the 2217 Model 120 Release 1, which was available around mid-1995, was but a "concept car." It lacked significant functionality. It could at best support only one SNA device (i.e., no SNA multidrop or multiple links), and despite its claim of being a SNA-centric solution lacked surrogate DLUR capability to ensure that traffic from any SNA device could be effectively transported across an APPN network. Even worse, it did not support HPR. This first release was to serve as a place-holder and demonstrate that SNA/APPN-centric multiprotocol solutions are indeed viable. This demonstration of viability could convince enterprises with a considerable investment in SNA/APPN to hold on until an HPR-based version could be delivered.

Subsequent models and releases, claimed to be available in 1996, are expected to remedy most if not all the weaknesses of the first release. If and when available, they will certainly support HPR. For enterprises whose 1997–S1998 traffic mix is expected to be still dominated by SNA/APPN traffic, such an HPR-based multiprotocol solution could be enticing. HPR, as was discussed in the previous chapter, does have some persuasive features. These include: agile and very low-fat, Layer 2 routing; dynamic alternate routing; state-of-the-art anticipatory congestion control; and elimination of intermediate node processing, in addition to all of APPN's strengths such as plug-and-play networking and COS-based multicriteria path selection.

With the 2217 approach of supporting all the other protocols over HPR, all of these features will automatically be available to the other protocols. Thus, NetBIOS and IPX will immediately gain the advantages of dynamic alternate routing. If the implementation-specific,

SNA TG-like parallel link capability of HPR is also made available on the 2217, as would be logical, all the protocols will enjoy the throughput-enhancing properties of load-balancing. HPR TGs will also ensure protection against single-link failures. However, as was seen in the previous chapter, much of the lure of HPR can disappear if HPR is going to be run across a frame relay WAN, as is likely to invariably be the case with most North American networks. This "revenge of frame relay" aspect could diminish the 2217's edge and appeal.

The major hurdle facing the 2217 is that it lacks true credibility and momentum. IBM's track record over the last six years in conceiving and delivering competitive multiprotocol LAN/WAN solutions has been abysmal. The star-crossed 6611 router, which IBM still gamely continues to market despite its cringe-inducing reputation unfortunately tends to cast a pall over other IBM bridge/router like solutions. The 2217 will feel the brunt of some of this 6611 backlash.

Cisco and Bay now have the mind share of IBM-centric enterprises. They can demonstrate tried and tested multiprotocol bridge/router solutions. Moreover, they support DLSw, APPN NN routing, surrogate DLUR, and RFC 1490. They are committed to offering HPR NN-based routing in 1996. Thus, the choice, even for an enterprise that is likely to have a 1997 WAN traffic mix consisting of 80% SNA, is going to be between a 2217 and an HPR-capable bridge/router. Unless IBM can dramatically improve its image and credibility when it comes to multiprotocol networking, most of these enterprises will have no choice but to consider a bridge/router solution. In many cases, the enterprises with high volumes of SNA traffic will indeed end up using HPR—not per the 2217 paradigm, but with bridge/routers replete with HPR NN routing that will permit the SNA traffic to be routed within the context of HPR alongside the other protocols.

8.5 REFLECTIONS

Multiprotocol networking over an IP-only or SNA/HPR-only backbone is indeed viable. There are a variety of technologies such as DLSw(+), tn3270, SNA-to-IP gateways, and even AnyNet that can be successfully used to implement IP-only backbones that are capable of providing unconstrained support for SNA/APPN interactions. The choices are considerably more limited when it comes to implementing an SNA-only backbone that will support TCP/IP interactions. On the whole, the only mainstream solutions for IP-over-SNA end up being AnyNet, Novells SNA Links, and a few SNMP-over-SNA solutions offered by a few hub vendors.

If an enterprise expects its 1997–1998 WAN traffic mix to be dominated by TCP/IP or SNA/APPN traffic, a single protocol backbone-based around the predominant protocol is certainly an attractive solution. If the backbone is to be TPC/IP-centric, there are essentially two distinct ways in which to realize this objective. The first is to opt for a bridge/router-based solution given that today's bridge/routers are inherently TCP/IP biased and based. The other approach is to restrict the backbone just to IP traffic irrespective of how the LAN/WAN network is implemented. In general the bridge/router-based approach should prove to be the more flexible and optimal solution in that it does not preclude a small amount of non-IP traffic from being routed across the WAN alongside the IP traffic.

The demand for SNA-HPR-only networks is nowhere near as strong as that for IP-only solutions. On the whole the overall trend is toward IP-oriented, SNA-HPR-capable multiprotocol networks. IBM's AnyNet technology, which is based on its 1992 Networking Blueprint vision, does offer some interesting options for realizing either SNA-over-IP, IP-over-SNA, or even LAN-over-SNA solutions. The 2217, which was an attempt to popularize HPR-centric multiprotocol networking so that SNA's heritage of value-added networking could be leveraged into the 21st century, appears to be too little, too late. Enterprises with large volumes of SNA traffic that wish to avail themselves of HPR's indubitably attractive features are likely to do so using HPR-capable bridge/routers, rather than opting for an HPR-only backbone.

CHAPTER 9

IBM's Nways ATM Solutions

A QUICK GUIDE TO CHAPTER 9

Bridge/router-based multiprotocol LAN/WAN networking was IBM's Waterloo vis-à-vis networking. Much to IBM's relief, the advent of ATM and switched networking has provided it with a new, and as yet level, battlefield on which it can mount an all-out assault to regain at least some of the lost ground and possibly its honor. Consequently, IBM has made it abundantly clear that ATM is now the end-all and be-all of its hardware-oriented networking initiatives.

It has gone as far as to claim that ATM's significance to IBM over the next 15 years is likely to be equivalent to what the S/370 has been for the previous 15 years. Given the phenomenal success of the S/370 family over the last three decades and its traditional role of being the lifeblood of IBM's revenue stream, this is quite a comparison. To IBM's credit, it has managed, at least to date, to be true to its fierce and vociferous commitment to ATM.

At present, at least on paper, IBM has a unique, full-spectrum, and persuasive ATM story that embraces products, technologies, an ATM-centric architecture as well as a Networking Blueprint framework for switched networks. IBM already has a comprehensive ATM product repertoire that addresses every conceivable aspect of ATM-based networking. Its product offerings range from 25Mbps ATM adapters for PCs all the way up to

51Gbps WAN switches targeted at carriers intending to provide public ATM networks.

Moreover, IBM also has the potential for delivering "soup-to-nuts" total system solutions by grafting its information processing prowess to its ATM technology. IBM's ability to synthesize data processing, application software, ATM hardware, and total system management à la SystemView to create bespoke, turnkey systems is unsurpassed and unlikely to be matched, at least in the near future, by any other networking vendor.

This chapter examines, in-depth, every facet of IBM's ATM story, and where appropriate draws parallels with comparable offerings from other vendors. In line with the overall structure of IBM's ATM, this chapter is divided into three broad sections. Sections 9.1 to 9.4 deal with the product offerings: the ATM adapters, the campus products, and the WAN switches each having their own section. Section 9.5 looks at the Networking BroadBand Services (NBBS) architecture, which can be thought of as an SNA for ATM networks, while Section 9.6 describes the nascent Switched Virtual Networking (SVN) framework.

The heart of all of IBM's ATM switches is the PRIZMA chip. The PRIZMA technology is described in Section 9.1.2. With its high-end, high-capacity 2220 broadband switches, for example, the Model 700, IBM intends to pursue the public ATM network market. IBM's prospects of succeeding in this market, in which it has historically never even been a bit player, is discussed in Section 9.4.2.

IBM, right now, is totally obsessed with ATM. IBM is craving after and actively pursuing ATM with wholehearted gusto and little, if any, restraint. It has managed to convince itself that ATM is indeed the magical excalibur that will somehow enable it to recapture and hold on to its fast eroding networking domain and credibility.

With ATM-based broadband switching technology, IBM intends to reverse the embarrassing setbacks it continues to suffer in today's burgeoning LAN/WAN internetworking market dominated by multiprotocol bridge/router, FRAD, and LAN switching technology and vendors. ATM is perceived as the next big networking wave that IBM hopes to master and ride for a very long time, given that it hardly managed to get splashed by the multiprotocol internetworking wave that has yet to ebb.

IBM for the last few years has been essentially betting its entire networking farm and reputation on ATM. At IBM's very first ATM roll-out in July 1993, Lois Gerstner and other senior executives unequivocally stated that ATM's significance to IBM over the next 15 years is likely to be equivalent to what the S/370 had been for the previous 15 years. That was quite a comparison. The S/370 is the phenomenally successful and unparalleled mainframe family that for the last 25 years has, indubitably, been the sustaining lifeblood of IBM. It is also the defining technology that is indelibly associated with IBM. To compare the impact of ATM to that of the S/370 could, and should, not have been an idle boast. If it was, it would border on being sacrilegious and would forever damn IBM's credibility within the marketplace.

True to its commitment to this nascent technology, IBM today, at least on paper, has a unique and persuasive ATM story that embraces products, technologies, and at least one ATM-centric architecture. The breadth and depth of IBM's ATM offerings, as well as the potential total system application solutions it can provide by grafting its information processing expertise to its ATM technology, is unsurpassed and unlikely to be matched, at least in the near future, by any other networking vendor. Thus, there really could be some merit to IBM's comparison of ATM to the S/370. The only bone of contention is when the ATM wave is likely to start surging. IBM had assumed that this was likely to be around 1995. Today the indications are that while there are certainly some early pioneers, the ATM wave as originally envisaged in 1993 by IBM is unlikely to gather enough momentum till at least 1997. This delay, which was discussed in Chapter 3, could take some wind out of IBM's ATM sails—given that it had been hoping to make some sizable ATM sales by 1996.

IBM's market-maker and front-runner role to date is atypical when it comes to ATM, to say the least. In the past, IBM has been a follower, rather than a leader, when it has come to endorsing and implementing open, industry-standard, Layer 3 and below, data transmission technology. Historically IBM would let others not only test the waters, but frolic in it for a good bit of time, before it would tentatively commit to any "not-invented-here" data communications standards. One only has to think of the time it took for IBM to get around to supporting: X.25, Ethernet, token-ring, T1/T3, LAN protocol routing, frame relay, FDDI, or ISDN. But with ATM things are indeed different. IBM has already rolled out a comprehensive ATM product repertoire even though the overall market is still only at the technology evaluation phase. One can only hope that IBM's faith in ATM is indeed rewarded and that ATM proves to be as profitable for it as the S/370 continues to be.

IBM's currently available ATM products already embrace the entire networking spectrum. They stretch from PC/workstation ATM adapters all the way up to high-capacity ATM WAN switches that can be used by network carriers to implement large-scale public ATM networks. Campus ATM switches integrated into multimedia hubs, ATM-capable LAN switches, and ATM mainframe interfaces fill in the middle between the PC adapters and the WAN switches. Figure 9.1 provides a high-level view of IBM's current ATM repertoire of offerings. Cisco's intention to start marketing ATM adapters by late 1995 does eliminate what was IBM's hitherto unique distinction of being the only vendor that addressed the total ATM spectrum in terms of product offerings. Cisco, nonetheless, still has a long way to go before it can truly match IBM's theoretical potential for providing total system solutions consisting of both computing and networking elements rather than network solutions.

IBM's comprehensive ATM product set is augmented by and consolidated through an IBM-specific broadband networking architecture that was rolled out in mid-1993. This architecture is known today as *Networking Broadband Services* (NBBS). (Prior to IBM encountering a naming conflict in late 1994 NBBS was known as BroadBand Network Services [BBNS].) NBBS can be thought of as an SNA-type architecture for ATM-oriented networks. It is in fact positioned by IBM as being the next iteration of the SNA type family of packet-switching architectures. This typecasting of NBBS as a packet-switching, as opposed to a cell-switching, architecture by IBM is intentional. *Though fully supporting ATM cell-switching, NBBS also concentrates quite heavily on variable-length packet switching à la IBM's PTM.* Thus it tends to treat cell-switching as a subset of packet-switching.

NBBS specifies a range of optional value-added facilities that augment the services provided by ATM. The facilities specified by NBBS in general set out to increase the efficiency, robustness, and flexibility of future, multimedia broadband enterprise networks. NBBS, like its distinguished predecessor SNA, is primarily targeted at enterprises that intend to build their own in-house, private ATM-oriented networks. Given that NBBS, at least as yet, is only supported by IBM ATM products, such NBBS-based private networks would have to be built around IBM offerings.

This essentially harks back the original IBM-centric private network paradigm as represented by SNA in the late 1970s. In those days given IBM's stature and clout, plus the lack of any other compelling alternatives, most enterprises were willing to pursue a networking strategy that involved them getting locked into IBM products and technology. *This does not mean that IBM ATM products cannot interoperate with ATM products from other vendors*; far from it. IBM's ATM products, for example, the 2220 switches, can interoperate with switches from other vendors when the interactions are based on ATM Forum's

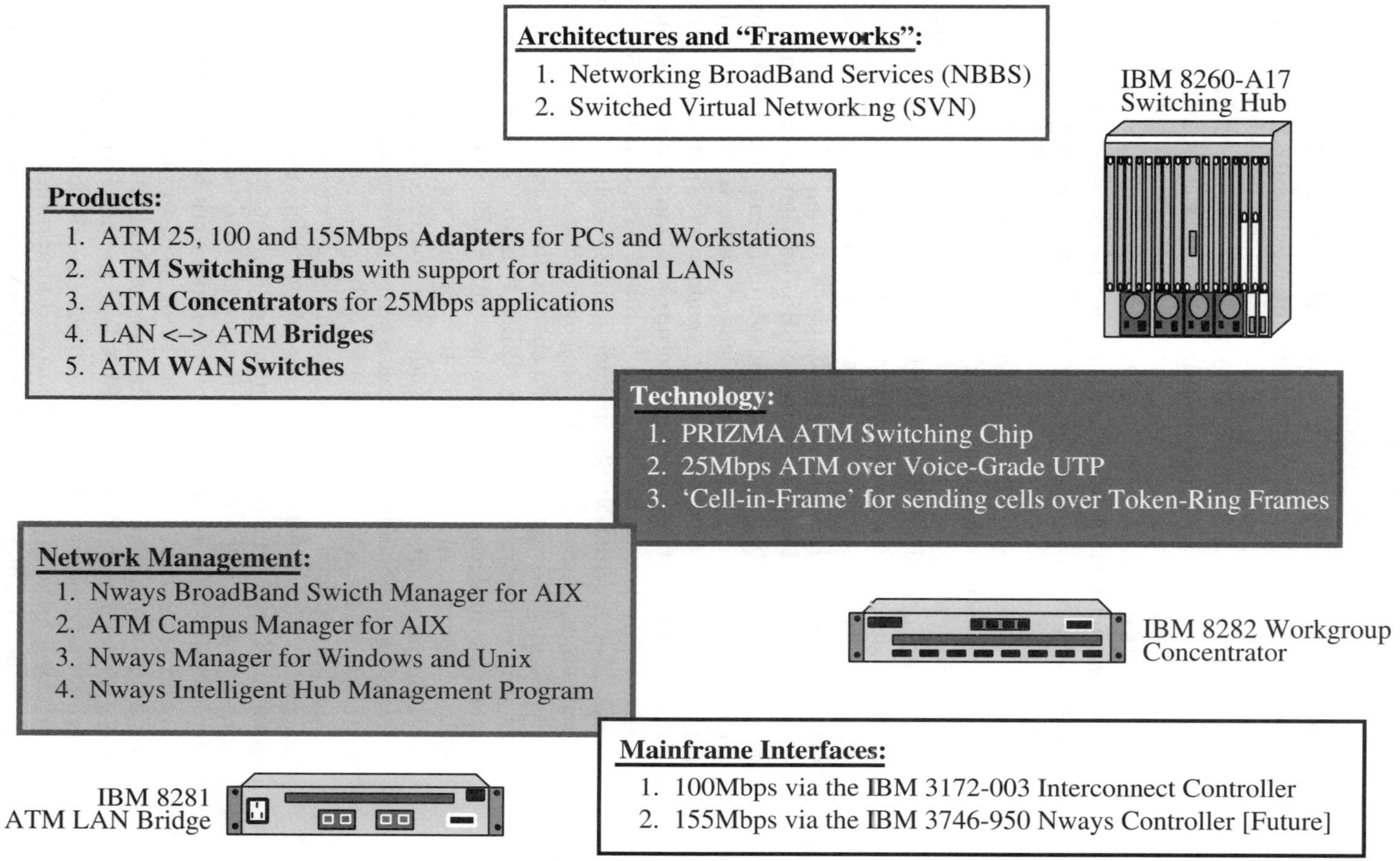

Architectures and "Frameworks":

1. Networking BroadBand Services (NBBS)
2. Switched Virtual Networking (SVN)

IBM 8260-A17
Switching Hub

Products:

1. ATM 25, 100 and 155Mbps **Adapters** for PCs and Workstations
2. ATM **Switching Hubs** with support for traditional LANs
3. ATM **Concentrators** for 25Mbps applications
4. LAN <-> ATM **Bridges**
5. ATM **WAN Switches**

Technology:

1. PRIZMA ATM Switching Chip
2. 25Mbps ATM over Voice-Grade UTP
3. 'Cell-in-Frame' for sending cells over Token-Ring Frames

Network Management:

1. Nways BroadBand Swicth Manager for AIX
2. ATM Campus Manager for AIX
3. Nways Manager for Windows and Unix
4. Nways Intelligent Hub Management Program

IBM 8282 Workgroup
Concentrator

IBM 8281
ATM LAN Bridge

Mainframe Interfaces:

1. 100Mbps via the IBM 3172-003 Interconnect Controller
2. 155Mbps via the IBM 3746-950 Nways Controller [Future]

FIGURE 9.1 Bird's-eye view of IBM's current ATM repertoire.

User Network Interface (UNI) or Network Node Interface (NNI), frame relay, HDLC, or Circuit Emulation. Interoperability issues will only occur when dealing with the value-added services provided by NBBS.

Interoperability is not an issue if standard ATM or frame relay is all that is required. It is the value-added services that present the problems; therefore enterprises have to make a hard choice. *If multi-vendor interoperability is deemed to be important, they have no choice but to restrict their networks to the lowest common denominator services covered by the frame relay and ATM industry standards.* If on the other hand, they do want additional value they may have no choice but to sacrifice the possibility of multivendor interoperability.

Today there is an understandable hesitancy to commit to an entire networking fabric that is totally contingent on networking products from a single vendor. Moreover, IBM's credibility and credentials vis-à-vis networking is not what it was in the heyday of SNA. Hence, the prospects for NBBS while it remains strictly IBM-only are not very rosy. If, however, IBM by hook or by crook manages to get a few other ATM switch vendors to support NBBS, its significance and appeal would dramatically improve. In today's ultra-competitive networking arena, where most of the market leaders perceive IBM as an underdog, the chances of other leading vendors wanting to endorse and embrace IBM's networking architecture would appear to be slim.

IBM still has a enviably huge installed base of private SNA networks. If supporting NBBS, in some form, is seen as a way to gain a tangible competitive edge in winning over some of these networks from IBM and other vendors, there is a chance that some of the larger vendors may contemplate adopting NBBS. It is worth noting that all of the leading bridge/router vendors now offer APPN NN-based routing and are committed to upgrading this facility in 1996 to conform to HPR, despite APPN and HPR both being patented technologies that can only be implemented on a license-based agreement with IBM. Consequently, there is still some hope for NBBS.

In late August 1995, IBM introduced an overarching blueprint, referred to as *Switched Virtual Networking* (SVN), that included NBBS as a subset. SVN at least in its initial form is more a Networking Blueprint-type high-level framework as opposed to a detailed bit-and-bytes architecture. SVN is an ATM-centric framework for the implementation and management of switched, as opposed to routed, networks. It postulates the notion of networks that are built entirely around ATM and LAN switches. The goal of SVN is to provide a scalable switching fabric for any-to-any, end-to-end connectivity with one-hop routing (i.e., everything appears to be adjacent) and guaranteed quality of service (e.g., response times). SVN will also endeavor to protect existing investment in networking equipment. The core of SVN networks will be a backbone made up of ATM switches à la IBM's 2220 Nways BroadBand Switches. Around this ATM backbone will be peripheral switches, à la IBM's 8271 or 8272 LAN switches, that will provide end user access to the overall switching fabric. NBBS and SVN are discussed in further detail later in this chapter.

IBM's capability to provide a "soup-to-nuts" ATM solution, replete with NetView- and SystemView-based system management tools, and capped by NBBS and SVN blueprints could prove in the end to be pivotal. Particularly so, given that IBM in many cases will also be providing the PCs, Unix workstations, LAN servers, mainframes, and possibly even the software applications being served by the ATM network. Most enterprises know of the trials and tribulations of multivendor solutions particularly when it comes to complex, leading-edge technology that is being deployed to support mission-critical applications.

This potential for a single-vendor solution, particularly with IBM's still somewhat impressive 30-year-plus track record in helping enterprises realize far-flung, highly reliable networks, could as yet give IBM the edge it needs to succeed with ATM. There is, however, one dark cloud in this otherwise rosy scenario: whether IBM, after its repeated faux pas in multiprotocol internetworking and LAN switching, still has enough credibility to be considered as a viable and reliable vendor for next-generation networking.

9.1 AN OVERVIEW OF IBM'S ATM PRODUCT SUITE

IBM's total ATM product repertoire that is either available, has been announced, or has been previewed in the form of statements of direction, as of November 1995, is listed below. This product repertoire has been divided into appropriate functional categories to highlight product relationships and emphasize IBM's coverage of the entire ATM networking spectrum.

- Desktop Offerings
  - 25, 100, and 155Mbps *TURBOWAYS ATM* adapters for PCs and RS/6000 workstations
  - IBM *TURBOWAYS 8282 ATM Workgroup Concentrator* for supporting PCs using the 25Mbps ATM adapters
- Campus Offerings
  - IBM *8281 ATM LAN* bridge for providing interoperability between traditional LANs and ATM networks

  - 155Mbps ATM up-links for the IBM 8271 Ethernet and 8272 token-ring LAN switches
  - IBM *8260 Nways Multiprotocol Intelligent Switching Hub* Models A10 and A17: 10- and 17-slot multimedia (i.e., token-ring, Ethernet, and FDDI support) hubs with an *ATM subsystem*
  - IBM *2220 Model 200* advanced multiprotocol router that will be an IBM 6611-like multiprotocol bridge/router with built-in ATM support and running the Proteon multiprotocol software à la IBM's 2210 low-cost bridge/router

- WAN Offerings
  - IBM *2220 Nways BroadBand Switches* Models 300, 500/501, 700, 800 etc.

- Mainframe Interfaces
  - IBM 3172 Interconnect Controller Model 003 with a 100Mbps TURBOWAYS ATM Adapter
  - IBM 3746 Model 950 Nways Controller Model 950 with a 100 or 155Mbps ATM interface

- Network Management
  - IBM *Nways BroadBand Switch Manager for AIX* for managing the 2220 WAN switches
  - IBM *ATM Campus Manager for AIX Version 1* for managing ATM campus networks
  - IBM *Nways Manager for Windows and Unix* for managing hubs including the 8260 Switching Hub, LAN switches, the 8281 ATM LAN Bridge, and the TURBOWAYS 8282 ATM Workgroup concentrator
  - IBM Nways Intelligent Hub Management Program (IHMP) for managing 82xx hubs

9.1.1 Nways and TURBOWAYS

The term *Nways* appears quite frequently in the above list. This term was introduced in June 1994 when some of the above ATM products were announced. Since then it has been used consistently to prefix all of IBM's ATM switches, such as the 2220 and 8260. Consequently, the term Nways has been thought by many as being a synonym for IBM ATM products. This is not the case. Products with no current ATM capability such as the 2217 MultiProtocol Converter, the 8238 stackable token-ring hubs, and the 2210 multiprotocol bridge/router also sport the Nways moniker. Thus the prefix Nways alone does not indicate the presence or absence of ATM capability.

IBM has three criteria that need to be satisfied in order for a new product to be assigned the Nways tag. First and foremost the product has to be multiprotocol-capable and, as a result, be capable of operating in multivendor environments. Secondly, the functions it performs must be associated with Layers 2 to 4 of the OSI Reference Model. (Rather than specifying this criterion in terms of the OSI Reference Model, IBM invariably does so by using its 1992 Networking Blueprint as shown in Figure 8.8. In the context of the Blueprint, a product that is to be called Nways must perform functions that fall between the CTS layer and the top of the physical layer.) Lastly, but by no means least, the product has to be deemed to be strategic.

Hence, per IBM guidelines, the term Nways will only be used with new products that are strategic and multiprotocol, and that perform functions associated with Layer 2 to 4. Obviously all of IBM's ATM and LAN switching products meet these three criteria. Therefore they are entitled to use the Nways prefix. The 2217 and the 2210 also satisfy these guidelines. Thus, the Nways prefix is in their official titles. The term Nways also appears on most of the new management products for ATM and campus networks, for example, Nways BroadBand Switch Manager for AIX. This would appear to violate the second criterion for Nways eligibility given that most management applications can be thought of as spanning Layers 2 to 7. In this context, however, Nways typically indicates that these products are capable of managing products that sport the Nways tag.

The bottom line here is that Nways per se is not just restricted to ATM-capable products—though most of IBM's *strategic* ATM products will have Nways or TURBOWAYS in their title. The need to be deemed strategic in order to be called Nways can obviously lead to some intriguingly puzzling oddities. The 8281 LAN Bridge, though performing the crucial role of providing interoperability between traditional LANs and ATM networks, has never had Nways in its title. At best, it is sometimes referred to as a member of the Nways family.

The 8281 is inordinately expensive for the functions it provides. There is a high possibility that IBM will provide more cost-compelling alternatives to the 8281 in the future. The absence of Nways in its official title may be a clue that this may indeed be more of a tactical solution, as opposed to a long-term strategic one. If this is indeed the case, the use of the Nways term could actually backfire on IBM. The market may see the absence of it in certain, supposedly strategic products, as a tacit warning from IBM as to the longevity of that product. If that happens, IBM will be forced to call all of its future networking hardware products Nways, which would obviously devalue the worth and significance of this "brand name."

Compared to the somewhat convoluted significance of the term Nways, that of TURBOWAYS is simple and straightforward. TURBOWAYS is used to designate the members of IBM's ATM PC/workstation family. Unlike Nways, TURBOWAYS, at least to date, has always indicated the presence of ATM. The TURBOWAYS moniker is also assigned to the 8282 ATM Workgroup Concentrator. This is appropriate and logical since the one and only role of this concentrator is to support 25Mbps TURBOWAYS adapters.

9.1.2 PRIZMA—The Heart of IBM's ATM Switches

Packet Routing Integrated Zurich Modular Architecture (PRIZMA) is an integrated ATM packet switch chip that was developed by IBM's Research Laboratory in Zurich, Switzerland. All of IBM's current ATM switches are based on this PRIZMA chip. The current version of the PRIZMA chip has 16 input ports and 16 output ports, that is, a 16 x 16 switch. Each of these ports have a full-duplex throughput rate of at least 400Mbps. At 400Mbps per port, the current PRIZMA chip has a total aggregate data throughput rate of 6.4Gbps (i.e., 400Mbps x 16). With this throughput, IBM claims that this chip can sustain an ATM cell-switching rate of 15 million cells per second.

The PRIZMA chip currently supports a maximum cell length of 64 bytes. The current PRIZMA chip is built on a 15mm die and holds 2.4 million transistors. This transistor capacity is more or less in line with other contemporary chip technology. For example, IBM's 66MHz PowerPC 601 chip has 2.8 million transistors on an 11mm die, while the original 66MHz Pentium had 3.1 million transistors on a 16mm die.

IBM ATM switches that were available in 1995, for example, the 2220 Models 300 and 500, did not use the full 400Mbps throughput rate of the PRIZMA chip. Instead, the PRIZMA chip was being used at 266Mbps, full-duplex, per port. (This is akin to PC processor chips, such as the Pentium or PowerPC 601 and 604, being run at different clock speeds, with most of these speeds being well below the maximum rate supported by that chip.) IBM switches available in 1996 are expected to have 400Mbps per port PRIZMA chips. It is also likely that IBM will offer a means whereby the slower switches could be field-upgraded, most likely for a fee, to run at the higher port speed.

Given the full-duplex capability of these ports, IBM sometimes refers to the PRIZMA port speeds in terms of an I/O rate. This I/O rate is double that of the throughput rate and reflects the simultaneous two-way data transfer that can take place. Consequently the 266Mbps rate is occasionally quoted as being a 532Mbps FDX I/O rate. The so-called throughput rates currently quoted for the 2220 switches are invariably based on this I/O rate. For example, the Model 500/501 combo has a total of 14 adapter slots, where each adapter is connected to a port on a PRIZMA chip running at 266Mbps. The aggregate throughput for this 14-adapter 500/501 would thus be 3.7Gbps (i.e., 266Mbps x 14). While this number is mentioned, it is not unusual for IBM to also refer to the 500/501 as having a 7.4Gbps capacity, which happens to be the full-duplex I/O rate (i.e., 3.7Gbps x 2). These throughput rates for the 2220 will be further discussed in Section 9.4.

The PRIZMA chip was developed from ground up with an eye toward a smooth evolutionary growth path that would deliver even more throughput and support higher port capacities. Hence, the cranking up of the port speeds to 400Mbps in 1996 are not the only enhancements in store. IBM is already talking about a 32 x 32 version of the PRIZMA chip. Even at a 400Mbps port rate this chip would have a total aggregate data throughput rate of 12.8Gbps. IBM, however, is expected to be able to support around 800Mbps per port with this chip. That would give it a total aggregate throughput rate of around 25.6Gbps.

Some of the higher end 2220 models, for example, the so-called Model 700, targeted at network carriers building public networks, are envisaged to have 32 or more slots. Such high-capacity switches would obviously benefit from having 32 x 32 switching chips rather than being forced to use multiple 16 x 16 chips in tandem. Though yet to be formally announced, IBM previewed the Model 700 as a part of its original 2220 announcement in June 1994. It was quoted as having an I/O rate of 25.6Gbps—which just happens to be the I/O rate of a 32 x 32 switch was being driven at 400Mbps per port (i.e., 400Mbps x 32 x 2). A 2220 Model 800 with a 51.2Gbps I/O rate was previewed at the same time. This 51.2 Gbps I/O rate would correspond to a 32 x 32 switch with 800Mbps ports.

That, interestingly, is still only half the throughput that IBM claims this family of chips will be able to sustain. When SVN was unveiled in August 1995, IBM made references to the PRIZMA chip evolving all the way up to a 50Gbps (i.e., 51.2Gbps to be technically pedantic) throughput switch-capable of switching 118 million ATM cells per second. Whether there will be a widespread demand for such switch speeds in 1996 or 1997 is debatable. It is also not clear whether IBM will make such switch speeds available during that timeframe given the ATM's slow ramp-up to date.

What is noteworthy is that IBM, in theory at least, will have this switching power when there is a market need for it. Such aggregate throughput switching rates will most likely be needed initially on switches being deployed to support large public networks. The jury is still out on whether IBM will succeed in becoming a switch vendor to public carriers for two reasons. This is an arena in which IBM has historically never had much joy. Moreover, IBM with its part ownership of the worldwide Advantis network, which is already in the process of offering ATM services, could be perceived by many of the carriers as an overt competitor as opposed to just being an equipment vendor. This issue is discussed further in Section 9.4.

9.2 THE TURBOWAYS ATM ADAPTER FAMILY

The TURBOWAYS family of ATM adapter cards for PCs and RS/6000 workstations currently consists of 25Mbps, 100Mbps, and 155Mbps offerings. The IBM 8282 Workgroup Concentrator, which, at least at present, is a mandatory prerequisite when using 25Mbps PC adapters, is also included within this family. Table 9.1 provides a snapshot of the characteristics of the TURBOWAY adapters that were available in November 1995. IBM has stated that it intends to offer 622Mbps TURBOWAY adapters in the future.

All the adapters in this family include an on-board Segmentation and Reassembly (SAR) chip for slicing-and-dicing data into 53-byte ATM cells when transmitting, and for reassembling cells into data blocks when in receive mode. The 100Mbps adapters, and most likely the 155Mbps adapters as well, also have an Intel i960 RISC processor for performing non-SAR related functions such as connection setup, bandwidth allocation, and on-board diagnostics. (The 155Mbps adapter has FLASH memory, as opposed to Erasable-Programmable Read Only Memory [E-PROMs], to facilitate the adding of enhancements, and obviously also any software fixes, via down-line loadable microcode upgrades.)

The 25Mbps and 155Mbps adapters can be used with UTP wiring that meets UTP Category 3 specifications, provided the wiring length does not exceed 100 meters. They can also be used with STP. STP wiring was initially recommended for the 100Mbps adapters. This could change, especially now that there are IBM and other offerings (e.g., Cisco's ZeitNet card for Sun Microsystems Unix workstations) that readily support 155Mbps over Category 5 UTP wiring.

All the adapters are full-duplex-capable and provide support for ATM Adaptation Layer 5 (AAL 5) services for connection-oriented and connectionless data exchanges. SVC support conforming to the ATM Forum UNI 3.0 specification is already available on the 25 and 100Mbps adapters. This immediately ensures that these adapters, or at least the 100 and 155Mbps adapters, will freely interoperate with non-IBM ATM switches that support Version 3.0 SVCs.

TABLE 9.1 A Snapshot of IBM's TURBOWAYS ATM Adapters as of November 1995

TURBOWAYS ATM Adapter	*Workstation BUS*	*Operating System*	*NIC/IF*	*Announce/ Availability*	*One-Off Price*	*Unit Price in ≥5 Packs*	*True Per-Port Price*
25Mbps[1]	PC ISA [min. 386 PC]	DOS 6.2 or Windows 3.1	NDIS 2.0.1	6/28/94 Now	$405	$395	$895
25Mbps[1]	PS/2 MCA	OS/2 & LAN Server	NDIS 2.0.1	6/28/94 Now	$405	$395	$895
25Mbps[1]	PCI	OS/2	NDIS 2.0.1	3/14/95 Now			
100Mbps	PS/2 MCA	NetWare 3.12 4.0.1 —>	ODI	6/28/94 Now	$1,995	$1,795	$1,795
100Mbps	PS/2 MCA	OS/2 & LAN Server	NDIS 2.0.1	6/28/94 1Q95	$1,995	$1,795	$1,795
100Mbps	*IBM RS/ 6000*; MCA	IBM AIX/ 6000 V3.2.5		3/29/94 Now	$1,995		$1,995
155Mbps	*BM RS/ 6000*; MCA	IBM AIX	NDIS ODI	6/20/95 Now	$1,995		$1,995
155Mbps	PCI	OS/2	NDIS	3/14/95 Now			

[1]These 25Mbps adapters have to be used in conjunction with the 8282 Workgroup Concentrator. The last column of this table indicates the true per-port cost with the price of an 8282 port factored in.

None of these adapters at present support voice or video. They only support data—and even that via LAN Emulation as opposed to a native ATM interface. LAN emulation, as has been discussed in earlier chapters, enables an ATM adapter to effortlessly masquerade as a bona fide token-ring or Ethernet NIC card vis-à-vis existing PC and workstation software. Thus, any software (e.g., Novell NetWare, OS/2, Communications Manager/2, etc.) that works with conventional LANs would also immediately work with the ATM adapters. Therefore, ATM can be introduced at the desktop level without requiring any new software or even modifications or updates to the existing LAN networking software. This nondisruptive, transparent migration to ATM using known and proven PC/workstation software is what makes LAN Emulation so attractive and significant.

The TURBOWAYS adapters in 1996 will provide LAN emulation for both token-ring and Ethernet. Independent of LAN media type,

LAN emulation, since it is trying to emulate a NIC, also tends to be dependent on the type of NIC interface as well as the Operating System being used on the PC/workstation. The NIC and Operating Systems currently supported by these IBM adapters are listed in Table 9.1. Note that the Operating System support across the board is still patchy on the 25 and 100Mbps adapters with neither as yet supporting the full complement of popular OSs as represented by DOS/Windows, OS/2, Unix, and NetWare. Also note that the TURBOWAYS LAN emulations, at least as yet, do not reach out to embrace other less popular Network OSs such as Banyan Vines. IBM's initial focus on just data transfer, and that via LAN emulation, is essentially par for the course in todays market. Native ATM interfaces for data and support for the ATM Forum's standard for Multiprotocol Over ATM (MPOA) and IP over ATM will only be available from most vendors in 1996.

The PC/workstation buses supported by the current TURBOWAYS adapters, as shown in Table 9.1, is somewhat parochial given the significant bias toward IBM's once favored MCA bus. This will have to change particularly now that IBM is no longer making PCs with MCA buses. Most non-IBM 155Mbps adapters now support the PCI bus and IBM will have no choice but to follow suit soon. TURBOWAY adapters with PCMCIA interfaces are another possibility. It may also consider a PCI bus version of the 100Mbps adapter. There is also a good possibility that it would consider a 100Mbps card for at least EISA bus PCs given that 100Mbps Fast Ethernet EISA NICs are now available. At this juncture it is also worth noting that IBM's 100 and 155Mbps adapter prices, as shown in Table 9.1, are a bit on the high side compared to non-IBM offerings. Non-IBM 155Mbps adapters with PCI buses are already available with list prices below $1,000. Hence, there is likely to be some serious price reductions from IBM over the next year.

All the TURBOWAYS adapters include on-board self-diagnostic software. They are delivered along with a configuration program to facilitate the installation of the adapter and the necessary device drivers. The 155Mbps adapters, which include an SNMP subagent, can be directly monitored and managed from SNMP Managers à la IBM's NetView for AIX.

9.2.1 25Mbps ATM

IBM was the original proponent of 25Mbps ATM. When IBM first started talking about 25Mbps ATM in 1993 it was considered by most ATM cognoscenti as being heresy. ATM at that juncture was only being envisaged as being viable and applicable at speeds above 45Mbps. IBM's rationale for 25Mbps was simple and pragmatic. IBM wanted 25Mbps to be a low-cost entry point into ATM, arguing that 25Mbps adapters had to be less expensive than 100 or 155Mbps adapters. Consequently, price/performance became the win-or-lose, live-or-die criterion for 25Mbps.

IBM's initial prices for 25Mbps ATM in June 1994 were somewhat persuasive and tended to justify its claim that this was the right entry-level approach to ATM. The TURBOWAYS 25Mbps adapters in quantities of five or more carried a list price of $395 per adapter. This figure was, however, misleading. These 25Mbps adapters could only be used with a Workgroup Concentrator, à la IBM's 8282, since IBM, as well as non-IBM, switches do not support direct 25Mbps inputs; that is, ATM switches do not have 25Mbps ports. Thus, the cost of a Workgroup Concentrator port has to be factored in when looking at the true cost of a 25Mbps adapter.

Today an 8282 port, whether on an eight-Port unit or on a 12-Port unit, costs around $499. Thus, the true cost per PC for 25Mbps ATM is $895. By today's standards that is expensive and unrealistic. Already, 155Mbps ATM adapters, albeit only for PCs with the relatively new PCI bus, are available for around that price. Moreover, 100Mbps Fast Ethernet adapters are now readily available at under $250 per adapter while token-ring-compatible 100Mbps AnyLAN (a.k.a. 100 VG) adapters retail for under $200. The new hubs required to support 100Mbps LANs have per-port costs that are typically below $150. Hence, 100Mbps to a PC can be realized these days for around $400 per PC.

To exacerbate this unfavorable price comparison, a 25Mbps TURBOWAYS supports only data—and that via LAN emulation. So all it is, at present, is a way of increasing the data-only bandwidth available to the desktop. If added data-only bandwidth is the only requirement, LAN switching as described in Chapter 4 or 100Mbps LANs are a much more cost-compelling option than 25Mbps ATM. IBM can obviously try to restore the luster and lure of 25Mbps by dramatically reducing the price of both the adapters and the 8282. This seems likely to happen fairly soon.

A price cut alone will not be enough. At the same time it will have to add voice and video support, and back this up by offering some tantalizing desktop multimedia applications such as multmedia e-mail and videoconferencing. Providing the multimedia capabilities, in particular for the applications, may not be as easy as cutting price. However, price cuts and multimedia capability will still have to be weighed against the inevitability that the prices of 155Mbps ATM adapters will also start to fall, quite sharply, starting in 1996. The only redeeming feature that 25Mbps will still have over 155Mbps is that it can be used with existing PCs with the old ISA buses, whereas 155 invariably requires PCI or MCA.

Despite the price/performance-related obstacles it faces, 25Mbps ATM is currently on a roll. Having initially rejected it the previous year, the ATM Forum in mid-1995 approved 25Mbps as bona fide ATM standard—albeit under the technically accurate 25.6Mbps heading. The 25Mbps consortium put together by IBM in 1994 is also thriving and there are now nearly 30 high-profile members, including: HP, National Semiconductor, Madge, Olicom, Whitetree Network Technologies, Fujitsu, Centillion (now a part of Bay Networks), and others. H-P, Whitetree, and National Semiconductor among others now offer 25Mbps ATM solutions alongside IBM.

The 8282 is an integral component of IBM's 25Mbps TURBOWAYS story. It has one and only one role in life. That is to act as an interface between PCs using 25Mbps adapters and ATM switches, given that there are no ATM switches today that sport 25Mbps ports. A 8282 base unit has eight UTP/STP ports. A four-port extension module can be added to the base unit to increase the port capacity to 12. Each 8282 port can only support one PC.

The 8282 is not an ATM or LAN switch. Hence, it cannot do any port-to-port switching between its 25Mbps ports. An 8282 is connected to a stand-alone ATM switch (e.g., Cisco LightStream 100 ATM Switch or IBM 2220) or to ATM switch module within a multimedia hub (e.g., IBM 8260). The current interface between the 8282 and the ATM switch is via a single, full-duplex 100Mbps ATM UNI-compliant Transparent Asynchronous Transmitter-Receiver (TAXI) interface. This 100Mbps TAXI connection is realized across a multimode fiber cable. Figure 3.8 depicts how an 8282 acts as an interface between PCs with 25Mbps ATM adapters and an ATM switch. Since the 8282 is incapable of performing any port-to-port switching, any such switching required has to be done by the ATM switch with the data going to and fro across the 100Mbps ATM interface.

There is a high possibility that IBM or other vendors, in 1996, will offer an ATM switching version of the 8282—possibly with more ports. There is even a likelihood that such a 25Mbps switch may be offered as an adjunct to a token-ring LAN switch in order to facilitate migration from token-ring LANs to 25Mbps ATM. Independent of 25Mbps switching, there is likely to be versions of the 8282 from IBM or others that will support 155Mbps ATM interfaces in place of the current 100Mbps.

9.3 IBM'S ATM-BASED CAMPUS OFFERINGS

The 8260 is at present the focal point of IBM's ATM offerings for the campus. The 8260 per se is the multimedia, intelligent family of chassis-based hubs that IEM OEMs from Chipcom (which is now a part of

3Com). To make an 8260 ATM-capable, IBM adds an ATM subsystem to it. Today, ATM subsystems are available for the 17-slot 8260 Model 17 and the 10-slot 8260 Model 10. When an ATM subsystem is installed in one of these hubs the model number is prefixed by an "A" to indicate that the hub is now ATM-capable.

The ATM subsystem for an 8260 Model A10 or A17 consists of three discrete components. These three components are as follows:

1. ATM Backplane for the 8260.
2. One, or optionally two, ATM Switch/Control Point Modules—each of which occupies two full-size slots within the 8260 hub. (The optional second ATM Switch/Control Point Module is only installed when a fault-tolerant, hot-backup configuration is required.)
3. One or more one-slot ATM Concentrator Module that contains the actual ATM ports. Today, there are two types of ATM Concentrator Module: a two-port 155Mbps Concentrator Module, and a four-port 100 Mbps TAXI Concentrator Module. The former will support multimode fiber, UTP-5, or STP connections, while the latter, as yet, is restricted to multimode fiber. It is likely that IBM will offer other Concentrator modules with more port options and permutations in the future.

The ATM subsystem resides, unobtrusively, alongside standard 8260 token-ring, Ethernet, FDDI modules occupying other slots within the 10- or 17-slot chassis. Given this coexistence between the ATM subsystem and the traditional LAN modules, the ATM backplane does not replace the standard 8260 backplane. Instead, the ATM backplane is installed behind the standard 8260 backplane. The LAN modules continue to use the standard backplane oblivious to the new ATM backplane. The ATM backplane is only used for data transfers involving the ATM Concentrator Modules, the ATM Switch/Control Point Module(s), and the optional 8281 ATM LAN Bridge Module. Existing Model 10 or 17 chassis can be field-upgraded to contain an ATM backplane so that they can support the ATM modules.

Despite coexisting within the same chassis, data transfer between the 8260 ATM modules and the LAN modules is only possible via an IBM 8281 ATM LAN Bridge or something comparable such as a NetEdge Systems' ATM Connect. The IBM 8281 functionality can be obtained either as a module that slots into an 8260 A10 or A17 chassis, or as a standalone device. The 8281 is described further in Section 9.3.1.

The 8260 ATM Switch/Control Point Module consists of two full-size blades (or cards) packed as one two-slot module. One of these blades is referred to as the *base card* and contains a PRIZMA ATM switching chip rated as having an aggregate throughput that exceeds 8Gbps. (This could be a new 32 x 32 chip running at 266Mbps per port, given that 8260s, unlike the original 2220 with the 16 x 16, 266Mbps chips that started shipping in late December 1994, did not start shipping until mid-1995.) The other blade is referred to as the *control point card* and contains a high-speed processor as well as FLASH memory that holds the ATM Control Point code executed by this card.

This Control Point code, somewhat analogous to an APPN Control Point, is responsible for discovering and maintaining network topology, route selection, and call setup. The 8260 ATM Switch/Control Point Module supports PVCs, SVCs per ATM Forum UNI 3.0, and switch-to-switch interactions-based on an UNI 3.0 extension per the ATM Forum P-NNI framework. It also supports the interconnection of local ATM networks over an ATM WAN, including end-to-end SVCs. Point-to-point and point-to-multipoint connections are supported both for PVCs and SVCs. Installing two Switch/Control Point Modules, which will take up four slots within the chassis, provides a redundant configuration to guard against the malfunctioning of one of the modules.

If only one Switch/Control Point Module is installed in an A17, the remaining slots can be used to hold up to 14 Concentrator modules. If the 100Mbps TAXI modules are being used, this would enable an A17 to support up to 56 100Mbps ATM ports. In the case of the 155Mbps modules the maximum port capacity will be 28. It would also be possible to mix 100 and 155Mbps modules within the same chassis. A maximum of eight Switch/Control Point Modules could be installed in an A10 chassis. If just one Switch/Control Point Module is installed, the A10 will be able to support up to 16 155Mbps ports, or 32 100Mbps ports. Figure 9.2 depicts a representative campus ATM network built around 8260 A17s.

The ATM capability offered by the 8260 is obviously not unique to IBM. For a start 3Com, now that it has acquired Chipcom, can market the ATM-capable 8260s under the Chipcom logo. Bay Networks with their 16-slot Model 5000 Switching Hub and Cisco with their five-slot Catalyst 5000 also offer comparable ATM-capable multimedia hubs. Other competitive offerings are available from Cabletron, Newbridge, U-B Systems, and Fore Systems.

9.3.1 The 8281 ATM LAN Bridge and Other ATM Campus Offerings

The 8281 ATM LAN bridge functions provide for interoperability between conventional LANs, whether token-ring or Ethernet, and ATM networks. This interoperability can be in two distinct forms. The 8281 will permit unhindered LAN-to-LAN interactions across an ATM back-

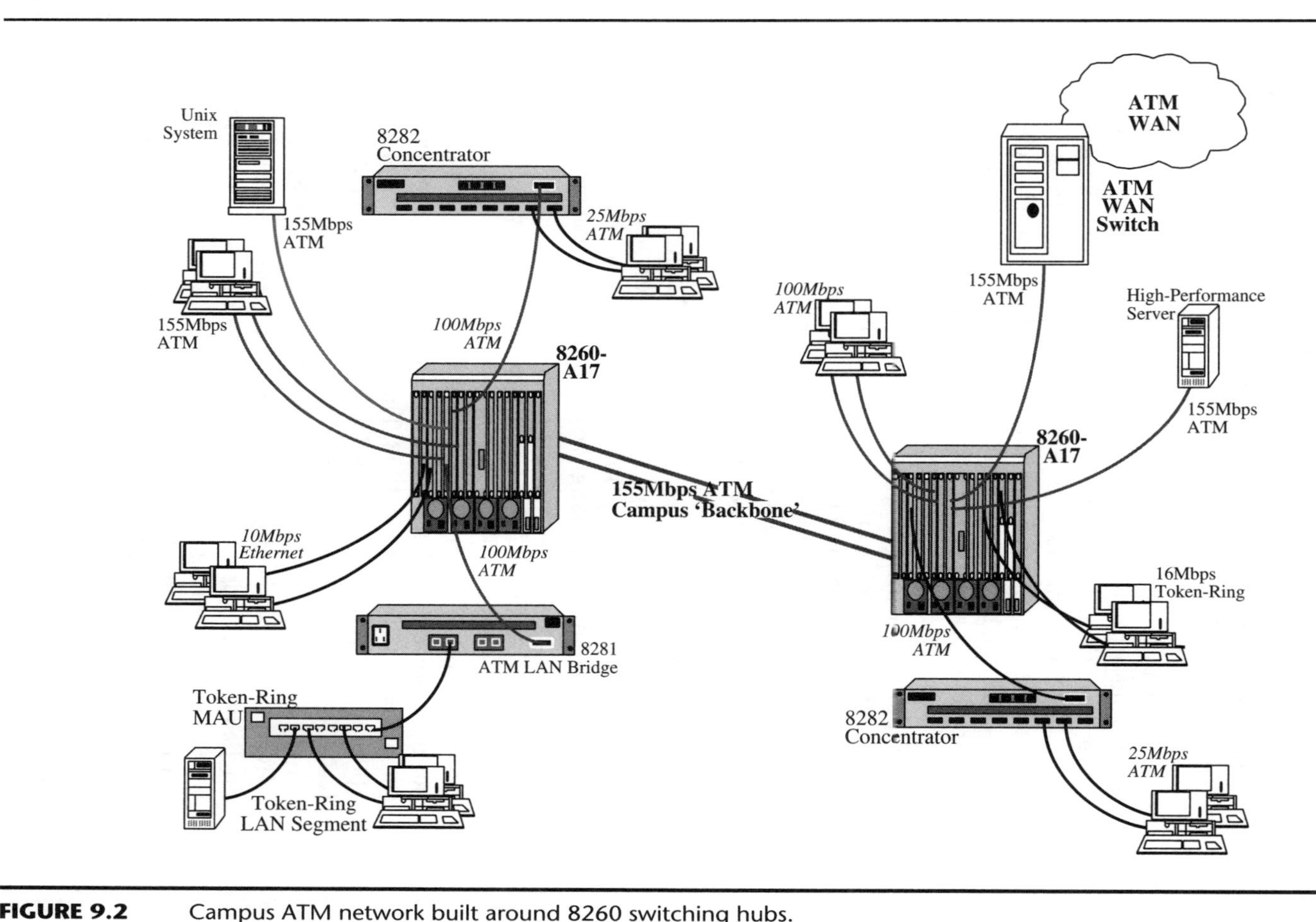

FIGURE 9.2 Campus ATM network built around 8260 switching hubs.

bone. In this mode of interoperability the 8281 is functioning as an ATM-based LAN-to-LAN MAC layer bridge. The 8281 thus becomes the ATM equivalent of a conventional LAN bridge that does bridging over a frame relay network. The 8281 relies on LAN emulation to achieve its LAN-to-LAN over ATM bridging.

The 8281 also permits a LAN-attached device to talk to a device connected to an ATM network. Thus, with an 8281 ATM LAN bridge a token-ring-attached PC could exchange data with a PC-based LAN server that is using a 155Mbps TURBOWAYS adapter. The server with the 155Mbps adapter could be connected to a 155Mbps Concentrator Module on a 8260 hub, while the token-ring client PC would access the 8281 via its token-ring. The 8281 in this case could either be a module within the 8260 hub or a stand-alone unit with a 100Mbps TAXI connection to a 100Mbps Conncentrator module in the 8260.

Figure 9.3 depicts how an 8281 can be used to provide LAN-to-ATM interoperability between a PC with a token-ring NIC and another PC with an ATM adapter. This mode of interoperability could be thought of as LAN-to-ATM Translation Bridging. When doing LAN-to-ATM bridging the 8281 will automatically and dynamically discover the location of the ATM partner being sought and establish the connection on an as-needed basis.

In addition to its LAN-to-LAN over ATM and LAN-to-ATM bridging prowess, the 8281 can also be used as a very expensive LAN-to-LAN bridge. When acting as a LAN-to-LAN bridge, the 8281 will support either transparent bridging when supporting Ethernet LANs, or SRB when dealing with token-ring LANs.

The 8281 ATM LAN bridge functionality is now available in two distinct forms: either as a stand-alone unit, or as a two-slot module for the 8260 A10 or A17. In either form the 8281 can support two or four LAN ports. At present, all LAN ports have to be of the same type. This restriction is likely to be lifted in 1996. The stand-alone version has an optional 100Mbps ATM port that is used to connect it to an ATM switch for LAN-to-LAN over ATM or LAN-to-ATM bridging. This optional ATM port could be omitted if the 8281 is to be used as a LAN-to-LAN bridge. (Given that a two-LAN-port 8281 has a current list price of $8,999, why anybody might want to contemplate using it as a LAN-to-LAN bridge, when two port bridges are available for less than one-fourth of that price, is a beguiling mystery.)

The 8281 Module for the 8260 does not have an ATM port. Instead, it is used with the ATM ports available on the 100Mbps or 155Mbps ATM Concentrator Modules. Interactions between the 8281 Module and the Concentrator Module will be via the 8260's ATM Backplane. See Figure 3.4.

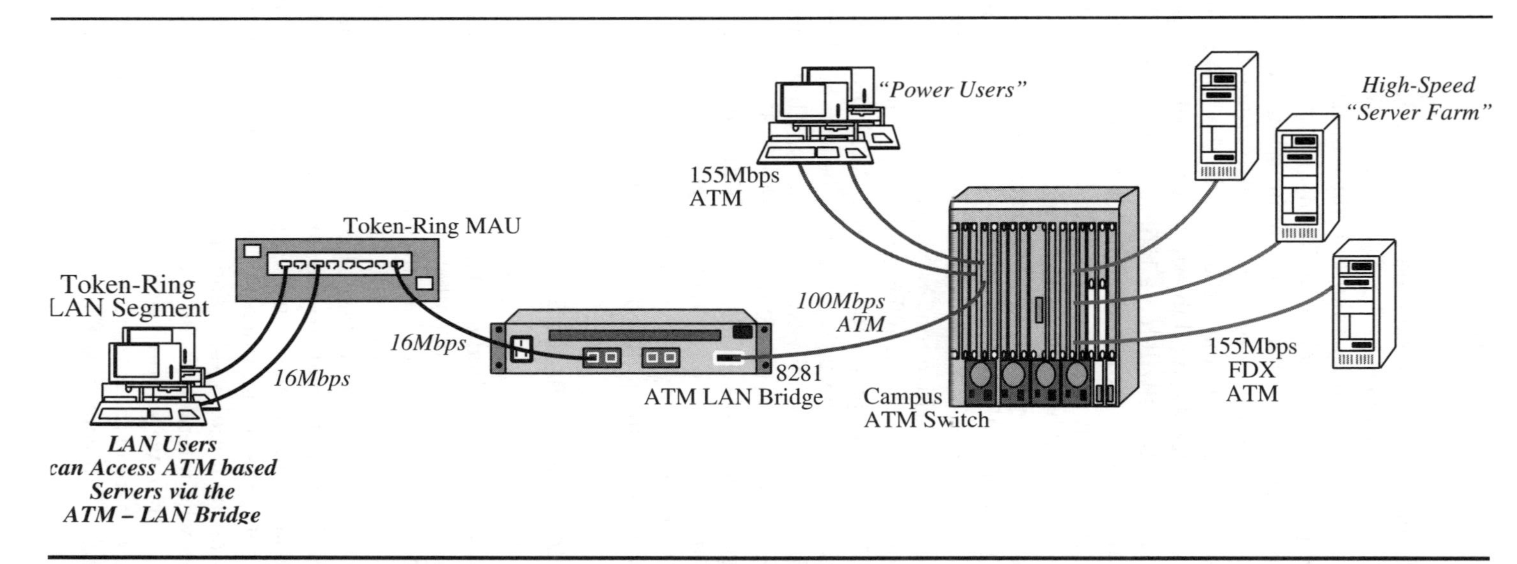

FIGURE 9.3 Unhindered and transparent LAN-to-ATM interworking through a bridge.

The other campus-oriented ATM-capable offerings from IBM are its 827x LAN switches and a new generation of multiprotocol bridge/routers that will finally take the place of the star-crossed IBM 6611 bridge/router. The 8271 is a family of stand-alone Ethernet LAN switches, developed by IBM in conjunction with Cisco/Kalpana, that today comes in 8, 11, 12 or 16 LAN port versions. The 8272, also developed with Cisco/Kalpana expertise, is the token-ring version of the 8271. It is currently available in 8, 12 or 16 LAN port versions. Both these LAN switches will have 155Mbps up-links in 1996. The configurational and networking possibilities of LAN switches with ATM uplinks have already been discussed in Chapters 3 and 4.

The new IBM bridge/routers—the 2220 Model 200 and the two-LAN-port, four-WAN-port 2210—will also support optional ATM 155Mbps uplinks. These new ATM-capable routers, in theory, could be used in place of the stand-alone 8281 to realize LAN-to-ATM interoperability. Most leading non-IBM bridge routers, in particular from Cisco, Bay Networks, and CrossComm, already offer similar ATM up-link capabilities.

All of the campus-oriented IBM ATM products discussed above, from the 8260 to the 2210, can now be managed from the new, integrated, SNMP-based Nways Manager, which will run under Windows or AIX. In addition, the 8260 and the 8281 can be comprehensively managed from IBM's ATM Campus Manager for AIX. This ATM Campus Manager focuses on providing access to ATM-specific entities such as ATM topology maps, PVC and SVC connections, and ATM Virtual Path/Virtual Channel link management.

9.4 THE 2220 NWAYS BROADBAND SWITCHES

The 2220 WAN BroadBand Switches are the core, as well as the pride, of IBM's ATM story to date. Three models of the 2220 WAN switches were available as of November 1995: the Models 300, 500, and 501. The Model 300 is a six-slot entry-level switch. The Model 500 is an eight-slot switch, with the 501 being strictly a six-slot expansion chassis for the Model 500. A 500/501 combo will thus have 16 slots.

The 300 and the 500/501 are targeted at mid-size to large enterprises that wish to implement their own private, in-house frame relay or ATM WAN networks. In addition to these end user switches, IBM has also talked about large, 32-slot and above, 2220 models aimed at network carriers who intend to build public ATM networks. Two of these larger switches that were previewed somewhat sketchily at the June 1994 announcement of the 2220 range were referred to as Models 700 and 800.

Each 2220 contains at least one Switch Module with a PRIZMA

ATM switching chip. The chip currently used with Models 300 and 500/501, as was discussed earlier, is the 16 x 16 version running at 266Mbps per port. This leads to the following aggregate throughput and full-duplex I/O rates:

> Model 300: up to 4 data slots. Aggregate Throughput = 4 x 266Mbps to 1.06Gbps; I/O rate = 2.1Gbps
>
> Model 500: up to 8 data slots. Aggregate Throughput = 8 x 266Mbps to 2.1Gbps; I/O rate = 4.2Gbps
>
> Model 500/501: up to 14 data slots. Aggregate Throughput = 14 x 266Mbps to 3.7Gbps; I/O rate = 7.4Gbps

The Model 700 is claimed to have a 12.8Gbps aggregate throughput and 26.6Gps I/O rate, while the corresponding figures claimed for the Model 800 are 25.6Gbps and 51.2Gbps.

Each slot in a 2220 can house either a data adapter or a voice adapter. In general, and particularly in the future, there is unlikely to be any restriction on how many adapters of each type can be slotted into a given 2220 chassis. Right now there is a restriction on the Model 300 that it may only support up to four data adapters (hence, the use of four adapters in the aggregate throughput and I/O rate comparisons performed above). There is no such restriction with the 500 or 500/501 combo, where each of the 8 or 14 slots may be taken up by a data adapter. Figure 9.4 shows generalized schematic of the internal structure of a 2220 Nways switch.

The actual functioning and the exact persona exhibited by a 2220 switch is dictated by its control software. This control software for the 2220 is called the *Nways Switch Control Program* (NSCP). NSCP is the 2220 equivalent of what ACF/NCP is for an IBM 37xx communications controller. NSCP, running within a given 2220, is responsible for the local operation, on-line maintenance, and management of that switch. NSCP implements three functional components of the NBBS architecture on each switch. These three components are: Access Services, Transport Services, and Network Control. These functions are described in detail in Section 9.5.1. An NSCP implementing functions specified by NBBS in 2220s is analogous to that of ACF/NCP implementing the functions of an SNA Type 4 node in 37xx's.

The parallels between NSCP and ACF/NCP, however, do not extend to the way that individual 2220 and overall network configurations are specified to a NSCP. In marked contrast to ACF/NCP, NSCP definitions can be done on the fly from a local or remote OS/2 workstation acting as a 2220 console. This is a welcome departure from the ACF/NCP paradigm

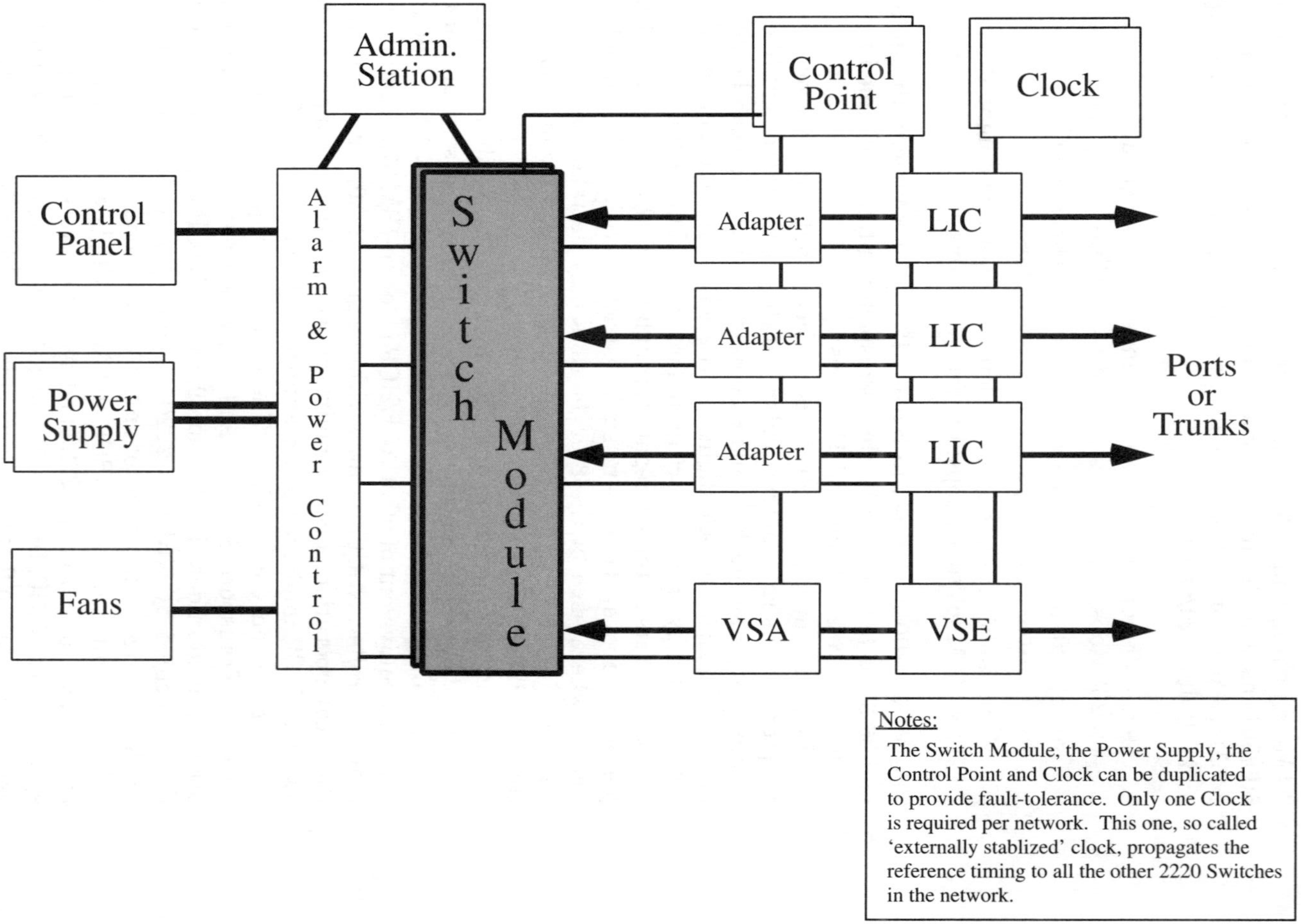
Admin. Station
Control Point
Clock
Control Panel
Power Supply
Fans
Alarm & Power Control
Switch Module
Adapter
Adapter
Adapter
VSA
LIC
LIC
LIC
VSE
Ports or Trunks
Notes:
The Switch Module, the Power Supply, the Control Point and Clock can be duplicated to provide fault-tolerance. Only one Clock is required per network. This one, so called 'externally stablized' clock, propagates the reference timing to all the other 2220 Switches in the network.

FIGURE 9.4 Generalized schematic of the internal hardware architecture of an IBM 2220 Nways switch.

that required extensive manual definitions that then needed to be compiled, as if it were a COBOL program, to produce a load module that then had to be loaded into a 37xx.

Overall management of a 2220 network, plus the updating of either the configuration or the NSCP software running on a particular 2220, is realized using the Nways BroadBand Switch Manager for AIX software that runs on one or more RS/6000s acting as network management consoles. At least for the time being, an Nways BroadBand Switch Manager for AIX interacts with NSCPs running on various 2220s using the OSI Common Management Interface Protocol (CMIP) and the Common Management Interface Services (CMIS) as opposed to SNMP. This, more than anything else, reflects IBM's late-1980s commitment to standardize on CMIP/CMIS in the days when IBM along with quite a few others were beginning to believe that OSI was indeed finally going to be a reality and, moreover, be the networking standard of the future.

Now that TCP/IP has somehow miraculously managed to sideswipe OSI from the picture altogether, IBM, in line with the rest of the industry, is an avid believer and follower of SNMP. Hence, there is a possibility that down the road IBM may replace this original CMIP/CMIS management scheme with an SNMP-oriented scheme. What really is noteworthy is that with the 2220s IBM elected to go with an open public-domain scheme as opposed to one based on SNA Management Services (SNA/MS) as would have definitely been the case in the past.

At this juncture it is also worth noting that all the 2220 switches in addition to their ATM prowess also have considerable frame relay capability. A 2220 can act as a frame relay DTE (i.e., terminal attached to a FR network) or as a frame relay DCE (i.e., a bona fide FR switch implementing a part of the WAN). Thus, for a start, 2220s in DTE mode can be interconnected to each other across a public or private frame relay WAN. More interestingly and significantly, 2220s can be used as very high-capacity and high-throughput switches to implement private and public frame relay networks.

Compared to a 3745, a 2220 certainly has the power and surprisingly the port capacity to be a daunting frame relay switch. With some of the port configurations already available, a Model 300 could act as a frame relay switch capable of supporting 240 56Kbps ports while a 500/501 combo could support 840 ports. Many of the 2220s that were in use in 1995 were being used as frame relay switches. Given that a 2220 is capable, at least in theory, of simultaneously acting as an ATM and FR switch such hybrid switching may become popular over the next few years. Enterprises planning to implement in-house private networks now have the option of first using 2220s as frame relay switches and then migrating a few ports at a time to ATM.

9.4.1 2220 Adapters, LICs, and Interoperability

The central switch module that contains the PRIZMA chip is the heart of a 2220 and is responsible for forwarding all traffic, in the form of ATM cells, between the various adapters. While they may not perform any ATM switching per se, there is, however, little doubt that the very soul and power of a 2220 rests within the adapters. A 2220 data adapter can be configured, via software, to perform one of three very different roles. These three different roles that can be performed by a data adapter are as follows:

1. Port Adapter that supports user premise data equipment
2. Trunk Adapter that supports high-speed connections between switches
3. Control Point Module that like its counterpart in a 8260 ATM subsystem performs network control functions such as topology management, path selection, and call setup

Figure 9.5 depicts the key trunk and port types that will be supported by the 2220 in 1996. Independent of the roles they are likely to perform, the 2220 data adapters come in four distinct flavors differentiated both by their bandwidth capacity and the number of VC connections they can support. The four adapter types are: Low Speed Adapter (LSA) Type 1, LSA Type 2, High Speed Adapter (HSA) Type 1, HSA Type 2. LSAs, whether Type 1 or 2, can support port or trunk connections at speeds from 2400bps to 2.048Mbps.

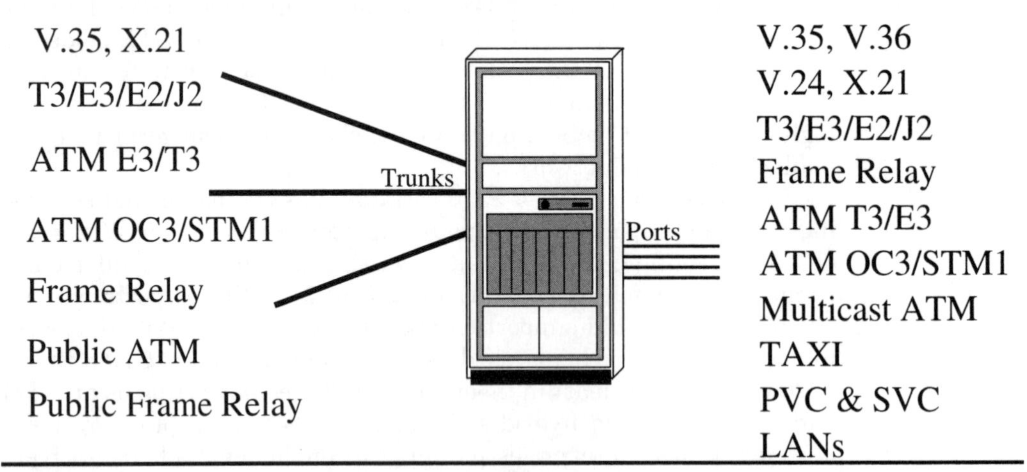

FIGURE 9.5 Key trunk and port types that will be supported by the IBM 2220 Nways switches in 1996.

Support for higher speeds require an HSA. The current HSAs can only support speeds up to 51.8Mbps. IBM will either increase the speed rating of the HSA, or introduce yet another adapter type to accommodate the 100Mbps, 155Mbps, and 622Mbps speeds commonly associated with ATM. Per IBM announcements, 155Mbps ATM trunk and port interfaces compliant with the necessary ATM Forum UNI or NNI specifications will be available in 1996. *These UNI- and NNI-based interfaces will make it possible for 2220s to be deployed in multivendor ATM environments.*

It should again be stressed at this juncture that IBM switches such as the 2220 can and will work in two distinct modes: a totally standards-compliant mode, or a value-added mode where NBBS and SVN functions (e.g., Packet Transfer Mode) will be overlaid on top of the standards-based offering. Multivendor interoperability should hopefully not be an issue when these switches are working in standards-compliant mode. Interoperability should only become an issue if value-added functions are being contemplated. The bottom line here is that enterprises will have to balance the importance of interoperability with those of the value-added services. If multivendor interoperability is a must then the value-added services will have to be forgone, or at least restricted to partitions within the network consisting of switches from the same vendor.

The Type 1 and Type 2 classifications for LSAs and HSAs bear no relationship to the bandwidth handling characteristics of an adapter. The bandwidth capacity classification is done purely in terms of the high-speed and low-speed designations. The Type 1 and Type 2 classifications are used instead to indicate either the number of connections supported by an adapter, or the size of the network that can be supported by a 2220 Control Point if the adapter is being used to run Control Point code. A connection in this context could be a ATM SVC or PVC, a frame relay SVC or PVC, or a port-to-port virtual circuit carrying traffic included with HDLC, SDLC, or ISDN Link Access Protocol-D (LAPD) frames. In the case of non-data traffic (i.e., voice or video) a connection would include Circuit Emulation Service (CES) channels used to carry CBR data.

A Type 2 adapter when used as a port or trunk adapter can support more connections than a Type 1 adapter. A Type 2 adapter when used as a Control Point module can support larger networks, with more trunks, than a Type 1 adapter. A dedicated Type 1 adapter, irrespective of whether it is an LSA or a HSA, is required to act as a Control Point Module and execute Control Point code. On the other hand, a Control Point can coexist within a trunk adapter that is supporting up to 1,000 connections, on an LSA or HSA Type 2 Adapter. Table 9.2 summarizes the various capacities of the four types of 2220 data adapter.

None of the 2220 adapters, whether supporting data or voice, contains any physical ports that can be used to realize either port or trunk

TABLE 9.2 **The Port, Trunk, and Control Point Capacities of the Four Types of 2220 Data Adapters**

Role	*LSA Type 1*	*LSA Type 2*	*HSA Type 2*	*HSA*
Port Adapter	up to 300 connections	up to 800 connections	up to 300 connections	up to 800 connections
Trunk Adapter	up to 900 connections	up to 1,000 connections	up to 900 connections	up to 1,000 connections
Stand-alone CONTROL POINT	Network up to 128 Nodes 128 Trunks	Network up to 400 Nodes 1,750 Trunks	Network up to 128 Nodes 128 Trunks	Network up to 400 Nodes 1,750 Trunks
Trunk and CONTROL POINT		Trunk up to 1,000 Connections Network up to 250 Nodes 1,000 Trunks		Trunk up to 1,000 Connections Network up to 250 Nodes 1,000 Trunks

connections. Instead, the actual ports for these adapters are provided via Line Interface Couplers (LICs), as shown in Figure 9.4. (This term has its origins with the 37xx, where the term *LIC* is used to describe the cards that support the external ports.) An LIC occupies the same 2220 slot as the adapter it is serving and is slotted in directly behind that adapter.

There are a variety of different LICs available today ranging from one that supports one T3 port to one that is capable of supporting up to 60 low-speed ports. Table 9.3 lists the LICs that are currently available. The repertoire of 2220 LICs will continue to grow as IBM adds support for increasingly higher link speeds. It is also highly likely that IBM will attempt to increase the port density of the entire 2220 range by offering LICs with more ports per LIC, for example, an LIC with four T3 ports as opposed to one that would immediately quadruple the number of T3 ports that can be supported by a given 2220 Model. (Recently IBM did a similar port capacity increasing exercise on the 6611 to make it more price competitive on a per-port basis.)

When it comes to voice traffic, Voice Server Adapters (VSA) take the place of LSAs/HSAs. In addition to VSAs, the 2220 also has so-called Voice Server Extension (VSE) modules that can be used to increase the number of voice channels that can be supported by a VSA. A VSA supports voice processing functions such as voice compression,

TABLE 9.3 Some of the 2220 Line Interface Couplers Available in November 1995

Line Interface Couplers (LICs) for High Speed Adapters

- LIC Type 513—one T3 at 44.736Mbps for a Port or Trunk
- LIC Type 523—one E3 (34.37Mbps), E2 (8.448Mbps), or J2 (6.312Mbps) for a Port or Trunk
- LIC Type 530—one High Speed Serial Interface (HSSI) at 51.84Mbps for a Port or Trunk

Line Interface Couplers (LICs) for Low Speed Adapters

- LIC Type 511—up to 60, V.35/X.21 (up to 256Kbps) or V.24 (up to 19.2Kbps) lines via two Ports.

  Each LIC 511 Port connects—over 1 cable—to one or two, separately 19" rack-mounted, Line Connection Boxes (LCBs).

  Each LCB supports up to 15 low-speed lines. Thus, each 511 Port can support up to 30 low-speed lines—i.e., two LCBs. With BOTH 511 ports active, 60 low-speed lines can be supported.

  This is the ONLY LIC that does not support trunks.
- LIC Type 512—four X.21 or V.35 (up to 2.048Mbps) Ports for Trunks or Ports.
- LIC Type 514—four T1/J1 Ports for Trunks and Ports. Channelization, where each port can support one 1.536Mbps channel or up to 24 64Kbps channels, is supported. Fractional T1/J1 is also supported.
- LIC Type 515—four Port E1 (2.048Mbps) variant of 514
- LIC Type 516—four Port E1 variant of 515 but with 120Ω RJ-48 connectors.
- LIC Type 522—four Port V.36 variant of 512

silence removal, fax/modem detection (i.e., precluding data traffic from being compressed as if it were voice), and digital echo cancellation. A VSA can process up to 20 64Kbps voice-channels.

A VSA is capable of operating in many different modes. Some of these modes include: voice compression with a 5:1 compression ratio, silence removal with a 2:1 compression ratio, as well as a combined voice compression and silence removal mode with a potential 9:1 compression ratio. In addition, a VSA can offer clear-channel pass-through mode where all value-added processing such as compression is bypassed. This is the mode used when the VSA detects fax/model data signaling and automatically turns off all of the compression. VSAs support touchtone dialing and include a built-in digital echo canceller that can compensate for up to 32ms of local delay. A VSE1 increases the number of channels that can be supported by a VSA to 80. A VSE2 increases the number of channels that can be supported by a VSA to 140. Table 9.4 shows the basic channel capacity of a VSA without VSEs and how the VSEs increase channel capacity in the various operational mode.

TABLE 9.4 Channel Capacity of VSAs without VSEs and How VSEs Increase Capacity

VSA Configuration	*Using All VSA Modes*	*With Only Compression and Silence Removal Active*	*Echo Cancellation Only Active*
No VSEs	8 Channels	20 Channels	20 Channels
With a VSE1	32 Channels	80 Channels	80 Channels
With a VSE2	56 Channels	140 Channels	140 Channels

9.4.2 2220s in Private and Public WANs

With this breadth of data and voice capability and capacity, the 2220 switches can, in theory, be used in two distinct roles. They can be used to implement private ATM networks. In this scenario, an enterprise just buys bandwidth in the form of high-speed links (e.g., 45Mbps T3 or 155Mbps OC-3 from a carrier) and then uses 2220s to build a full-function frame relay, ATM, or hybrid ATM/frame relay WAN using these high-speed links. Facilitating the implementation of such broadband private WANs is the primary role envisaged for 2220s.

The other role for 2220s is to act as the high-performance, high-capacity switches that are used to implement the switching fabric within the public ATM or hybrid ATM/Frame networks that will be operated by network carriers such as AT&T, Sprint, MCI, WillTel, and so on. In this scenario, IBM becomes the switch provider to the relevant network carriers. This alone is an interesting twist to the whole unfolding ATM scenario vis-à-vis IBM.

IBM is essentially backing two horses when it comes to ATM WANs. Ideally, IBM, given its legacy and penchant for private enterprise networks (e.g., SNA), would like mid-size to large enterprises to implement their own private, in-house ATM WANs using 2220 mid-range switches such as the Model 300 or the 500/501 combo. Given that these switches can also adroitly perform frame relay switching, such enterprises considering private networks now even have the choice of first implementing a private frame relay network to put the 2220s through their paces. Once satisfied with the integrity and the reliability of the 2220s as WAN switches, the enterprises can transition the WANs in their entirety to ATM over time or elect to settle for a hybrid ATM/frame relay network.

If on the other hand, customers decide that they would rather outsource the considerable responsibilities of running an in-house frame relay or ATM WAN and go with a public service, IBM is hoping that it could also satisfy some of this market by being the vendor supplying the appropriate high-capacity switches, for example, the Models 700 and 800, to the appropriate carriers. How successful IBM will be with this dyadic and beguiling notion of trying to sell the same family of ATM switches to end users as well as to carriers is still open to debate. IBM historically has never had the appropriate product repertoire or the necessary goodwill to be successful in the network carrier or service provider segment.

Now with the 2220 range, IBM, at least on paper, certainly has the necessary horsepower and capacity to interest the carriers. IBM, however, is also seen by these carriers as being a direct and dangerous competitor. For a start, IBM's active promotion of private frame relay and ATM networks explicitly impinge on the popularity and the demand for public networks. Moreover, IBM with its connection to Advantis, which happens to be a provider of bandwidth and services worldwide, is a direct, head-to-head competitor that is trying to woo prospective customers away from other carriers. Given these factors it is likely that IBM will face considerable opposition in trying to penetrate this lucrative but highly selective market sector.

To counter this, IBM intends to make its ATM switches compelling to the carriers by concentrating on providing value-added, system solutions. These could include: special facilities for tightly integrating cable TV-related services with ATM; innovative but necessary directory lookup services, and customized billing and accounting capabilities for ATM users. This is the area where IBM's background and prowess in information processing will provide them with an indubitable differentiating factor over other networking-only vendors.

9.5 NETWORKING BROADBAND SERVICES

Networking BroadBand Services architecture can be best thought of as an SNA for high-speed, broadband networks. The services and functions specified by NBBS complement and support those provided by ATM. The NBBS services and functions essentially sit above the AAL, which in the context of the OSI Reference Model occupies the lower portion of Layer 2. NBBS attempts to fill the gap between the base level, "plain vanilla" functionality as specified by the current frame relay and ATM standards, and the feature-rich, value-added, industrial strength facilities sought by the large enterprises contemplating private broadband networks.

NBBS is the offshoot of a pioneering fast-packet research project conducted by IBM Research at Hawthorne, NY, starting in 1988. This research project and the prototype switches emanating from it were known as PARIS (Packetized Automatic Routing Integrated System). Within the context of this PARIS project, IBM initially developed a gigabit switch that was not ATM-compliant. NBBS, as it is today, evolved from work done between Hawthorne and IBM Network Systems in Raleigh, NC, during 1989 to 1994.

The primary value-adds that NBBS strives to provide enterprises fall into the following categories:

- *Guaranteed Quality of Service:* Trying to ensure guaranteed QOS by proactive bandwidth management, congestion control, and heuristic bandwidth provisioning is one of the primary goals of NBBS. (Bandwidth provisioning in this context is the acceptance or rejection of new voice or video calls based on their bandwidth requirements relative to the current bandwidth availability on the end-to-end path for those calls.) Guaranteed QOS is invariably a prerequisite for satisfactorily supporting CBR traffic. Without such guarantees voice conversations and video applications would be susceptible to prolonged periods of unacceptable silence or slow-motion jerkiness. Some of the QOS criteria supported by NBBS include:

    1. Desired end-to-end delay (i.e., network latency) characteristics—real-time or non-real-time
    2. Limit on the variation of the arrival time of successive cells—the so-called jitter factor
    3. Need for dynamic, nondisruptive path switching in the event of node or link failures
    4. Desired security level for the connection—is there a need for external data encryption?
    5. Acceptable cell or packet loss ratio—percentage of cells/packets that may be discarded over a given period of time
    6. Feasibility of dynamically adjusting the bandwidth requirements originally stated at call setup based on actual usage relative to the cell/packet ratio deemed acceptable for that connection using heuristic techniques
    7. Maximum number of hops that should ideally be used in the end-to-end connection
    8. Requested priority when seeking the establishment of a new call
    9. Feasibility of path reallocation after the initial path has been established to accommodate new connections with higher priority

- *Set or Group Management:* Ability to create multiple virtual subnetworks within an overall NBBS network to either reserve certain sets of resources for specific groups of users, enhance data transmission security, or support VLANs. The types of initial VLAN partitioning envisaged by NBBS and the Switched Virtual Networking framework that goes above it include: partitioning via protocol (e.g., all devices using IPX), partitioning all the devices attached to certain ports on one or more switches, as well as partitioning based on media. Other VLAN partitioning schemes, such as partitioning users based on their organizational or functional affiliation, are likely to follow soon.

- *Proactive and Continuous Bandwidth Management:* Given NBBS's overriding desire to deliver guaranteed QOS to its users it has to be careful how many connections it accepts for a given end-to-end path through the network. *To avoid packet reordering delays at the destination and to ensure consistent cell delivery delays, ATM switches inevitably prefer to use fixed path routes for all connections but particularly so for CBR traffic.* A key element of bandwidth management is to select the optimum path for each new CBR call so as to guarantee its end-to-end QOS requirements, always bearing in mind that the bandwidth of a given path, even if it is 155Mbps or 622Mbps, is still finite.

  Reserving bandwidth, à la TDM, on each path to make sure that QOS criteria can be easily satisfied is obviously anathema in any broadband network—especially one that is ATM-based. However, continually allocating new connections to a given path without regard for the total amount of bandwidth that may be required at peak periods, especially when "bursty" data traffic is also involved, is foolhardy. If connections are allocated in this unconstrained manner, the switches will have no option but to discard cells to cope with the higher volume of traffic when one or more of the connections start using the bandwidth at peak levels. This will not always be acceptable since the cell discard ratio for a given connection can be a part of its QOS requirements.

  Before allocating a new call to a given path, a switch should determine what other fixed paths calls (i.e., CBT connections) have been previously allocated to that same path. This decision will be based upon the maximum bandwidth being requested by the new call, the cell loss ratio it is willing to accept and the previously allocated bandwidth for the path in question. A new call requiring fixed path routing will only be accepted if the switch determines that there is sufficient bandwidth on that path to accommodate the new call—relative to the cell loss ratio deemed to be acceptable. By allocating fixed path calls on such a basis, a switch can ensure that traffic related to these fixed path calls is not delayed or disrupted due to overutilization of the subject path.

This path selection mechanism is not a bandwidth reservation scheme. All the bandwidth of the path will be totally available to all the traffic types sharing that path—through statistical multiplexing. The path selection mechanism makes sure that the guaranteed throughput requirements of all of these traffic types can be safely satisfied. Nonetheless, basing this new call acceptance or rejection criteria on the peak bandwidth requested by that call could result in nonoptimum allocation of the bandwidth.

Hence, straight off the bat, NBBS uses what it refers to as the *equivalent capacity* of a call, rather than peak bandwidth required by that call, as its call acceptance or rejection criteria. NBBS dynamically calculates the equivalent capacity for a call based on four criteria. These are the call's peak bandwidth rate; accepted cell loss ratio; what NBBS deems to be that call's mean bandwidth rate; and what NBBS estimates to be the average length of the traffic bursts that will occur on that call. The latter two criteria used to calculate the equivalent capacity are heuristically calculated by NBBS as opposed to being specified as one of the call setup parameters. NBBS will estimate the values for these two criteria based on its ongoing experience of other similar connections as well as on the type of tariffing requested for that call. (The tariff notion in this context is similar to using something akin to frame relay's CIR concept to determine the amount of guaranteed bandwidth a user is willing to pay for a call.)

In addition to accepting or rejecting new CBR-related calls based on an equivalent capacity, NBBS also has the ability to dynamically adjust the bandwidth associated with a given call based on its actual traffic patterns. Such bandwidth fine-tuning is only done if it was deemed to be acceptable in the initial QOS requirements (as listed in the QOS section above), and provided that any bandwidth adjustments do not cause the connection to fall outside the bounds of its guaranteed QOS.

- *Dynamic, nondisruptive path switching:* As discussed relative to guaranteed QOS.
- *Dynamic priority-based preemptive call rerouting:* The desirability of using fixed path routing to support voice and video CBR connections is discussed above. There could be situations where a new CBR call should ideally be allocated to a path that is now deemed to be at its capacity supporting previously allocated fixed path calls. This rerouting feature caters to such eventualities. Whether a call can be rerouted once it is set up is determined by the QOS parameters of that call. If call(s) can be rerouted, NBBS will move them to another path in order to accommodate higher priority calls

on the original path. NBBS realizes such rerouting using its nondisruptive path switching capability normally used for dealing with node or link failures.

9.5.1 The Anatomy of NBBS

The NBBS architecture consists of the following three primary functional components:

1. *Access Services* govern how users interface to and use an NBBS-based network.
2. *Transport Services* provide the end-to-end connections across an NBBS-based network.
3. *Network Control / Control Point Services* dynamically control, monitor, manage, and allocate the various physical resources that make up an NBBS-based network.

The above three NBBS functional components are implemented within each node that supports NBBS. For example, the 2220 Nways Broadband Switch Control implements these NBBS functional components within a 2220 switch. Figure 9.6 depicts the overall functional structure of an NBBS-based network while Figure 9.7 provides an exploded view of how the NBBS functional components are layered within a given node.

In much the same way as any other network, NBBS networks consist of nodes that are interconnected via links. A link in the context of NBBS, however, is a unidirectional transmission mechanism. Consequently all links are deemed to occur in the form of pairs—one unidirectional link in each direction between any two nodes so as to ensure full-duplex communications between those nodes.

NBBS also postulates the possibility of a node being made up of multiple subnodes that share one or more of the Network Control/ Control Point functions. Typically the functions that would be shared by the subnodes include a common network topology database and a spanning tree algorithm-based routing scheme. This spanning tree-based routing scheme is used to exchange network control and network availability information on a Control Point-to-Control Point basis among the NBBS nodes.

Each subnode is considered to be a switching element and has one or more full-duplex links associated with it. Having this notion of subnodes enhances implemetational latitude by permitting multiple disparate switching elements to be implemented within a single physical switch without requiring each switching element to have its own sepa-

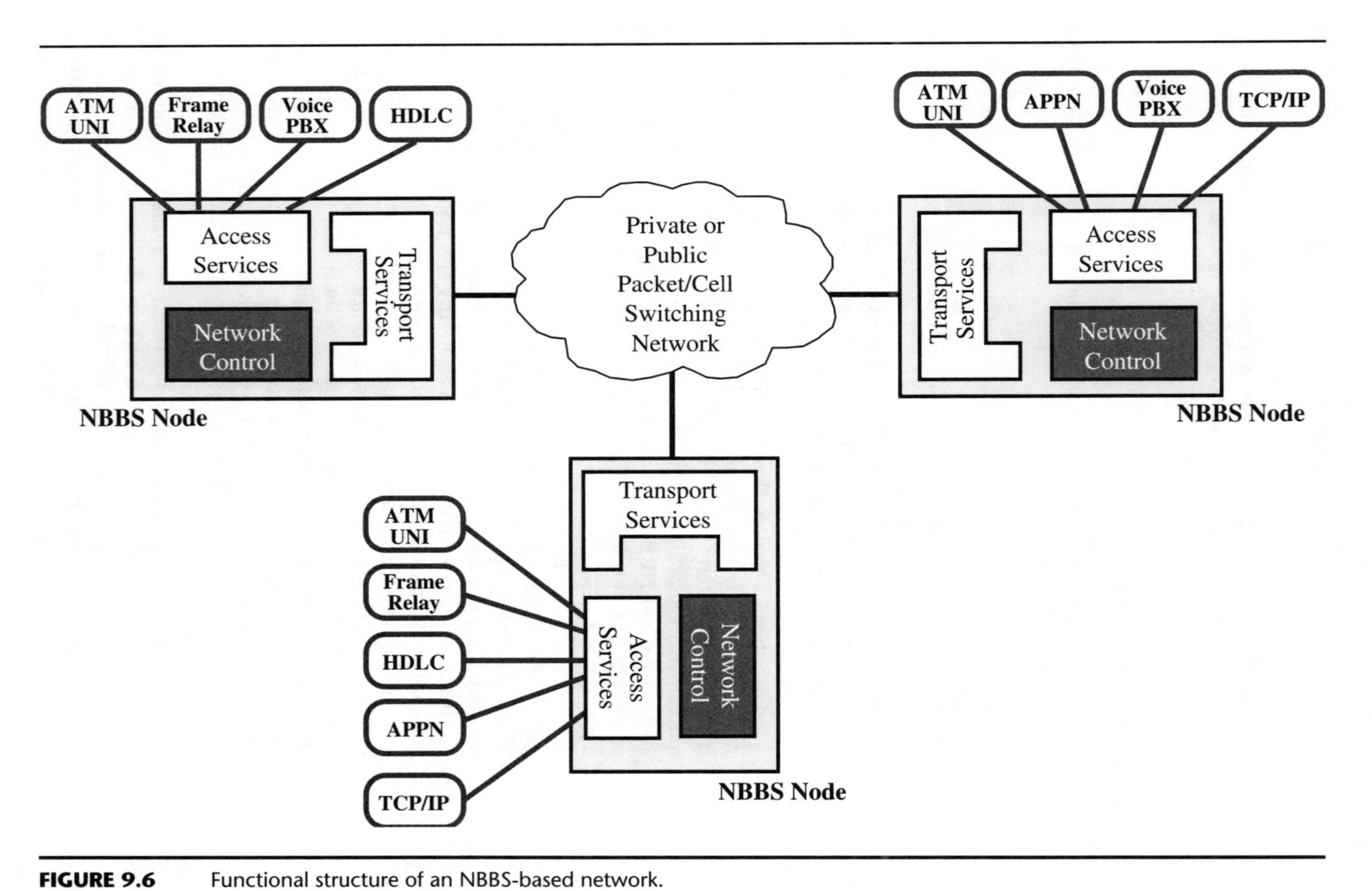

FIGURE 9.6 Functional structure of an NBBS-based network.

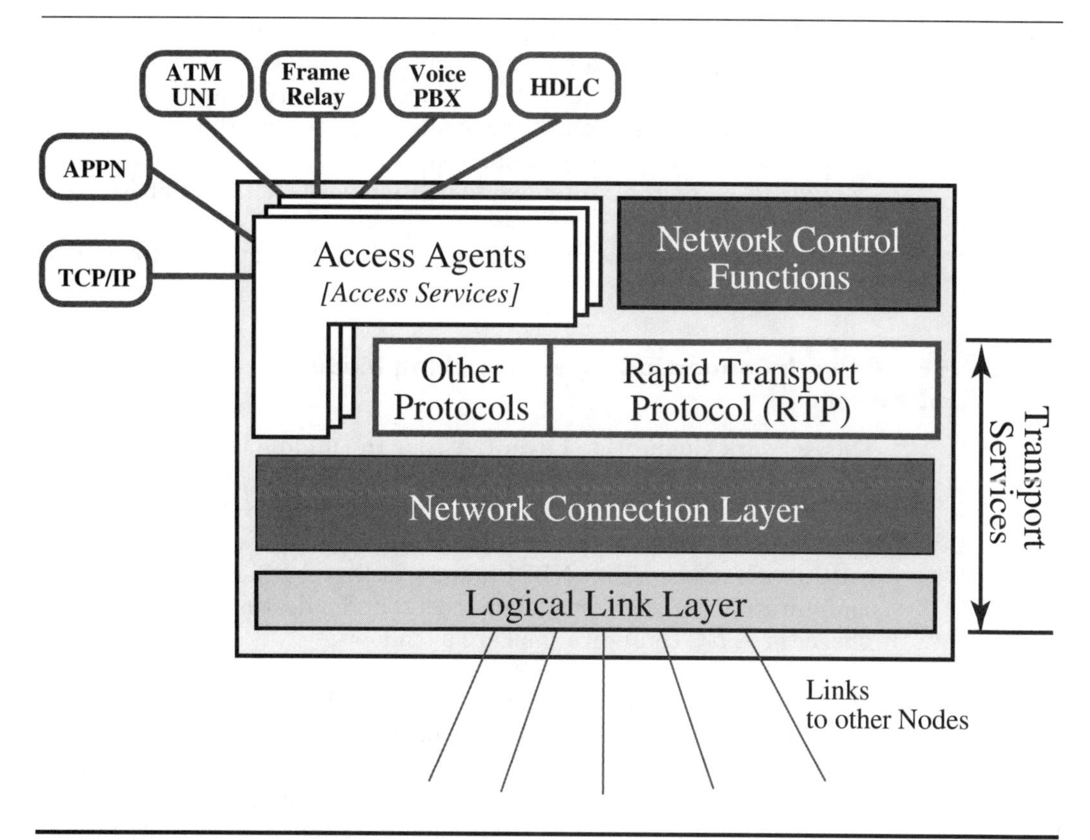

FIGURE 9.7 Functional layering of the NBBS components within a node.

rate network control services and network topology database. (This is a definite improvement over SNA, which never catered for the possibility of having multiple subnodes, such as a Type 4 and a Type 2, within a given device sharing a common path control function.)

NBBS nodes (or subnodes) communicate with each other by exchanging packets. (Note that as discussed at the start of this chapter IBM positions NBBS as a packet, as opposed to a cell, switching architecture and treats ATM cells as a subset of packets.) Nodes transmit and receive packets through their LLC function—which as shown in Figure 9.7 is the bottommost layer of the NBBS Transport Services component. The higher NBBS services consider the LLC as providing them with a virtual packet pipe between logically adjacent subnodes. NBBS specifies the following three different types of LLC interfaces:

1. *Bit-level interface* is targeted at T1 to T3 low-speed links and optionally includes a preemptive resume facility that enables high-priority, real-time traffic (e.g., voice) to interrupt and go ahead of lower-priority non-real-time traffic (e.g., data) that was in the process of being transmitted.
2. *Byte-level interface* intended to be used with OC-3 and above high-speed links.
3. *MAC level interface* for exchanging NBBS packets over token-ring or Ethernet LANs.

9.5.2 NBBS Network Connection Layer, Switching Modes, and Packets

The Network Connection Layer (NCL) is the middle layer of the NBBS Transport Services component. It is responsible for providing the layers above it with end-to-end Network Connections (NCs) in the form of virtual packet pipes between origin subnode(s) and a destination subnode(s). (Cf. SNA VRs.) NBBS supports three different types of NC configuration between origin subnode(s) and a destination subnode(s). These three NC configurations are as follows:

1. *Point-to-point* NC connects one origin subnode to one destination subnode. A point-to-point NC is deemed to consist of two independent unidirectional paths, one in each direction. The two paths that make up a given point-to-point NC do not have to traverse the same physical path (i.e., go through the same intermediate nodes). NBBS even reserves the right to reroute one path of a NC independent of the reverse path.
2. *Point-to-multipoint* NC connects one origin subnode to multiple destination subnodes. This type of NC is deemed to be strictly unidirectional, with packets only being allowed to be transmitted by the origin subnode.
3. *Multipoint-to-multipoint* (or multipoint) NC connects multiple subnodes to each other. Such a multipoint-to-multipoint NC can be implemented in the form of a series of unidirectional spanning trees or as a single bidirectional tree.

In addition to providing NCs, the NCL is also responsible for transporting packets on an end-to-end basis. It does so by using the services of the LLC layer beneath it. In the context of end-to-end packet transport, the functions performed by the NCL layer include: transmission scheduling, packet switching, congestion control, and packet buffering.

NBBS supports an eclectic collection of packet switching modes. The ones currently supported are as follows:

1. *ATM switching mode:* ATM cells, in their original form, are switched unadulterated in anyway across an NBBS network with the NBBS switches reading and interpreting the standard Virtual Path Identifier (VPI) and Virtual Channel Identifier (VCI) fields that appear in the 5-byte header of all ATM cells. Any ATM cells entering an NBBS are always automatically switched using this mode to obviate the need for any header translation and to maximize switching performance. When an ATM cell enters an NBBS network, NBBS reserves the right to replace the 4-bit Generic Flow Control (GFC) field that appears at the start of all 5-byte ATM headers prefixing ATM cells flowing across an UNI interface. This is not deemed to violate the integrity of the ATM cells since the GFC only has local significance between an ATM end-point and the ATM switch to which it is attached. In NBBS scenarios, an NBBS switch (e.g., 2220) in most cases will be the ATM switch serving end-points. The GFC is not carried through the network, and the 5-byte ATM headers prefixing ATM cells flowing between switches, across the so-called NNI interface, does not include a GFC field.
2. *HPR's Automatic Network Routing (ANR) switching mode:* HPR's ANR functional component was described in detail in Chapter 7. This routing mode uses the routing information that appears in the ANRF that appears in the header of each packet. The ANRF, which is comparable to the RIF used in SRB, specifies the intended route of the packet in the form of a series of ANR labels. This series of ANR labels successively identify the next hop (e.g., link) that needs to be traversed by that packet in order to reach its destination. With ANR switching, each switch determines the next hop on the route by reading the first ANR label that appears within the ANRF. It then removes that label before switching that packet across the specified hop.

   The ANRF's datagram-like self-describing end-to-end routing information permits NBBS to use this mode of switching to realize a connectionless service. NBBS uses this ANR connectionless packet switching mode to establish end-to-end connections, exchange network control messages, and sustain previously established connections. With the exception of using this ANR mode as a connectionless service, NBBS in general, like its predecessor SNA, is primarily connection-oriented.
3. *Label-Swap switching mode:* This is a switching mode analogous to that used by APPN NN routing, frame relay, and ATM. In this

mode, each packet is prefixed by a label that uniquely associates it, at each intermediate node, with the end-to-end connection (i.e., session) to which it belongs. The label in essence serves as a pointer that is used to access a lookup table that contains information about all the connections passing through a given intermediate node. At each intermediate node the connection information pertaining to a given packet is looked up, using the label as the pointer, to determine the next hop on to which it should be switched. At this juncture the old label may be replaced by a new one that will act as the pointer to the appropriate connection at the next node. The new label to be used would be included as a part of the connection-related information maintained by each node. With label-swapping, the connection information at each node, as well as the inbound label used to access it and the new label that will replace it on the next hop, have all got to be set up when the end-to-end connection is being established.

4. *Remote-Access-to-Tree switching mode:* This switching mode is used to support point-to-multipoint and multipoint-to-multipoint connections. Such switching that involves multiple destinations is said to be realized using multicast trees. With a *multicast tree* a switch will forward a source packet on all the outbound hops (i.e., links) that appear on a given tree. NBBS realizes remote-access-to-tree switching through a synthesis of ANR and label-swapping. The goal here is to permit a source subnode to multicast along a tree without having to be a part of that tree per se. To do this, the source subnode uses ANR switching to reach an access node that happens to belong to the relevant multicast tree. The last ANR label in the ANRF when it reaches the access node will identify the specific multicast tree. Label-swapping will be used from this juncture to switch the packet through the outbound hops specified in the multicast tree. NBBS uses this switching mode to implement the spanning tree maintained between the Control Points in all the nodes as well as to implement any other tree-based routing scheme necessary to support multicast or VLAN functions.

With the exception of the ATM switching mode, the other three switching modes also support a Copy Option as a part of their switching function. This option enables an intermediate node to retain copies of packets that it is forwarding on to other nodes. (This is akin to "promiscuously" reading all frames that flow through a LAN without heed to their actual destination address.) This copy option in effect enables NBBS to create a type of point-to-multipoint connection where the end-point destinations as well as all the intermediate nodes along the route each get a copy of a packet. NBBS uses this copy option during connection setup.

NBSS assumes that the switching function at intermediate nodes will be realized by a hardware-based switching element, for example, a PRIZMA chip. It also assumes that the switching element is capable of supporting all of the switching modes—which obviously include standard ATM switching. A short field (i.e., around 3 bits) at the start of all NBBS packets, including ATM cells, specifies the switching mode to be used to forward that packet through the network. This field also indicates the format of the rest of the NBSS header. (Compare FID field at the start of all SNA/APPN THs.)

NBBS switches and networks always support both variable-length NBBS packets as well as standard, 53-byte ATM cells. The generic format of an NBBS packet is shown in Figure 9.8. The overall length of an NBBS packet will only be limited by the maximum advisable packet length for the various links that the packet will have to be transmitted to reach its final destination.

The NCL header that appears at the start of each NBBS packet contains such information as: the switching mode to be used; a switching information field (e.g., ANRF); whether the copy option of the switching mode is active; and the packet discard/packet delay priority of the packet per its QOS that would determine its fate if there is congestion at intermediate nodes.

9.5.3 NBBS Access Services

Access Services can be thought of as NBBS's equivalent of SNA LUs. Access Services provide the interface between End Users or End Points and an NBBS network. This interface is specified and realized in terms of NBBS *access agents*. An access agent enables a particular type of End User or End Point to gain access to the NBBS network. Consequently there are different access agents corresponding to the various disparate types of End Points already supported by NBBS. NBBS can only support a given End Point type if it has a specific access agent for that type of End Point.

NBBS access agents have already been defined to support, among others, the following End Point interfaces: ATM UNI interface; frame relay DCE interface; Voice PBX interface; HDLC interface (with idle removal); clear-channel interface; TCP/IP interface; APPN NN interface; and the Fiber Channel Standard interface

An access agent essentially has two interfaces: one at either end. The interface at the top provides the necessary adaptations and emulations to ensure that the End Point type it is servicing can easily and

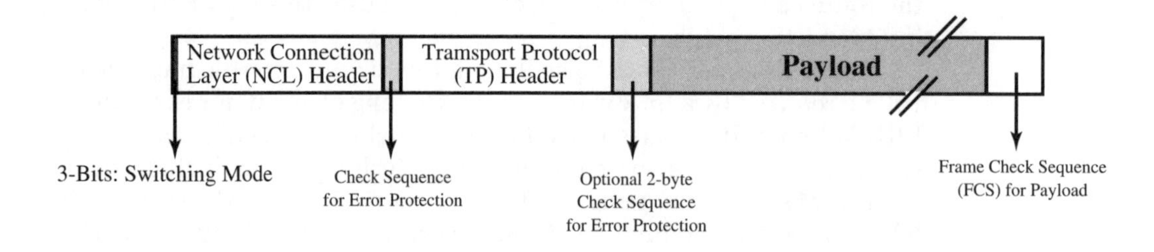

FIGURE 9.8 Overall structure of a switching-centric network as envisaged by SVN.

seamlessly interact with the agent—using its native protocol. Thus, for example, the APPN NN access agent would appear to a bona fide APPN NN (e.g., AS/400) as if it were another full-function APPN NN. The interface at the bottom of an access agent will mesh with the Transport Services layer of the NBBS subnode it is resident within.

NBBS further defines an access node as consisting of three functional subcomponents. These three subcomponents are the *protocol agent*, *directory agent,* and *connection agent.* The protocol agent is responsible for the top interface that accommodates the End Point being serviced (e.g., APPN NN emulation). The directory agent is tasked with interpreting the address scheme being used by the external End Point and mapping it to a scheme understood and used by NBBS. The connection agent uses standard NBBS protocols to establish, sustain, and tear down the end-to-end connections required to permit End Points from being able to freely communicate across an NBBS network.

9.5.4 NBBS Transport Services

All end-to-end communications across an NBBS network are realized through the use of the NBBS Transport Services. As shown in Figure 9.7, both the Access Services (i.e., all the Access Agents) and Network Control/Control Point components in a given (sub)node gain access to the network in order to communicate with entities in other (sub)nodes via the Transport Services Layer—consequently, the notion that any entity communicating across an NBBS network presents itself to that network in the form of a transport (layer) user.

The NCL and the LLC Layer that were described earlier form the bottom two sublayers of the Transport Services component in a given (sub)node. A series of Transport Protocol stacks identified by NBBS are then layered on top of the NCL. NBBS, at present, supports five Transport Protocols, which are frame relay, HDLC (with idle removal), Continuous Bit-stream Operation (CBO), Pulse Code Modulated (PCM) voice, and HPR's Rapid Transport Protocol (RTP). In general, the NBBS Transport Protocols strive to offer, at a minimum, the following optional facilities: message segmentation and reassembly, sequence checking, and packet retransmission in the event of a packet loss or error.

HPR's RTP was described in detail in Chapter 7. The Network Control/Control Point functions of NBBS and the Access Agents rely on RTP for nearly all of their interactions. NBBS positions RTP as a reliable (i.e., connection-oriented), full-duplex transport protocol that supports variable-length packets and guarantees in-sequence delivery. As is the case with HPR, NBBS recognizes that RTP's selective retransmission facility can be deactivated if all that is required is a fast, unreliable transport scheme where higher-level software will take care of any retransmissions that may be necessary.

Information required by the Transport Protocols are included in the Transport Header, which comes directly after the NCL Header in NBBS packets—as shown in Figure 9.9. It is assumed that NBBS will use the standard RTP Transport Header (THDR), as used by HPR, when using RTP. This THDR, depending on the Control Vectors included, could be in excess of 20 bytes in length. With the other Transport Protocols, NBBS typically uses a 1-byte Transport Header. The primary purpose of this 1-Byte TH is to convey packet sequence numbers. However, in the case of voice traffic, this 1-Byte TH may also include information pertaining to voice compression and silence removal.

9.5.5 NBBS Network Control/Control Point Services

The NBBS Network Control functions dynamically control, monitor, manage, and allocate the various physical resources that make up an NBBS-based network. NBBS refers to the entity that performs this set of functions as an NBBS CP. The functions performed by an NBBS CP is a superset of those performed by an APPN or HPR CP.

The network control functions performed by an NBBS CP include the following:

1. *Directory Services:* Maintaining information on how various resources and entities can be reached within or across the network. It will also maintain such information for End Points being serviced via Access Agents. This service also includes name and address resolution and translation functions.
2. *Network Topology Database Maintenance:* Ensuring that all NBBS nodes have an up-to-date awareness of the active topology of the NBBS network in terms of the NBBS nodes that are active and the

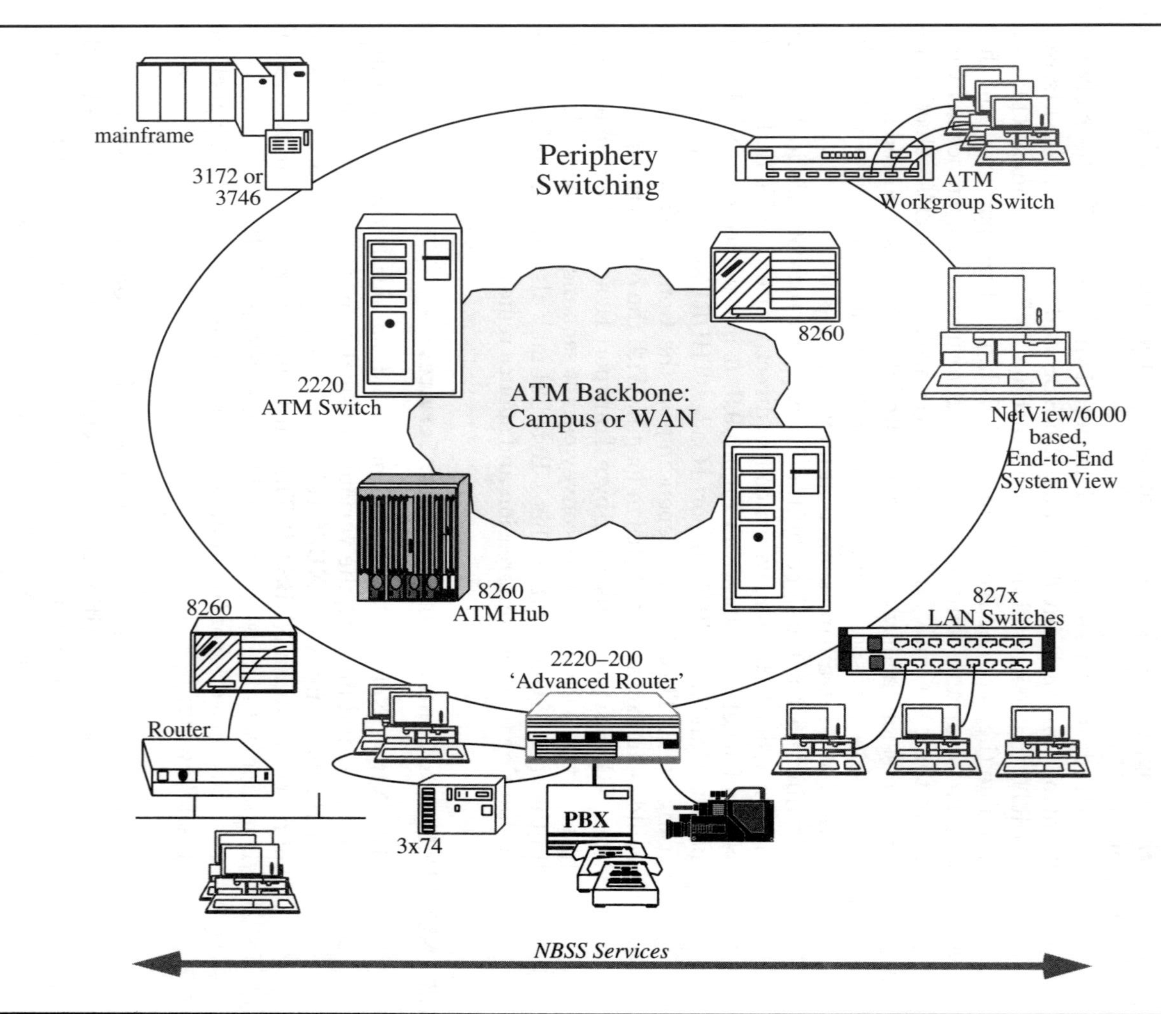

FIGURE 9.9 Generalized format of an NBBS packet.

operative links between them. Just as with APPN, the NBBS CPs will notify each other whenever they detect a change in the active topology.

3. *Optimum Path Selection:* Choosing the optimum end-to-end path for each new connection being established based on its QOS requirements and the current bandwidth allocation levels of all the potential paths across the network.
4. *CP-to-CP Spanning Tree Maintenance:* NBBS CPs interact with their counterparts in other nodes via a spanning tree-based routing scheme. It is via this tree, for example, that a CP notifies its peers of any changes to the active network topology.
5. *Congestion Control:* Ensuring that unexpected congestion does not impact the QOS requirements of the active connections. When faced with congestion, NBBS will be forced to discard packets. It will, however, try to do so within the acceptable packet loss ratio QOS associated with a given connection.
6. *Proactive and Continuous Bandwidth Management:* As discussed near the start of Section 9.5.
7. *Set or Group Management:* As discussed near the start of Section 9.5.
8. *Node Initialization:* In much the same vein as an APPN CP, an NBBS CP on its node being activated will set about exchanging identification (cf. XID), CP-capabilities, and control information with the CPs in logically adjacent NBBS nodes.
9. *Multiprotocol Switched Services:* This is a new set of functions introduced into NBBS in August 1995 by the new IBM SVN framework. SVN is discussed in detail below. Multiprotocol Switched Services (MSS) appears to still be in the process of evolving. It currently embraces the notion of distributed routing, ATM LAN emulation (i.e., the ability of an ATM adapter to masquerade as a LAN NIC), Broadcast Management (e.g., destination discovery within the context of a VLAN), and the whole issue of VLANs.

    SVN's notion of *distributed routing* deals with how already existing bridge/routers or those that were planned to be deployed can now be eliminated from an SVN/NBBS-based switched network. SVN plans to distribute the Layer 3 routing functions, hitherto performed by bridge/routers, to the periphery of the network—with such functions possibly done by LAN servers or even individual PCs/workstations.

    In order to cope with migration scenarios, the SVN/NBBS switching network could present the appearance of a peer-router to those bridge/routers already installed within the network being migrated. This router emulation will most likely be realized via an NBBS Router Access Agent. (Compare to APPN NN Access Agent.) The SVN/NBBS distributed routing functions will initially provide support for routing between devices using LAN emulation, VLANs, and IP-over-ATM applications. In the future this support will be extended to include the ATM standard MPOA.

9.6 IBM'S SWITCHED VIRTUAL NETWORKING FRAMEWORK

Switched Virtual Networking is IBM's ATM-centric overarching blueprint for the implementation and management of switched, as opposed to routed, networks. *SVN postulates the notion of networks that are built entirely around LAN and ATM switches.* The goal of SVN is to provide a scalable switching fabric for any-to-any, end-to-end connectivity with one-hop routing (i.e., everything appears to be adjacent) and guaranteed quality of service (e.g., response times). SVN also endeavors to protect existing investment in networking equipment—with the possible exception of non-IBM bridge/routers.

SVN, which was unveiled at the end of August 1995, is the latest chapter in IBM's already voluminous ATM story. It attempts to synthesize what has preceded it, and slots in actual products against networking roles and functions hitherto described by NBSS. The core of SVN networks will be a backbone made up of ATM switches à la IBM's 2220 Nways Switches. Around this ATM backbone will be peripheral switches, à la IBM's 8271 or 8272 LAN switches, that will provide end user access to the overall switching fabric. With IBM now reselling Bay Network's Centillion token-ring switch, possibly as a stop gap solution for the star-crossed 8272, it is possible that SVN peripheral switching, if not ATM switching, in the future will also embrace non-IBM switches.

The key precepts of SVN are as follows:

- ATM, as has been repeatedly stated by IBM, is indeed the ultimate holy grail of contemporary networking. The goal of SVN is to make this a reality—as soon as possible.
- Large centralized collapsed-backbone-type routers and networks built around such routers are now passé. SVN networks will be switch-centric, and will rely on distributed routing as described by NBSS's new MSS component above.
- Routing functions will be distributed, with some routing functions pushed all the way to the desktop (i.e., being performed by PC/workstation software).
- NBSS will acquire MSS, which covers the functions associated with Distributed Routing, LAN Emulation vis-à-vis ATM, Broadcast Management, and Virtual LANs.

ATM is the very soul of SVN. With SVN, IBM categorically states that ATM is the endgame in contemporary networking. The big question that is still left unanswered, however, is when ATM is likely to be a reality to the bulk of IBM's transaction-processing oriented, SNA-based commercial sector customers.

SVN obviously has to advocate a phased approach when it comes to switching. It starts off with periphery switching, which revolves around LAN switching. It then builds upon this switching base to create an IBM-centric blueprint for end-to-end interactions when ATM does indeed become the basis of next generation networking. Given frame relay's current popularity, particularly in North America, SVN does also acknowledge the possibility of FR-based backbones, albeit as an interim to or at worst a localized adjunct to ATM. It does, however, concede that frame relay may continue to be the optimum means for attaching branch offices, for example in an banking environment, to a SVN backbone. Figure 9.9 depicts the overall structure of a switching-centric network as envisaged by SVN.

SVN, in addition to pinning all of its hopes on ATM, postulates, possibly with a twinge of sour grapes given IBM's singular lack of success in this arena, that large centralized routers and router-based networks are passé. With SVN, IBM intends to distribute and disperse routing functions such as destination identification, optimum route selection, and protocol transformation. Where possible some of these routing functions will be pushed all the way back to the desktop.

This immediately raises the question as to how these distributed routing functions will be provided at the desktop. Ideally, it should be in the form of an Operating System independent piece of software. While IBM has already stated that it will provide this functionality on its ATM TURBOWAYS adapters, this alone will not be enough. The TURBOWAYS adapters will only address PCs/workstations running native ATM. SVN, by definition and design, also deals with LAN switching. PCs/workstations using LAN switches will not have new NICs. Some are also likely to be using non-IBM NICs.

Distributed routing and other MSS functions will also have to be available in software form to cater to scenarios involving LAN switching, as well as non-IBM ATM adapters. Only being able to provide such distributed desktop routing with OS/2 will not be enough by a long shot. The IBM world, in line with the rest of the world, is rapidly standardizing on Windows(95). Thus at a minimum this distributed routing and MSS functions should be available for PCs running Windows(95) and using LAN switching.

A model for switched networks à la SVN is not a new or unique concept. If anything, most of the other leading networking vendors had already beaten IBM to the punch with their models for such ATM-oriented networking. These include: CiscoFusion, Bay's Bay Networks

Switched Internetworking Services (BaySIS) and Cabletron's Secure-Fast Virtual Networking.

9.6.1 The Structure of SVN

The SVN framework currently consists of four elements:

1. Periphery switching
2. Backbone Switching
3. Networking BroadBand Services
4. System Management

Periphery switching functions will be provided by a gamut of IBM products, including: 827x LAN switches, 8260 Switching Hubs, 3746-950s Nways Controllers, 3172 Interconnect Controllers, a yet to be unveiled ATM workgroup switch, and the Proteon multiprotocol software-based, ATM-capable 2220 Nways Switch Model 200 Advanced Router, which will make obsolete the star-crossed 6611 bridge/router.

The backbone switching functions will be provided by 8260 Switching Hubs and 2220 Switches. NBBS will provide SVN with advanced network control functions and MSS as discussed earlier. MSS embraces distributed routing, LAN emulation within ATM (i.e., the ability for LAN adapters to look and act like LAN adapters), broadcast management (e.g., destination discovery), and VLANs. The 2220-200 router, the 8260 ATM Hub, 827x LAN Switches, the future ATM workgroup switch, and IBM's TURBOWAYS ATM adapters for PCs and workstations will, in time, all include MSS functionality.

At present, SVN has a fairly modest notion of what a VLAN could be. SVN VLANs can consist of users grouped together, irrespective of their actual physical location, per one of the following criteria:

- By specific ports of LAN switches and hubs. (Most LAN switches already support this type of VLAN.)
- By Layer 2 MAC addresses. (This again is already supported by some LAN switches.)
- By Layer 3 protocol (e.g., IP, IPX) as well as by Network Addresses. (Some LAN switches, particularly Ethernet switches, support such VLANs.)
- By the type of LAN Emulation being used over ATM, such as token-ring or Ethernet emulation.

In time VLAN technology will evolve to permit the criteria for grouping a set of users together to be more sophisticated. Ideally, VLANs will permit user partitioning along the lines of organization or functional affinity, independent of the physical proximity of the users being grouped together. Thus it would be possible to create VLANs that embraced all the members of, say, the Human Resources department or all the hardware engineers working on a particular product. SVN does not address this type of VLAN as yet.

The last, but by no means least, element of SVN is tightly integrated, end-to-end, SNMP-based System Management, which will be built around NetView for AIX (né NetView/6000). Given that SVN includes Virtual LANs, this System Management function will also address the myriad issues related to defining, configuring, and monitoring VLANs. NetView/6000 is discussed in detail in the next chapter.

9.6.2 Facets of SVN That Need Further Refinement

To IBM's credit, SVN is without doubt a bold step in the right direction. Given IBM's reliance on and oft-avowed commitment to ATM, it did not have a choice but to put forward a grand unification framework for switching-centric networks such as this. There are, however, three crucial areas that SVN has to tackle in more detail if it is to be truly credible and useful. These three areas that still need further refinement are: multivendor interoperability; how existing bridge/router-based networks are going to be systematically migrated to SVN; and the whole thorny but crucial issue of seamlessly assimilating SNA/APPN traffic into a switched network.

Multivendor interoperability is the Achilles heel of all of these SVN-like, ATM-oriented switching frameworks. For that matter multivendor interoperability is also the bane of even the more modest VLAN schemes currently offered by LAN switches—the problem being that there are no industry or even de facto standards, as yet, to facilitate such interoperability. Thus, at present, SVN alas is a bit analogous to SAA of yore. Enterprises have to make a leap of faith and trust IBM to deliver all the necessary goods to make a SVN-based switched network real.

In essence, with SVN, as it is today, enterprises have to commit to adopting a predominantly IBM-centric networking fabric. While some enterprises, given their loyalty and trust in IBM, may have been willing to contemplate this even a few years ago, this is definitely not the case now. IBM's credibility and credentials when it comes to contemporary networking is somewhat tarnished. Over the last five years or so IBM has had trouble delivering compelling and competitive networking products.

IBM may have turned the corner with its ATM product repertoire but it is still too early to judge exactly how solid and competitive these products are compared to the other products in the field. IBM has even had trouble delivering its 8272 token-ring switch on time. SVN has to surmount this credibility issue. IBM-centric networks, à la SNA networks, will in future be a minority. *Enterprises will opt for eclectical and egalitarian networks based on the best of breed products from multiple vendors—and do so through a systems integrator if necessary*. The only way that SVN can become credible is to comprehensively address the interoperability issue. But this cannot be done in isolation. Other major vendors also need to commit to supporting SVN.

Bridge/routers and multiprotocol internetworking have captured the mind share and the networking budgets of the IBM community. Cisco, Cabletron, Bay Networks, Hypercom, and 3Com are playing an increasingly significant role in "IBM" networking, now particularly so that IBM relies on 3Com (who acquired Chipcom) for some of its 8260 technology; on Cisco (who acquired Kalpana) for its 827x LAN switching technology; and on Bay Networks from whom it OEMs the Centillion token-ring switch, albeit as a stopgap while it brings the 8272 to market. Hence, SVN has no choice but to tackle how its functions will be provided on switched networks made up of peripheral and backbone switches from multiple, essentially competing vendors.

Now to IBM's credit, and possible relief, Bay Networks, on the back of the Centillion OEM deal, has made some token gestures that they will look at the possibility of supporting SVN. With the Centillion switch likely to play a major role in any peripheral switch environments installed by IBM over the next year, such support would definitely add credibility to SVN. There is obviously plenty of precedence for this type of multivendor cooperation. Cisco, Bay, 3Com, and others now actively support APPN and plan to support HPR. There is, however, one crucial difference between APPN/HPR and SVN: The bridge/router vendors supporting APPN/HPR do not have a comparable SNA routing technology of their own. (Cisco abandoned its initiative to develop a competitor to APPN, which was known as APPI, way back in 1993.)

In the case of SVN, all of the major players, as mentioned earlier, have their own frameworks such as Fusion, BaySIS, and so on. Hence it will be intriguing to see how SVN plays out against these other schemes. There is little doubt that in the long-term all these schemes will get around to providing some level of multivendor interoperability, but only after standards are developed for this through the ATM Forum or some other forum.

The lack of a migration path for bridge/router-based networks is in a sense an extension to the multivendor problem. Most router-based networks in existence today tend to be non-IBM-based. Just dismissing routers as being passé may not be adequate. Also, supporting exiting routers via some type of NBSS Access Agent that makes a switched SVN network look like a peer router to an existing router network may not be enough. Many enterprises are likely to adopt ATM by first using it between bridge/routers. Thus, at least for the next few years, bridge/routers will provide the backbone ATM switching functions. Peripheral switches, such as LAN switches, will be connected to these core bridge/routers. Against this background, SVN cannot just advocate a distributed different routing as being the strategic direction for the future. It also has to provide credible and pragmatic schemes for systematically assimilating and migrating existing bridge/router networks into SVN-based networks.

The other key issue that SVN has to address with certitude is that of how SNA traffic will be assimilated into an SVN network. To be fair, SVN already makes some vague references to SNA support centered around APPN on the mainframe, ATM interface to the mainframe, and HPR across the WAN. Unfortunately, this not enough. Despite IBM's relentless pressure on mainframe-centric SNA enterprises to embrace APPN, the reaction to date has been mainly apathetic—most likely due to the absence of generalized DLUR support. Now with nearly all major bridge/routers, with the ironic exception of IBM, offering surrogate DLUR support, the tide may start to reverse. But that is still a very big "if." Plain subarea-oriented SNA still rules supreme when it comes to mainframe-centric networks.

HPR is also a promise as yet—and alas many think of it as being just another flavor of APPN. Though full HPR (i.e., ANR and RTP) support is now available on the latest ACF/VTAM and ACF/NCP, it is still not available on any peripheral devices such as PCs, AS/400s, 3174s, and so on (The current OS/400 just supports ANR.) Moreover, as was seen in Chapter 7, there is now some consternation as to the efficacy of HPR's supposed state-of-the-art anticipatory congestion control mechanism at speeds beyond 45Mbps. With 155Mbps being considered by most as the starting point for WAN ATM, this concern with congestion control will have to be stamped out if HPR is going to be a viable option for supporting SNA traffic over high-speed ATM WANs.

The ATM interface to mainframes is the last thorny issue that IBM has yet to grasp with conviction. Today, the 3172-003 offers a 100Mbps ATM interface to mainframes, based on a standard OS/2 TURBOWAYS 100 adapter. However, as has been discussed before, most ATM cognoscenti refuse to accept 100Mbps as being true ATM. Bowing to this pressure, IBM has already indicated that the 3746-950 Nways Controller will provide a 155Mbps ATM interface sometime in late 1996. This supposed 155Mbps support vis-à-vis mainframes is a somewhat beguiling statement. Today, the fastest speed that the fiber-based ESCON II channels can run at is 136Mbps. Thus, just providing a 155Mbps interface on 3746-950s alone is not going to be enough.

IBM must also increase the throughput of its channels—possibly with ESCON III. There has yet to be any clarification of this very crucial issue. Until such time, there will always be some question marks as to the true viability of 155Mbps mainframe ATM interfaces.

9.7 REFLECTIONS

IBM has a multibillion dollar vested interest in trying to make ATM into a resounding success as quickly as possible. It sees ATM as its savior vis-à-vis networking. With ATM, IBM is hoping to regain its former role of being a leading supplier of commercial sector networking solutions. A compelling ATM story backed up by strong and competitive products should permit it, in theory, to offset the continuous setbacks it suffered over the last few years when it came to bridge/router-oriented multiprotocol LAN/WAN networking.

True to this goal, IBM already has a unique and full-spectrum ATM product repertoire, replete with the NBBS architecture and the SVN framework. NBBS, following the lead of its mentor SNA, sets out to provide value-added, industry-strength networking within the context of ATM. NBSS and SVN are not the only ways that IBM can add unparalleled value to ATM-based networking. IBM alone of the current crop of ATM vendors has the ability to deliver total system solutions from PCs with ATM adapters all the way to mainframes with ATM interfaces. Moreover, it has the ability to deliver the software necessary to make an ATM network tick, such as directory services and usage-based billing programs. To cap it, unlike many of its ATM competitors, IBM, despite its faux pas of late, does have an impressive three-decade-plus legacy of delivering and sustaining complex, mission-critical systems centered around far-flung networks.

Nonetheless, it is not out of the woods yet. Its ATM products have yet to demonstrate their true mettle. High-speed, ATM Forum compliant interfaces will not be available on the flagship 2220 till late 1995. Adoption of ATM by its customer base has also been much slower than what IBM had initially expected. Together, this means that IBM's ATM lineout is not really going to get a true workout till 1996 or later. Hence, the bottom line here is that the omens look promising but all it is at present is indeed promise. Multivendor interoperability and 155Mbps mainframe interfaces are the other gadflies. In time, IBM will, no doubt, overcome these issues. But time is of the essence, if IBM is to reestablish its networking credibility and credentials through its mastery of ATM technology. In 1996, IBM will have to stop thinking about ATM in terms of theory and step up to the line to start showing how ATM can be truly exploited, in practice, to further the cause of commercial networking.

Chapter 10

Total Enterprise Management: An Elusive Goal

A QUICK GUIDE TO CHAPTER 10

Network Management is the glue that holds a network together and prevents it from unraveling out of control at the least provocation. Nobody now doubts or underestimates the importance and value of incisive network management vis-à-vis the successful operation of a production system. Vast strides have been made in the field of network management in the last decade.

The ability of contemporary SNMP-based Network Management Systems to vividly portray the real-time status of a multivendor, multiprotocol network using compelling graphical displays is nothing short of awesome. The nearly 20-year-old mainframe-resident NetView/390, the undisputed doyen of network management, is no laggard either. It now supports graphical displays and offers a raft of value-added services such as automated network asset management (i.e., inventory control by electronically reading serial numbers). It is, however, worth noting that IBM's current strategic and flagship network management platform is the SNMP-oriented and Unix-based NetView for AIX, rather than NetView/390.

Despite these indubitable advances, most network professionals still claim that realizing a sound and all-encompassing network management mechanism is still the most elusive goal in contemporary networking. This elusiveness becomes further exacerbated in networks involving both SNA and non-SNA traffic.

A fundamental cause of this frustration is that it is still very hard to oversee and manage a far-flung and heterogeneous network from a single management platform. For a start, the prowess of SNMP managers does not extend to managing conventional SNA devices such as 3174s. Similarly, NetView/390 is not the optimum platform through which to monitor and control a large network of multiprotocol bridge/routers, LAN switches, and multimedia hubs.

Enterprises with 3,000 or more SNA end users really do not have any other choice but to resort to a cooperative management scheme involving both NetView/390 and a SNMP-manager. Section 10.3 elaborates on such cooperative management schemes and the ways to build bridges between the two disparate systems using techniques such as NetView Service Points. This section also looks at the latest advances in NetView-compatible multiprotocol, multivendor management such as bridge/routers, hubs, and FRADs replete with integrated Service Points and the ability to embellish SNMP-based topology maps with actual SNA LU, PU, and link names.

The other major problem of today's management schemes is that they are much too network-centric. There is no reason why fault, performance, configuration, and change management is just restricted to networking devices. Management, in future, should embrace the complete electronic fabric of an information dissemination system including applications, PCs/workstations, and LAN servers. This is the brave and bracing frontier of the so-called Total System Management initiatives. Sections 10.1 and 10.2 cover the overall requirements and parameters for Total System Management. Section 10.4, using IBM's SystemView blueprint as a framework, expounds on the possibilities and dynamics of future System Management schemes.

In the end, when all has been said and done, and push has come to shove, it is the effectiveness and the incisiveness of the network management scheme in use that will make or break a network. The good news on this front is that these days both the customer and vendor communities, without any equivocation, understand and value the importance of sound and thorough network management. Network management is now rightly seen as the all-powerful gravitational force field that will hold a network together and prevents it from unraveling at the least provocation. To their credit, most enterprises, especially so in the IBM world, currently devote as much effort and energy to selecting the optimum network management scheme and developing the necessary processes to manage the network as they do to selecting and implementing the various devices that will make up the actual network.

Unfortunately, these network-management-conscious enterprises, at least for the time being, face an uphill struggle. Solid, all-seeing, all-powerful, integrated network management, especially for multivendor, multiprotocol LAN/WAN environments with a considerable amount of SNA/APPN traffic, is still a frustratingly elusive goal. This despite the enormous strides that have been made in the last few years by today's popular, highly graphical, typically SNMP-based, management systems such as H-P's OpenView, IBM's NetView for AIX (né NetView/6000), Sun Microsystem's SunNet Manager, Cabletron's Spectrum, and of late Cisco's CiscoWorks.

The areas where these systems have made the most amount of tangible progress during this time are: ease-of-use, procedural consistency, multiprotocol management, and multivendor management. Most now include plug-and-play features such as *auto-discovery*. This facility enables them to automatically and on the fly discover the network topologies being used by key protocols (e.g., IP) when connected to a network and given the address of one of the nodes using the relevant protocol.

The graphical interfaces of all of the Unix-based and SNMP-driven network management schemes are awesome at displaying a topological map showing the high level activity status of a network, overlaid on top of accurately rendered geographic maps if necessary. They also excel at displaying consolidated statistics such as traffic volumes, transient errors (e.g., retransmissions due to line hits), pending error reports, processor utilization, and so on. They invariably offer a gamut of compelling schemes through which to present such data, including speedometer type rate displays, histograms, X and Y coordinate graphs, and color-coded logs.

This operator-friendly, graphical, point-and-click paradigm, as yet does not alas typically extend all the way down to managing all individual devices on the network—particularly so in multiprotocol networks involving SNA/APPN traffic. Trying to individually monitor or control a specific device on the network could lead to the graphical interface suddenly becoming nothing other than a window with a textual command line interface, or a window full of bewildering text messages and numbers. In some cases, even worse, the device that needs to be monitored or managed may not be supported at all by the network management scheme and will appear to be a "black hole." *The primary problem is that SNMP despite its galloping popularity is still not the universal management protocol of all networking devices, especially those with an SNA heritage.*

10.1 THE TWO PRIMARY CHALLENGES OF ENTERPRISE MANAGEMENT

The inability to manage traditional SNA entities using SNMP leads to the first major management-related hurdle now confronting nearly all enterprises that are transitioning from an SNA-only network to a multiprotocol, multivendor network. IBM's NetView/390 running on mainframes is still the most practical and optimum way to manage mid- to large-size SNA environments with, say, over 3,000 simultaneously active end users (i.e., SNA LUs). In addition, most traditional SNA devices, such as 3x74s, 37xx's, 4700 Financial Systems, or even 3172s supporting SNA traffic can only, as yet, be monitored and managed using NetView/390. NetView/390's network management capability is, however, not totally restricted to managing SNA entities.

NetView/390 does offer some SNMP- and CMIS/CMIP-based management capability provided that TCP/IP or OSI software, respectively, is coresident on the mainframe. NetView/390, nonetheless, cannot in any way shape or form compete with Unix/SNMP-based network management systems, à la OpenView or NetView/6000, when it comes to managing inherently SNMP-centric devices such as intelligent hubs, multiprotocol bridge/routers, LAN switches, switching hubs (e.g., IBM 8260), or ATM switches. These SNMP-managed devices, moreover, are now the staple means of implementing contemporary, SNA/APPN-capable multiprotocol LAN/WAN networks.

Even IBM does not deny SNMP's pivotal role vis-à-vis management today and in the future. *For the last few years it has been adamant and unequivocal that SNMP-based NetView/6000, rather than NetView/390, is now the strategic focal point for IBM-centric enterprise management.* In the case of multiprotocol networks with large numbers of SNA users, IBM categorically advocates a dyadic scheme whereby both NetView/390 and NetView/6000 are used in tandem.

In such so-called cooperative management schemes, NetView/390 using SNA/MS protocols will manage the SNA entities (e.g., LUs, SDLC links, SNA nodes). In parallel, NetView/6000 using SNMP and OSI CMIP over TCP/IP (CMOT) will, in theory, manage everything else including T1 TDM multiplexors, PBXs, and even applications running on PCs/workstations. Various conduits such as IBM's AIX SNA Manager/6000 will provide a degree of interoperability between these two very disparate management systems.

Figure 10.1 depicts the overall structure of a multiprotocol, multivendor LAN/WAN network that is being cooperatively managed by NetView/390 and a SNMP-based management platform. This figure also introduces IBM's notion of *segment managers*. A segment man-

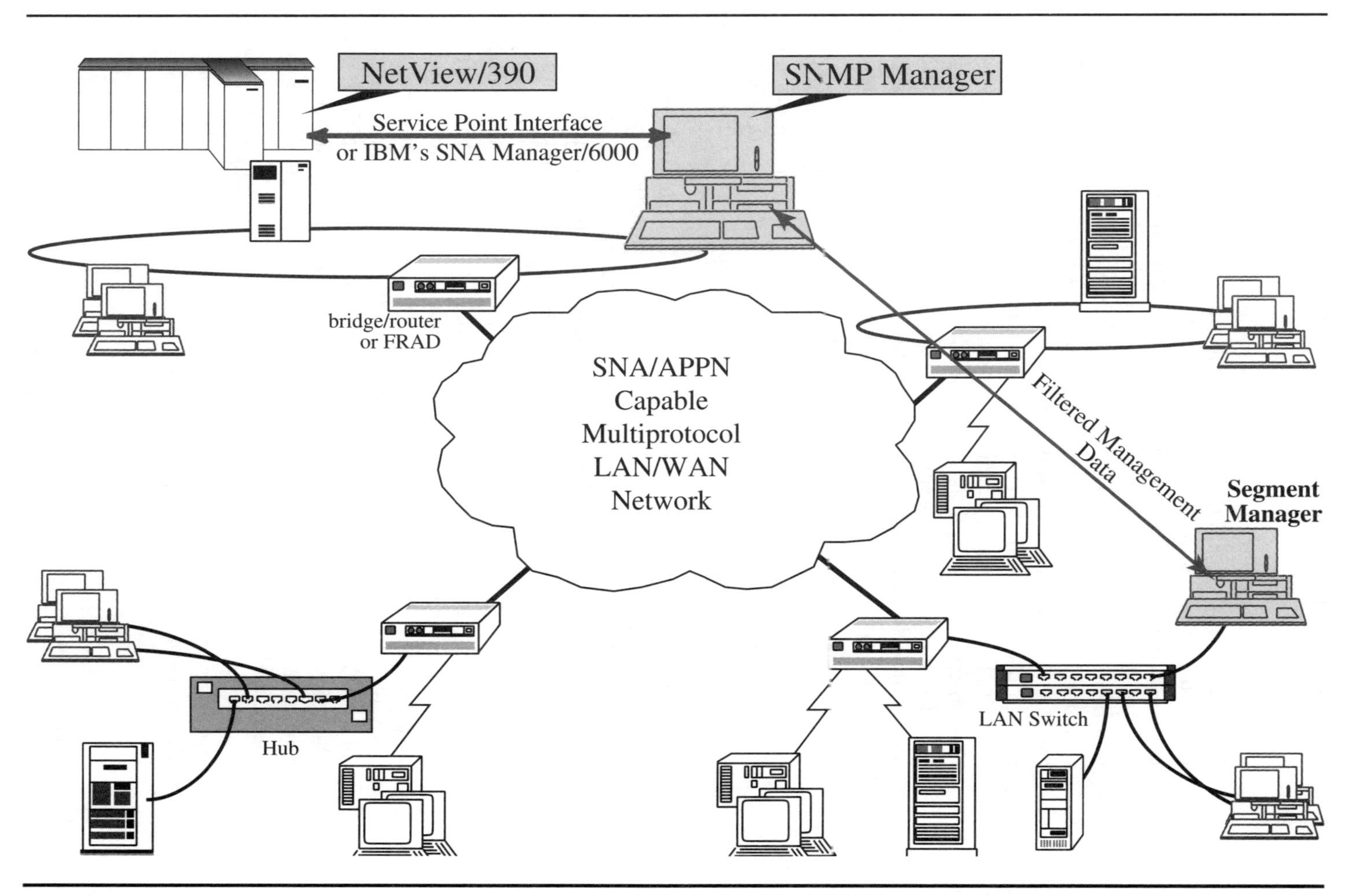

FIGURE 10.1 Cooperative management of a multivendor, multiprotocol network using NetView/390 and an SNMP manager in tandem.

ager attempts to detect and resolve problems as close to the source of the problem as possible. With segment managers it will no longer be necessary for all management information to be sent to a central point—in particular to a NetView/6000 for analysis and action.

A segment manager, for example, on each critical LAN segment, will be able to provide much of the management functions locally. It will use heuristic techniques, built on rules-based and event-driven Correlation Engines. In essence these segment managers will be mini-, though peer-oriented, NetView/6000 systems. They will not require a dedicated workstation, and moreover will be able to run on non-IBM Unix platforms as well as even non-Unix platforms. IBM's Systems Monitor for AIX is an example of a segment manager à la IBM. It offers powerful, and user customizable, event filtering and threshold analysis capabilities, so that only critical alerts and consolidated performance data has to be forwarded to NetView/6000.

The dual-platform-based management scheme, replete with downstream segment monitors that filter the management information conveyed to the central SNMP-platform, as shown in Figure 10.1, is IBM's current and strategic network management paradigm for multiprotocol networking that involves mainframes and SNA/APPN traffic. Given this dual-platform notion advocated by IBM, Figure 10.2 goes on to depict the various management platform permutations that can be used to realize contemporary enterprise management schemes.

NetView/390 and NetView/6000, despite any protestations from IBM to the contrary, are not the only means of realizing dual-platform-based management schemes, such as that shown in Figure 10.1. For example, Sterling's NetMaster could be used on the mainframe in place of NetView/390. In general, however, NetView/390 tends to be the preferred and popular option for the mainframe side—and is by far the uncontested market leader. This is definitely not the case when it comes to NetView/6000. Though, indubitably powerful, competent, and quite popular, NetView/6000 is by no means the automatic choice for SNMP-based platforms, even in heavily IBM-oriented environments. OpenView and SunNet Manager continually and actively jostle with NetView/6000 for the SNMP management role.

Consequently, one of the major IBM-oriented network management requirements of today is that of implementing cohesive and comprehensive cooperative management systems for SNA/APPN-capable multiprotocol LAN/WAN networks using NetView/390 (or an equivalent) and any of the popular SNMP management platforms. The challenges of and the options for implementing such cooperative management schemes are discussed in Section 10.3.

The other major challenge facing enterprises is that of total system, as

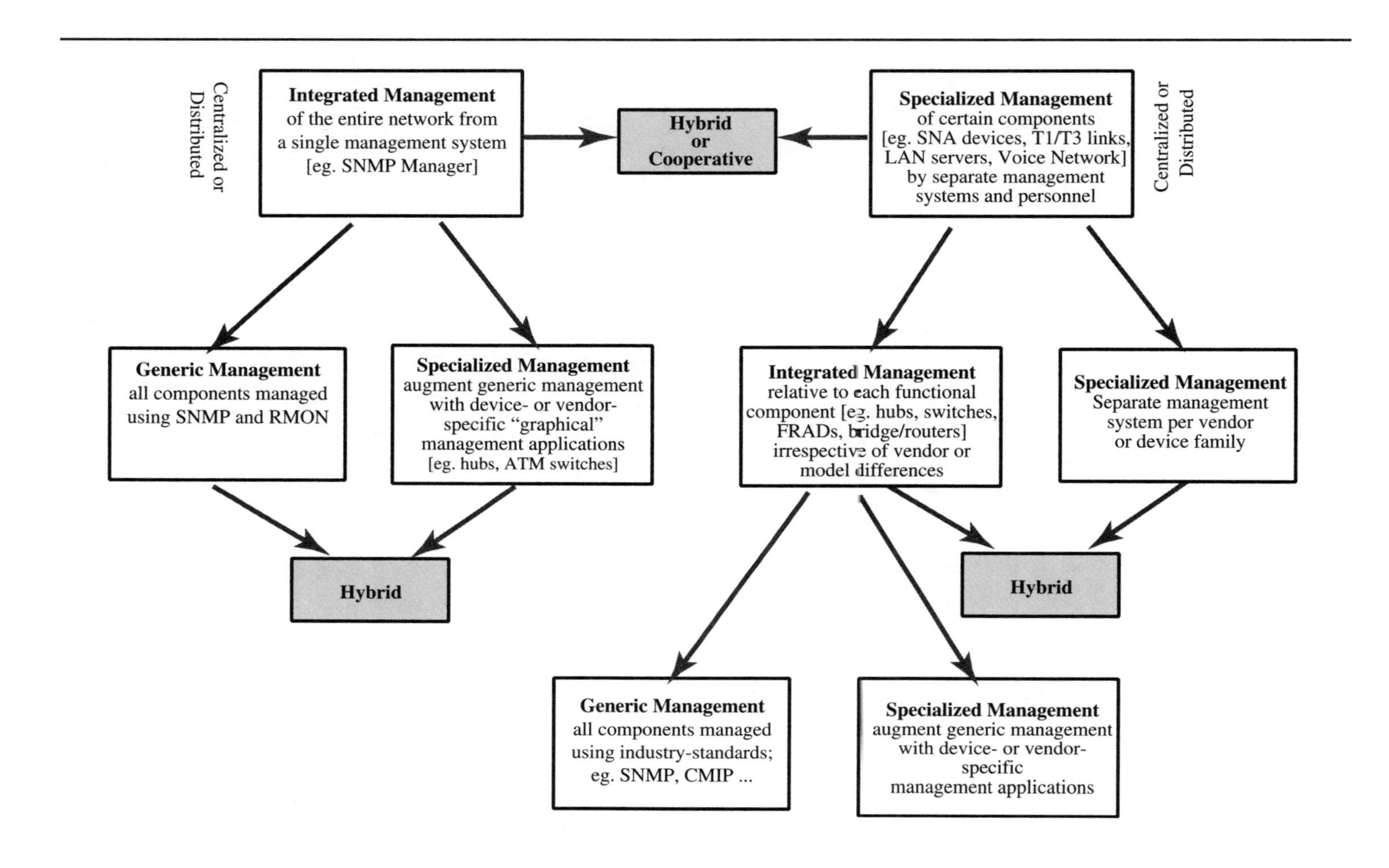

FIGURE 10.2 There are multiple permutations to realize a contemporary enterprise management scheme.

opposed to just network, management. Today the ability to receive nearly instantaneous alerts of problems being encountered by networking devices and the ability to gather copious statistics on networking related activity is mundane, and taken for granted. Unfortunately, such monitoring and management capability is invariably restricted to the devices being used to implement the networking infrastructure. This is particularly true when dealing with non-SNA entities. Devices not actually involved in creating the network fabric tend to be outside the scope of most Network Management schemes irrespective of the importance of those devices when it comes to the successful functioning of the network.

LAN servers are a good case in point. Though not involved in maintaining the network fabric, LAN servers are an integral and vital component of today's LAN/WAN networks. Nonetheless, incisive and proactive LAN server management is still in its infancy—particularly when compared to the scope and depth of conventional network management. The possibilities when it comes to LAN server management are considerable and can include such obvious features as automatic alerts to an appropriate Help Desk either when disk space or processor utilization exceeds preset thresholds, or when a user experiences an error when trying to store or retrieve a file. Such LAN server management schemes are only now beginning to appear on the market. In most cases they are just in the process of being tightly integrated with SNMP-based management platforms. But LAN servers are but the tip of this iceberg.

The ability to closely monitor and comprehensively manage entities, in time, needs to extend to cover everything attached to and served by the network. This is the ultimate and ambitious goal of Total System Management. Total System Management, in reality, even goes much further. It addresses issues such as software distribution (i.e., change management), capacity management, operations management (e.g., shift scheduling) and even business-aspect management (e.g., automating lease and maintenance contract payments). Since September 1990, IBM has been slowly trying to make such all-encompassing Total System Management a reality under the banner of its SystemView initiative. While some progress has been made to date, truly integrated all-seeing Total System Management is still a long ways away. Much of the so-called SystemView offerings, in general, still focus on what are still essentially network-centric management issues.

Interestingly, IBM at present is the only party actively promoting and working on Total System Management. DEC and AT&T who in the early 1990s had also expressed a keen interest in this field have since had a change of heart. IBM obviously has more than a little vested interest in making System Management a reality. Despite its diminishing role as a major network provider, IBM now more than ever is in a unique situation. It is the only vendor that still has the ability to deliver total systems from S/370- or Unix-based mainframes all the way down to PCs, with a LAN/WAN or ATM network in the middle, made up of boxes that all carry the same logo. In addition, it can supply most of the key central applications, such as database (e.g., DB2) and transaction processing systems (e.g., CICS), to boot.

By being the only true total system provider, IBM now certainly has the onus of providing its customer base with the necessary means to manage such systems, irrespective of whether it supplies all or just some of the boxes and applications. Thus IBM definitely has an obligation, not to mention a motivation. For a start, Total System Management, given its scope and its innate appeal, is likely to be highly lucrative—especially if IBM ends up, even initially, having the monopoly on this technology. In addition, a clear leadership role in such a key and high-profile technology, will without doubt, enable IBM to reclaim some of the account control and credibility it has lost of late due its inexplicable faux pas in the multiprotocol internetworking arena.

This said and done, developing a meaningful and useful Total System Management scheme is not trivial and does require considerable fiscal and human resources. Fortunately IBM does have such resources to back up its motivation, to make SystemView-based System Management a reality. The SystemView initiative is discussed in detail in Section 10.4. If IBM, in time, can make real even 70% of the initial goals of SystemView, what is today loosely referred to as *enterprise management* could be considered to have finally come of age and worthy of its name.

10.2 THE KEY REQUIREMENTS OF ENTERPRISE MANAGEMENT

Before delving into the possible options for managing multivendor, multiprotocol networks with SNA/APPN traffic, and what Total System Management promises to deliver, it is useful to take a pragmatic snapshot of today's pressing enterprise management requirements. These management-related requirements can be categorized as follows:

- *Multiprotocol Management:* Provide, at a minimum, fault and configuration management for multivendor, multiprotocol networks carrying both SNA/APPN and non-SNA networks, and made up of non-SNA (e.g., bridge/routers) and SNA (e.g., 3745) devices. Such networks are most likely to be managed through a dual-platform scheme involving NetView/390 and an SNMP manager.

  Though not imperative for their functioning, it is desirable in most instances to have some degree of interoperability between these two disparate management platforms to enable each to have some access to information being processed by the other. This, if nothing else, ensures that a network operator using a console attached to one platform is not totally isolated from what is happening to the other side of the network. IBM facilitates bidirectional interactions between NetView/390 and non-SNA systems, via a mechanism known as a NetView Service Point. Figure 10.3 illustrates the use of NetView *service points* to interface non-SNA devices to NetView/390.

A service point implementation that will run on most Unix systems is available from IBM in the form of the AIX NetView service point program. This Service Point can be gainfully used to interface most all SNMP-based management platforms with NetView/390. This, though the simplest, is not the only way by which management information can be exchanged with NetView/390. Other schemes such as integrated service points within bridge/routers and FRADs, IBM's AIX SNA Manager/6000, Cisco's CiscoWorks Blue, and Cabeletron's Spectrum/BlueVision are discussed in Section 10.3.

- *LAN Workgroup Management:* Ability to manage LAN servers as well as LAN-attached PCs and workstations. A key feature of this support will be that of alert generation in the event of component failures within a workstation or server. Hard disks, memory, monitors, keyboards, serial ports, and floppy drives are some of the components that could be covered by this feature. Remote system configuration and remote operation, a facility particularly applicable for servers, will also be possible. Capacity and performance management will also be features. Thus it would be possible to remotely monitor CPU utilization of a server or a client, the disk utilization of a server, or the I/O statistics of a server.

  IBM's NetFinity software product, that can be interfaced to most SNMP-management systems, is an example of this class of LAN Workgroup Management product. (NetFinity supports the fast emerging Desktop Management Interface standard specified by the Desktop Top Management Task Force.) Novell and Microsoft among others have comparable offerings.

  Till recently, management of MAU, as opposed to intelligent hub-centered, token-ring LANs and IBM's SRBs were only possible using IBM's OS/2-based LAN Network Manager (LNM) program. The need for LNM is rapidly disappearing. Enterprises are replacing their MAUs with intelligent hubs or token-ring switches, and the aging SRBs with bridge/routers or token-ring switches. Intelligent Hubs, bridge/routers, and token-ring switches are all geared to be managed via SNMP rather than LNM.

- *Change Management:* At a minimum, a facility whereby software or data files can be systematically distributed to (and installed on) multiple remote systems from a central site on a scheduled and controlled basis. Such a change management scheme can be profitably used to download data into LAN servers from a corporate data center on a nightly basis, or to remotely install new system software on a particular product (e.g., 3174 or AS/400) over a weekend. IBM's NetView Distribution Manager, which is now available in both mainframe and Unix versions, is an example of such a software and data distribution product.

- *Performance Management:* Satisfactory network performance is a prerequisite in order to have happy network users. Thus being able to monitor network performance on an ongoing basis and being able to easily compare it against expectation and predictions is an important part of contemporary network administration and operation. Performance monitoring should be possible at an individual device, user, or link level. In addition, it should be possible to generate an Alert if performance of a particular entity drops below a prespecified threshold. If and when there is a major deviation in performance, facilities should be available to enable a network manager to quickly pinpoint the cause of the problem. Most SNMP managers are now beginning to offer performance management applications.

  Many 3270 control unit class products, such as 3174s, SNA LAN gateways, and SNA emulator support SNA's RTM feature. RTM permits accurate end-to-end (i.e., keyboard to mainframe application to screen) response times, experienced by individual users (i.e., LUs) to be measured and logged over a long period of time. RTM statistics can be extracted from various devices (e.g., 3174) and tabulated using the mainframe resident NetView Performance Monitor (NPM).

- *Asset (or Inventory) Management:* The ability to dynamically and automatically maintain an electronic inventory of all the devices attached to a network. Such an inventory would include, at a minimum: device type, model number, serial number, latest hardware update level, and some amount of user-defined device identification information. Some SNA devices, most notably the newer 3174s, have the ability to automatically supply NetView/390 with this level of information when they are powered up or when explicitly interrogated by NetView/390. This SNA defined capability, is referred to as VPID—Vital Product Identification.

Ideally, Asset Management should not be restricted to hardware devices. It should also cover software running on various devices so that enterprises can maintain an accurate and up-to-date inventory of the software in use. SNMP managers and non-SNA devices are now beginning to offer nascent electronic asset management schemes. NetFinity, for example, can provide a software inventory, albeit on a LAN workgroup basis. The notion of software inventories can be further extended to include software license management—particularly so for software running on PCs and workstations. License management will ensure that all software is at the right level, legitimate, and, in the case of a single copy of a program being shared off a server, make sure that any maximum user limit has not been exceeded. Such license management tools are also now beginning to appear on the market.

- *Trouble Ticketing Mechanism:* Automated system for tracking, logging, and systematically escalating unresolved network/system problems. Ideally, such a trouble ticketing facility should be an integral component of the management platform. Most SNMP management platforms now offer some form of built-in, automated trouble ticketing scheme.

- *Automated Operation:* With networks getting more complex and larger, both in terms of the number of users being served as well as their geographic spread, there is a growing demand for tools that will permit many routine management functions to be performed automatically. Example: Attempt to automatically reactivate a link or node on receipt of notification of a transient link or node failure prior to seeking human intervention. Management automation is invariably realized using a scripting language that essentially permits an event-driven program to be created. NetView/390, true to its mainframe orientation and 15-year-plus product life, offers extensive task automation facilities. NetView/390 task automation scripts can be easily and quickly developed using CLISTs (i.e., control or command lists) or IBM's REXX scripting language.

- *Capacity Management:* This facet of management deals with monitoring and recording resource utilization and using this data to empirically predict how the system and network resources need to be upgraded (or possibly even down graded) to satisfy usage trends. In the past, the notion of capacity planning was mostly restricted to determining and controlling the growth of mainframe resources such as processor cycles (i.e., MIPS), memory, disk space (i.e.,

DASD), tape storage, and so on. Today, such capacity planning disciplines are getting directed at networks and LAN servers so that network administrators and operators can forecast and budget for new services (e.g., bandwidth) and equipment.

Capacity management involves two distinct processes. The first is the automated and ongoing collection of pertinent resource usage statistics. The second is the analysis of this accumulated data to extrapolate usage trends and determine how resource capacities need to be altered to keep up with the usage trends. IBM's Teleprocessing Network Simulator and NPM, which have both been around for at least 15 years, are examples of mainframe-centric, network capacity management tools. Today, IBM offers other options such as NetDA/2. Capacity planning tools for bridge/router-based multiprotocol networks are available from a variety of vendors, including some of the main bridge/router vendors.

- *Usage Accounting:* Being able to bill departments or individual users for their actual network usage on a monthly or quarterly basis, even if the bills were just for internal accounting purposes, is a cherished and widely used feature of SNA networking—especially in large corporations. This charge-back capability that permits administrators and managers to determine how network bandwidth has been utilized is the SNA equivalent of producing itemized telephone usage bills. Enterprises obtain the raw data—the number of bytes sent and received by a given SNA LU—necessary to produce these bills from the capacity planning statistics automatically collected by the mainframe-resident NPM.

  (N.B.: NPM, at least at present, can only collect SNA usage statistics for SNA/APPN traffic that goes through a 37xx. If a 37xx is replaced with a 3172 or equivalent, NPM will not be able to gather SNA/APPN usage statistics, at least for the time being. There is a possibility that IBM, in the future, will provide a Network Performance Analyzer [NPA] feature for 3172s to ensure that NPM can work with 3172s. If and when such a feature is available, vendors that make 3172 equivalents [e.g., Cisco, Bus-Tech, etc.] may OEM this code from IBM or try to develop their own version. Consequently, for the time being, the loss of NPM data is one of the few compelling reasons for persevering with a 37xx despite the temptation of replacing it with a much lower-cost 3172-type gateway.)

  Ironically, bridge/routers in general shy away from the notion of providing accurate network usage statistics per user (e.g., MAC address) on a protocol basis. There are two primary reasons for this. The first has to do with performance and memory. Keeping track and tabulating the source, destination, length, protocol, and time of day of packets will, indubitably, negatively impact the performance and the sacrosanct packets-per-second forwarding rates of bridge/routers. Storing this data, until it is extracted on a nightly basis across the network by a SNMP-based management platform, will also require considerable amounts of memory—assuming that bridge/routers would not want to start maintaining hard disk drives for the sole purpose of keeping such data.

In addition to these two technical reasons, there is also a philosophical issue involved. Keeping track of network usage, especially in today's world of unrestrained e-mail exchanges and Internet surfing, is seen by many as a tool that will be used by Big Brother to keep an eye on how and for what a network is being used. The bottom line here is that, at present, obtaining network usage statistics per user, à la NPM, in the context of multiprotocol networking is difficult, if not impossible. In time there will be a backlash against this as enterprise administrators (e.g., CFOs) insist on knowing exactly how the millions of dollars a year being spent on networking are being used—and by whom. *Eventually, bridge/routers, LAN switches, and ATM switches will have to provide network accounting data, even if it means that this will slow down their performance by as much as 30%.*

- *Application Management:* This is a nascent but increasingly significant facet of Total System Management. It has always been a key pillar of the SystemView vision. Over the next few years, application management will be one of the fastest expanding growth areas in the management arena. The possibilities for, and the advantages of, being able to automatically monitor error conditions on user applications, in much the same way as that for networking hardware, are immense, patently obvious, and long overdue. Application management will typically be based on SNMP-based platforms as opposed to NetView/390.

  With application management, as envisaged by IBM and others, it would now be possible to have an alert dispatched to a designated Help Desk, if a Lotus 1-2-3 user encounters a catastrophic I/O error while trying to open a spreadsheet, or is unable to save a spreadsheet on a file server because the server has run out of disk space. Similarly, alerts could be generated if a query application was unable to service a query; if a distributed database application had to perform a roll-back operation because an update could not be applied; or if a transaction processing application encountered an catastrophic error while processing a request.

This type of application management compellingly complements LAN Workgroup management. Already some database software vendors such as Legent have included hooks in their software to permit Alerts to be generated and routed to a designated SNMP-management system. Application management will also build upon today's centralized software distribution (i.e., Change Management) tools to permit applications to be remotely installed, configured, and operated.

- *Total System Management (Part I):* In addition to application management, which is without doubt a key requirement, the first phase of Total System Management will have to deal with seamlessly integrating the management of network resources with that of managing other system resources. The primary goal is to provide a mechanism whereby both network and system resources can be incisively monitored and managed using a consistent and intuitive user interface and the same set of management disciplines.

  Ideally, at the first few levels of overall system availability monitoring and management, there should be no difference between ascertaining the operational status of a modem or a mainframe-attached disk drive. In essence, today's network topology maps that indicate the activity status of various networks components using color-coded icons need to be extended to embrace system components such as LAN servers, mainframe channels, databases, and even PBXs. Obviously, all of this information does not need to be available or displayed on the same management console.

  Total System Management, even more so than today's network management schemes, will advocate and rely on the deployment of multiple, distributed, typically SNMP-based management platforms. It will also build upon the notion of segment managers discussed above to enable hierarchical management schemes to be implemented with different groups of management consoles being responsible for different parts of the overall systems. The underlying goal, however, is to empower the management infrastructure so that it embraces all the resources in the system, including applications and databases, rather than just being restricted to the networking components.

  The nonnetworking entities that will be included in the first phase of Total System Management will include:

  - Applications running on PCs, workstations, Unix systems, minicomputers, and mainframes
  - LAN servers
  - Individual PCs and workstations

- Unix systems
- Databases, whether resident on mainframes, Unix systems, minicomputers, or LAN servers
- Mainframe resources, in particular channels, disk drives, tape units, processor cycles, and memory utilization
- PBXs and the in-house telephone network

- *Total System Management (Part II)*: The second phase of Total System Management will move beyond just managing physical entities and applications to also embrace operational, business, and fiscal issues that have a direct bearing on the successful functioning of the total system. Some of the functions that will be addressed by this phase include the following:
  - Automated system operations tasks such as scheduling personnel for the three-shift operational coverage required by nonstop, 24 x 7 systems
  - Automatic tracking and possibly even electronic payment of recurring monthly costs such as equipment lease, leased line, and frame relay network access payments
  - Automated tracking and reminders as to when maintenance contracts or licenses have to be renewed

  The so-called business aspect of Total System Management will be further addressed in Section 10.4 within the framework of SystemView.

10.3 SNA/APPN COGNIZANT MULTIPROTOCOL, MULTIVENDOR MANAGEMENT

The inexorable move from SNA-only networks to SNA/APPN-capable multiprotocol LAN/WAN networks is having a dramatic affect on how enterprises that rely on SNA mission-critical applications have to restructure their network management schemes. Managing SNA-only networks is straightforward and proven. Mainframe-resident NetView/390, which is now nearing its 20th birthday, provides comprehensive fault, performance (i.e., RTM and NPM), change (e.g., NetView/DM), asset (e.g., SNA VPID collection), and capacity (i.e., NPM) management facilities—albeit in some instances in conjunction with other programs such as NetView/DM or NPM.

NetView/390s manage SNA entities using SNA/MS-defined commands and requests such as Network Management Vector Transport (NMVT) and Request Maintenance Statistic (REQMS). These interactions are typically exchanged across the SSCP-PU session that is established between ACF/VTAM and each and every SNA node whenever a node is activated. NetView/390, which is defined and installed as an authorized ACF/VTAM network management application, gains access to the data being sent across SSCP-PU sessions through ACF/VTAM. (Refer to Appendix A for details on SSCPs, PUs and SSCP-PU sessions.)

The problem now being faced by enterprises is that NetView/390 cannot by itself manage non-SNA devices such as bridge/routers, intelligent hubs, LAN switches, ATM switches, or LAN servers. All of these non-SNA devices, which are now the bread-and-butter of contemporary networking, typically use SNMP as their management protocol. A few such as IBM's 2220 ATM switches currently use OSI CMIP/CMIS. Consequently, enterprises now have no choice but to standardize on SNMP-based management platforms, such as OpenView, SunNet Manager, and NetView/6000, as their primary means of managing the overall networking fabric. Some of these SNMP centric platforms, such as NetView/6000, provide a degree of support for CMIP even if it is in the form of CMOT. This enables them reach out to at least some of the CMIP/CMIS-oriented products.

RISC/Unix workstations are the preferred and popular habitats of these SNMP managers. However, some such as OpenView and NetView can also be run on Intel-powered, DOS/Windows PCs. IBM has postulated the possibility and the potential need to run such managers on massively parallel Unix supercomputers, such as IBM's SP/2, when they are tasked with managing very large and complex networks.

10.3.1 Trying to Manage SNA/APPN Devices from SNMP Platforms

The one downside of standardizing on an SNMP-based managed scheme is that SNMP per se cannot be used to monitor, activate/deactivate, and manage traditional SNA devices such as 3x74s, 37xx's, or even PCs that only contain an SNA stack. The reason for this is very simple. Most traditional SNA devices do not include built-in support for TCP/IP and hence SNMP. Given the lack of SNMP support, a SNA device also does not contain a Management Information Base (MIB) software and control block that is a prerequisite for SNMP-based management.

There has been much activity over the last few years to define an SNMP MIB that would be implemented within SNA devices to facilitate SNMP-based management. The problem here is that most SNA equipment vendors, including IBM, have not as yet included this MIB in all of their new SNA software. In addition to including this MIB, they would also have to in some form bundle in TCP/IP software along with the SNA software in order to permit SNMP-based access to the MIB. Even if these hurdles could be overcome, this MIB would still only be found within SNA devices running the latest and greatest software. This still would not fix the problem for the huge installed base of SNA devices that are no longer receiving any new software upgrades. Such out-of-support devices include thousands of stable, active, and productive plug-compatible 3274s whose original suppliers are no longer in business.

The bottom line here is that in much the same way that NetView/390 cannot manage non-SNA devices, SNMP managers cannot manage SNA devices. *It is, however, important to note at this juncture that SNMP managers can provide a degree of Layer 1 and 2 management for some SNA devices.* This is particularly true if the SNA devices are either LAN-attached or are being supported through bridge/routers or FRADs using SDLC-to-LLC:2 conversion. In these scenarios the SNMP manager will still not be able to directly talk to or monitor the SNA device per se. However, it will have visibility and control of the physical interface to which the SNA device is attached, for example, a port on a hub in the case of LAN-attached devices, or a serial port on bridge/router or FRAD in the case of link-attached devices. At a pinch, an SNMP manager can provide a first-cut device active or inactive status for SNA devices based on the activity status of the physical port.

SNMP managers, with active help from bridge/routers or FRADs, could also provide a degree of MAC-address-oriented monitoring and management for LAN-attached SNA devices as well as those supported via SDLC-to-LLC:2 conversion. Thus it is possible today to display up-to-the-minute accurate topology display maps of DLSw(+) and RFC 1490 networks showing the activity status of the various SNA devices participating in these networks in terms of their Layer 1 and 2 status. These DLSw or RFC 1490 maps could be overlaid at the SNMP console on topology maps for other protocols such as IP and IPX.

IBM, Cisco, and Cabletron, for a start, can now further embellish these DLSw(+) and RFC 1490 type maps being displayed on SNMP consoles. (In the case of IBM the maps could be just of the SNA network as supported by the mainframe using SNA lines. NetView/6000 gets these SNA-only maps from NetView/390 using AIX SNA Manager/6000.) The Cisco and Cabletron capability works independently of NetView/390. The feature they offer can best be thought of as the correlation of SNA device MAC addresses with the actual SNA link, PU, and LU names as defined to ACF/VTAM and ACF/NCP.

With this feature, they use a mainframe-resident ACF/VTAM application to gain access to the ACF/VTAM and ACF/NCP definition files that are stored in "SYS1.VTAMLST" (or equivalent) library. They then use an LU 6.2 session between this ACF/VTAM application and the SNMP-manager to download this SNA definition data from the mainframe to an application running alongside the SNMP manager. This application reads this definition data and correlates SNA names with the MAC addresses defined for them. It would typically do this by looking through the so-called Switched Node definitions that occur for all LAN-attached SNA devices. When displaying SNA-related maps, these true SNA names will now be displayed alongside the SNA devices. *Though an elegant and very useful feature, this name correlation facility does not ensure that an SNMP manager can see beyond Layer 2 of an SNA device.*

There is one scenario where an SNMP manager, by itself, can provide fairly comprehensive management for an SNA/APPN network. This is when all the SNA/APPN traffic is being routed across the WAN using APPN NN support on bridge/routers. In such a configuration, the NNs in the bridge/routers will have session-layer (i.e., Layer 5) awareness of what is happening within the APPN network in terms of end-to-end, databearing LU-LU sessions. In addition, each NN will have true visibility of and access into all of the End Nodes attached to it through the CP-CP sessions that are established between all APPN nodes. (SNA nodes will have to be supported through the surrogate DLUR function found in bridge/routers.)

The only problem here is that most enterprises with mainframe-centric SNA networks are unlikely to opt for APPN NN routing on bridge/routers as their means of integrating the SNA traffic with non-SNA traffic. As many as 80% of these enterprises are expected to select DLSw(+) or RFC 1490 in preference to APPN NN routing. Hence, this APPN NN-based scenario for standardizing on an SNMP-only management scheme is not going to be universally applicable.

The crux of the issue here is that enterprises have to make a hard choice. If their SNA network is relatively small, and possibly also in decline, they could standardize on SNMP and forget about NetView/390. They would then have to rely on ACF/VTAM and the Layer 1–2 visibility available to the SNMP manager as their sole means of monitoring and managing their SNA entities. At this juncture it is worth noting that even when NetView/390 is present, it is ACF/VTAM that is still the all-powerful control point of an SNA network. NetView/390, which is an ACF/VTAM application, eases and facilitates network management. However, NetView/390 still relies on ACF/VTAM for all of its interactions with the actual SNA devices. Thus, in the absence of NetView/390, ACF/VTAM operators can still activate, deactivate, and monitor the status of the SNA network. It just will not be as slick, automated, and graphical as it can be with NetView/390.

SNMP-only based management, however, is not going to be adequate or practical if a multiprotocol LAN/WAN network is still supporting 3,000 or more simultaneously active SNA end users. In such networks, where SNA is still likely to be a major component of the overall traffic, enterprises really have little choice but to implement a dual-platform-based management scheme involving both NetView/390 and SNMP-management, as shown in Figure 10.1.

10.3.2 Building Bridges Between NetView/390 and SNMP Platforms

When NetView/390 and an SNMP manager are being used in tandem to manage an SNA/APPN traffic-bearing multiprotocol network, it is often useful to have some degree of interoperability between these two management schemes. In most instances such interoperability is sought because the NetView/390 and the SNMP management consoles are not co-located and operated by the same group of people. Given their historic mainframe-centricity, the NetView/390 consoles are likely to be in the main machine room and be operated by system operators who are responsible for managing the mainframe and ACF/VTAM. Given that such system operators already have their hands full and do not typically have any background in non-IBM management, it is also possible that an SNMP console would not be installed in the machine room.

Instead, the SNMP consoles are likely to be in network control rooms and at Help Desks. The network control rooms might in many cases also house the central-site bridge/routers, hubs, modems, and multiplexors. These SNMP consoles will be operated by network operators who are trained and experienced in multiprotocol LAN/WAN networking. Though they are likely to have access to a NetView/390 they, given a choice, prefer to devote their attention to the SNMP consoles. In such split-function, dichotomic management setups, it is useful for NetView/390 operators to have some access to the management data available to the SNMP operators and vice versa.

This type of cooperative management between the two platforms would enable a NetView/390 operator who detects a problem on the SNA side of the network to immediately establish the status of the multiprotocol network in order to determine if the SNA problem is due to a malfunctioning of the multiprotoctol network. With cooperative management, the NetView/390 operator would be able to do this directly from the NetView/390 console and without having to bother the network operators sitting at the SNMP consoles. In some instances, using the Resource Object Data Manger (RODM) and the Graphic Monitor Capability, the NetView/390 operator may even be able to display the status of the non-SNA network in graphical form—albeit using information supplied to it, via a Service Point by its SNMP partner.

Cooperative management will permit a SNMP operator to display SNA-related activity status topology maps (e.g., SNA-only or DLSw(+)) alongside topology maps for other protocols. Thus, the SNMP operators will have an overall picture of the entire network, including all the devices using SNA. The SNA name correlation feature discussed above makes such maps even more SNA-oriented and operator-friendly. If IBM's AIX SNA Manager/6000 is being used, an SNMP operator would also be able use the SNMP console as a surrogate NetView/390 console. Thus, with SNA Manager/6000 the same console can double up as an SNMP and NetView/390 console. Any commands that can be issued from a NetView/390 console or any management data that can be displayed on a NetView/390 console could now also be issued or displayed at the SNMP console.

SNA Manager/6000 is not the only way to establish a bidirectional bridge between NetView/390 and an SNMP manager. SNMP managers as well as non-SNA devices, as shown in Figure 10.3, can interface with NetView/390 via a Service Point. A NetView Service Point is a formal mechanism defined by SNA/MS, in the mid-1980s, to enable NetView/390 to monitor and manage non-SNA entities. The Service Point capability was originally offered by IBM as a PC-based program known as NetView/PC. NetView/PC included an application interface (i.e., API) as well as full SNA node with its own PU. NetView/PC interacted with NetView/390 using a SSCP-PU session established between ACF/VTAM and this PU. Typically, a DOS program had to be written to the NetView/PC API to interface the non-SNA device to NetView/390. The non-SNA device would be attached to the PC executing NetView/PC via a LAN or a serial port. The NetView/PC interface application would be responsible for performing all of the translations required to ensure that each side actually understands what the other side is saying and doing.

With NetView/PC, the interface program would have to intercept events forwarded to it by the non-SNA device and then convert it to NMVT form. In the opposite direction, NetView/390 operators would issue explicit commands to the non-SNA device by typing the appropriate command as a text string appended to the NetView/390 RUNCMD. When the RUNCMD is executed, NetView/390 would deliver the text string to NetView/PC over their SSCP-PU session. The NetView/PC program would then hand over this command to the interface program. The interface program would now be responsible for ensuring that the non-SNA device executes that command. It would typically do this by using some form of programmatic operator interface—an interface, or port, that enables programs to issue operator commands.

This Service Point capability even in its NetView/PC form was a powerful and effective way to provide bidirectional management inter-

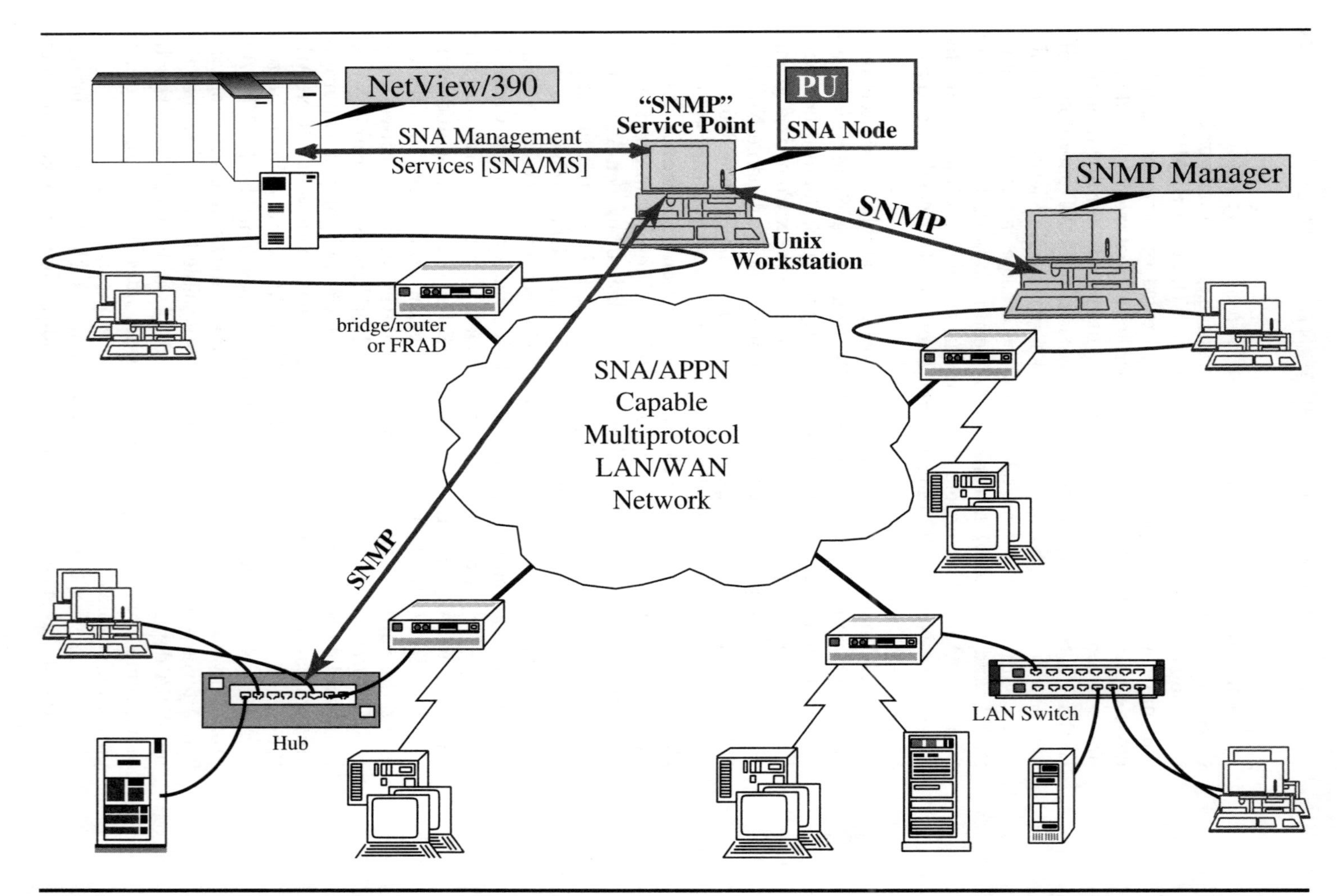

FIGURE 10.3 Using an SNMP MM-Specific net view service point to interface an SNMP manager or SNMP device to NetView/390.

actions between NetView/390 and non-SNA devices. This Service Point technology has moved on considerably since NetView/PC's debut around 1987. NetView/PC has now been complemented by IBM's Unix-based AIX NetView Service Point that executes in nondedicated mode on a workstation running IBM's AIX or Sun Microsystems SunOS. AIX NetView Service Point has a built-in SNMP-gateway to minimize the interfacing that has to be done when trying to attach an SNMP-based system to NetView/390. In general, using AIX NetView Service Point running on an Unix workstation is a straightforward, proven, and relatively cost-effective means of interfacing SNMP-managers with NetView/390.

However, vendors such as Cisco, Cabletron, and Sync Research, among others, have gone further. Instead of relying on an external Service Point such as NetView/PC or the AIX version, they build in the appropriate Service Point software into their products. In general, implementing an integrated Service Point within a contemporary networking device is not too difficult. The main requirement is that of implementing a SNA Type 2 or 2.1 node within the node. This SNA node is required to realize the SNA-based SSCP-PU, or now LU 6.2, interface between the device and NetView/390. Proven, efficient, and compact code to implement SNA nodes is now readily available, thanks to the work done over a decade ago to port SNA nodes to PCs.

Ironically, implementing an integrated Service Point within a non-SNA device, such as a bridge/router, actually makes that device into an SNA device—since it will now contain bona fide SNA node replete with an SNA PU. Hence, any device with an integrated Service Point looks to NetView/390 as a native SNA device. However, all that NetView/390 sees is the SNA node subcomponent. It could use native NetView/390 commands (e.g., VARY NET, ACTIVATE, etc.) to activate or deactivate that node. However, that will not have any bearing on the actual functioning of the non-SNA device, which will still only accept SNMP-based management commands. Thus, despite the built-in SNA node, NetView/390 will still only be able to interact with the non-SNA device via the Service Point.

Once the SNA node is in place, developing the interface between it and the native management software of the device (e.g., bridge/router) is all that is really required. In many cases this interface software can be developed using the NetView PC template. In essence, this interface software will have to convert selected SNMP events to NMVTs and accept commands sent to it, via RUNCMD, by NetView/390.

In the case of bridge/routers, hubs, and FRADs with integrated NetView Service Points, NetView/390 will be able to interact with these devices without having to use an external Service Point, or having to go through its partner SNMP manager. SNA name correlation and Integrated Service Points are examples of the lengths which vendors such as Cisco and Cabletron are now willing to go to facilitate cooperative NetView/390 cum SNMP management of multiprotocol LAN/WAN networks. Figure 10.4 shows a multiprotocol LAN/WAN network built around bridge/routers or FRADs with integrated Service Points.

10.4 THE SYSTEMVIEW TOTAL SYSTEM MANAGEMENT INITIATIVE

SystemView was a strategy put forward by IBM in September 1990. Given that the now defunct SAA was IBM's overarching framework for the so-called next-generation computing model, SystemView was presented as an SAA strategy. In those days when SAA appeared to be as important to IBM as ATM appears to be today, calling it an SAA strategy was meant to give it credibility, respectability, and a guaranteed future. Ironically, as discussed in earlier chapters, SAA has ignominiously drifted into oblivion. Nonetheless SystemView lives. Moreover, it was essentially relaunched by IBM in mid-1995 as the SystemView Series. (This version of SystemView was known for awhile by its internal code name—SystemView Karat.)

SystemView was and still is an ambitious and even slightly audacious total system management framework for the complete life-cycle of a system. It deals with systems involving heterogeneous computing platforms, various equipment from multiple vendors, and multiprotocol networks. It attempts to boldly embrace every aspect and phase associated with realizing and sustaining such complex systems. The SystemView framework covers planning, capacity management, and the myriad issues of coordinating changes. It goes on further to deal with practical, day-to day operational issues such as budget tracking as well as problem escalation tracking.

The total breadth and depth of integrated, consistent, and all-seeing management as envisaged by SystemView has yet to become a total reality. NetView/6000 today, however, is providing a firm but flexible foundation upon which IBM can start building the complete repertoire of tools and applications that will be necessary to make SystemView tangible. At present, in addition to NetView/6000, NetFinity, NetView/390, NetView/DM, Nways Manager, ATM Campus Manager, Trouble Tickets for AIX, NPM, Performance Toolbox for AIX, and others, IBM markets about another 150-odd management-related products for mainframes, AIX, and PCs. With the combined functionality of all of these products, IBM probably now addresses about 60% of the original spectrum of functions and facilities postulated by SystemView.

While 60% may sound extremely promising and encouraging, the

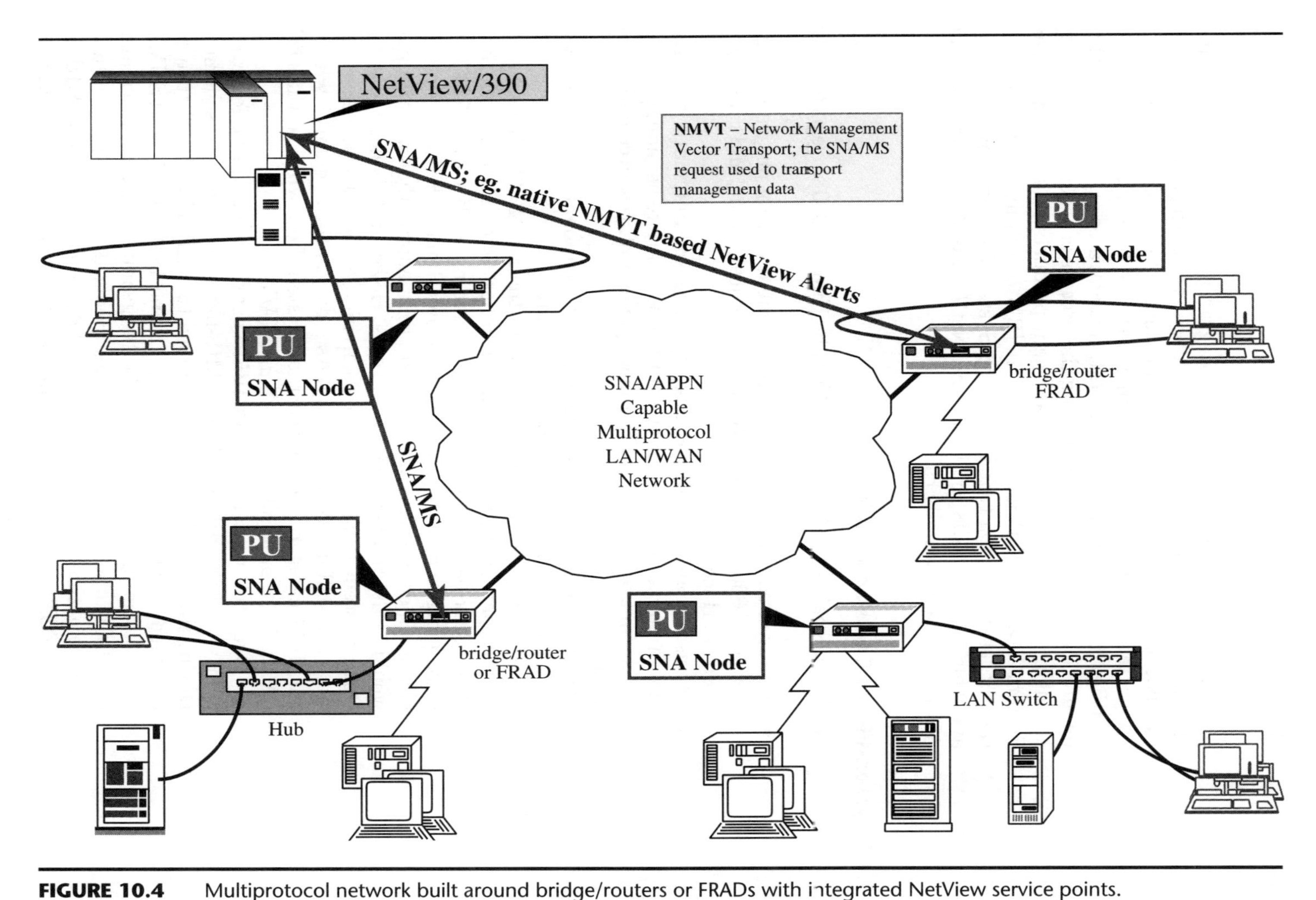

FIGURE 10.4 Multiprotocol network built around bridge/routers or FRADs with integrated NetView service points.

problem here is that there is little integration, consistency, and cohesion across is multitude of products. A primary precept of SystemView was not only to provide total system coverage but to do so in a tightly integrated manner using a common look and feel. The latest SystemView Karat initiative, to its credit, concentrates on addressing this integration aspect and streamlining the process of ordering, installing, and using the myriad IBM management products that fall under the SystemView and NetView banners. If IBM continues at this current rate, at least 75% of a true integrated SystemView offering should be available around 1998, built around NetView/6000, or NetView for AIX as it is now formally called.

10.4.1 SystemView's Goals and Structure

The initial goals stated for SystemView by IBM in 1990 were as follows:

- Improved usability of system management control programs (e.g., NetView/.6000) through intuitive, graphical, iconic, and consistent user-interfaces
- Consistent and, where possible, dynamic and automatic resource definition mechanisms for the entities that are to be managed
- Consistent access to, and the management of, resource definition data
- SystemView architecture and protocols compliant with Open System industry standards and architectures

  (It should be remembered that in the late 1980s there was a very strong sentiment that mainframes were history and that RISC/Unix-based Open Systems including IBM's own RS/6000 were going to rule the roost. Within this climate IBM paid lip service to Open Systems while all the time hoping that mainframes would be around for another 20 years. However, this need to show Open System loyalty lead to an ironic decision whose repercussions are still in evidence today. As a part of the SystemView roll-out, IBM categorically stated that SystemView will use the OSI CMIP/CMIS protocols rather than SNA/MS.

  Over the next few years IBM spent considerable effort and money to start building CMIP/CMIS-based management schemes. During this period, IBM even claimed that APPN will in future be managed using CMIP/CMIS. Then around 1992, TCP/IP and SNMP came storming to the fore on the burgeoning Internet wave and blew OSI totally out of the picture. Being the astute pragmatists they sometimes can be, IBM hastily abandoned CMIP/CMIS in favor of SNMP. However, new products that were in the process of being developed during the 1990–1993 period such as the 2220 ended up using CMIP/CMIS as opposed to SNMP or SNA/MS.)

- Tools to facilitate automated management tasks
- The provision of architected management APIs to ensure customer and third-party participation and integration. To this end there is now a SystemView consortium of vendors working with IBM to develop SystemView-compliant products as well as products to augment and complement SystemView offerings. It should be noted that there is also a NetView/6000 Consortium for vendors developing tools and applications for the NetView/6000 platform.
- The provision of relevant, enterprise-wide support applications to facilitate and sustain SystemView related activities

The SystemView strategy is defined in terms of three interrelated elements referred to as *dimensions*. These three dimensions are: *end-use dimension*, *application dimension*, and *data dimension*. Each dimension provides relevant guidelines, standards, architectures, and APIs for implementing and integrating SystemView-compliant total system management applications. Figure 10.5 depicts the overall SystemView structure as originally presented by IBM.

The end-use dimension deals with the guidelines for building a consistent and intuitive user interface for SystemView compliant applications. This user interface, as is to be expected, is heavily oriented toward iconic, object-based, point-and-click technology. Any of the graphical user interfaces found on today's SNMP managers are likely to live up to, or even exceed, this user interface as envisaged by SystemView in 1990. The criteria defined in this dimension were responsible for ensuring that SystemView met its goal of improving the usability of management applications.

The data dimension concentrates on the requirements for the consistent definition of resources, the access of such data, and the sharing of such definition data among multiple management applications. The application dimension specifies the services and interfaces that will be necessary to support the various tasks that will be required to realize Total System Management.

SystemView identified six management disciplines that were thought to be key in a total system management scheme. These six SystemView disciplines are listed below. Table 10.1 provides a detailed list of the various tasks associated with these six SystemView disciplines.

1. Business Management: Asset management, resource registration, financial administration related to system components as well as planning and scheduling system operation-related functions

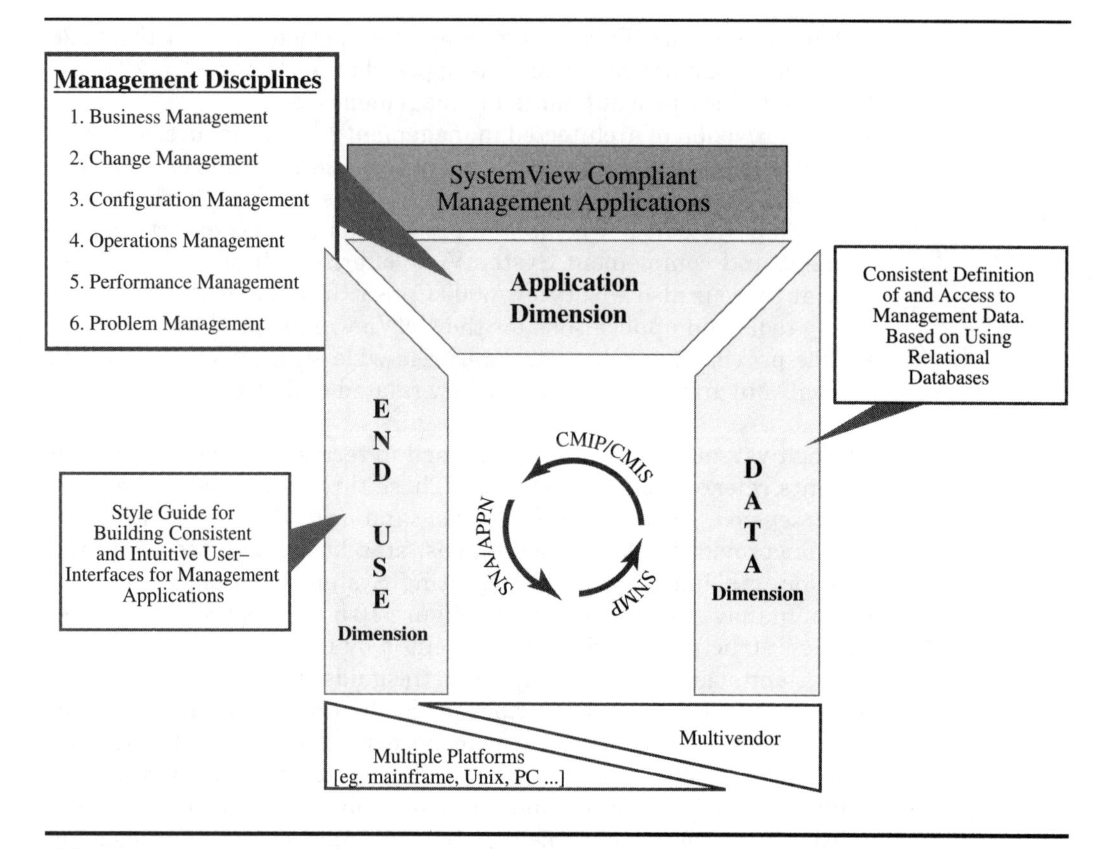

FIGURE 10.5 The overall structure of the system view framework for total system management.

2. Change Management: This discipline dealt with all hardware, software, and configurational changes affecting the system. It covered planning, scheduling, and change distribution. It also addressed how change management should be controlled, tracked, and logged.
3. Configuration Management
4. Operations Management
5. Performance Management: Including capacity planning and the collection of performance data for system tuning
6. Problem Management

10.4.2 The Goal of SystemView Series

The primary goal of the latest SystemView initiative, that is, SystemView Series (or Karat), is to change the overall basis on which

IBM-based management is selected and used. SystemView Series sets out to change the basis of IBM-centric management from being product-oriented to becoming process-oriented, irrespective of whether system or just network management is being considered. Today, with IBM's 150-odd repertoire of management products, the onus is on network managers and administrators to decide and select the optimum product set required to meet their needs. They have to evaluate the features and functions of these various products and see how these fit in with their management needs. SystemView Series sets out to turn this product-centric approach on its head.

With SystemView Series, network managers stop concerning themselves as to what a particular management product can do. Instead they concentrate on the management process (or task) they wish to tackle. Currently, SystemView Series identifies around 20 such tasks. These include: Problem Management, Print Management, Performance Management, Software Distribution, Database Management, Security Management, Administration Management, and Workload Management.

Once a process is selected, SystemView Series will not only automatically select the necessary products but will also install them. In time, product selection and installation will also include non-IBM management products. True to a basic precept of SystemView, the products thus installed will all have a common and consistent, heavily icon-oriented look-and-feel user interface. There will be a heavy reliance on object-oriented technology to achieve this user interface as well as some, but initially not all, of the processes.

SystemView Series, will, however, not initially be a single product. It cannot be, given the diverse management platforms with which IBM currently deals. There will be at least four "flavors" of SystemView Series, one each for: AIX, OS/2, mainframe MVS, and AS/400. Other flavors, including versions for non-IBM platforms, are also being considered.

The AIX and OS/2 versions will be delivered on a CD-ROM. Each CD-ROM will contain all the software for the various products required to implement the repertoire of processes supported by SystemView Series. Using a user-friendly SystemView Series installer, a system administrator will select the desired management processes. System View Series will then determine which products, off the CD-ROM, need to be installed to realize these processes. An enterprise, obviously, will only want to pay for the processes that they want to use, rather than for all of the SystemView Series processes or even worse for all the software on the CD-ROM.

To accommodate this, SystemView Series will display an access code once the processes have been selected and it has determined the products required. The administrator will call up IBM with the access code, which will determine how much the customer will be billed for the software. IBM, once it has determined the cost of the software requested by the enterprise, will provide the system administrator with an authorization code. When a valid authorization code is entered, SystemView Series will then install the necessary software and all of it, in theory, will be ready to run. This type of authorization code-based software or font selection off a CD-ROM has been widely used in both PC and Macintosh communities.

SystemView Series, at least in the first year or so, will not be a fully rounded solution completely true to the SystemView vision. For a start, the emphasis, understandably, will still be network management-centric rather than being Total System Management. The bottom line is that SystemView Series, in 1996, will not be the ultimate panacea for total system management. Total System Management à la SystemView is still some ways down the pike.

10.5 REFLECTIONS

Incisive and in-depth management, whether network- or system-centric, is in the end the all-powerful gravitational force that will hold that network or system together over the long haul. Today, nobody doubts the significance or worth of good management schemes and practices. Powerful, graphical, standards-based, and cost-effective SNMP-based management systems are understandably in vogue. Nonetheless, all-seeing, all-powerful management is considered by many to still be an elusive utopian dream. The primary reason for this frustration vis-à-vis management is the lack of a universal network management protocol that can be used to manage all devices. In time SNMP, or its descendants such as SNMP II, might end up becoming this universal management panacea. But this is not likely to happen for at least another five years. In the meantime, enterprises that are still heavily reliant on SNA mission-critical applications now face the challenge of managing SNA/APPN traffic across an SNMP-managed, multivendor, multiprotocol network.

In most cases enterprises with large, active populations of SNA/APPN users will be forced to settle on a dual-platform scheme involving NetView/390 and one of the many powerful and compelling SNMP management platforms. It is possible to run such a dual-platform management scheme without any interoperability between the platforms. However, most enterprises will opt to have some level of interoperability between these platforms. Today, there are quite a few ways in which to build "bridges" between NetView/390 and SNMP systems, including: SNA Manager/6000, Service Points, SNA Name Correlation, and even bridge/routers replete with integrated Service Points.

Unfortunately, multiprotocol, multivendor management, though vital, is still at best only half the challenge of overall enterprise management. Enterprise management has to extend beyond the network cloud to embrace all the entities that make up the total system served by the network. The types of alert processing, status monitoring, and overall management functions that are today taken for granted with network management now need to be extended to embrace applications, LAN servers, databases, mainframe channels, DASD, and PBXs. Total System Management needs to go even further. It has to address such issues as software distribution, capacity management, operations management, and business management.

Total System Management will, indubitably, be the exciting new management frontier for the remainder of the 1990s. SystemView, with its ambitious and even audacious framework of what all-encompassing, complete-life-cycle Total System Management is all about, is likely to serve as the blueprint for mastering this new frontier.

CHAPTER 11

Putting It All Together

A QUICK GUIDE TO CHAPTER 11

This is the last chapter of this book; two appendixes on SNA and APPN, respectively, are all that follow—apart from the glossary, list of acronyms, and the bibliography. Hence, the express goal of this chapter is to synthesize and reinforce all of the key concepts, methodologies, and technologies discussed in the previous ten chapters.

To this end, this chapter includes many real-life, before-and-after examples of various network reengineering initiatives in what were hitherto very IBM-centric networking environments. These real-life examples focus on token-ring switching initiatives to increase campus-level LAN throughput, and parallel network consolidation efforts. In essence, these examples thus demonstrate the outcome of the first phase of the network reengineering process vis-à-vis the IBM world. This is really the state of the art in this arena at the end of 1995.

While some enterprises are indeed evaluating and prototyping ATM, particularly at the campus level, few if any have really reached the point of being able to claim that they have completed the second phase of the reengineering process. That is not likely to happen for another few years. Consequently, ATM-centric real-life examples, from the IBM world, are scarce and currently nonrepresentative in their scope. Most of the prototype ATM implementations are but variations of the campus-level configurations shown in Chapters 3 and 9.

In keeping with the theme that the network reengineering process has two distinct phases, this chapter consists of two main sections. The first section, 11.1, deals with the issues pertaining to realizing an effective, SNA/HPR-capable multiprotocol LAN/WAN network. The opportunity of replacing costly 37xx's with relatively inexpensive enterprise gateways à la the 3172 is a key decision that needs to be made during the network consolidation phase. To facilitate this decision process, this section includes a detailed treatise on the pros and cons of getting rid of 37xx. It also lists the scenarios in which it would be inadvisable to contemplate replacing a 37xx.

The second part of this chapter addresses challenges as well as the possibilities of the ATM-oriented phase of the reengineering process. It deals with deploying ATM at the campus level and then building upon this base to branch out toward desktop ATM, multimedia ATM, and finally WAN ATM.

This section also provides an introduction to IBM's beguiling notion of cell-in-frame technology. Cell-in-frame could in theory permit existing token-ring NICs to masquerade as ATM adapters of sorts and transmit and receive 53-byte ATM cells—albeit with the cells encapsulated within some amount of standard MAC/LLC framing. The rationale for cell-in-frame is that it can serve as a no-cost, data-only entrée to desktop ATM.

Enterprises that hitherto relied heavily on IBM-technology-based networks, particularly for their mission-critical WANs, are now at a major crossroads. In the majority of cases this is the first networking-related crossroads that they have encountered in the last two decades. During that period, which is more than a few lifetimes in this fast-paced industry, IBM's hold on commercial sector networking, thanks in large part to SNA, was awesome and nearly mystical. IBM single-handedly dictated the shape, scope, tempo, and direction of commercial sector networking. Though most SNA-based networks bristled with non-IBM plug-compatible equipment, the overall networking parameters were still under the control of IBM.

The delay in the introduction of LANs into SNA networks is a good case in point of IBM's sway over this entire market. In the early and mid-1980s non-SNA enterprises gleefully rushed to embrace 10Mbps Ethernet as an optimum way to interconnect the PCs that were beginning to crop up on desktops overnight. For sound technical reasons, such as its lack of deterministic access, IBM eschewed Ethernet in preference for 4Mbps token-ring. Though unveiled in 1986, full-scale SNA-over-token-ring was not viable for another couple of years. In addition, token-ring solutions from day one have always been considerably more expensive than comparable Ethernet solutions. While token-ring does have some value-added features not found in Ethernet (e.g., deterministic access and priority mechanism), the significant price differential between the two technologies, especially today, is hard to justify.

The SNA community, however, really had no choice. They had to wait for token-ring technology to be readily available and then pay a premium for it. The repercussions of this delay in the availability of token-ring LANs, and their higher costs, can still, unfortunately, be seen and felt today. SNA shops are typically 18 months behind non-SNA shops when it comes to innovatively exploiting LAN technology to realize high-speed, multiprotocol networking. Without doubt some of this laggardness is due to the overriding concern of preserving mission-criticality. The rest can be attributed to the late start, as well as the impact of IBM's current lack of leadership in LAN and multiprotocol networking. Now that most have made a significant and in many cases irrecoverable investment in token-ring technology, IBM glibly endorses and promotes Ethernet, as was discussed in Chapter 4.

To cap it off, and to add insult to injury, IBM now uses Ethernet, as opposed to token-ring, to connect operator consoles and network management stations to its flagship 2220 Nways BroadBand ATM WAN switch. Token-ring is not the only example of IBM's role in controlling the potential popularity of networking technologies when it comes to commercial sector networking. X.25 is another notable case in point. IBM's fawning endorsement of frame relay, including native built-in support within ACF/NCP, is in marked contrast to its disdain for X.25. Consequently, X.25, especially in North America, never really played a major role in Fortune 1000 networks.

IBM's uncontested domination of commercial sector networking is now at an end. *The SNA-only networks that for so long were the cornerstone of commercial networking are now anachronisms*. SNA/APPN-capable multiprotocol networks that will eventually dovetail into ATM-centric multimedia, broadband networks are now the wave of the future. Nobody, not even IBM, doubts or denies this. Contemporary multiprotocol networking is dominated by bridge/router and FRAD technology. In this arena, IBM is not a market leader and at best is a second-tier supplier behind the new heavyweights in this industry such as Cisco, Bay Networks, 3Com, Motorola, Hypercom, and Cabletron. With ATM, IBM hopes, desperately, to recapture some of the ground, credibility, and revenues it has lost vis-à-vis networking over the last few years. Widespread ATM to the degree that IBM hopes for, and certainly needs if it is to make any kind of mark, however, is not going to happen till after 1997.

IBM's current supporting, as opposed to leading, role vis-à-vis networking also means that there is no longer an SNA-like framework around which to model future enterprise networks. The Open Blueprint, described in Chapter 8, is still too academic and theoretic; it lacks conviction as a pragmatic model for today's dynamic and pell-mell networking world. IBM's SVN and NBBS, as discussed in Chapter 9, are at present heavily IBM-centric and lack sufficient hooks for facilitating multivendor interoperability or adoptability. Consequently, in their current form, they are unlikely to enjoy the phenomenal success of SNA given that enterprises, today, are understandably weary of what appears to be proprietary solutions.

Instead, the standards that are going to dictate the shape of future networks are going to be those such as: frame relay, ATM, RFC 1490, MOPA, and DLSw. The Internet IETF with their RFC mechanism, the ATM Forum, and the frame relay Forum, as opposed to individual vendors, will for at least the next decade be the purveyors and arbitrators of the standards that will mold the shape of future networks. Individually, none of these standards will, alas, provide the solution for realizing a complete end-to-end multiprotocol, multimedia LAN/WAN network that is likely to consist of, at a minimum, intelligent hubs, LAN switches, bridge/routers, ATM switches, LAN Servers, SNA control units, ATM adapters in PCs/workstations, SNA LAN gateways, mainframe gateways, and PBXs.

As was said at the start of Chapter 1, the ultimate onus of synthesizing these new networks using the appropriate standards now falls on network managers and network administrators—possibly with some help from system integrators. In reality this may not be as bad as it first appears. In the end it may be a small price to pay for the freedom and flexibility of being able to readily pursue optimum and cost-effective technological solutions without in anyway being locked into a particular vendor or overarching architecture.

11.1 THE FIRST STEPS IN REENGINEERING IBM NETWORKS

Consolidating the dreaded parallel SNA and non-SNA networks, while at the same time enhancing the overall networking bandwidth are the two highest priority tasks, and the logical first steps, in reengineering IBM-centric networks. LAN Switching, 100Mbps LANs, and frame relay, particularly so in North America, will be the primary options for enhancing networking bandwidth. These three technologies will be the precursors to ATM in the IBM world. ATM, in many cases, will first get introduced into IBM-oriented environments in the form of 155Mbps connections between LAN switches or high-end bridge/routers. Transitioning to a full-scale, multimedia ATM-centric network will be the next logical step in this reengineering process. Most enterprises are unlikely to be ready to wholeheartedly embark on this phase of the reengineering process until 1997 or later.

LAN switching will be the most cost-effective and least disruptive way of increasing bandwidth at the campus level whether it be to individual desktops, LAN segments, or the backbone. Fast Ethernet or AnyLAN, both of which run at 100Mbps, by themselves or in conjunction with LAN switching, are another way to significantly increase campus-level bandwidth, albeit now with the added cost and disruption of replacing existing NICs with new ones. Full-Duplex NICs, including 200Mbps Fast Ethernet NICs, can be installed just on the LAN servers, when using LAN switching to further enhance the performance of all interactions involving the LAN servers.

Figures 11.1, 11. 2, and 11.3 show actual, real-life, before-and-after examples of how token-ring Switching has already been used to improve LAN performance. The reduction in latency and the increase in the throughput, in each instance, are shown within these three figures. Figure 11.4 shows how a token-ring switch with SRB can be used to provide a fault-tolerant, alternate path configuration between a multiprotocol WAN network and a 3745 front-ending a mainframe. *The LAN switches shown in these figures as well as the other LAN-centric equipment such as hubs and MAUs would be managed using SNMP management platforms.*

The viable and predominant technologies for parallel network consolidation, as discussed Chapters 3 and 5, are going to be:

1. TCP/IP encapsulation of SNA/APPN traffic à la DLSw
2. Natively encapsulating SNA/APPN traffic within frame relay frames per RFC 1490/FRF.3
3. IP-only WAN network with SNA/APPN interactions being supported via: tn3270(e), IP-to-SNA Gateways, or even AnyNet SNA-to-IP conversion
4. APPN/HPR NN-based routing by multiprotocol bridge/routers such that SNA/APPN traffic can be routed alongside non-SNA traffic
5. HPR-only WAN network with non-SNA traffic routed across this WAN via an IBM 2217-ilk AnyNet gateway approach

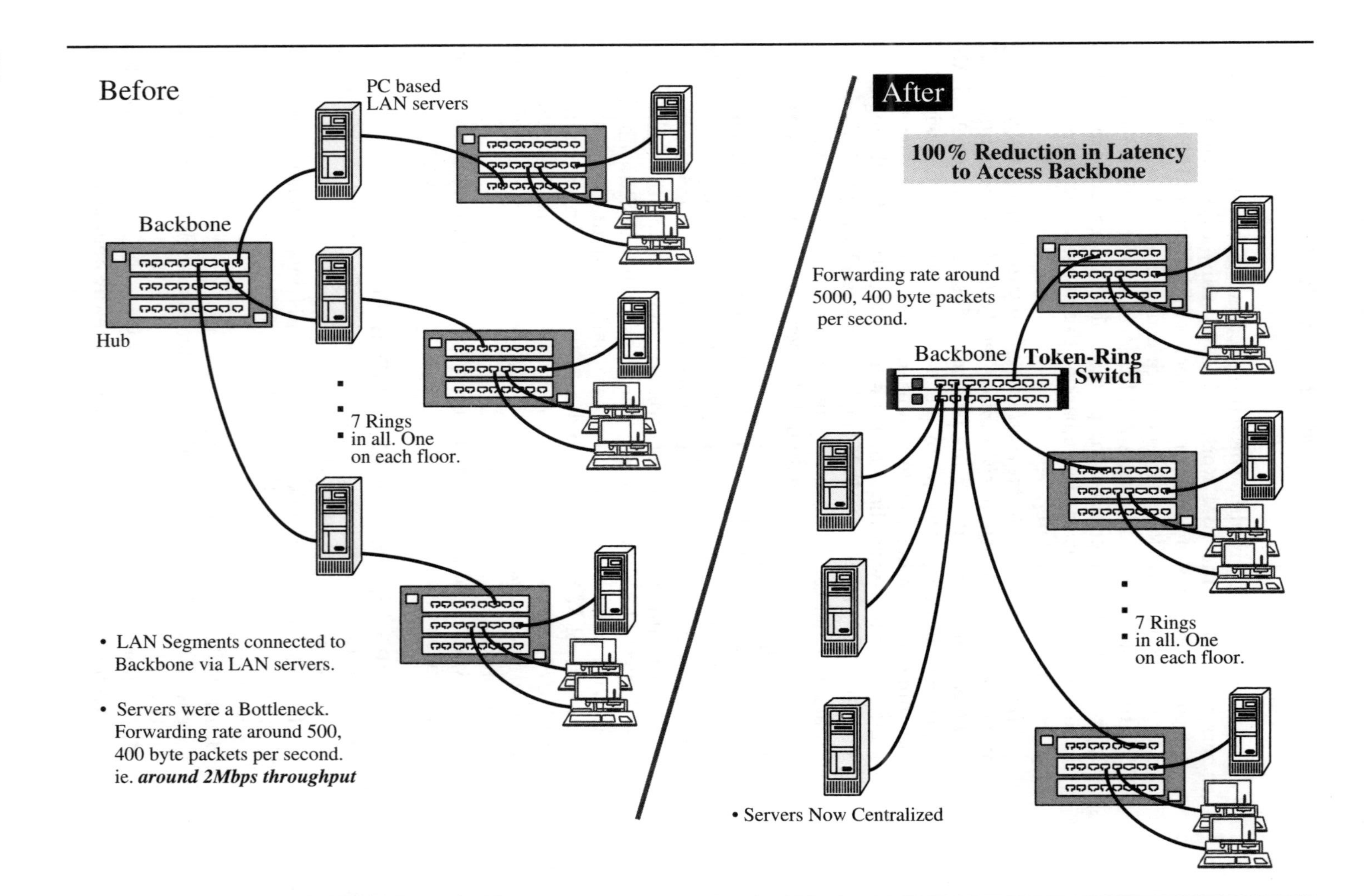

FIGURE 11.1 Real-life example of using a token-ring switch to dramatically increase LAN performance.

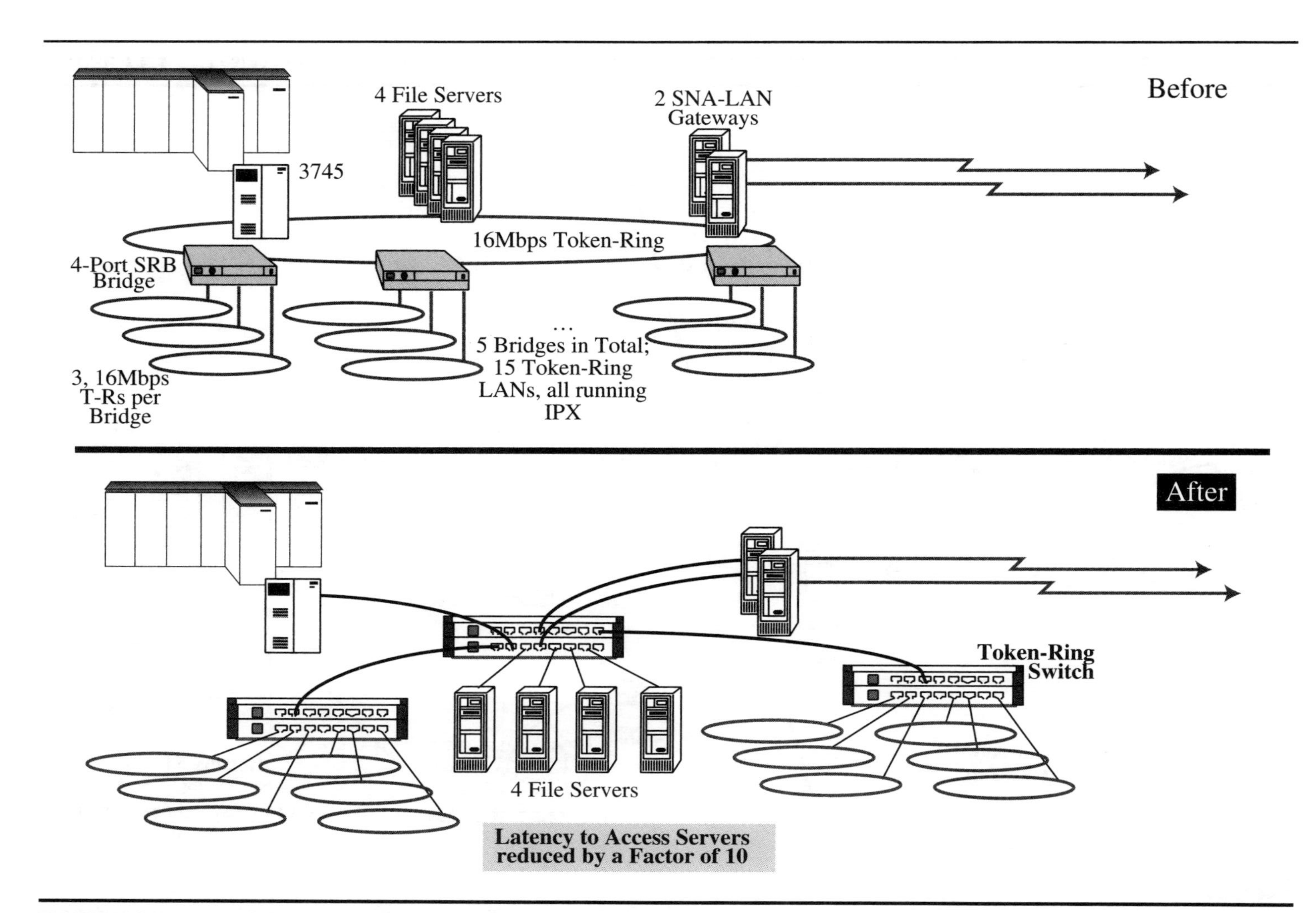

FIGURE 11.2 Another real-life example of using token-ring switches.

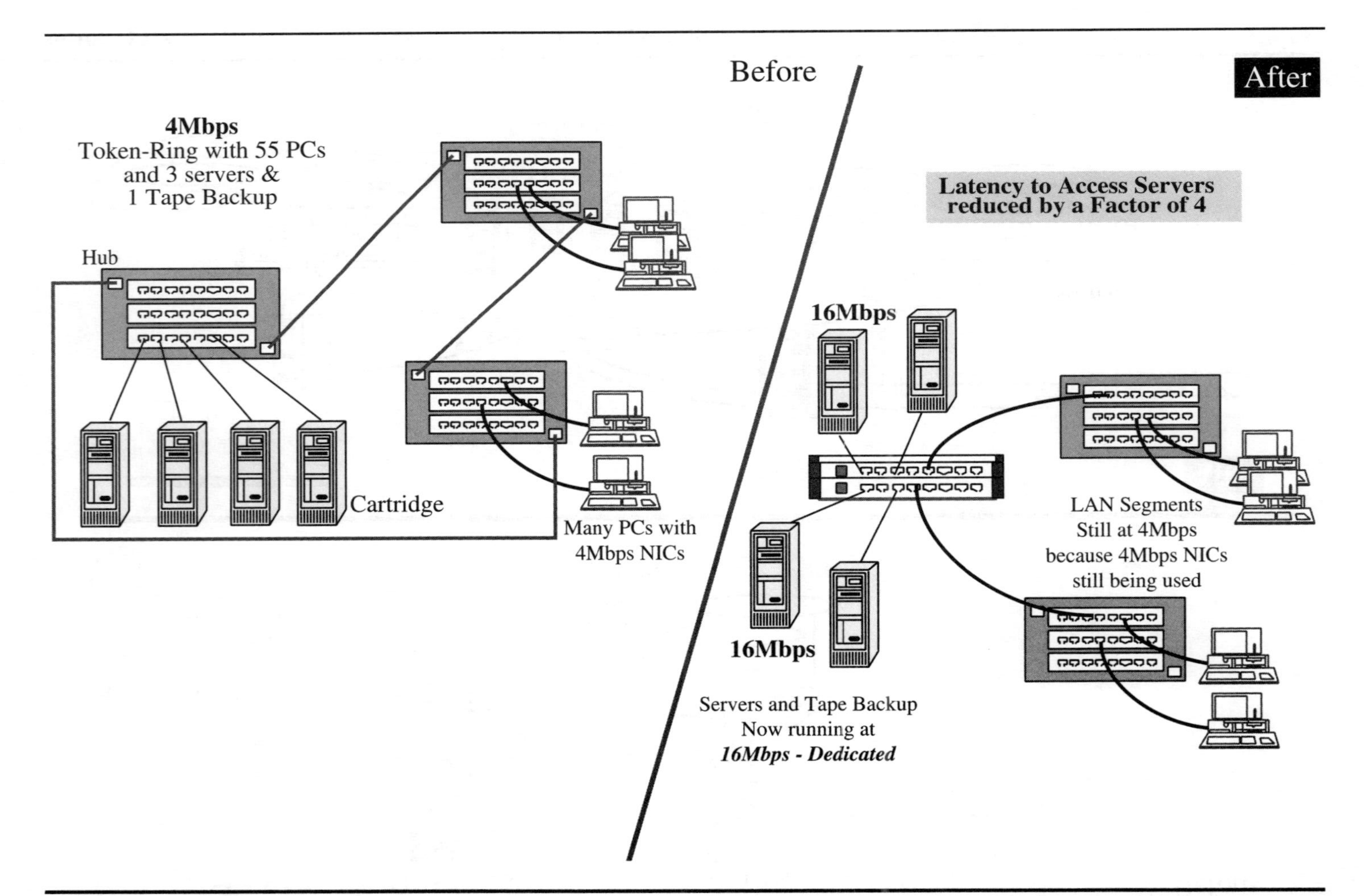

FIGURE 11.3 Real-life example of segmenting a large LAN and reducing the average delay to access servers by a factor of 4.

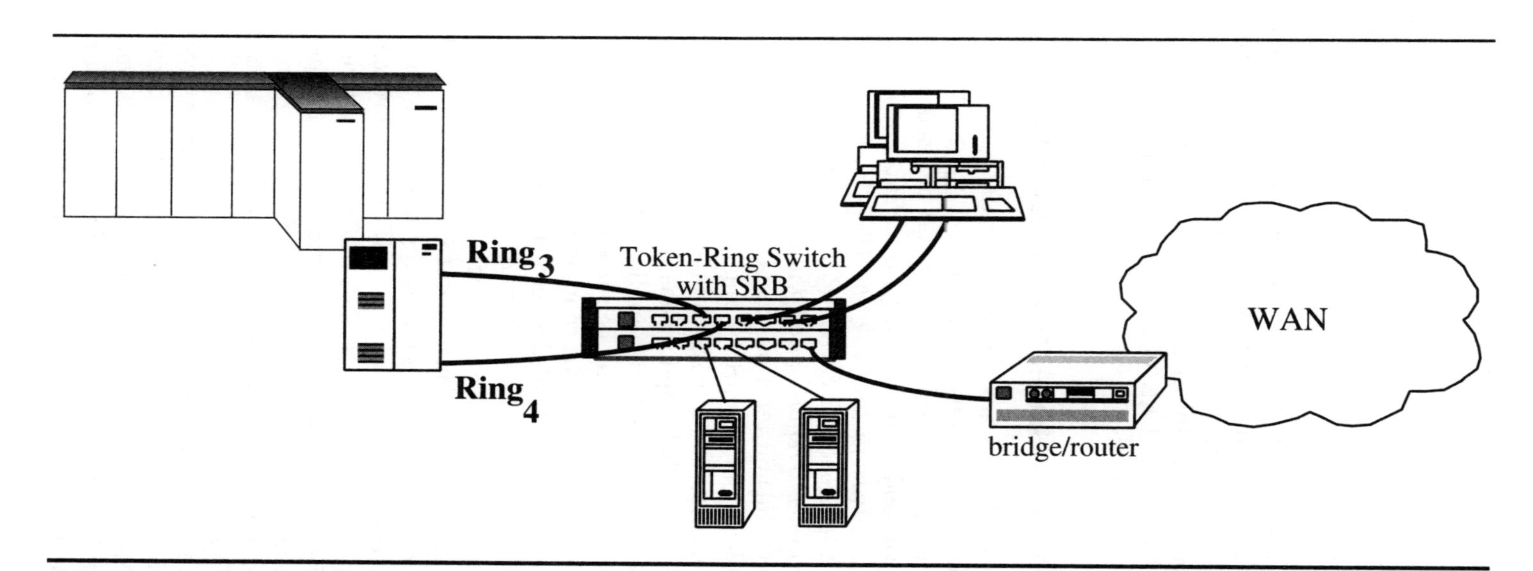

FIGURE 11.4 Using a token-ring switch with source-route bridging to provide an alternate route to a 37xx as a means of fault tolerancy.

6. SDLC-to-LLC:2 and BSC-to-SNA-to-LLC:2 conversions to support SDLC and BSC link-attached SNA devices such as automated teller machines

The first three technologies listed above are non-SNA techniques that accommodate SNA/APPN traffic. The next two listed technologies, on the other hand, are SNA/APPN/HPR-oriented. The last listed technology permits link traffic to be freely incorporated into a consolidated multiprotocol LAN/WAN network implemented using one of the other five technologies.

11.1.1 Parallel Network Consolidation Alternatives: A Reprise

The WAN traffic mix envisaged for the 1997–1998 time frame should be the metric used to determine whether a non-SNA (e.g., DLSw) or SNA-oriented (e.g., 2217) technology should be used to realize the parallel network consolidation. *Typically, an HPR-centric, 2217/AnyNet solution would only be justifiable if the 1997–1998 WAN traffic mix is going to consist of at least 80% SNA/APPN traffic—with this SNA-heavy traffic mix expected to remain much the same for quite a few years.* This last caveat as to the mid-term trend in traffic mix is crucial in this instance.

More and more non-SNA applications are now coming on line to complement and augment SNA-based mission-critical applications. At the same time, SNA applications, albeit it slowly, are being replaced by non-SNA, typically Unix, applications. Consequently, over the next five to six years there could be some marked shifts in WAN traffic mix compositions with the SNA component of this traffic, in terms of percentage, either steadily or dramatically declining year to year.

An IP-only network should really only be contemplated if an enterprise is already in the throes of trying to replace their SNA applications with non-SNA applications, or out-sourcing the SNA applications. If, on the other hand, there is still going to be some amount of SNA interactions across the WAN, a strictly IP-only approach may not be optimum, especially if bona fide SNA traffic from native SNA devices (e.g., 3174, AS/400) have to be transported across the network. Instead, a DLSw(+) approach that permits SNA traffic to freely coexist within what is a IP-centric network may be preferable to trying to arbitrarily ban all SNA traffic from the WAN.

On a similar vein, Desktop DLSw that even precludes SNA traffic from traversing a LAN is on the whole only applicable for supporting mobile SNA users. In the case of LAN-attached PCs/workstations, DDLSw invariably becomes a cute technology in search of a real problem. Also note that desktop TCP/IP encapsulation based on the current DLSw standard that requires two active TCP connections per end-point can suffer from serious scalability problems. All of these issues pertaining to IP-only networks and DDLSw were discussed in detail in Chapter 4.

The majority of the 22,000-odd enterprises, worldwide, with SNA networks today are likely to have a 1997–1998 WAN traffic mix where the SNA/APPN component only accounts for about 70% or less of the total volume. In many cases this SNA/APPN component of the traffic mix is also likely to be on a downward trend. DLSW(+), RFC 1490, or HPR/APPN NN are all potentially viable means for successfully implementing multiprotocol LAN/WAN networks that have this type of WAN traffic mix characteristics.

Of these, HPR/APPN NN routing would be most applicable when a network either has four or more mainframes all running applications that need to be remotely accessed, or if it already has a fair amount of existing APPN traffic due to APPN being used between distributed AS/400s. Outside these two scenarios it is unlikely that APPN NN routing adds enough tangible advantages to merit its selection over RFC 1490 or DLSw(+). *Using HPR/APPN NN, over frame relay, in a network with at most a couple of mainframes or AS/400s, where the bulk of the traffic is SNA/3270 rather than LU 6.2, could prove to be futile and unnecessarily expensive.*

In this type of network, APPN nor HPR has much opportunity to add any real value to the network, as was explained in Chapter 8. Key networking features such as Layer 2-based processing, elimination of intermediate node processing, dynamic alternate routing, and congestion control is going to be provided by frame relay, rather than HPR. Invoking HPR or APPN NN routing over frame relay in this type of network will just increase the amount of processing that has to be done by the bridge/routers.

Moreover, given that APPN and HPR are IBM licensed technology, most bridge/router vendors charge a premium price for APPN/HPR to offset the annual license fee. Hence, APPN/HPR should only be used in situations where this added cost is easily justifiable, for example, in a network with large numbers of geographically dispersed mainframes. In other scenarios, particularly when only one or two mainframes or AS/400s are involved a RFC 1490 or even a DLSw(+) solution is likely to be more efficient and cost effective.

Making the right choice between DLSw(+) and RFC 1490 can be vexing—especially since bridge/router vendors despite offering comprehensive support for RFC 1490 tend to more vociferously promote DLSw(+) because of its TCP/IP orientation. If IP is the predominant traffic type in a network, it may be best to opt for DLSw(+), putting aside all other considerations, since this would then enable IP to be standardized as the underlying routing mechanism for the entire network. However it should be carefully noted that DLSw, if not DLSw+, does have some serious limitations. These are scalabity, the inability to reroute around failed destination bridge/routers, and the relatively high header-to-pay load ratio that impacts overall efficiency.

RFC 1490, though not affected by these limitations per se, does have the unfortunate restriction of only being feasible with frame relay-based WANs. In general, in frame relay WAN-centric networks, with only a few SNA/APPN or NetBIOS destinations, where 40% or more of the WAN traffic is SNA/APPN, RFC 1490 is likely to be a more straightforward and efficient solution than DLSw(+). If the WAN is not frame relay-based, DLSw(+) will be the automatic and uncontested choice, provided of course that other options such as APPP NN routing have already been eliminated.

11.1.2 Real-Life Parallel Network Consolidation

At this juncture, it would be instructive to look at the mechanics and structure of some representative consolidated, SNA/APPN-capable multiprotocol, multivendor LAN/WAN networks. The best way to start would be to look at a classic, even iconic, SNA leased-line network with the remote sites containing SNA devices such as 3174s and AS/400s. The remote sites as well as the corporate headquarters where the mainframe is resident also have multiprotocol LANs with TCP/IP- and IPX-based interactions. Figure 11.5 shows such a classic SNA network which of late has been infiltrated by multiprotocol LANs. An effective consolidated multiprotocol LAN/WAN network to meet the needs of this enterprise could, obviously, be implemented using DLSw(+), IBM 2217s, RFC 1490 or APPN NN, with SDLC-to-LLC:2 being used to cater for the SDLC-attached devices.

Given that this is a single mainframe home-run (i.e., all SNA links home in on the mainframe) configuration with just plain vanilla SNA end nodes, APPN NN routing is not going to be of any consequence. If the non-SNA traffic across the WAN is minimal and expected to be so for a few years to come, a 2217-type AnyNet Gateway solution would be a good fit. Figure 11.6 depicts how the network shown in Figure 11.5 could be consolidated into a multiprotocol network using a 2217-type AnyNet Gateway solution. If on the other hand, non-SNA traffic accounts for at least 30% of the WAN traffic mix, then the choice will have to be between DLSw(+) or RFC 1490. If this is a North American network, and frame relay proves to be the most cost-effective way of obtaining 56Kbps range WAN bandwidth, then RFC 1490 is likely to be the best option.

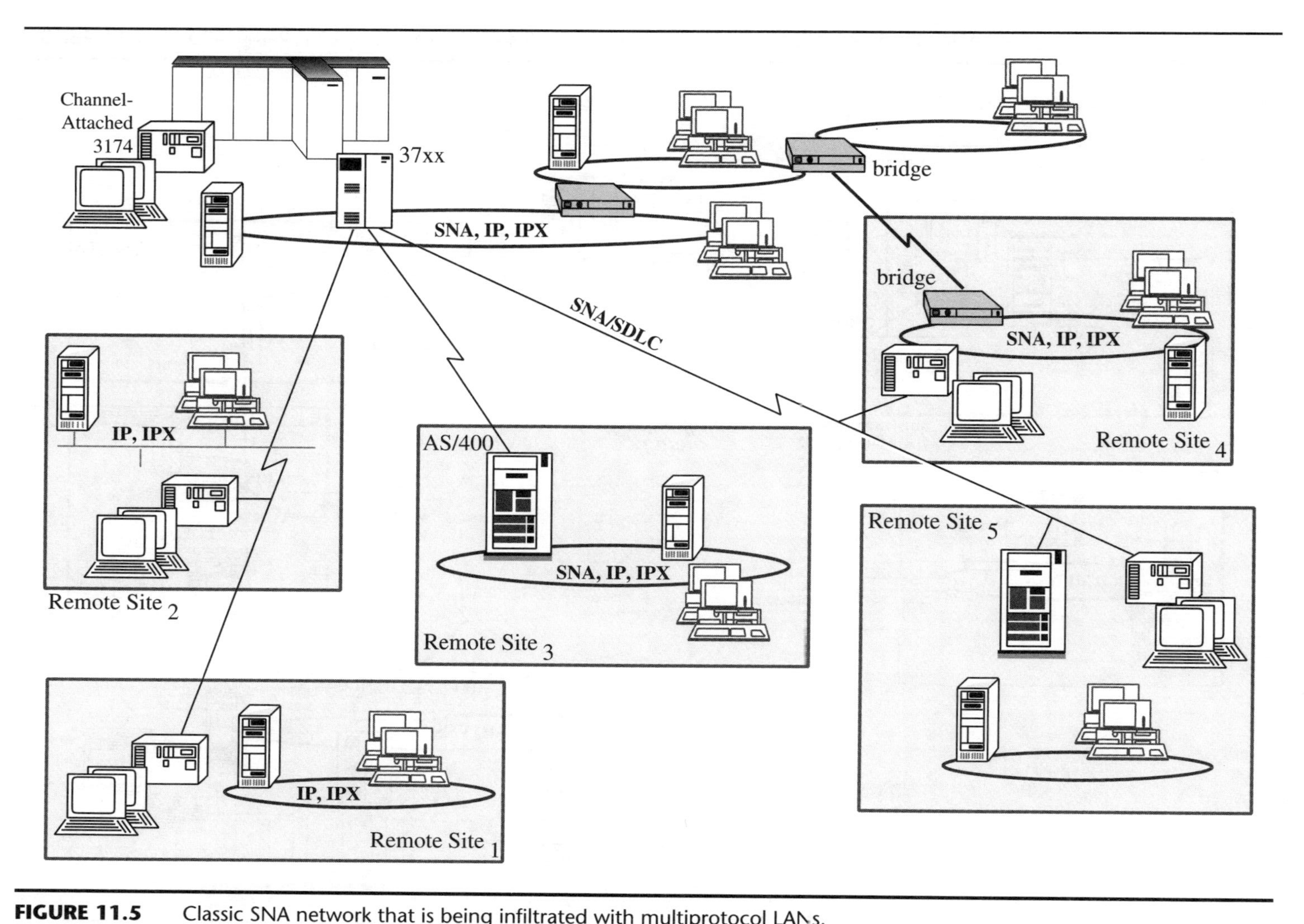

FIGURE 11.5 Classic SNA network that is being infiltrated with multiprotocol LANs.

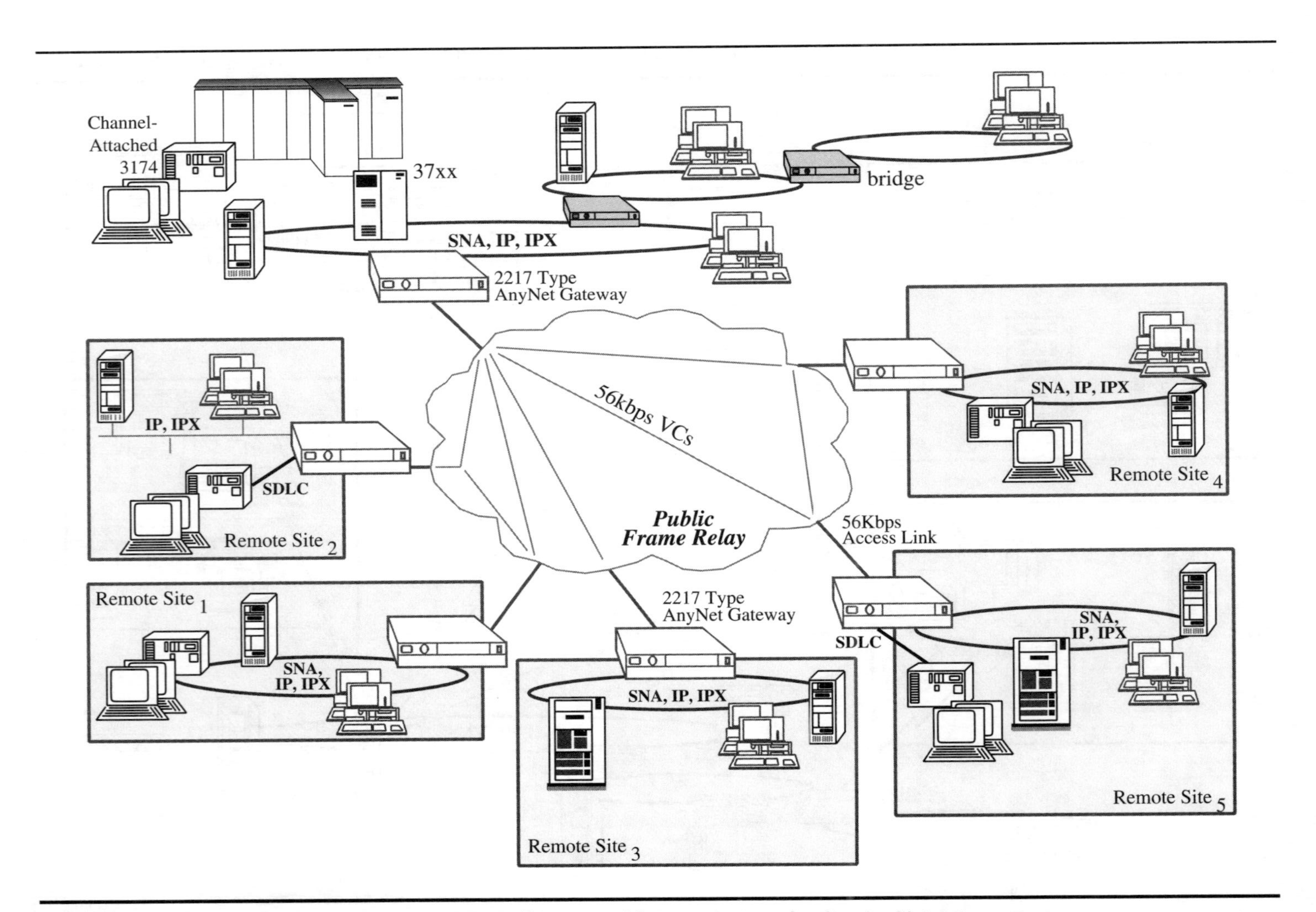

FIGURE 11.6 Converting the classic SNA network into a multiprotocol network using AnyNet gateways.

Figure 11.7 shows how this multiprotocol LAN/WAN network could be implemented using a RFC 1490 solution. Note that the networks shown in both Figures 11.6 and 11.7 have the same topology and structure. The only difference between them is the type of traffic being transported across the WAN. In the 2217-based example shown in Figure 11.6 all the WAN traffic will be HPR/SNA-based even though multiprotocol interactions are being supported across the network. In the RFC 1490-based WAN, each interaction will be conveyed across the WAN in terms of its own native protocol—suitably encapsulated within frame relay frames with RFC 1490 headers at their start. Figure 11.7 shows the possibility of using dial-back between the FRADs to guard against access link failures, and the use of two access links at the mainframe site again as a means of mitigating the impact of link failures. In keeping with contemporary trends, the LANs at the corporate HQ shown in Figure 11.7 have been interconnected using a LAN Switch.

The consolidated networks shown in Figures 11.6 and 11.7 immediately raise the question as to the need for a 3745-type gateway in this type of LAN-centric configuration. In this type of network it would be feasible to replace the 37xx with a 3172, a 3172 equivalent, a channel-attached intelligent hub (e.g., Cabletron), or even a channel-attached bridge/router (e.g., Cisco). Figure 11.8 shows the same network as Figure 11.7 but now with the 37xx gateways replaced with a 3172. Considerable capital, hardware maintenance, and monthly ACF/NCP software maintenance (e.g., $600/month) cost saving could be achieved by replacing a 37xx with any other type of mainframe gateway. However, these cost savings do come at a price, attesting yet again that there are no free lunches in this market. The downside of replacing a 37xx even in this type of relatively small and noncomplex network is as follows:

- Increases the mainframe cycles expended by ACF/VTAM to support each SNA session by about 20%. (See Chapter 2.)

  In general, this is not as bad as it sounds. Cost of processor cycles has come down dramatically in the last six years, thanks to the pressure from RISC/Unix systems. They also continue to go down—particularly now that IBM is offering air-cooled mainframes based on the less expensive but slower CMOS semiconductor technology as used by microprocessors (e.g., Pentium and PowerPC) in parallel to those based on the more expensive but faster Bipolar technology. In addition, most mainframes typically have about 20% unused, excess capacity.

  The additional ACF/VTAM resource consumption is very unlikely to eat up all of this spare processor bandwidth. Moreover, and ironically, as enterprises move mission-critical applications from

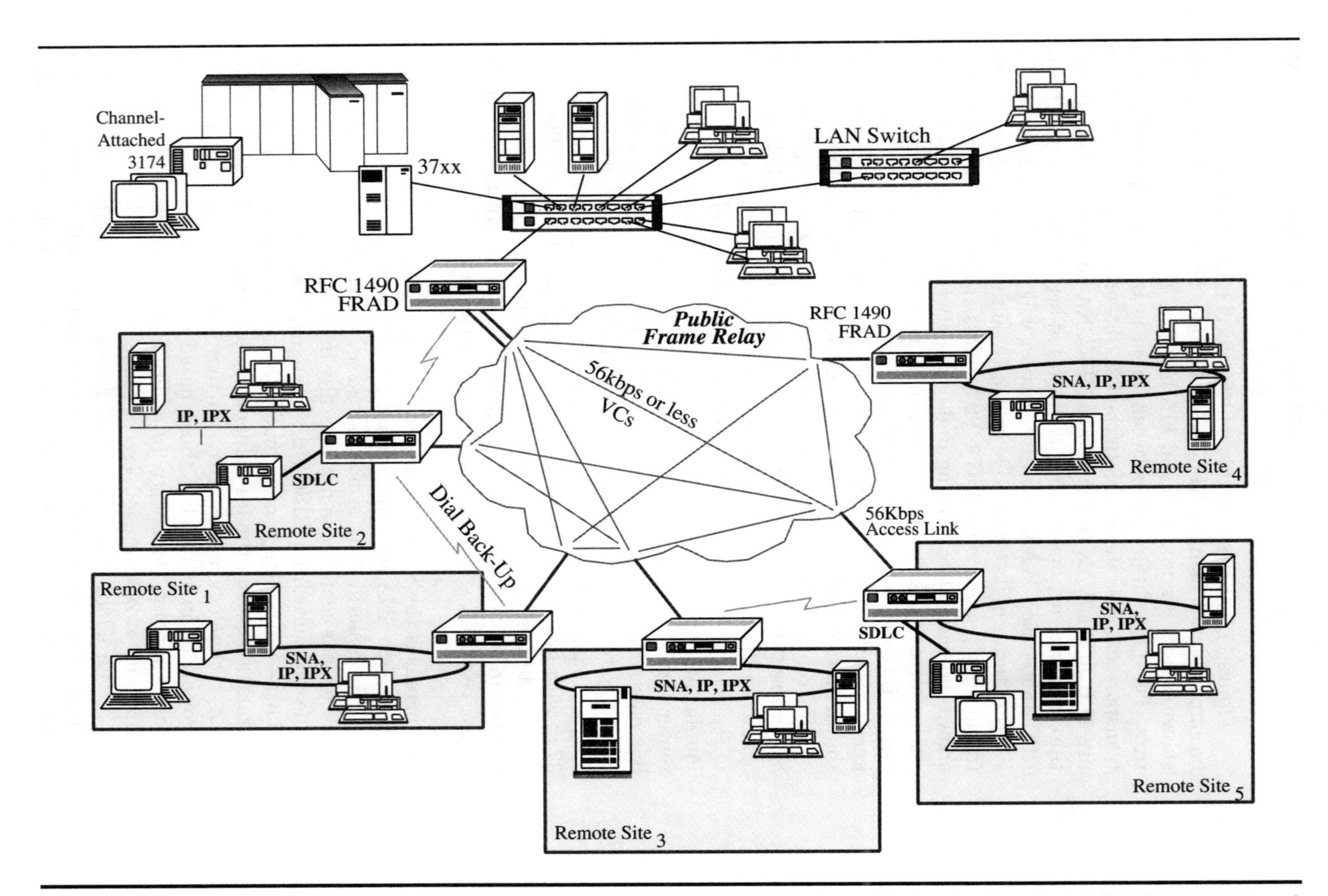

FIGURE 11.7 Upgrading the network shown in Figure 11.5 into a contemporary multiprotocol network using RFC 1490 FRADs and LAN switches.

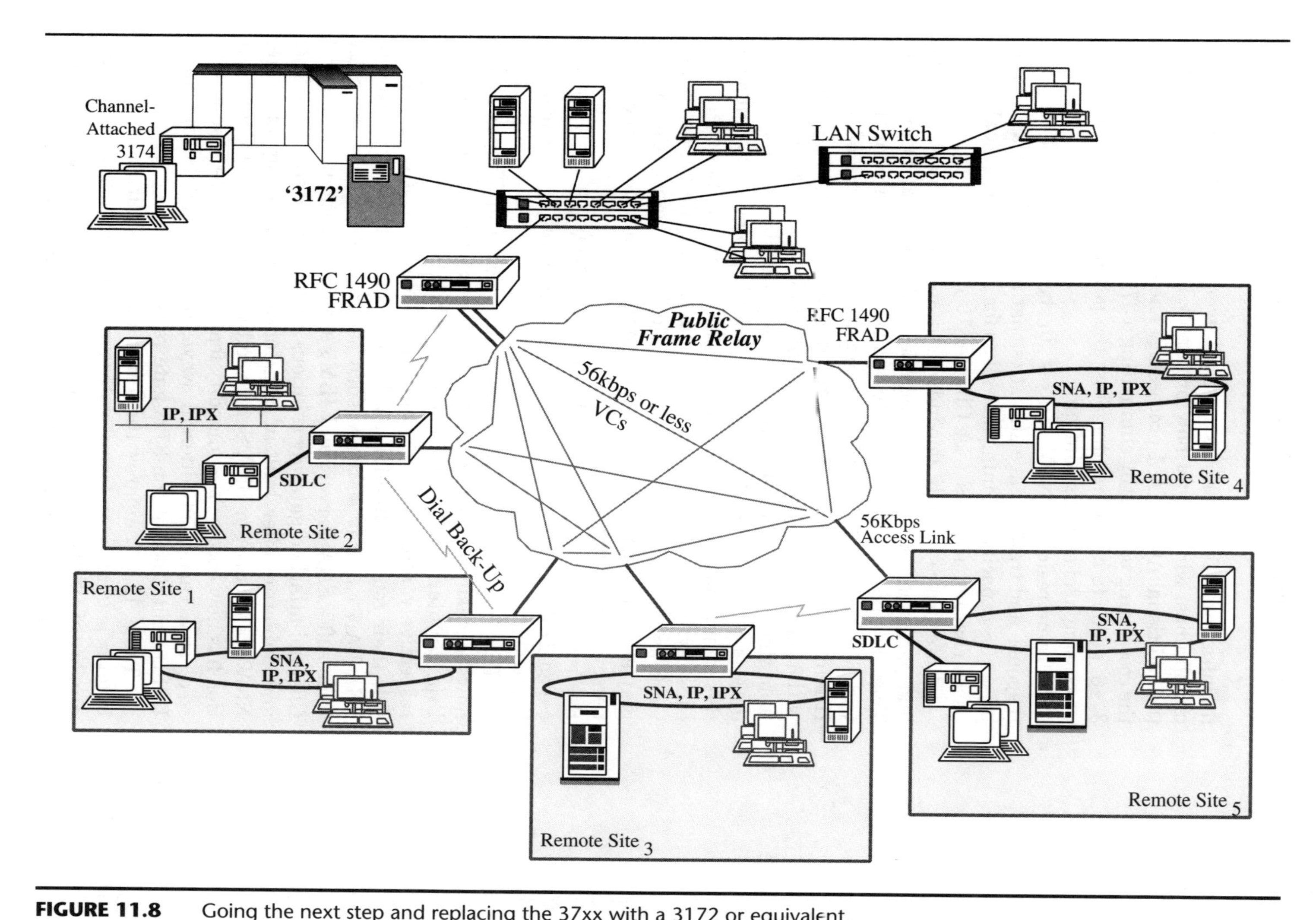

FIGURE 11.8 Going the next step and replacing the 37xx with a 3172 or equivalent.

mainframes to Unix systems it creates even more excess capacity on mainframes, which could now profitably be used to accommodate the additional ACF/VTAM processing. The bottom line here is that the cost saving possible by eliminating a 37xx, particularly a 3745 or 3746, needs to be balanced against the potential cost of the added processing load on the mainframe.

- Mainframe-resident NPM ceases to be able to collect SNA response and traffic usage statistics that are used by many enterprises for capacity planning as well as for generating actual usage-based charge-back bills. (See Chapter 10.)

  Enterprises heavily dependent on such SNA network usage and performance data may understandably balk at losing NPM. There is a possibility that a future release of software for the 3172 will support NPM. There is also a chance that some of the non-IBM gateway suppliers, Cisco in particular, will try and come up with their own solution for providing NPM with the requisite data from their non-37xx gateways. It should also be noted that NPM data is restricted to SNA traffic. In multiprotocol environments this is now only of limited interest since network managers need capacity, response time, and network usage data for the entire network, as opposed to just the SNA traffic. Hence, network managers may elect to start looking for multiprotocol solutions rather than persevering with NPM.

- Restriction in the number of downstream SNA nodes (i.e., the so-called DSPUs) that can be natively supported by the non-37xx gateway.

  This limitation, as was discussed in detail in Chapter 2, can now be easily overcome by using PU Concentration facilities. Bridge/router and FRAD vendors, such as Cisco and Hypercom, respectively, already offer built-in PU Concentrator Gateways inside their routers. In addition, PU Concentration is supported by SNA LAN gateway software such as Novell's NetWare for SAA, Microsoft's SNA Server, and IBM's Communications Manager/2. Channel-attached gateway solutions such as those from Bus-Tech and Cabletron offer built-in PU Concentration using integrated NetWare for SAA or SNA Server software. Hence, PU Concentration has now become a commodity feature that can be readily realized, in a variety of different ways, to overcome the maximum DSPU limitations of non-37xx gateways.

Table 11.1 compares the functional capabilities of *some* of today's leading mainframe gateways and shows the maximum number of DSPUs that can be supported across each gateway prior to invoking PU Concentration. Also note that Cisco, for a start,

TABLE 11.1 Comparison of Some Representative Channel-Attached Mainframe Gateways (Part 1)

	Representative Mid-Size IBM 3745 Model 21A	*IBM 3746 Model 950*	*IBM 3172 Model 003*	*Cisco CIP for 7000 Family Bridge/Routers*
Minimum H/W Costs (Purchase Price)				
Basic, Min. Chassis	$151,950	$51,400	$9,970	$19,000
Token-Ring Adapter	$22,060	$17,500	$1,060	$4,000
Bus-and-Tag Channel Interface		$15,370	Not Available	$5,250 $28,000
Min. H/W Cost for Bus-and-Tag Configuration	$189,380	N/A	$16,280	$42,000
ESCON II Channel Interface	Note 1	$21,700	$9,500	$40,000
Min. H/W Cost with ESCON II I/F	$215,510[1]	$90,600	$20,530	$63,000
Minimum S/W Cost	$4.5–$24.8K per YEAR	$0	$6,060	$2,000 (+$5,000 for APPN NN)
Minimum H/W Maintenance/Annum	$5,000	$6,232	$1,172	$4,200
Minimum S/W Maintenance/Annum	included in monthly license			
Max. # Channel Interfaces	8[2]	10	2	8
Max. # LAN Interfaces	8[2]	20 Token-Ring	4	16
ATM Interface	No	Future	Yes, 100Mbps	Yes, 155Mbps –>
FDDI Interface	No	No	1	Multiple
IP Support	Yes	Future	Yes	Yes
APPN Support	Composite Node	Network Node	Pass-through	Network Node
Frame Relay Support	Yes	Future	Yes	Yes
Max. # Downstrean PUs prior to PU Concentration as LAN Gateway	9,999 –>	At least 9,999 –>	1,020 with all 4 LAN adapters	4,000
Value-Added	All of ACF/NCP functions including NPM		TCP/IP Offload	Full-Function, Multiprotocol Router + TCP/IP Offload
Proven Solution	Beyond Doubt	Not Avail. till Dec. 1995	Yes	New to Market

Notes:

[1]Need 3746-900 Expansion Unit.

[2]Without 3746-900 Expansion Unit

TABLE 11.1 Comparison of Some Representative Channel-Attached Mainframe Gateways (Part 2)

	Bus-Tech 3172–BT–1	*Bus-Tech 3172–NT–1*	*Bus-Tech ENC*
Minimum H/W Costs (Purchase Price)			
Basic, Min. Chassis	$11,730	$17,720	$11,635
Token-Ring Adapter	$1,060	$1,060	$1,000
Bus-and-Tag Channel Interface	$5,250	$5,250	Included in Base
Min. H/W Cost for Bus-and-Tag Configuration	$18,040	$24,030	$12,635
ESCON II Channel Interface	No	No	Future ($13,000)
Min. H/W Cost with ESCON II I/F	N/A	N/A	$19,135
Minimum S/W Cost	$4,995	$6,059	$4,995
Minimum H/W Maintenance/Annum	$1,599	$2,199	$1,599
Minimum S/W Maintenance/Annum	$250	Per Call or $1,000	$250
Max. # Channel Interfaces	2	2	2
Max # LAN Interfaces	4	4	6
ATM Interface	Future	Future	Future
FDDI Interface	Yes	Yes	Yes ($1,295)
IP Support	Yes	Yes	Yes
APPN Support	No	No	No
Frame Relay Support	No	No	Future
Max. # Downstream PUs prior to PU Concentration as LAN-Gateway	2,032[1]	250[1]	2,032[1]
Value-Added	Built-in Novell NetWare for SAA SNA-LAN Gateway + TCP/IP Pass-Through + tn3270 Server	Built-in Microsoft SNA Server + Opt. TCP/IP Pass-Through + tn Server	Built-in Novell NetWare for SAA SNA-LAN Gateway + TCP/IP Pass-Through + tn3270 Server (also Microsoft SNA Server)
Proven Solution	Yes (over 400 installations)	New to Market	Getting There (100 instalations)

Note:

[1]Always with PU Concentration

TABLE 11.1 Comparison of Some Representative Channel-Attached Mainframe Gateways (Part 3)

	Cabletron LCAM-BC/ES in MMAC	*Cabletron CIM-9@110-BT/ES in MMAC-Plus*	*CNT's Brixton 6600 Integrated Gateway*
Minimum H/Costs (Purchase Price)			
Basic, Min. Chassis	$4,750 (with Redundant Power)	$15,000 (with Redundant Power)	$36,000
Token-Ring Adapter	Included with Channel i/f	$13,995	$3,900
Bus-and-Tag Channel Interface	$18,995	$21,995	Included in Base
Min. H/W Cost for Bus-and-Tag Configuration	$23,745	$50,990	$39,900
ESCON II Channel Interface	$29,995	$31,995	$4,000
Min. H/W Cost with ESCON II I/F	$34,745	$60,990	$43,900
Minimum S/W Cost	$0	$0	$19,500
Minimum H/W Maintenance/Annum	1st year included in Base Price	1st year included in Base Price	$3,192 in Base Price
Minimum S/W Maintenance/Annum	1st year included in Base Price	1st year included in Base Price	$2,340 in Base Price
Max. # Channel Interfaces	6	12	2
Max # LAN Interfaces	6	12	4
ATM Interface	Yes	Yes	Future
FDDI Interface	Yes	Yes	Yes
IP Support	Yes	Yes	Yes
APPN Support	Yes	Yes	No
Frame Relay Support	Yes	Yes	Yes
Max. # Downstream PUs prior to PU Concentration as LAN-Gateway	128	256	Note 1
Value-Added	NetWare for SAA, MicroSoft SNA Server, tn3270 server, TCP/IP Offload (future)	NetWare for SAA, MicroSoft SNA Server, tn3270 server, TCP/IP Offload (future)	tn3270 server, TCP/IP Offload, SNA Sessions Switching between multiple mainframes
Proven Solution (close to 100)	Getting There	New to Market	New to Market

Note:

[1]1,000 concurrent SNA sessions using Brixton's 'PU5/PU4' (i.e., mainframe emulating) software

already supports full APPN NN routing on their 3172 compatible gateway. Cisco has stated that this APPN capability will be upgraded in 1996 to be HPR compatible.

The possibilities of eliminating 37xx gateways and the issues relating to using non-37xx gateways are discussed further in the next section.

Figure 11.9 shows a representative example of dreaded parallel networks, where an enterprise is using an SNA network, a bridged network for NetBIOS and IPX traffic as well as a router-based network for supporting DECnet traffic between DEC VAX systems. Figure 11.10 illustrates how these parallel networks could be consolidated into a single multiprotocol network using DLSw. In this instance, APPN NN would also be an option given that multiple mainframe sites are involved. Obviously, RFC 1490 would also work. Using APPN NN routing would eliminate the two-hop routing that RFC 1490 and DLSw(+) would be forced to use in this scenario to compensate for their lack of support for SNA Routing. Why two-hop routing is necessary and how APPN NN routing eliminates it are discussed in detail in Chapters 2 and 7.

Figure 11.11 depicts a 200-plus branch office environment where SNA and TCP/IP are used between the branch offices and a central corporate site. Note , that this frame relay-based consolidated network also supports voice—albeit using separate 64Kbps channels on fractional T1 links for voice and data, rather than multiplexing both across the same VC. It is, however, not obligatory to keep the voice and data on separate VCs. Some of today's FRADs will permit voice and data to be multiplexed over the same set of VCs and access links.

Figures 11.1 to 11.11 graphically illustrate the types of contemporary, multivendor, multiprotocol networks that are now possible with LAN switching and parallel network consolidation technology. The sizes of the networks shown in these figures were intentionally kept small so that they could be depicted within relatively small diagrams without too much clutter and confusion. Subject to a few caveats such as DLSw's current scalability problem, all the techniques shown in these figures could in general be scaled up to deal with much larger networks.

It should also be noted that in a big network it would be possible, and in some cases even profitable to use different consolidation techniques to deal with different parts of the network. Chapter 7 already addressed the possibility of building hybrid networks where DLSw(+) or RFC 1490 are used in conjunction with APPN NN. It would also be possible to use IBM 2217s in conjunction with DLSw(+) or RFC 1490 given that the output of a 2217 consists entirely of LU 6.2 traffic. Figure 11.12 shows how DLSw(+), APPN NN, and 2217s could also be used within one network.

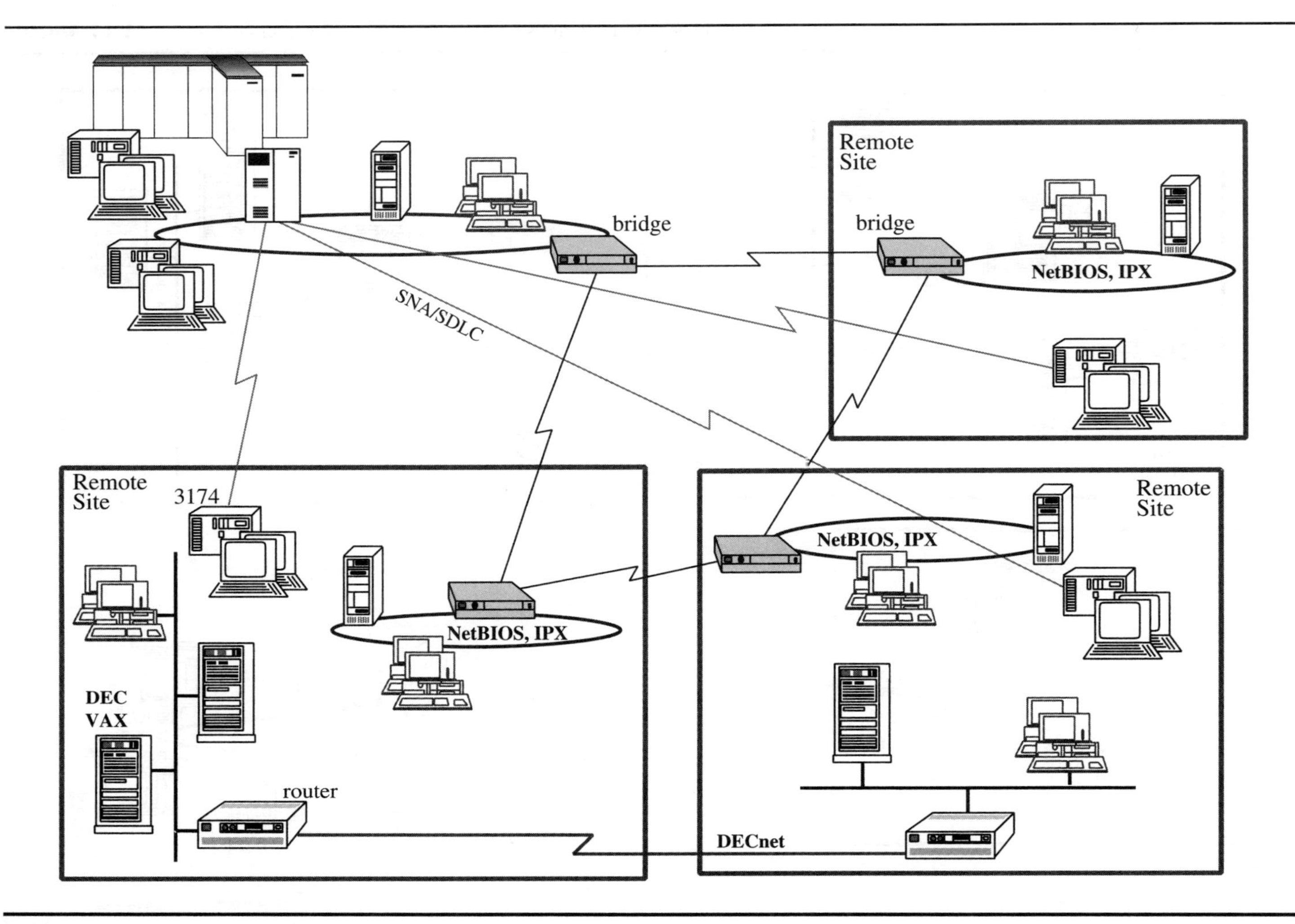

FIGURE 11.9 Class case of the dreaded parallel networks.

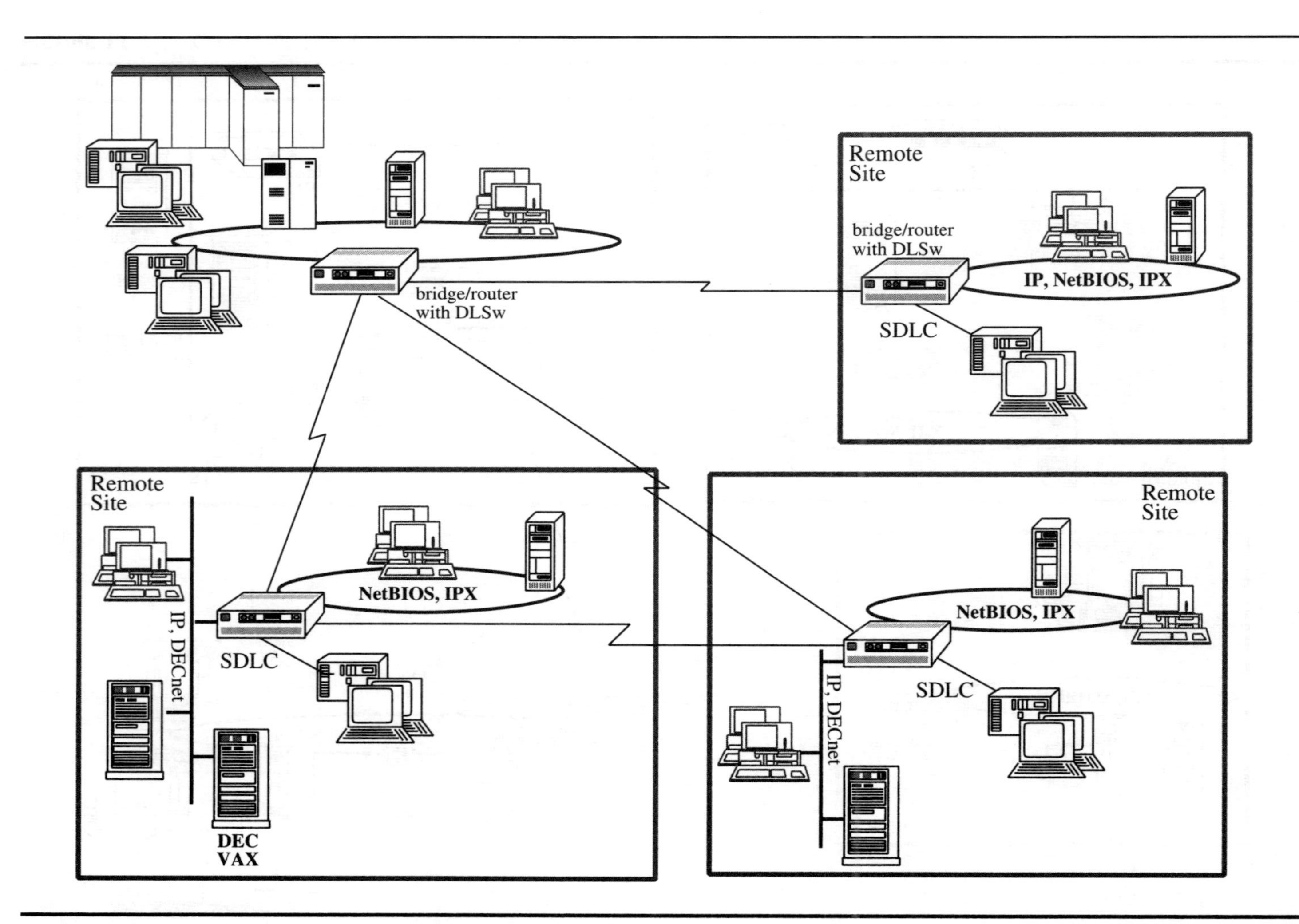

FIGURE 11.10 A classic DLSw-based, SNA/APPN-capable multiprotocol network.

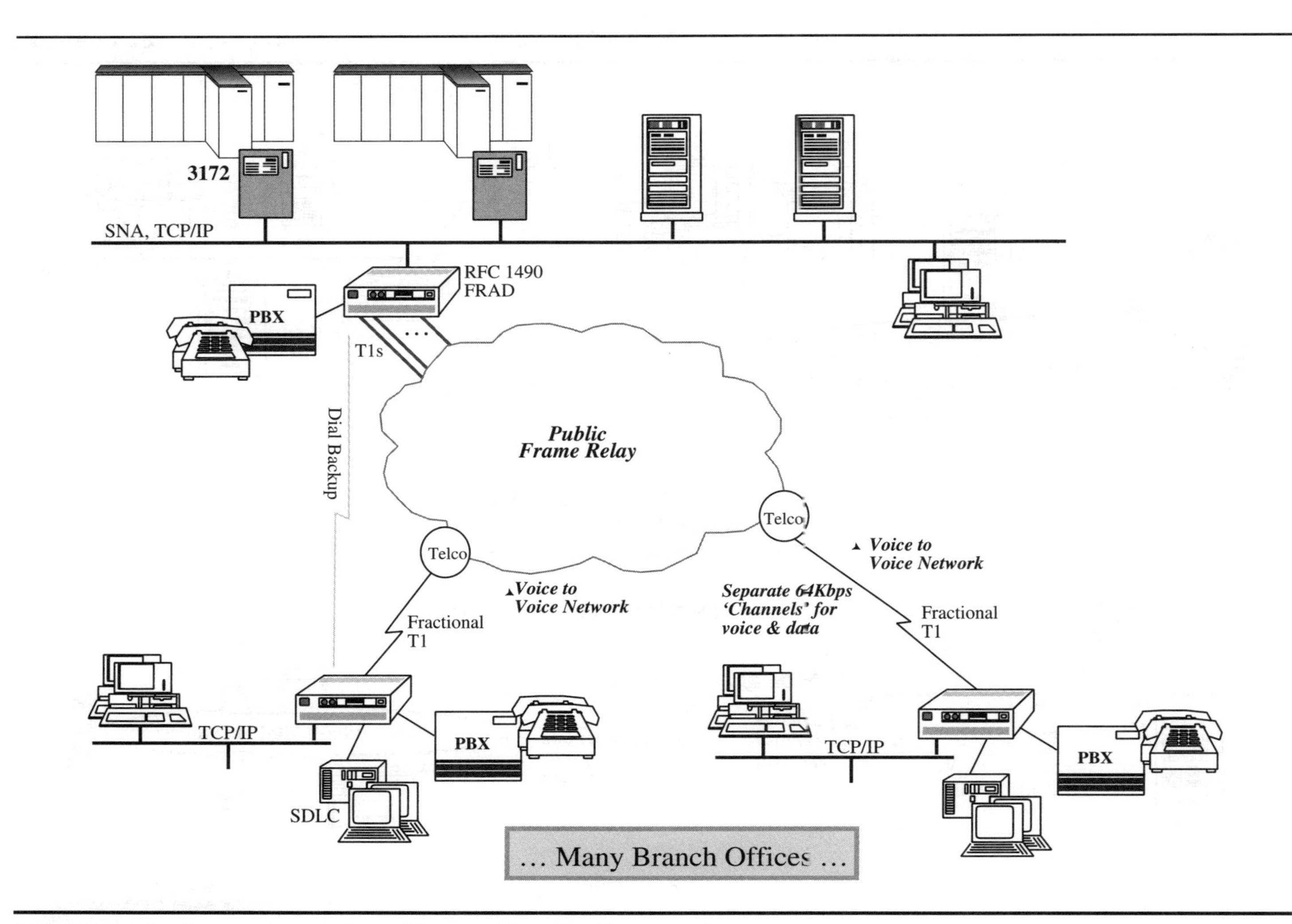

FIGURE 11.11 Frame relay-centric, multiprotocol branch office network.

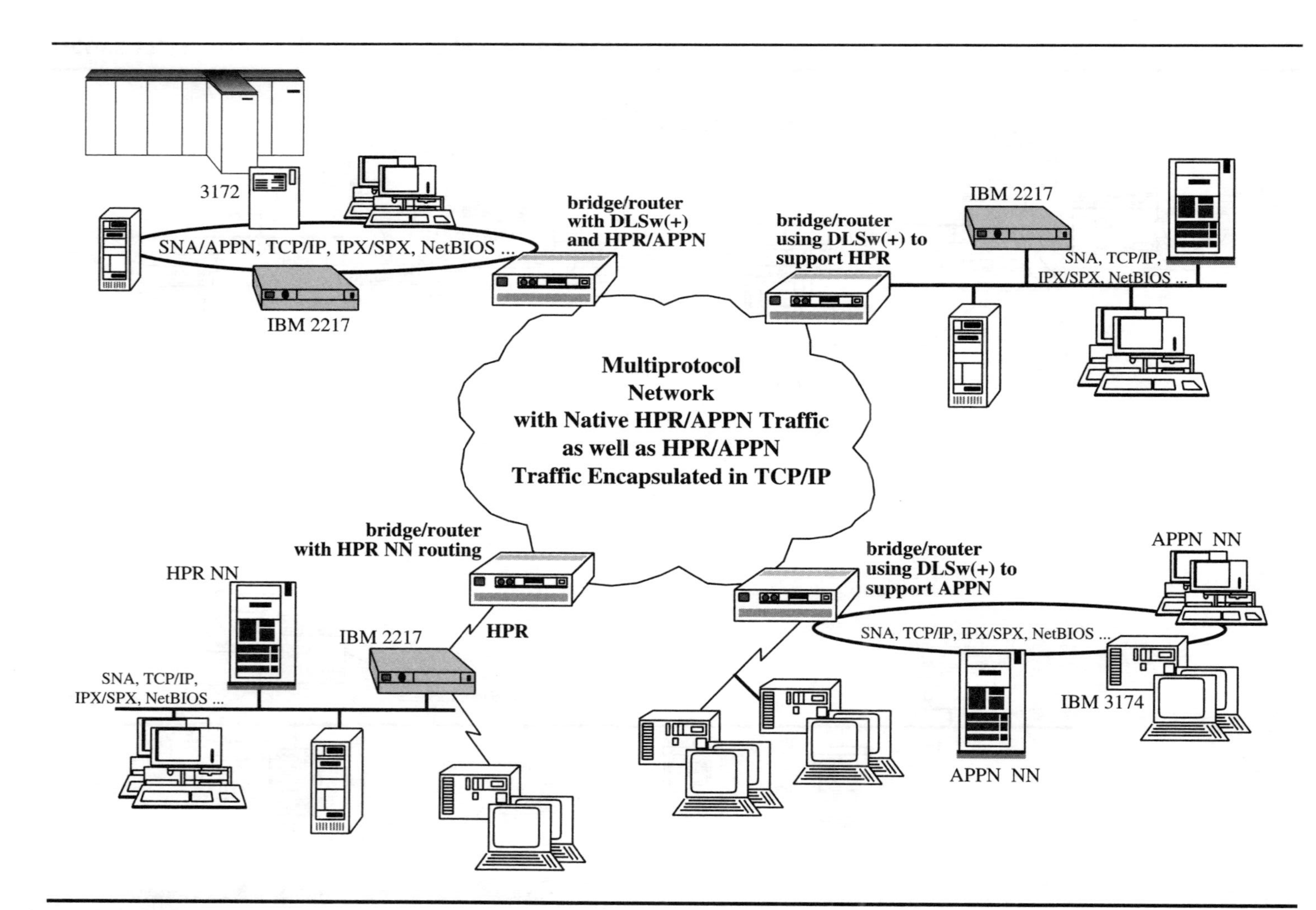

FIGURE 11.12 Hybrid network where DLSw(+), APPN/HPR, and even 2217s are used within one network.

The management of the networks shown in Figures 11.6 to 11.12 would invariably revolve around using a dual-platform management scheme involving both NetView/390 and a SNMP manager. The issues involved in implementing this type of dual-platform management scheme were discussed in Chapter 10.

11.1.3 Evaluating the Possibilities of Replacing 37xx's with Enterprise Gateways

IBM 37xx communications controllers, starting with the 3705 that was available in 1974 to the current 3745/46s, have traditionally been the gateways between mainframes and SNA only WANs. Supporting WANs and performing sophisticated SNA functions such as SNA subarea routing and SNA Network Interconnection (SNI) were and still are the indubitable fortes of 37xxs. They excel and are peerless at handling huge numbers of low-speed (i.e., up to 256Kbps) serial links.

However, as was shown in Figure 11.8 and discussed above, the need to use 37xx's as mainframe gateways is now starting to dissipate. There are the following three overriding reasons for this:

1. *Disappearing need for serial ports at the mainframe site:* There are two major trends that together are rapidly eradicating the need for serial ports on 37xxs, namely:

   - *Migration from SDLC-attached devices to LAN-attached devices:* Most new SNA devices, deployed over the last few years, have been LAN-attached, rather than being link-attached. In addition, the predominant client type in most of toady's SNA networks are LAN-attached PCs that are acting as SNA/3270 terminals. These LAN-attached PCs either run full-stack SNA or using a SNA-LAN Gateway à la NetWare for SAA to derive their SNA-functionality. These LAN-attached PCs have virtually usurped the 3270 terminals that used to be connected to SDLC or BSC link-attached 3x74 control units via coax cables.

     Ubiquitous support for SDLC-to-LLC:2 by bridge/routers and FRADs: The installed base of existing SDLC devices can now be easily and cost-effectively supported locally, at the various remote sites, by the bridge/routers or FRADs being used to implement the multiprotocol network. SDLC-to-LLC:2 conversion is an integral feature of DLSw that is now supported by all major bridge/router vendors. FRADs, in addition to supporting SDLC-to-LLC:2, also excel at supporting other link protocols such as BSC and Async. *BSC-to-SNA-to-LLC:2 conversion tech-*

nology is even available to eradicate the need for serial ports on 37xxs to support BSC traffic (e.g., 3270) that is being supported through ACF/NCP. (Serial ports will still be required if the emulation program [EP] is being run on 37xx's to support non-3270 BSC traffic, such as 2780/3780 BSC. This would only be the case if very old, pre-SNA applications that use IBM's Basic Telecommunications Access Method [BTAM] were still being run on the mainframes.)

Thanks to these two trends it is now possible to implement SNA/APPN-capable multiprotocol LAN/WAN networks where the mainframe gateway only has to support a LAN or ATM interface. All traffic from SDLC or 3270 BSC devices in the network would be presented to this LAN-only gateway as emanating from LAN-attached devices.

2. *In multiple mainframe environments bridge/routers can now provide support for SNA routing through their APPN NN routing capability:* Facilitating SNA Subarea-to-Subarea routing was a key role of 37xxs, particularly so in the case of link-attached, remotely deployed RCPs. Today, however, all leading bridge/routers offer APPN NN routing as well as providing surrogate DLUR support to ensure that this APPN NN routing can be used to freely accommodate all SNA traffic. Moreover, a channel-attached Cisco bridge/router can perform APPN NN, making it functionally equivalent in this respect to the first release of IBM's 3746-950, which only supports APPN NN routing. (It should also be noted that the initial version of the 3746-950 does not support IP.) Figure 11.13 shows a consolidated multiprotocol network involving multiple mainframes built around channel-attached and remote bridge/routers that are using APPN NN routing to integrate SNA/APPN traffic with non-SNA traffic.
3. *Move from SNA-only networks to SNA/APPN-capable multiprotocol LAN/WAN networks:* In such multiprotocol networks, bridge/routers or FRADs perform all the WAN-related functions. All that will be expected of the 37xx is to act as a SNA/APPN and TCP/IP gateway between the mainframe and the multiprotocol network. A 37xx is an expensive means of realizing such a gateway.

As a direct consequence of the above three factors the need for and the appeal of 37xx's have, understandably, begun to wane. What is required in most instances is a cost-effective gateway that can adroitly support both SNA/APPN and TCP/IP. A 37xx (with the exception of the current 3746-950), without doubt, can support SNA/APPN and TCP/IP. It, however, has trouble doing this cost-effectively. The phrase *a cost-effective 37xx* would be thought by many as being a classic exam-

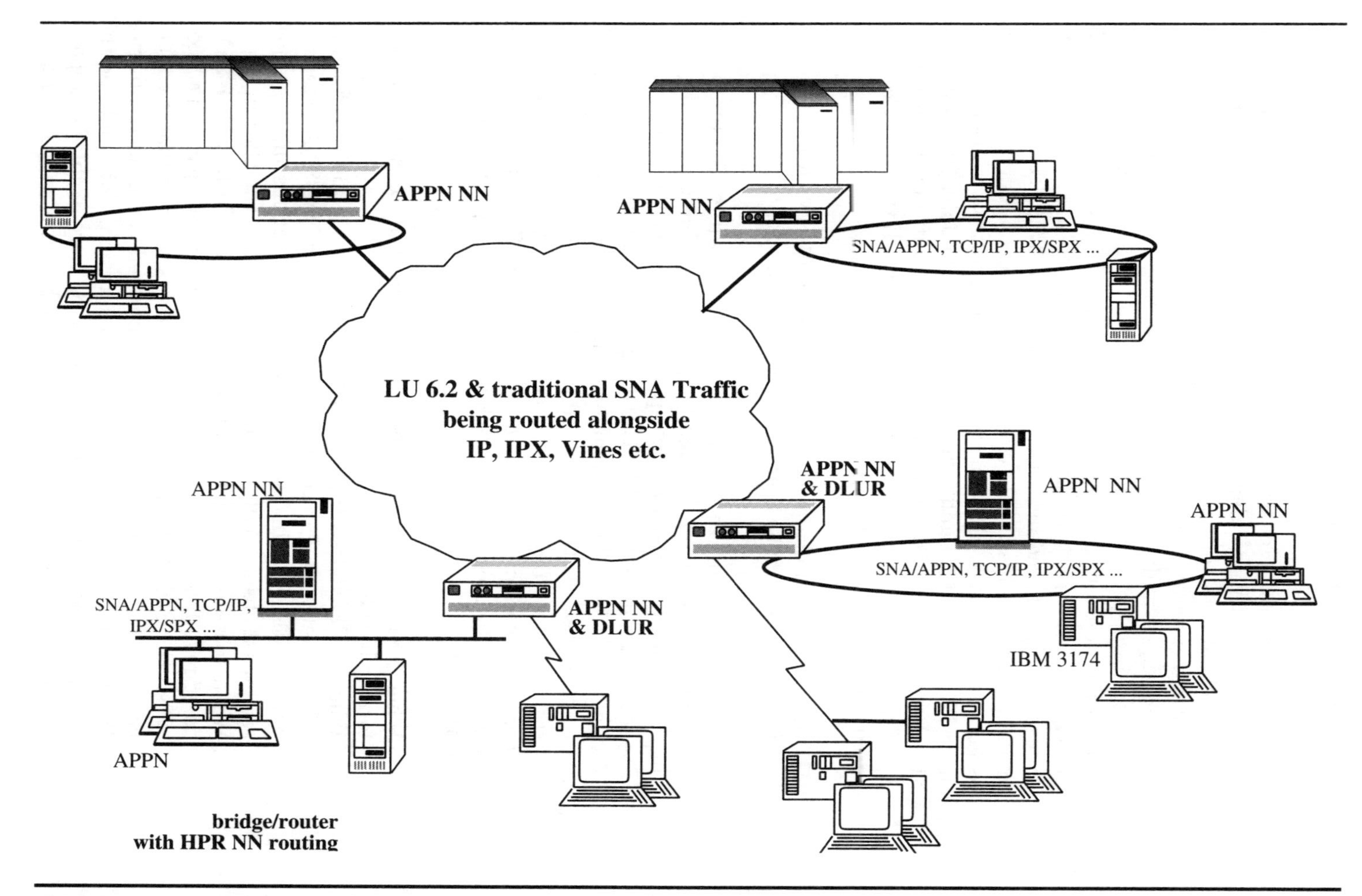

FIGURE 11.13 APPN network built around channel-attached bridge/routers.

ple of an oxymoron. For what they do, 37xx technology is now exceedingly expensive, particularly so when compared to the price/performance of today's bridge/routers, FRADs, hubs, and even other mainframe gateways such as IBM's own 3172.

The staggeringly steep $26,200 for a single high-speed 3745/46 port was discussed in detail in Chapter 6 in the context of why it is inadvisable to contemplate connecting a 3745 to a frame relay WAN through a serial port. A token-ring interface on a 3745 can cost in excess of $20,000, and in excess of $15,000 on a 3746. These high premium prices were justifiable when 37xx's provided a unique role. This is no longer the case. Much more cost-effective gateways are available from IBM (e.g., 3172) and others. Table 11.1 tabulates and compares the characteristics of some of today's leading mainframe gateways.

Two relatively serious shortcomings of non-37xx gateways were discussed above, these being:

1. Increase in mainframe cycles consumed by ACF/VTAM
2. Loss, at least for the time being, of NPM functionality

However, in many cases, enterprises may be willing to put up with these impositions given the significant and tangible cost saving that they can immediately realize by replacing a 3745 with a low-cost 3172. The maximum number of DSPUs that can be supported by these non-37xx gateways is no longer a major issue given the widespread availability of proven PU Concentration technology.

It is, however, imperative to note that there are scenarios where it would not be possible, or advisable, to replace a 37xx with a 3172-type alternative. The scenarios in which it is important to stick with 37xx's are as follows:

1. When SNI is being used: SNI Gateway functions can only be performed by a 37xx.
2. When ACF/NCP adjuncts such as Network Routing Facility (NRF), Network Terminal Option (NTO), or Non-SNA Interconnect (NSI) are being actively used: The functions of these now somewhat obsolescent products are described in the Glossary.
3. When there is a need to still support BSC traffic to/from BTAM-based mainframe applications using EP software and ports on a 37xx.
4. When the SDLC and BSC installed base is so huge that it cannot be cost-effectively or meaningfully supported via bridge/routers or FRADs performing SDLC-to-LLC:2 conversion: This could be the case if an enterprise currently has one or more 3745/46s supporting 500 or more SDLC ports—where each port is supporting quite

a few multidropped SNA/SDLC devices. In such environments, where the installed base of SDLC devices is likely to be in the thousands, using a 3745/46 may still prove to be the most cost-effective way of supporting these devices. *Migrating this SDLC network from SDLC to frame relay, relative to the 3745/46, could be a means of tangibly reducing the cost.* This conversion could be realized using either built-in SNA-over-FR support if available (e.g., 3174s, AS/400s and OS/2s) or RFC 1490-based mono-FRADs. SDLC to FR conversion is described in Chapter 6.

11.2 THE NEXT PHASE IN THE REENGINEERING PROCESS

Implementing ATM-centric, multimedia, broadband networking, as has been stated many times in the preceding chapters, is seen today by both the vendor and user community as the final goal in contemporary networking. In much the same way that SNA networks have provided the commercial sector with sterling service for two decades, ATM networks are envisaged, today, as being able to satisfy nearly all networking requirements for at least the next 15 years. IBM certainly thinks so as testified by its claim that ATM will be as significant to IBM in the next 15 years as S/370s have been in the past.

Obviously, just as with SNA, ATM will, of course, evolve during this period. The amount of bandwidth associated with such networks will even exceed the 622Mbps that is already envisioned. It is even possible that today's sacrosanct 53-byte cell length may even get increased by a factor of two or even four as switching technology gets increasingly faster and is able to switch a 128- or 256-byte cell in the time it takes to switch a 53-byte cell today.

While there is no argument whatsoever that ATM is indeed the endgame, there is still considerable consternation as to exactly when enterprises are going to start embracing ATM in earnest—particularly so when it comes to the innately conservative mission-critical community that makes up the bulk of the IBM-centric networking world. There had been much clamor that the 1994–1995 time frame would see the coming of age of ATM. While there were indeed early adopters of ATM during this period, especially for medical, scientific, and engineering imaging, the overall penetration of ATM technology across the market was relatively small.

Most large enterprises are still evaluating and at best prototyping ATM technology. Toward the end of 1995, the consensus was that widescale adoption of ATM is unlikely to happen, across the board, until 1997. This is significant and encouraging in that this is exactly in line with the ATM migration schedule proposed in Chapter 3. However, all 34,000-odd enterprises that have an SNA or APPN network today are not going to be embracing ATM in 1997 or 1998. The availability of voice support over frame relay, and relatively jitter-free full-motion video support with LAN switching, is also likely to minimize the urgency to adopt ATM. Neither frame relay nor LAN switching can in any way or form match the guaranteed QOS essential for high-quality voice or video applications. However, the availability of a degree of voice and video support outside ATM will enable some enterprises to prototype new multimedia applications prior to adopting ATM.

The adoption of ATM technology will be slow, phased, and systematic—exactly mirroring the way and pace at which all new technology is assimilated by this community, whether it be LANs, bridge/routers, or frame relay. *The bottom line here is that the move to ATM is likely to be as sedate and orderly as has been the move to consolidate parallel networks*. The IBM-centric networking community, saddled with the responsibility of providing high-availability mission-critical networking, is not noted for being early adopters of radically new networking technology. The risks vis-à-vis mission-criticality are much too great.

This delay in the widescale adoption of ATM until 1997–1998 does not and should not have a negative impact on the IBM-centric networking community. If anything, it is a welcome respite given that this community is still in the throes of coming to terms with parallel network consolidation. Now there is adequate time and plenty of motivation to tackle parallel network consolidation with gusto without continually having to look over one's shoulder to see what needs to be done on the ATM front. This delay also effectively eliminates the last valid excuse that a few have clung onto as to why they wish to delay parallel network consolidation. This excuse is that they wished to see what ATM had to offer in terms of seamlessly accommodating both SNA and non-SNA traffic.

The possibility of running parallel SNA and non-SNA networks across an ATM backbone, using one set of ATM Virtual Channels for SNA traffic and another for the non-SNA traffic, was discussed in Chapter 3. What was, however, concluded was that the easiest and most cost-effective way of interfacing remote SNA devices (e.g., LAN-attached PCs, 3174s, or AS/400s) to an ATM network would be through an ATM-capable bridge/router or FRAD. If bridge/routers or FRADs were being used to interface SNA traffic to ATM at the remote sites, the chances are that they would also be used to interface all the non-SNA traffic to ATM as well.

Using a bridge/router or FRAD to realize ATM connectivity in this manner is but an extension of parallel network consolidation. Relative to the bridge/routers or FRADs, ATM as opposed to say frame relay is now being used to transport the SNA and non-SNA traffic alongside each other. The bridge/routers or FRADs will still have to use some sort of consolidation technique, even if it is just plain bridging over ATM, to accommodate the end-to-end routing of the multiprotocol traffic. If such parallel network consolidation in one form or another is inevitable even in the context of ATM, enterprises may as well devote time and energy to successfully realize this first step without slavishly waiting for ATM. In parallel, they might as well explore LAN switching and 100Mbps solutions as cost-effective interim means of enhancing bandwidth at the campus level.

11.2.1 ATM at the Campus Level and Beyond

ATM in general, particularly in the context of commercial enterprises, is likely to first happen at the campus level, as opposed to across WANs. ATM, at 155Mbps and beyond, will be the ultimate panacea for eradicating congestion on LANs and turbocharging intracampus PC/workstation-to-server, PC/workstation-to-PC/workstation, and server-to-server interactions. In most cases, however, ATM will get introduced at the campus level as a high-speed interconnect mechanism between existing internetworking devices such as LAN switches, bridge/routers, and hubs.

As was discussed in Chapter 9, the technology and the products to comprehensively realize campus-level ATM, whether at the desktop or in terms of campus-level switches (e.g., IBM 8260s), is now available from IBM and others. A major bone of current contention is the viability and the need for ATM at the desktop. Today, ATM adapters for PCs/workstations only really support LAN emulation. They do not have native ATM interfaces for data, voice, or video. Consequently, ATM to the desktop, today, is just a means of delivering more bandwidth to individual PCs/workstations. With full-duplex 155Mbps adapters the amount of bandwidth that can be provided to a desktop is considerable and awesome.

However, 155Mbps ATM adapters typically support only new, high-speed buses such as PCI. The bulk of the installed base of PCs and workstations, across the market, does not posses such buses. Instead, most of the PCs that are currently gracing enterprise desktops have ISA or Extended Industry Standard Architecture (EISA) buses. It is unlikely that such PCs will ever be able to fully enjoy 155Mbps ATM. Instead, 100Mbps ATM or Ethernet/AnyLAN may be their limit. This causes a dilemma.

Having to upgrade PCs/workstations in order to be able to utilize 155Mbps can double or triple the overall cost of desktop ATM. A 155Mbps ATM adapter may cost $1,000, but the new PCI bus-based

PC to support that adapter may cost another $3,000. When thousands of PCs are involved across the board, an upgrade of this nature can prove to be very expensive and disruptive. Moreover, most have not budgeted for the PC upgrade given that ATM evangelists, in general, did not spend too much time talking about this aspect of desktop ATM.

If PC upgrades are out of the question for the time being, the only options open are to consider using either 25 or 100Mbps ATM, or 100Mbps Ethernet/AnyLAN. The per port cost of 25Mbps, as shown in Chapters 3 and 9, is no longer compelling, especially since today's 25Mbps ATM adapters support only data—and that using LAN emulation. If anything, 25Mbps data-only ATM even if it goes down to $400 per port is downright expensive compared to today's prices for 100Mbps Ethernet. Note that with LAN switching and full-duplex adapters, 100Mbps Ethernet could be made to deliver 200Mbps FDX bandwidth in the context of PC-to-Server interactions.

There is a possibility that IBM will use a so-called cell-in-frame technology to try to stimulate interest in desktop ATM in general and in 25Mbps ATM in particular. With cell-in-frame, IBM will provide 16Mbps token-ring users with new MAC-level software drivers for their existing NICs. This new software will slice-and-dice the data being sent to the 16Mbps token-ring adapter into ATM-compatible 53-byte cells. The token-ring adapter will now transmit these cells—albeit packaged inside some amount of token-ring MAC/LLC framing. Hence the name *cell-in-frame* to denote that the cells occur only within standard token-ring frames.

Obviously, cell-in-frame (CIF) is somewhat superfluous and redundant to be used in the context of conventional, shared-media LANs. Instead, the only way to gain some advantage of CIF would be to connect all the PCs using this new CIF software to a new breed of LAN or ATM switch. Essentially the switch will have to have special CIF ports that understand the format of CIF and can extract the ATM cells out of the frames. Once the ATM cells are extracted this switch will be able to forward these cells to their eventual destination as if they were any other ATM cell.

CIF can only be rationalized and justified on one and only one criterion. *It does provide a semblance of desktop ATM using existing NICs and wiring*. The ability to use existing NICs was the beauty and the lure of LAN switching. CIF uses that very same rationale—albeit this time in the context of low-speed entry-level ATM. Just as with LAN switching, CIF will also only work when the relevant PCs are connected to a CIF Switch via UTP/STP wiring.

CIF has yet to be unveiled by IBM. It is still at the "heavily confirmed rumor" stage. However, the technology behind it is feasible and demonstrable. All that CIF is doing in the end is segmenting the typical 2,000-byte token-ring frame into a series of 53-byte frames. A key issue that has yet to be resolved is the amount of framing that needs to be used around each cell. Standard MAC/LLC framing will add over 20 bytes of framing per cell. This will cause CIF to have a header-overhead-to-payload ratio of close to 50% (i.e., 1 byte of header per every 2 bytes of payload).

In reality, a standard MAC/LLC frame may not be required. It is even possible, that despite the name, the frame could be dispensed with all together. The use of a nonstandard frame, or even no frame, would be feasible since each PC using CIF will be directly attached to a CIF-specific port on a switch over dedicated media. The CIF frames only go between a PC and a CIF port. Given this self-contained, one-to-one relationships between CIF PCs and ports, any frame or cell format could be used within reason to realize CIF.

CIF at best is a short-term interim measure. It is more like a "suck-it-and-see" trial offer for ATM. In time, possibly post-1998, high-speed 155Mbps ATM to the desktop will become a reality. Not overnight, but on a gradual basis. If CIF does spark the imagination of the masses, there is a likelihood that 25Mbps ATM may enjoy some success, despite its price, as a means of getting true ATM rather than having to compromise on CIF. Prior to 155Mbps desktop ATM becoming widespread, ATM at the campus level will have become commonplace.

11.2.2 The Next Frontier of Networking

By the start of the 21st century, 155Mbps ATM will be the de facto fabric for campus backbone. Whether 622Mbps ATM will be pervasive by then is open to question given current slow ramp-up to ATM. (When evaluating potential technology penetration time lines for ATM it is always salutary and sobering to recall that X.25, ISDN, frame relay, and parallel network consolidation all took longer to make their indelible mark than had been initially predicted and expected.)

By 1998, public ATM WAN services should be competing against frame relay services. By 1996, IBM, Cisco, Newbridge, and Northern Telecom, among others, will have ATM WAN switches that could be used to implement either private or public ATM-cum-frame relay networks. As was discussed in Chapter 3 and 9, there will be some large enterprises that opt to implement their own in-house ATM networks as the optimum way of guaranteeing QOS, maintaining desired security levels, and possibly even containing costs. The current indications are that most enterprises will in time elect to use public ATM services rather than incur the costs and headaches of trying to administer and manage a largescale in-house network.

The bottom line is that ATM will be a reality. ATM will usher in a new age of broadband, multimedia, multiprotocol, multivendor networking. The migration to ATM will be slow and systematic. Widescale, widespread ATM, particularly in the mission-critical community, is now unlikely to be here much before 1997–1998. Even then, it will mainly be at the campus level in the form of a high-speed backbone for interconnecting bridge/router, hubs, and LAN switches. In the meantime, frame relay and LAN switching will be the predominant networking technologies that will help enterprises enhance their networking bandwidth and possibly even integrate some voice and video traffic alongside their data traffic.

While waiting for ATM, most if not all enterprises will gainfully exploit today's promising frame relay, LAN switching, and bridge/router technology to successfully complete the first part of their networking reengineering data, that is, consolidating their parallel data networks and enhancing the overall network bandwidth. That alone will constitute a major and noteworthy achievement that results in a new generation of multivendor, multiprotocol, mid-range bandwidth networks.

These new networks will be capable of effectively and efficiently supporting both existing mainframe-centric mission-critical and new-wave Unix-centric client-server applications. With the data side thus adequately dealt with, the reengineering process can then set about in earnest to extend the bounds and fabric of this network so that it can also agilely accommodate voice and video traffic. The overall technology (i.e., ATM), viable products, and even some architectures (e.g., NBSS) for this exciting phase of the reengineering process rather than being just "pie-in-the-sky" are already available in nascent, tantalizing, but tangible form. There really can be no more excuses for delaying network reengineering. The time is ripe and the technology is here. Capitalize on this unique opportunity to be a part of a bold and new networking frontier—and do not look back.

APPENDIX A

Essential SNA in a Nutshell

Systems Network Architecture (SNA) is not a product. It is not possible to call up IBM, or anybody else for that matter, and just order "SNA." SNA, as unequivocally stated by the last word in its title, always has been, and always will be, an architectural design specification à la the OSI 7-Layer Reference Model. The overall SNA specification, not counting APPN and HPR, runs to over 5,000 A4 printed pages and is contained within 16 separate IBM manuals. (See Bibliography.) Various functional subsets specified by the SNA architecture are implemented by or within products such as: ACF/VTAM, ACF/NCP, 3174 microcode, OS/400, PC/3270, NetWare for SAA, or even Cisco's PU Concentration facility.

The original SNA specification, then provocatively and possibly presciently named as *Single* Network Architecture, came out in mid-1974. IBM, which was in the midst of an antitrust embroilment changed the name to *system*, since *single* implied that this would be the only network architecture of any consequence. Little did they then know that SNA would indeed go on to become the single most important architecture in commercial networking for the next two decades. The author, who while working for IBM first heard of SNA as a result of this name change, has always jokingly maintained that the *single* only referred to that fact that the first three versions of SNA were only ever capable of supporting networks that contained one mainframe. The landmark support for multiple mainframes only came with the fourth version in 1977. The key phases of IBM's evolution from 1974 to date are described in Table A.1.

TABLE A.1 The Key Phases of SNA's Evolution Since its Inception in 1974

Phase	*Period*	*Salient Characteristics*
SNA 0	1974	One mainframe; one channel-attached 37xx; leased lines.
SNA 1	1975	One mainframe; two 37xxs –> one channel-attached, other remotely attached over one leased SDLC link.
SNA 2	1975	One mainframe; multiple channel-attached 37xxs –> each of which could support one remote 37xx switched links to peripheral nodes.
SNA 3	1977	Multiple mainframes.
SNA 4	1978	Fully interconnected mesh networks. No restrictions on the deployment of remote 37xxs. Transmission Groups
APPC	1982–1983	LU 6.2 and Type 2.1 peer nodes.
SNI	1984	Free and transparent interoperability across autonomous SNA networks.
Peer	1987–1989	Tight integration between the T2.1 peer mode and the traditional hierarchical mode of operation to permit transparent interoperability.
Dynamic	1989–1991	Dynamic resource definition and network reconfigur ation. Also XRF for Fault Tolerant networking.
APPN	1992–1994	Mainframes and 3746s can act as true APPN NNs.
HPR	1995 –>	Advent of HPR as the heir elect to both SNA and APPN.

The core SNA specification, incisively and with considerable panache, deals with the functions and protocols necessary to implement efficient, highly reliable, secure, and feature-rich mainframe-centric distributed data processing environments. Extensions to this core specification deal with issues such as peer-to-peer networking, program-to-program communications, and data distribution. The SNA specifications diligently eschew mentioning any product names or product types. All functional definitions are specified in product independent form. Implementations of SNA functionality within various products by IBM over the years have set the tone for what class of product implements what subset of SNA. The specifications per se do not make any such recommendations. The SDLC, contrary to a widespread belief, is also not part of the SNA specification. It was specified independently of SNA.

A.1 SNA CHARTER VIS-À-VIS END USERS

The term *end user* as it relates to networking was first coined by SNA. SNA categorically states that there can only ever be *three possible sources or destinations* for any and all the data that flows across a network. These three universal sources or destinations of all networking data are referred to as *end users*. The three universal data sources and destinations are: *application programs, terminal operators, and I/O devices*. (*Application program* in SNA's context of an end user essentially refers to all software in general as opposed to just high-level, application-specific software.) Thus, per SNA, an end user is one of these entities.

The various data transfer possibilities between these three classes of end user are as follows:

- program to I/O device
- program-to-terminal operator
- program-to-program
- terminal operator-to-terminal operator
- I/O device-to-I/O device
- terminal operator-to-I/O device

The above six end user pairings in actuality cover all conceivable data transfer configurations that could ever be devised for any type of information processing system. Thus, SNA's definition of end users is intrinsically sound and totally comprehensive.

The mission of SNA was to define a data communications fabric that provided for efficient, reliable, and secure data transfer between end users. Such data transfers would take place irrespective of the physical location of the various end users relative to each other, as well as the exact physical and functional attributes of the various end users. Thus, SNA set out to provide generic and universally applicable data transfer mechanisms that were not in any way tied to the capabilities or characteristics of a particular product or class of products. Thus, for example, when dealing with I/O devices, SNA does not specify one data transfer scheme for disk drives and another for tape drives.

The task of adapting the generic mechanism specified by SNA to the exact capabilities and requirements of a particular product are performed by the SNA Logical Unit (LU) function. (Compare NBBS Access Agents described in Chapter 9.) An LU thus always has two interfaces: a generic, SNA-defined one vis-à-vis the SNA network, and an SNA-independent product specific one that melds the SNA functions with that of a particular product. An LU 6.2 is the one possible exception to this. LU 6.2s only deal with application programs. Thus the product interface of an LU 6.2 is the API it presents to a program. This API, which is essentially a set of calls, can be the same whether a program is running on a mainframe, PC, AS/400, a Unix workstation, or on a printer. Hence, it does not have to be product-specific. SNA can and does define this interface, that is, the API. This API is referred to as the *LU 6.2 Protocol Boundary*. CPI-C is a standard and consistent product- and platform-independent implementation of this protocol boundary.

IBM's preoccupation with LU 6.2-based program-to-program communications can also be easily understood in the context of end users. (Advanced Program-to-Program Communications [APPC] is a marketing term for LU 6.2-based networking.) With the advent of inexpensive microprocessors, IBM realized that the days of totally hardware-based terminals and I/O devices were coming to an end. It correctly foresaw that all peripherals would contain one or more microprocessors and would be driven and controlled by software. If all peripherals were going to have software, there was going to be no need to support I/O device and terminal operator class end users. All data transfers vis-à-vis a network would originate from a program and be destined to a program.

Voila! Of the six end user pairings listed above, SNA, in theory, could forget about five of them and just devote all its attention to program-to-program interactions. *This was the dream of LU 6.2: a world where all end users were programs.* Unfortunately, things did not exactly work out the way that IBM had envisaged. The world did truly embrace microprocessors, and all contemporary peripherals are software-driven. However, partly due to IBM's lack of leadership in this respect in the mid-1980s, the SNA software used in these microprocessor-based peripherals was not based on LU 6.2.

Take PCs as a quintessential example of a software-driven peripheral. When IBM was first writing SNA software for PCVs it should have used LU 6.2. It did not. Instead, it wrote 3270 emulators for the PCs. This was expedient and ensured mindless, totally transparent backward compatibility with mainframe applications. But it derailed the cause of LU 6.2. *What IBM really should have done was to develop LU 6.2-based 3270 emulation.* LU 6.2 program-to-program interactions would have been used between the mainframe and the PCs. At either end, the program talking to the LU 6.2 would do the necessary protocol conversion between LU 6.2 data flows and 3270 datastream. Ironically, nearly a decade after it should have done it in the first place, IBM now does offer exactly this solution under the name APPC/3270 (A3270). But it is too late. The initiative was lost and IBM to date is still paying the price in terms of APPN's lack of success within the mainframe community. APPN only natively supports LU 6.2. Today, it can only support

3270 traffic using DLUS/DLUR technology. If IBM had migrated 3270 traffic to LU 6.2 in the late 1980s via an A3270 type solution, APPN would not be in the doldrums facing an uncertain future.

A.2 THE BASIC BUILDING BLOCKS OF SNA

All SNA networks consist of a set of functional entities, each with its own unique address, that are interconnected to each other via a path control network. These uniquely addressable functional entities are known as *network addressable units* (NAUs). Thus, each and every SNA network consists of a set of NAUs interconnected via a path control network. (In APPN, NAUs are known as network accessible units, since APPN unlike SNA does not rely on assigning fixed addresses to all of its functional entities.)

SNA only defines three types of NAU, which are:

1. System Services Control Point
2. Logical Unit
3. Physical Unit

SSCPs are the absolute masters of mainframe-centric SNA networks. *Traditional, mainframe-centric SNA networks must contain at least one SSCP*. Each mainframe in such a network would contain an SSCP. This is not negotiable. If a mainframe wants to be a part of an SNA network it must have an SSCP. The SSCP function within a mainframe is implemented by IBM's ACF/VTAM software. Today, SSCPs are also found within PU Concentrator Gateways. Consequently, SSCPs can now be found inside Hypercom FRADs and Cisco bridge/routers.

An SSCP is responsible for providing centralized *Directory* (i.e., name to address translation), *Configuration* (i.e., knowing the location of all the resources), and *Management Services*. The set of resources controlled by a given SSCP is referred to as the *domain* of that SSCP. The resources that make up a domain consist of LUs, PUs, path control network components, and physical links. If a network only has one SSCP (i.e., one mainframe) it is said to be a single domain. Networks with multiple SSCPs (i.e., multiple mainframes) consist of multiple domains. Thus, the term *cross-domain* is sometimes used to describe interactions that take place in networks with multiple mainframes.

The key functions performed by an SSCP are as follows:

- Knowing the SNA names assigned to all the LUs, PUs, and links. (Such names are assigned as a part of the ACF/NCP or ACF/VTAM definition process.) An SSCP automatically assigns SNA network addresses to these resources each time the network is activated. SNA end users only use names to specify the partners they wish to interact with. Humans never use network addresses when dealing with or using SNA. SNA only expects system control functions and system software to know and understand network addresses. This is a fundamental and major difference between SNA and TCP/IP. Given that end users do not know or use network addresses, the SSCP has to get involved each time an end-user-to-end-user session needs to be established so that it can perform the necessary name-to-address translations. This is the so-called Directory Services function.

- Keeping track of the location (e.g., at the end of what link) of all of its resources individually and with respect to each other
- Activating and deactivating the resources within its domain: An SNA resource cannot do any productive work until it is explicitly activated by an SCCP. When an SNA network is started up, every SSCP in the network issues explicit Activate PU (ACTPU) and Activate LU (ACTLU) commands to each and every PU and LU in its domain. DACTPUs and DACTLUs are issued when a network is being shut down.
- SSCPs have to actively intervene in the establishment of all sessions involving LUs and PUs in its domain. A PU can only talk to its controlling SSCP, that is, the SSCP in charge of the domain that the PU is resident within. A PU talks to its controlling SSCP via an *SSCP-PU session*. SSCPs are responsible for establishing SSCP-PU sessions with all the PUs in its domain. SSCP-PU sessions are initiated via the ACTPU command. LUs, in addition to talking to their controlling SSCP, also talk to other LUs. These LU-to-LU interactions are realized via *LU-LU sessions*. LU-LU sessions, in mainframe-centric networks, can only be set up with the help of one or more SSCPs. This need for SSCP intervention even applies to LUs within Type 2.1 peer nodes, if the T2.1 node is being controlled by a mainframe.
- Acting as the Management Focal Point for its domain and collecting all the relevant network management data: This data is typically obtained over SSCP-PU sessions. This is the reason, as described in Chapter 10, why NetView/390 has to be an ACF/-VTAM application program. Given that ACF/VTAM implements the SSCP function, all network management information first come to it. NetView/390 has to get this data from ACF/VTAM.
- Providing the operator interface through which system operators can monitor and manage the network

Next to the SSCPs, LUs are the most important NAUs within an SNA network. *LUs are the ports through which end users gain access to the SNA network in order to be able to communicate with other end users*. In today's terminology, LUs can be thought of as providing the *adaptation layer* between the SNA network and a given end user. LUs perform the following functions necessary to initiate and sustain end-user-to-end-user communications:

- Initiating and establishing the prerequisite LU-LU sessions across which all end user interactions are performed
- Methodical transfer of data across LU-LU sessions, replete with built-in DES algorithm-based end-to-end, LU-to-LU *cryptography* if needed
- Any necessary data transformations, that is, presentation services, to ensure that data being transmitted corresponds to the format and style expected by the receiving end user
- Data flow control and *pacing*. Pacing is SNA's primary mechanism for controlling congestion. With pacing, a data recipient can control the amount of data being sent to it. It does so using a *pacing window* scheme. A pacing window, which is a number such as 4, corresponds to the number of packets, of a certain maximum size, the recipient will accept within a given transmit cycle. Once the transmitter has sent the number of packets corresponding to the window size, it cannot send any more until it receives a pacing response from the recipient. It can then send another pacing window worth of packets.

  Pacing is performed, independently, on the data flowing in each direction of a session, with each side able to use its own window size. Traditionally, a fixed pacing window size for each end of a session was established during LU-LU session initiation. This scheme for obvious reasons is referred to as *fixed size pacing*. LU 6.2s use a variable window size pacing scheme known as *adaptive pacing*. With adaptive pacing, the pacing response issued by the recipient to give the go-ahead for the next round of packets transmittals specifies the window size for that transmit cycle. *Dynamic pacing* is used on SNA Virtual Routes where the pacing window size can be continually and dynamically adjusted between a minimum and maximum value to ensure maximum throughput relative to various degrees of network congestion.

An end user cannot gain access to SNA without going through an LU. SNA, however, does not specify how many end users a given LU may support. Most traditional, non-LU 6.2 implementations, use one LU per end user. Thus, a 3x74 control unit supporting coax-attached 3270 terminals would typically have at least one LU per coax port. An

LU 6.2, however, will typically support multiple end users—each of which will be an application program.

PU is the most misunderstood and misused term, without a shadow of a doubt, in all of networking. *An SNA PU does not refer to a physically tangible entity*. Thus, a 3174, a 3745, or a mainframe is not a PU—let alone a PU-2, a PU-4, or PU-5. Take as an analogy a driver operating a car. In the context of SNA the car is not a PU; the driver is.

An SNA PU is responsible for controlling, supporting, and managing the physical entities, in particular the data communications links, that are found within the network. *Thus, a PU rather than having a palpable physical form that can be touched and felt is instead invariably just a piece of software resident within an SNA node.* All SNA nodes, with the exception of Type 2.1 peer Nodes and APPN Nodes, contain a PU. (In T.21 and APPN nodes the PU functions are performed by the Control Point.) The PU within a given node has total responsibility for controlling and managing all of the SNA Path Control Network-related components used by that node—in particular all of the data communications links (e.g., SDLC link, channel-interface, LAN-interface) attached to that node. Only one PU can be resident within any given node.

The various types of SNA node have very different Path Control Network (PCN) capabilities and requirements. For example, Type 4 subarea nodes used within 37xx communications controllers need to be able to support large numbers of links as well as a wide variety of different link types: SDLC, frame relay, LAN, channel. In marked contrast, Type 2 nodes, implemented within typical peripheral devices such as a 3x74 control unit, usually only need to support one link. Type 5 nodes, found in mainframes, may have to support multiple links but they will all typically be channel connections.

To cater to the different PCN requirements of each node type, SNA defines a different PU type for each. Thus, a Type 5 node will contain a PU Type 5 (PU_T5), a Type 4 node PU_T4, and a Type 2 node a PU_T2. The misnomer of referring to SNA nodes and the devices within which they are installed as "PU Type this, that, and the other" came about as a result of this PU Type notation where a PU_T*x* is found with a Type *x* Node.

Some of the key functions performed by a PU are as follows:

- Activating, deactivating, and reconfiguring the PCN components (e.g., links) attached to its resident node under the control of an SSCP

  When a device containing an SNA node (e.g., 3174) is powered up, at least one of its links needs to be activated in order for the mainframe-resident SSCP to gain access to the PU and activate it with an ACTPU command. From an SNA architectural standpoint, a PU, however, only activates or decativates links at the explicit behest of an SSCP. But an SSCP can't reach a PU until at least one link is activated between them. To cater to this link activation scenario at device power-on, each non-Type 5 node (i.e., nonmainframe) contains an entity known as a *physical unit control point* (PUCP). At power-on, the PUCP masquerades as an SSCP and instructs its coresident PU to activate a specified link in order to open up a communications path with the mainframe.

- Monitoring and managing the PCN components of its node and notifying the SSCP of any status changes
- Providing SSCP with management information related to its node—including LU-related information such as the response time readings
- Activating or deactivating the node it is resident in at the behest of an SSCP
- Loading, where applicable and necessary, the control software associated with a Type 2 or 4 node. (This software download could either be over a link or from a diskette.)
- Obtaining, upon request from an SSCP, a complete storage dump for a Type 4 node and sending it to the mainframe for debugging purposes

An LU 6.2 consists of the following three distinct components:

1. A set of LU 6.2 functional characteristics
2. The LU 6.2 Protocol Boundary
3. A set of LU 6.2 Service Transaction Programs (STPs)

The quintessential functional characteristics of an LU 6.2 can be summarized as follows:

- Support for multiple, concurrent end users
- Formal support for multiple, simultaneous LU-LU sessions with other LUs, as well as Parallel Sessions, that is, multiple, active sessions between the same pair of LUs
- Activating or deactivating the node it is resident in at the behest of an SSCP
- Support of end-user-to-end-user *conversations*. Prior to LU 6.2, the extent of IBM's end-to-end connections were LU-LU sessions. LU-LU sessions, however, terminated at the LUs at each end. They did not embrace the end users. LU 6.2 conversations, on the other hand, do not end at the LUs but provide total end-to-end coverage. LU 6.2s,

by definition, only support application program end users. Thus, LU 6.2 conversations only occur between application programs.

- LU 6.2 sessions and conversations are intrinsically peer-to-peer. Consequently, T2.1 and APPN nodes are the optimum platforms for LU 6.2s, even though they can be implemented within other node types.
- LU 6.2s provide two levels of resource (e.g., transaction or database) *synchronization* support. The first is a mandatory, basic level, known as CONFIRM support. With CONFIRM, end user partners can explicitly signal the successful completion of a specified unit of work (e.g., complete processing of a transaction that involves the updating of a database). The CONFIRM protocol is executed, on an end-user-to-end-user basis, at the Application Layer, independent of all the lower level responses and acknowledgments being used by SNA and the Data Link Control protocol (e.g., LLC:2).

  The second, optional level of synchronization is known as *two-phase commit checkpointing*. This mode of synchronization is a prerequisite for transactions that involve the need to update distributed databases, such as debiting money from an account in one database and crediting it to an account in another. In such transactions involving changes to multiple databases, it is imperative that any changes made to a database can be completely undone if the transaction fails at some point. For example, in the above scenario if the transaction could not update the second database to credit the relevant account, it is obligatory that it reverses the debit operation on the first database. Otherwise, money withdrawn from the first account would disappear into a "black hole" causing much consternation at a later date.

  With two-phase commit, LU 6.2s take an active roll in ensuring that resources are checkpointed prior to any updates being made and that there is a failsafe mechanism whereby the status of a resource can be restored to the same state that it was in prior to a given update. This is one of the most powerful and endearing facilities of LU 6.2.

The LU 6.2 Protocol Boundary is defined in terms of a series of verbs. Some of the key verbs that make up this boundary are as follows:

ALLOCATE: Issued by an application to establish a conversation with a designated partner application

BACKOUT: Used in resource synchronization scenarios to roll back resources (e.g., database) to the prior checkpoint level

CONFIRM: Used in resource synchronization scenarios to seek "everything is OK" status from partners before committing to a new checkpoint. A backout will result if the response to a CONFIRM is negative.

CONFIRMED: A positive response to a CONFIRM.

DEALLOCATE: Deactivates conversations.

FLUSH: LU 6.2s always work in store-and-forward. Thus, there is no guarantee that data forwarded to an LU 6.2 by an application it is servicing is immediately sent across the network to its eventual destination. Typically, an LU 6.2 may wait until either a set of buffers is full or an application completes a send cycle and reverts to receive mode before it forwards the data it has received to that point across the network. An application can issue FLUSH to force data that has been previously forwarded to be sent across the network.

GET_ATTRIBUTES: Used by an Application to determine the characteristics of its underlying LU-LU session

POST_ON_RECEIPT: Requests an LU 6.2 to reactivate the application when it receives any data or status associated with a specific conversation

PREPARE_TO_RECEIVE: Signals that an application has finished a transmit cycle and is now entering receive mode

RECEIVE_AND_WAIT: Application is ready to receive any data that the LU 6.2 may have already received from the other side. If there is no data pending, the application will wait for data to arrive.

RECEIVE_IMMEDIATE: Application will accept any pending data from the other side. If there is no data as yet, the application will not wait for any to arrive. Instead, it will resume other processing activities and come back at a later point to see if any input has arrived in the interim.

REQUEST_TO_SEND: Used to seek change of transmit direction on a conversation

SEND_DATA: Used to transmit data across a conversation. Data forwarded via a SEND_DATA may not immediately be transmitted across the network. Instead, it can get queued up at the local LU 6.2, until a PREPARE_TO_RECEIVE or a FLUSH is issued.

SENT_ERROR: Used to convey error status

SYMCPT: Used in two-phase commit resource synchronization scenarios to establish a new checkpoint before any subsequent processing is attempted on a transaction

TEST: Used by applications to determine if a specified conversation has any change of status information posted against it,. The reception of a REQUEST_TO_SEND from the other side would be an example.

WAIT: Notifies local LU 6.2 that the application is entering a wait state pending data or status information on one or more designated conversations

LU 6.2 conversations come in two forms: b*asic conversations* and *mapped conversations*. Mapped conversations are geared for applications written in high-level languages (e.g., C) and are aimed at customer- and third-party-developed LU 6.2 applications. Mapped conversations permit any user-defined record types of any arbitrary format to be interchanged between applications. Thus, an application can send data to an LU 6.2 in any format it wants. Basic conversations, on the other hand, are a lower-level interface to an LU 6.2. With the basic conversations, the LU 6.2 will only accept data in one format—the so-called Logical Record format where each record needs to be prefixed with a 2-byte record-length field.

The LU 6.2 STPs refer to IBM-supplied utility programs that can be used to augment the basic networking functions provided by an LU 6.2. For example, the software modules used to realize LU 6.2 resource synchronization or to change the number of parallel sessions between two LUs are considered to be STPs. IBM also positions services such as SNADS, DDM, and SNA/FS as being LU 6.2 STPs.

A.2.1 SNA Nodes

Forget PUs. All SNA networks are built around SNA nodes. *An SNA nodes is the basic unit of functionality within SNA*. SNA functionality cannot be provided on a device without implementing an SNA node within that device. This is also not negotiable. SNA NAUs, (i.e., LUs and PUs) can only reside within SNA nodes.

An SNA node consists of one or more NAUs and a set of PCN components. Nodes are interconnected to each other to form an SNA network by means of physical links supported by the PCN. Figure A.1 illustrates the fundamental relationships between end users, NAUs, PCN and nodes.

SNA defines five types of node. These five node types are: Type 5, Type 4, Type 2, Type 2.1, and Type 1. (There never has been and never will be a Type 3.) The presence and absence of certain NAU types and the functional capabilities of the PCN components are the two factors that determine SNA node type. Type 1 and Type 2 nodes are known as

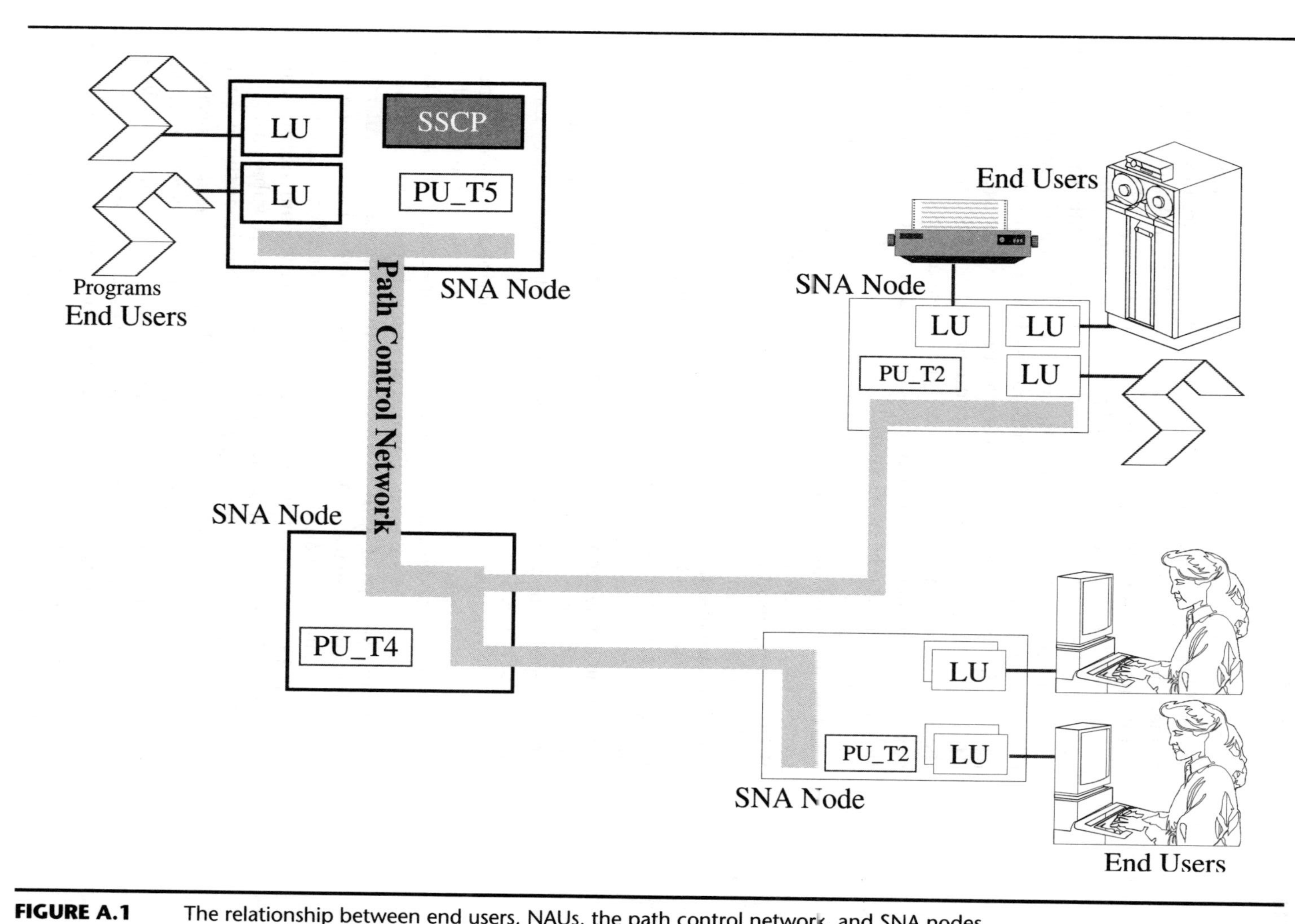

FIGURE A.1 The relationship between end users, NAUs, the path control network, and SNA nodes.

peripheral nodes, while Type 4 and Type 5 nodes are known as *subarea nodes*. Peripheral and subarea nodes are differentiated by their interconnection and message-routing capabilities. Figure A.2 depicts where the various SNA node types will occur within a typical SNA network.

Type 2.1 nodes are the original peer-to-peer nodes that were the precursors to APPN. Type 2.1 nodes and the LU 6.2s within them can operate independent of and outside the control of SSCPs. Consequently, LU 6.2s resident with T2.1 and APPN nodes that can function in peer-to-peer mode without SSCP intervention are known as *Independent LUs* (ILUs) to denote their SSCP-independent status. With the exception of the T2.1 node, the other four node types were all defined in 1974 and have retained their original composition and interconnection capabilities throughout the ensuing years.

Each Type 1 or Type 2 peripheral node in an SNA environment must be attached to a subarea node. A peripheral node that is not attached to a subarea node does not have an identity or a role within an SNA. A peripheral node cannot be directly attached to another peripheral node, nor can a peripheral node be concurrently attached to more than one subarea node at any one time.

A subarea node may be concurrently interconnected to several other subarea nodes via the PCN. A subarea node can also support multiple, simultaneously active SNA links. The subarea-node-to-subarea-node connections do not preclude, or limit, a given subarea from being able to have multiple peripheral nodes attached to it. A subarea node supports the peripheral nodes that are attached to it by means of a boundary function component.

A subarea node along with all the peripheral nodes that are attached to it is known as a *subarea*. If a subarea node does not have any peripheral nodes attached to it, that subarea node on its own is still considered to be a subarea. The notion of a subarea is an intrinsic and important concept of SNA. SNA addressing and message unit routing revolve around subareas. An SNA subarea is somewhat analogous to a TCP/IP subnet. *A subarea is always created around each subarea node of an SNA environment*. Thus, a typical SNA environment involving at least one S/370 host will be composed of one or more subareas.

NAUs resident within any SNA node can be either the originators or recipients of SNA message units. *The PCN components of peripheral nodes are, however, endowed with little or no message-routing capabilities and cannot in any form or shape perform SNA subarea-to-subarea routing*. In general such a component can only transmit or receive SNA message units to or from NAUs resident in its peripheral node and an adjacent subarea node.

Since peripheral node NAUs can only interact with other NAUs

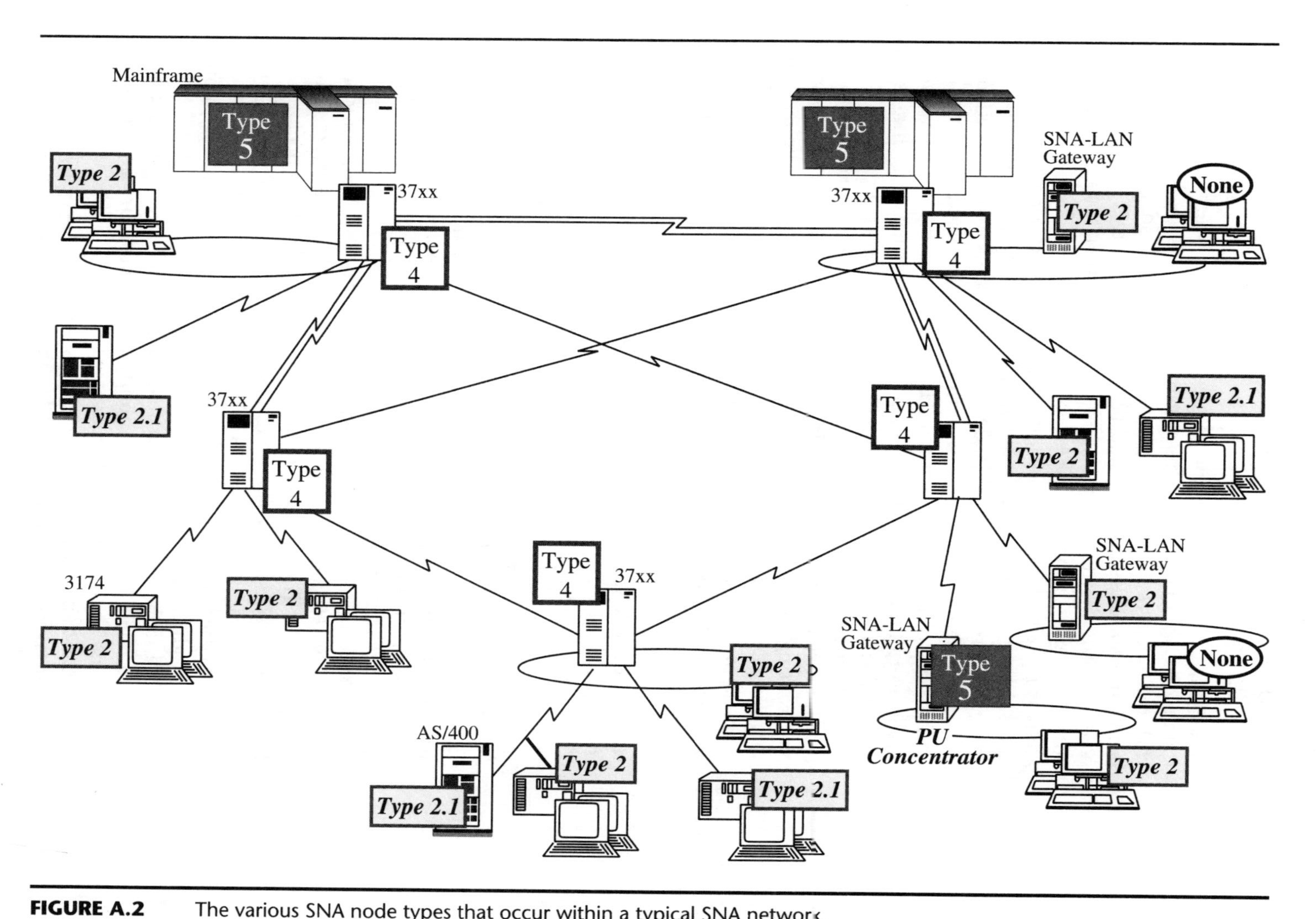

FIGURE A.2 The various SNA node types that occur within a typical SNA network.

through a subarea node they do not need to have an overall awareness of the unique SNA network addresses mandatorily assigned to each NAU within a network. Instead they use a simple local addressing scheme between themselves and the adjacent subarea node and let the subarea node perform the necessary address transformation between the local addresses and the global SNA network addresses. The use of local addressing simplifies the design and implementation of peripheral nodes. Thanks to local addressing, peripheral nodes can be designed with standard, preassigned local addresses that can be used unchanged within any SNA environment. In addition, these local addresses are fixed and do not need to be changed or updated to reflect any reconfigurations in the overall topology of an SNA network. This obviously facilitates the installation, reconfiguration, and maintenance of peripheral nodes.

The path control component of a subarea node, in contrast to its counterpart in a peripheral node, does have extensive SNA message-unit-routing capabilities including the ability to perform subarea-to-subarea SNA routing. The routing capability of a T2.1 node PCN component is similar to that of a peripheral node. *It can only forward or receive data from logically adjacent nodes. It cannot perform intermediate node routing*—the ability to receive message units from one node attached to it and then forward those to another node. This is where APPN NNs come in. A T2.1 node that is capable of doing intermediate node routing is called an APPN NN.

All SNA nodes, with the unique exception of T2.1 nodes, must contain a PU. T2.1 and APPN nodes do not contain a PU. Instead they contain a CP. A CP provides its node with Configurations and Directory Services. All SNA and APPN nodes are capable of supporting LUs. However, a traditional Type 4 node that is to be implemented with 37xx's will not contain any LUs since it does not have to directly support any end users. Instead the end users will be supported by the Type 2, 2.1, and 5 nodes that occur around a Type 4 node.

A Type 5 node has to contain an SSCP. An SSCP can only occur within a Type 5 node. It is the presence of the SSCP that makes a node a Type 5 node. The presence or absence of an SSCP is the primary distinction between the Type 5 and Type 4 subarea nodes. A Type 5 node cannot contain multiple SSCPs. Table A.2 tabulates all the pertinent characteristics of the five SNA node types and also lists the quintessential product implementations for each node type.

A T2.1 node can support multiple, concurrently active attachments. The forte of T2.1 nodes is their ability to establish direct peer-to-peer connections with other T2.1 nodes without any SSCP intervention. These peer-to-peer connections could be to other T2.1 nodes, subarea nodes, or APPN nodes. A T2.1 node, like a subarea node, can support

TABLE A.2 **Characteristics of SNA and APPN Node Types**

Node Type	*Classification*	*Contains*	*No. of Links*	*Routing*	*Quintessential Implementation*	*Status*
Type 5	Subarea Node	SSCP, PU_T5, LUs—0 to thousands	Many	Subarea-to-Subarea	ACF/VTAM (mainframes)	Active
Type 4	Subarea Node	PU_T4, LUs—0 to thousands, but typically 0	Many	Subarea-to-Subarea	ACF/NCP (mainframes)	Active
Type 2.1	Peer Node	CP, LUs—0 to thousands	Many	None	S/38, APPC/PC	Not Strategic
Type 2	Peripheral Node	PU_T2, LUs—1 to 255	1	None	3x74s, 3776, 4700	Active
Type 1	Peripheral Node	PU_T1, LUs—1 to 64	1	None	3767, 6670	Obsolete
APPN EN	APPN Node	CP, LUs—0 to thousands	Many	None	AS/400, CM/2	Strategic
APPN NN	APPN Node	CP, LUs—0 to thousand	Many	APPN NN-to-NN	AS/400, CM/2, bridge/routers	Strategic

multiple, simultaneously active SNA links. In marked contrast to a Type 1 or Type 2 peripheral node, a T2.1 node does not have to be attached to a subarea node in order for it to participate in an SNA environment. With T2.1 and APPN nodes it is now possible to construct SNA environments that do not contain any subarea nodes.

A T2.1 node that is, however, attached to a subarea node has the option of electing to appear and act as a bona fide Type 2.1 node, or to masquerade as a mere Type 2 node. When thus attached a T2.1 node is still supported by the subarea node's boundary function component, just as in the case with other peripheral nodes, irrespective of whether the node is acting as a bona fide T2.1 node, or merely as a Type 2 node. Being attached to a subarea node does not in any way preclude a T2.1 node from also being able to concurrently support direct peer attachments to other T2.1 Nodes. It is this peer-to-peer ability, even within the context of subarea networks, that is exploited by LAN-over-SNA solutions and the IBM 2217 to ensure that LAN-to-LAN interactions do not have to be relayed through a mainframe.

In mid-1987 the basic subarea network architecture as well as the T2.1 node addressing scheme were enhanced to facilitate the seamless integration of T2.1 nodes into traditional subarea networks. These changes were incorporated into ACF/VTAM Version 3 Release 2 and

ACF/NCP Version 4 Release 2. Thanks to this T2.1 integration facility, LU 6.2s in T2.1 or APPN nodes can now freely participate in subarea networks, without in any way having to deviate from their native, SSCP-independent mode of operation. They can have multiple sessions, including parallel sessions, with LUs in the subarea network. They can also participate in direct, T2.1-to-T2.1 or APPN-to-APPN sessions across the subarea network. In the case of such SSCP-independent peer-to-peer sessions, the entire subarea network irrespective of its configuration, complexity, and geographic dispersity appears to the T2.1 nodes as if it were just a simple, point-to-point link between physically adjacent T2.1 or APPN nodes.

A.3 SNA SESSIONS

Sessions are the lifeblood of SNA. NAUs can only interact with each other in the context of sessions; the notion of connectionless interactions is not condoned. End users exchange data with each other across LU-LU sessions. *Consequently, the only and whole purpose of any SNA network is to enable LU-LU sessions to be established and sustained.* SNA defines four types of session:

1. LU-LU sessions
2. SSCP-LU sessions
3. SSCP-PU sessions
4. SSCP-SSCP sessions

Apart from the all important LU-LU sessions, all the other three sessions types are SSCP-related. These SSCP-related sessions are there to help establish and support LU-LU sessions. Figure A.3 shows the various SNA sessions required to sustain a SNA network.

An LU cannot participate in any LU-LU sessions until it has an active SSCP-LU session with its controlling SSCP. When an SSCP activates a given node, it sets about trying to establish SSCP-LU sessions with all the LUs in that node by issuing ACTLUs to them. Once established, the SSCP-LU session stays continually active, even though there may be no data whatsoever going across it, until the LU or node is deactivated. LUs use their SSCP-LU session to request the SSCP to help them establish or terminate a LU-LU session.

SNA terminal operators typically type a "Logon <<application name>>" request (e.g., Logon TSO) to establish a LU-LU session with a given mainframe resident application. The logon request will be transmitted to the SSCP over an SSCP-LU session. The desired characteristics for the proposed session will be specified in terms of a mode name.

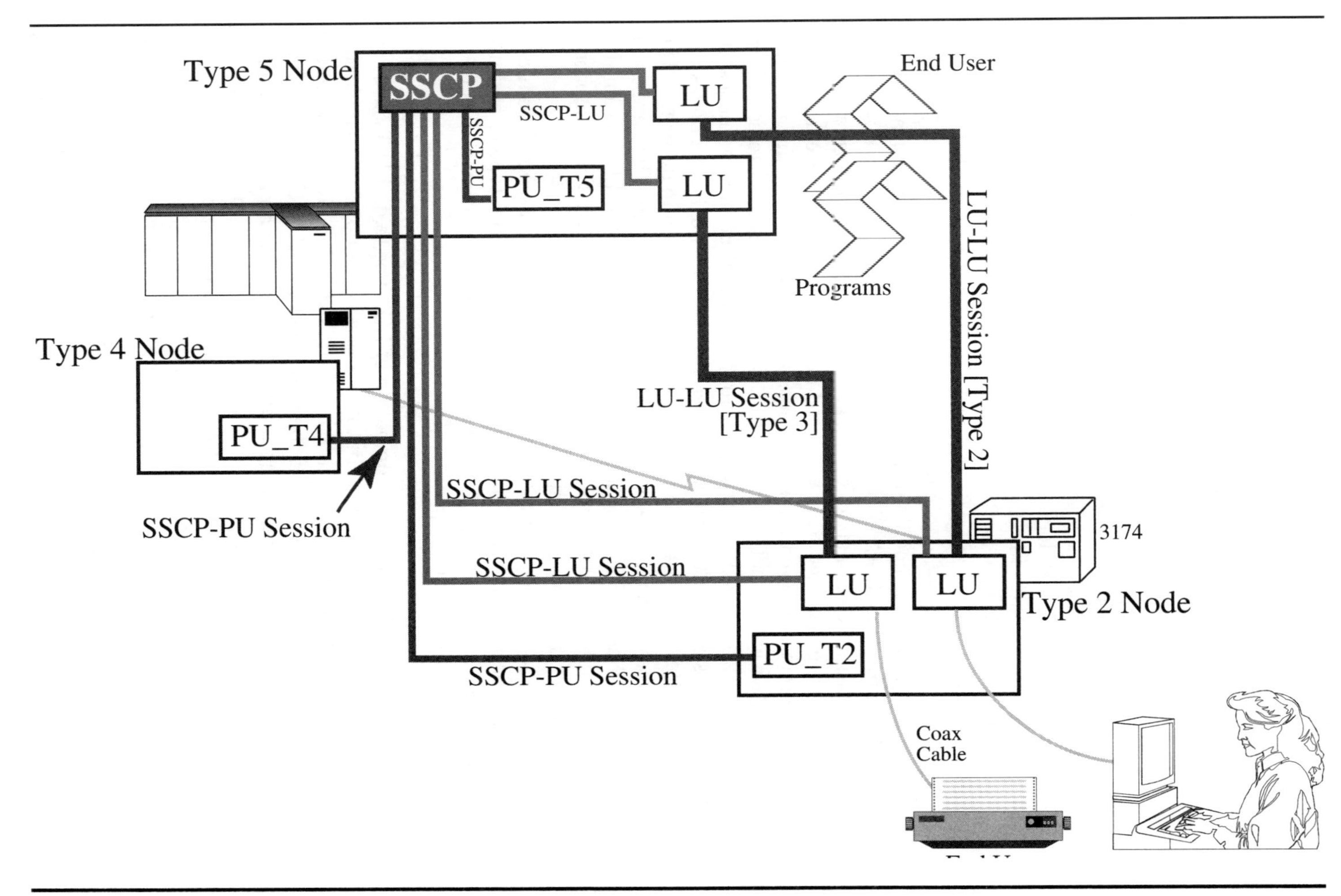

FIGURE A.3 The session infrastructure required to support two LU-LU sessions between a mainframe and one Type 2 peripheral node.

A *mode name* corresponds to an entry within a *mode table*. Each entry in the mode table will specify a different set of session characteristics: for example, full- or half-duplex mode, pacing window sizes, maximum permitted message unit size, message acknowledgment scheme, error recovery responsibility, and so on. The mode name, corresponding to the desired session characteristics, may be either specified as a part of the logon request (e.g., Logon Applid(TSO), Logmode(INTRACT6)), or associated with the logon issuing LU through a default mode name included as a part of its definition to ACF/NCP or ACF/VTAM.

Figure A.4 depicts the simplest, but nonetheless the most used, sequence of SNA requests and responses used to establish a LU-LU session in a single-domain (i.e., one mainframe) SNA network.

The PCN components of a node are activated and deactivated by an SSCP through an SSCP-PU session. The SSCP-PU session is also the conduit across which all network management requests and information (e.g., alerts) are exchanged. It is also used to load control software into a node or to obtain storage dumps of nodes. The SNA requests and responses associated with the T2.1 integration facility that is used to seamlessly support T2.1 and APPN nodes within the context of subarea nodes also flow across SSCP-PU sessions between the SSCP and Type 4 nodes.

SSCP-SSCP Sessions only occur within multidomain networks. They permit cross-domain LU-LU sessions to be established. They are essentially used as an SSCP-to-SSCP extension to an SSCP-LU session. Figure A.5 builds upon Figure A.4 to show how an SSCP-SSCP session is used to set up a cross-domain LU-LU session.

SNA defines seven very distinct types of LU-LU session. The fundamental distinction between these session types is the type of end user served by each. Table A.3 tabulates the salient characteristics of the seven LU-LU session types.

In general, with the exception of LU 6.2, there is no such thing as an LU Type in SNA. Like PU, the notion of LU Type (e.g., LU Type 2) is a misnomer, the reason being that it is a session as opposed to a particular LU that has a type associated with it. Table A.3 illustrates that an application program is one of the end users for LU-LU Session Types 1, 2, 3, 4, 6, and 7. Consequently, an LU supporting an application program (e.g., CICS, IMS, TSO) may need to support multiple different LU-LU session types: Type 2 for sessions with terminal users, Type 3 for 3270 printing, Type 1 for RJE support, and Type 6 for interprogram communications. Such an LU supporting different types of session obviously cannot have an LU type.

The type of session support by an LU in a peripheral node can also change. For example, take a 3270 printer coax-attached to a 3x74. Some applications use 3270 datastream (i.e., LU-LU Session Type 3)

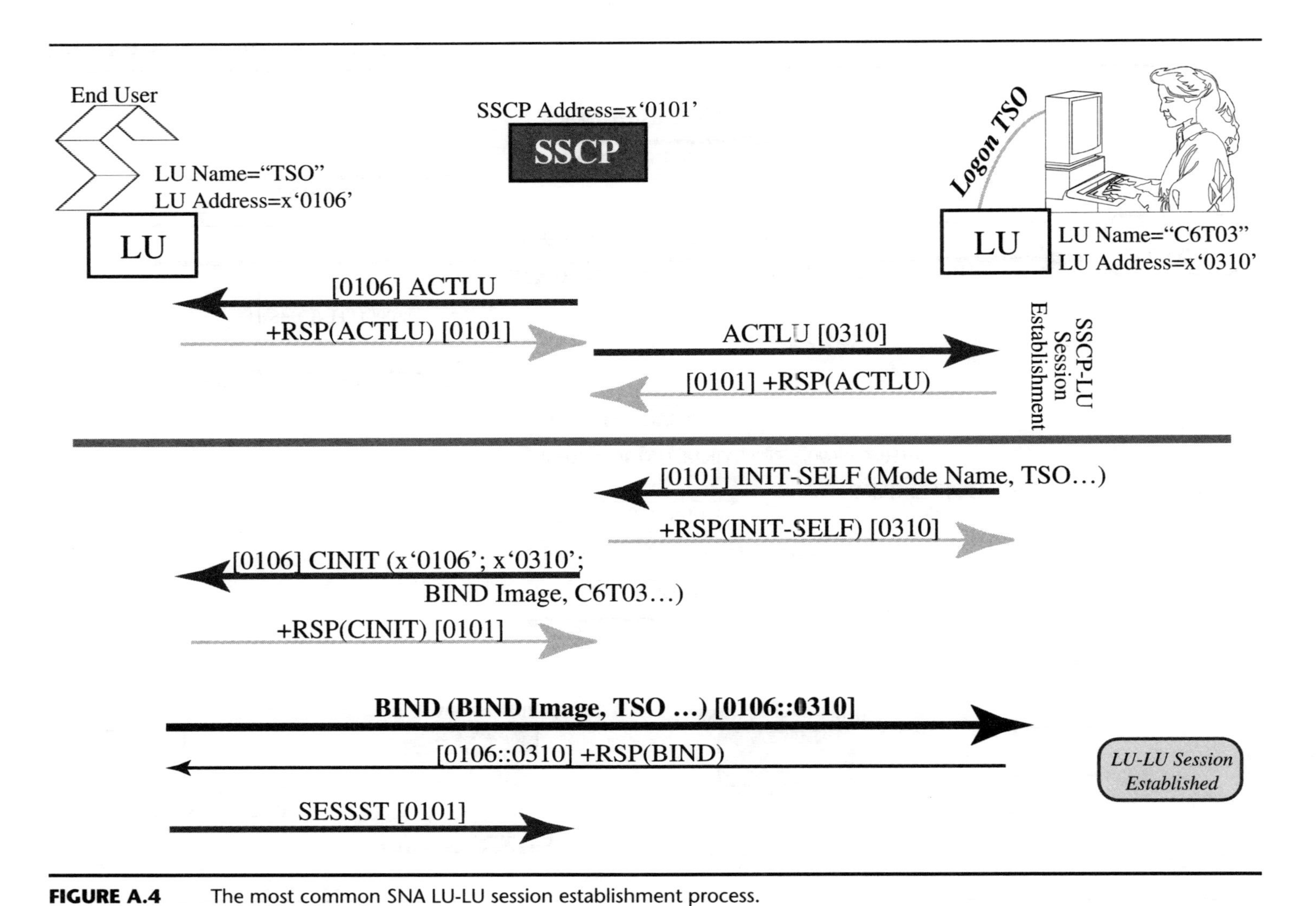

FIGURE A.4 The most common SNA LU-LU session establishment process.

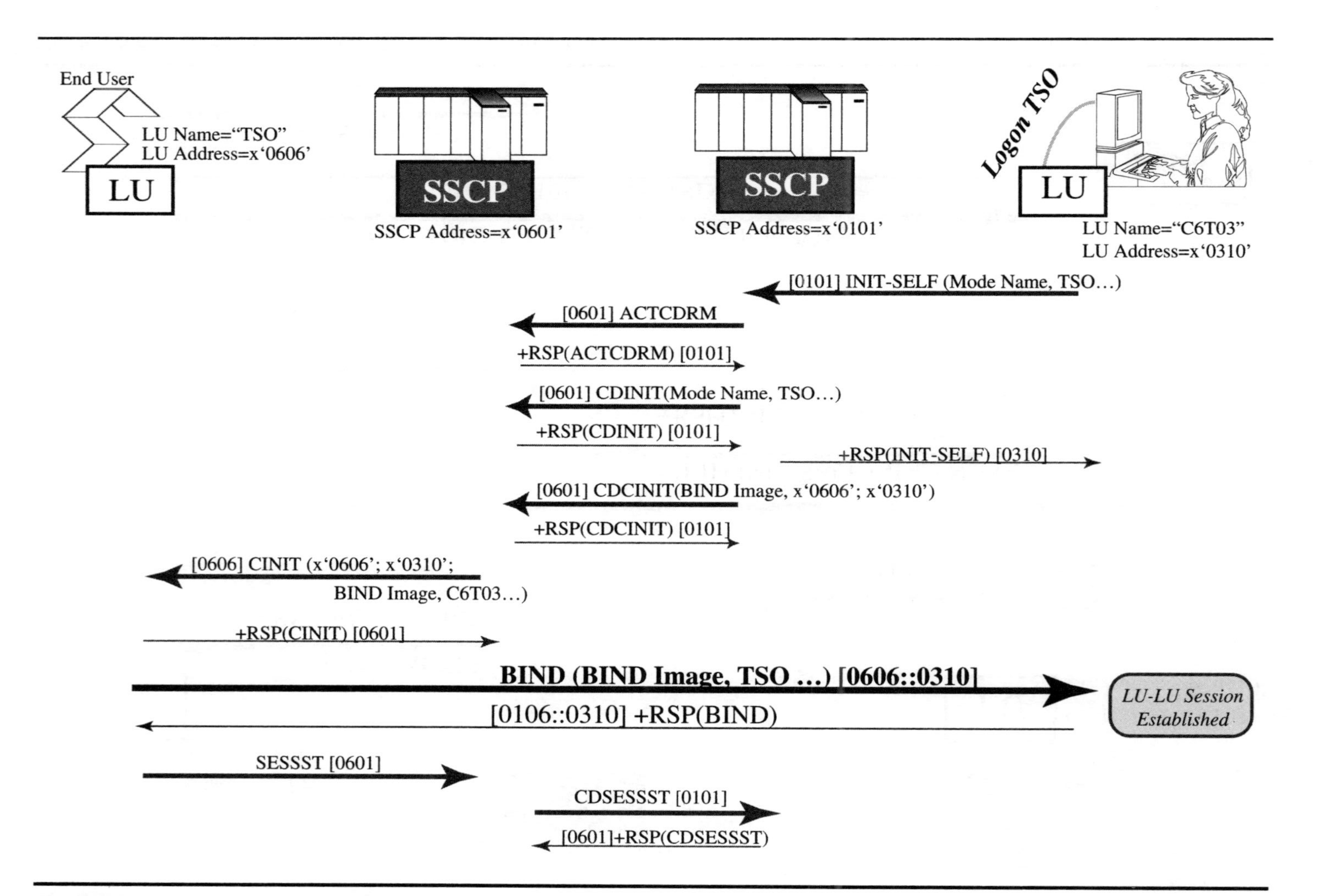

FIGURE A.5 A cross-domain SNA LU-LU session establishment process.

TABLE A.3 Characteristics of the Various SNA LU-LU Session Types

LU-LU Session Type	*The Two End Users Served by the LU at Each End*	*Disposition*	*Data Stream*
0	Any or unknown. SNA does not specify or provide end user services with this session types. Just provides a hands-off, clear-channel end-to-end between two LUs	Typically Master-Slave, but could be peer.	None Defined
1	Application program and I/O device. *(I/O device could be a 3270 printer. Cf. with Type 3.)*	Master-Slave; Application: Master	SCS
2	Application program and 3270 terminal	Master-Slave; Application: Master	3270
3	Application program and 3270 printer	Master-Slave; Application: Master	3270
4 (variant 1)	Application program and Word Processing Terminal or I/O Device	Master-Slave; Application: Master [Now obsolete]	SCS
4 (variant 2)	Two word processing terminals	Peer [Now obsolete]	SCS
6	Two application programs	Peer (Superseded by LU 6.2)	GDS
6.2	Two application programs	Peer	GDS
7	Application program and 5250 terminal (in AS/400, S/38 and S/36 environments)	Master-Slave; Application: Master	5250

to send data to such printer. Others use SNA Character Stream (SCS), that is, LU-LU Session Type 1. The same LU can thus support different session types. Hence, it would be misleading to refer to such an LU by a type. LU 6.2s, however, only interact with LU 6.2s. Thus, they support only LU-LU Session Type 6.2s. Hence, it is permissible to refer to an LU 6.2 as being an LU Type 6.2.

A.4 SNA ADDRESSING AND SNA MESSAGE UNITS

Unique SNA network addresses are assigned to all NAUs and links. Each SSCP and PU is assigned a single address. LUs may be assigned one or more addresses; multiple addresses being required to support parallel sessions. The PCN link stations found at each end of a link are also assigned individual addresses. SNA network addresses, in much the same way as IP addresses, are made up of two distinct parts: a *subarea address* component and an *element address* component.

Each subarea node (i.e., mainframe or 37xx) in an SNA network is assigned a numeric subarea ID. This becomes the subarea address for that node. It also becomes the subarea address component of the network addresses assigned to all the NAUs and link stations that are part of that subarea. Element addresses are assigned to NAUs and link stations on a per subarea basis. An element address is only unique within a given subarea.

Initially SNA used 16-bit network addresses. It also did not specify how these 16 bits should be divided up between the subarea address component and the element address component. (IP tackled this with its Class A, B, and C addressing schemes where each class used a different number of bits for its "netid" and "hostid" components.) How the 16 bits would be split between the two components was determined when an SNA network was being installed based on the "MAXSUBA=" operand, which specified the maximum number of subareas the network was likely to contain.

With SNA-4, in 1978, the addressing scheme was upgraded to support 48-bit addressing—with 32 bits allocated for subarea addresses and 16 bits allocated for element addresses. This was indeed an ambitious scheme, which permitted up to 4 billion subareas per network each with up to 65,536 element addresses. Though the SNA FID-4 and FID-F THs have 48-bit address fields, IBM has never implemented this 48-bit addressing, correctly stating that it would unnecessarily increase the amount of memory needed to hold address tables in 37xx's.

Instead, in 1985 it implemented a 23-bit scheme—8 bits for subarea and 15 bits for element—known as *extended network addressing* (ENA). This 23-bit scheme was implemented as a subset of the 48-bit scheme and still used the 48-bit fields with the THs. ENA was superseded in 1988 with a 31-bit scheme, with 16 bits for subarea and 15 bits for element, known as *extended subarea addressing* (ESA). This scheme, which caters to up to 65,500 subareas, is still the scheme being used today.

The network addresses of peripheral node NAUs are supplemented by local addresses. Local Addresses are only unique within a given peripheral node. All Type 2 nodes within a network will use the same 8-bit local addressing scheme. The boundary function in subarea nodes are tasked with performing network address to local address conversion.

NAUs in T2.1 and APPN nodes are not assigned fixed network or local addresses. Instead they use a dynamic addressing scheme, known as a *local form session identifier* (LFSID), that allocates a specific, but reusable, identifier to each session established over a given link. A single link attached to a T2.1 or APPN node can theoretically support up to 65,536 concurrent sessions relative to that node.

All message units that flow within an SNA network, independent of their source of origin, are classified as being either *requests* or *responses*. A request, in SNA, refers to any message unit that contains either end user data, or an SNA command (e.g., ACTLU), irrespective of the nature of the end user data or the SNA command. A *response* in SNA deviates somewhat from the standard usage of that term, especially when used in conjunction with the term *request*. *In the context of SNA, a response is not the reply to a request.* In SNA, the reply to a request will be in the form of another request.

A response, on the other hand, only denotes the successful delivery of a particular request to its designated destination, and the outcome of the preliminary evaluation conducted by its recipient, as to the viability of processing that request. Thus, a response is, in effect, just a positive or negative acknowledgment of the feasibility of processing a given request. Though serving in part as an acknowledgment of the successful delivery of a request, an SNA response should not however be thought of as being a substitute, or a duplicate, of a data link control protocol acknowledgment.

A data link control protocol acknowledgment only refers to the uncorrupted, and in-sequence, transmission of a data block across a single link. An SNA response is a higher-level, end-to-end interaction, possibly traversing multiple links, which in addition to confirming the safe delivery of a request, also indicates the feasibility of that request being processed. Thus an SNA response is not just the SNA equivalent of a data link control protocol acknowledgment. Each response, in an SNA environment, corresponds to a previously issued request. A request, however, may only have, at most, one response associated with.

A typical SNA message unit, with some small exceptions, consists of the following three discrete components:

1. A Request or Response Unit (RU)
2. A Request or Response Header (RH)
3. A Transmission Header (TH)

An RH is always 3 bytes long and has a single fixed format that is the same for any and all RUs that it prefixes. Bits within this header specify such things as: the SNA category of the RU (e.g., SNA command type or end user), whether the RU is a part of a chain or bracket, and pacing status. (Chaining and brackets are described in the Glossary.)

Whereas there is only one type of RH, there are seven very different TH formats. The TH contains the origin and destination addresses corresponding to each message unit. Figure A.6 depicts the seven THs used by SNA/HPR and specifies the exact conditions under which each type of TH is used. The formats of FID-2 and FID-5 THs are shown in Figure 7.7. FID-2 THs that are used by Type 2, T2.1, and APPN nodes are the most prevalent TH type found in most SNA/APPN networks.

All RUs must be prefixed by a RH. A RU prefixed with an RH is known as a *Basic Information Unit* (BIU). BIUs can be segmented to be sent across data links. Thus for example, a 2Kbyte-long BIU might get segmented into 256-byte segments when being sent across an SDLC link. A BIU, a segmented BIU or a TH by itself is known as a *Path Information Unit* (PIU).

A.5 THE SNA PATH CONTROL NETWORK

The PCN that is responsible for delivering all the message units that flow between the various NAUs in an SNA network consists of a set of PCN components and data link control (DLC) components. The PCN components select and manage the overall, end-to-end path through the network. The DLC components control the transmission of the message units over the physical data links that make up the end-to-end path chosen by the PCN components.

The PCN between subarea node consists of the following three hierarchically layered subcomponents:

1. Virtual Routes (VRs)
2. Explicit Routes (ERs)
3. Transmission Groups (TGs)

VRs are the basis of SNA routing. They are bidirectional, logical connections between two end-to-end subarea nodes that are supporting one or more sessions between them. The sessions being supported between two end-to-end subarea nodes over a given VR may consist of one or more session between either NAUs in the two subareas, NAUs in one subarea, and NAUs in a peripheral node attached to the other subarea, or NAUs in peripheral nodes attached to each subarea.

Each SNA session is assigned to a VR at session activation. A session can only be assigned to one VR at any given time. However, multiple sessions can and typically will share the same VR between two end-to-end subarea nodes. Each VR has assigned to it one of three transmission priorities: high, medium, and low.

Each VR is associated with an underlying ER. An ER can be associated with multiple VRs. An ER is a physical path between two end-to-

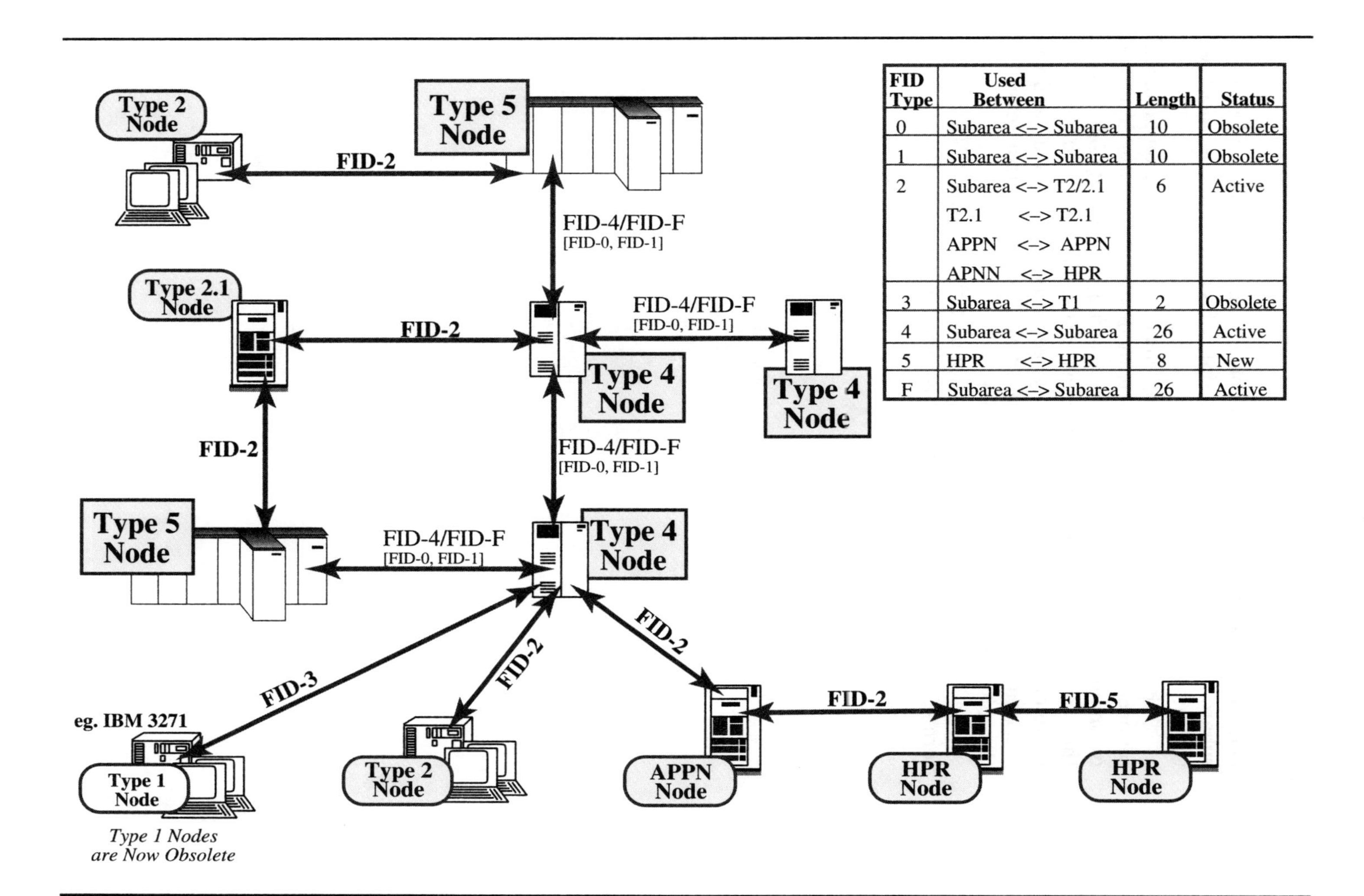

FID Type	Used Between	Length	Status
0	Subarea <-> Subarea	10	Obsolete
1	Subarea <-> Subarea	10	Obsolete
2	Subarea <-> T2/2.1 T2.1 <-> T2.1 APPN <-> APPN APNN <-> HPR	6	Active
3	Subarea <-> T1	2	Obsolete
4	Subarea <-> Subarea	26	Active
5	HPR <-> HPR	8	New
F	Subarea <-> Subarea	26	Active

FIGURE A.6 The different transmission headers used between the various SNA/HPR nodes.

end subarea nodes. An ER consists of a fixed set of TGs and a set of subarea node PCN components. Multiple ERs may be defined between the same two subarea nodes.

TGs are responsible for providing one or more bidirectional logical connections between adjacent subarea nodes. SNA TGs are described in Chapter 2. In the case of sessions involving peripheral node NAUs, their route through the PCN involves a so-called *route extension*. A route extension consists of a link between a peripheral node and the subarea node to which it is attached, as well as PCN components with the subarea and the peripheral nodes.

Refer to the Glossary for the explanation of other SNA-related terms. APPN-related concepts are discussed in Appendix B.

Appendix B

Thumbnail Sketch of APPN

In March 1987 when IBM launched its then very strategic and all-encompassing Systems Application Architecture grand vision for the next wave of computing, SNA was conspicuous by its absence. SAA networks were going to be based on something called *Low Entry Networking* (LEN) as opposed to SNA. LEN in those days was a marketing term to describe peer-to-peer networking in much the same way that the term APPC was being used to denote program-to-program communications. In the 1987 time frame, LEN in effect was IBM's way of talking about the then nascent Advanced Peer-to-Peer Networking (APPN). With SAA's unequivocal endorsement of LEN/APPN, over SNA, IBM was stating that APPN was indeed going to be groomed to graciously but convincingly inherit SNA's then glittering and expansive networking empire.

APPN, which in comparison to SNA came across as being nimble, contemporary, slick, and dynamic, was going to vanquish the then looming threat of OSI. APPN, which was far removed from "your father's SNA" was going to rejuvenate IBM centric networking and make sure that it continued to prosper well into the next millennium. That was nearly 10 years ago. APPN is still waiting in the wings watching the demise of SNA-only networks and the emergence of HPR—which in 1996 will eclipse it to become the eventual successor to SNA.

Despite the significance and prominence that IBM was going to be assigning to it within a year, APPN first saw the light of day in mid-1986 with no fanfare and little, if any, promotion. *Though an architecture like SNA, it was not unveiled as an architecture as was the case with SNA*. Instead, APPN's first outing was as an optional, peer-to-peer networking scheme on IBM's System/36 family of departmental minicomputers. The S/36, which was hugely successful in the small business world, was a predecessor of the AS/400.

APPN was included as a standard, no-charge facility within the S/36 operating system software: System Support Program Interactive Communications Feature. APPN in its current form, replete with so-called APPN End Nodes, came to being in June 1988 with the availability of the AS/400. Just as with the S/36, APPN was included as a standard facility within the OS/400 operating systems. Given that IBM had patented certain APPN protocols (e.g., the initial XID/CP-Capabilities exchange) and had originally intended to keep it as a closed architecture, the APPN architecture specification per se was not released until March 1993. *This initial lack of openness mortally wounded APPN*—and it has never fully recovered from it. SNA was the epitome of an open, public domain networking architecture.

During its first decade (1974–1984), all inclusive, elaborately detailed, and up-to-date SNA specifications were readily and inexpensively available from IBM. New SNA specs were typically available at least 18 months prior to any IBM implementations that corresponded to that spec. There were no patents or licenses. Other vendors could implement SNA with impunity. That was the real catalyst that propelled SNA's success. Due to its pervasive multivendor support it was the de facto networking standard for multivendor interoperability.

Given SNA's openness, the industry, apart from political and cultural ideology, did not require an OSI. SNA fitted that bill. But APPN was different. IBM initially eschewed non-IBM adoption, even when it relented it never abolished the license fees. Now IBM wonders why APPN never even came close to emulating SNA's success. But it does not seem to have learned. HPR, NBBS, SVN, and so on propagate the APPN proprietary model as opposed to SNA's original and winning "come and get it; it's good and it's free" philosophy.

B.1 THE BASIC COMPOSITION OF APPN

APPN is built on top of SNA T2.1 peer nodes. Its genesis can be traced back to a somewhat academic but landmark paper published by five IBM employees in the May 1985 IEEE *Journal on Selected Areas in Communications* (Vol. SAC-3, No. 3), entitled "SNA Networks of Small Systems." This paper, which introduced the term LEN, talked about a highly dynamic, peer-to-peer networking scheme that was based on the then relatively new T2.1 architecture. Given these roots it is not surprising that even today, the APPN architecture is based entirely on standard T2.1 and LU 6.2 concepts and constructs. APPN, however, eliminates the one glaring flaw in T2.1 nodes, which is their inability to do intermediate node routing.

The fundamental difference between APPN and T2.1 nodes boils down to CP-CP sessions. T2.1 and APPN nodes both have CPs. However, the scope and awareness of a CP in a T2.1 node is restricted just to the node it is resident in. A T2.1 CP does not talk to other CPs. There are no CP-CP sessions between T2.1 nodes. In contrast, all bona fide APPN nodes have CP-CP sessions either on a NN-to-NN or NN-to-EN basis. APPN does not support the notion of CP-CP sessions between ENs. All EN-to-EN interactions, vis-à-vis APPN, are conducted through one or more NNs.

LU 6.2 is the native LU scheme supported by APPN. APPN, at present, can only support non-LU 6.2 sessions using the DLUS/DLUR technology that was discussed in Chapter 7. *LU 6.2 is also the means by which APPN realizes all of its control functions.* Thus, APPN can be thought of as being an LU 6.2 application. Moreover, given that APPN directly supports only LU 6.2 transactions, it is an LU 6.2 application that only supports other LU 6.2 applications.

The two key LU 6.2 facilities used by APPN are: the LU 6.2 *General Data Stream* (GDS) and the notion of LU 6.2 *Service Transaction Programs* (STPs) designation. (GDS is described in Chapter 7, while STPs are introduced in Appendix A.) APPN uses APPN-specific, but homogeneous, *GDS variables* to convey all network operation-related interactions between APPN nodes. *These GDS variables are exchanged across CP-CP sessions.* Some of the key GDS variables used by APPN, which in effect constitute its command repertoire, include the following:

- CP Capabilities (X' 12C1'): Used at node activation to establish the facilities supported by an APPN CP.
- Register Resource (X' 12C3'): Used by APPN ENs to register their LUs with an APPN NN. See Chapter 7.
- Locate (X' 12C4'): Used as the overall GDS variable transport structure for the dynamic remote LU discovery process that is described in Chapter 7. Other GDS variables such as Cross-Domain Initiate and Find Resource are slotted within the GDS structure defined by a Locate variable.
- Cross-Domain Initiate (X' 12C1'): Used in conjunction with the Locate variable during the broadcast search process to locate an unknown LU.
- Delete Resource (X' 12C9'): De-register previously registered EN LU.
- Find Resource (X;pr 12CA;pr): Used in conjunction with the Locate variable.

- Found Resource (X;pr 12CB;pr): Used in conjunction with the Locate variable to denote the positive outcome of a broadcast search process.

An APPN Control Point is implemented as a standard LU 6.2. An APPN CP, thus uses standard LU 6.2 protocols and facilities—including the notion of STPs. An APPN CP uses APPN-specific STPs to realize all of its key functions. Thus there are separate STPs that perform the following APPN functions: CP-capabilities exchanges, dynamic LU registration, and broadcast searches. Given that CPs are in fact LU 6.2s, APPN CP-CP sessions are in reality just plain vanilla LU 6.2-to-LU 6.2 sessions. *CP-CP sessions are the basis for all of the powerful networking capabilities, including the peerless plug-and-play facilities provided by APPN.*

At the time APPN was being developed, LU 6.2, somewhat surprisingly, supported only half-duplex sessions. (LU 6.2 was revamped around 1993 to finally rectify this.) *To ensure full-duplex communications, a given CP-CP session is established in the form of two back-to-back, half-duplex LU 6.2 sessions*. Each CP establishes one of the two half-duplex sessions that make up a CP-CP session. A CP uses the half-duplex session that it established as its send path to the other CP. Giving each CP its own send path relative to the CP-CP session ensures that both CPs can have immediate, unhindered communications access to each other.

B.1.1 APPN Node Types

An APPN network consists of one, or more, APPN NNs, each of which can optionally support a set of one, or more, link-attached ENs. When multiple APPN NNs are present, they can be interconnected in any appropriate mesh-type topology. Such NN interconnection ensures that any particular NN that needs to interact with another NN can do so either via a direct connection (e.g., SDLC link), or via a path traversing one or more intermediary NNs. Intermediate node routing across NNs, in addition to being one of its fortes, is also a defining characteristic of APPN. The interconnected NNs together provide the APPN Path Control Network backbone that supports intermediate node routing, on-the-fly remote LU location, and the automatic adaptability to changes in the network topology. Figure B.1 depicts the structure of an APPN network and shows the relationship between NNs and ENs.

CP-CP sessions are only established between logically adjacent NNs. The NNs interact with each other to perform the necessary net-

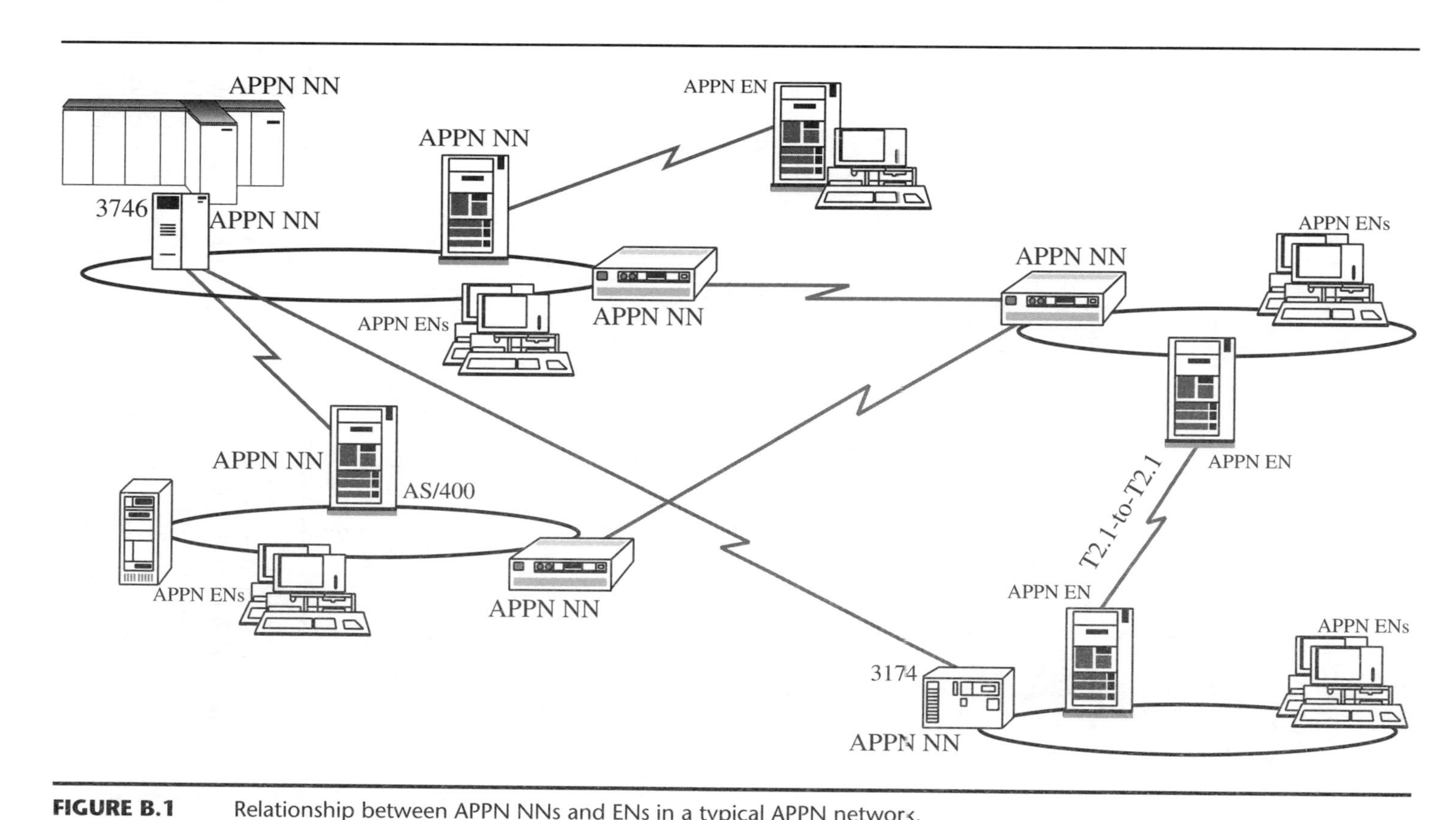

FIGURE B.1 Relationship between APPN NNs and ENs in a typical APPN network.

work control, broadcast search, and network topology update operations over these CP-CP sessions. There is, however, no need for end-to-end CP-CP sessions between all the NNs within a given network. CP-to-CP interactions that need to be propagated across the entire network such as broadcast searches and topology updates are forwarded hop-by-hop between adjacent NNs. The key functions performed by an NN are as follows:

- Performing intermediate node session routing for LU-LU sessions
- Acting as an NN Server for all attached ENs; specifically dynamically locating remote LUs, and calculating the optimum end-to-end path for proposed sessions based on a multicriteria scheme supported by APPN (i.e., directory services)
- Accepting LU registrations from attached ENs (i.e., configuration services)
- Maintaining an up-to-date network topology database and notifying adjacent NNs on changes to the active topology (i.e., configuration services)
- Acting as the management focal points for the network by soliciting and automatically collecting network management information—including alerts (i.e., management services)

Some NNs may be deployed purely to act as routing or EN server nodes. Such NNs, just as is the case with SNA Type 4 nodes, may not contain any LUs that support end users.

APPN ENs come in two distinct flavors: bona fide *APPN end nodes* and so-called *LEN nodes*. *APPN ENs have a CP that is capable of establishing a CP-CP session with a designated NN server*. It uses this session to perform LU registration and to solicit help in locating the unknown remote LUs. The APPN broadcast search mechanism is described in Chapter 7. An APPN EN may only have one active CP-CP session, with one NN server, at any given time. This is the equivalent of a SNA peripheral node only being able to have an active SSCP-PU session, with one SSCP, at any given time.

LEN Nodes, in contrast to APPN ENs, are just unadulterated T2.1 nodes that do not contain any APPN-specific extensions. In particular, they do not contain a CP that is capable of establishing a CP-CP session with an APPN NN Server. Thus, a LEN node treats the entire APPN network and the NN to which it is attached, as if they were just another adjacent T2.1 node. LEN nodes have absolutely no cognizance of the dispersed multinode APPN network and the NNs that form its core. From their perspective, all the remote LUs that they intend to interact with are all resident in the peer-attached, adjacent T2.1 node.

In the case of an APPN network, this adjacent T2.1 node will happen to be an NN. This NN thus has to act as a transparent gateway to the entire APPN network.

Given their lack of CP-to-CP sessions, LEN Nodes cannot register LUs with the NN or explicitly ask them to conduct a broadcast search. When an LU in a LEN node wants to establish a session with an LU that is not within its node, it and the CP in that node both assume that the remote LU is located in an adjacent node. The LEN node will have a manually created directory that specifies for each remote LU the link that should be used to reach the adjacent node containing that LU. In the case of APPN, all the remote LUs defined in the directory of a LEN Node will specify the link leading to the NN.

LU 6.2s in T2.1 nodes attempt to establish sessions with LUs in adjacent T2.1 nodes by issuing an unsolicited BIND request to the appropriate node over the link specified in its local directory. A LEN node attached to a NN will use this same mechanism to try to establish various LU-LU sessions. However, in the case of APPN the required destination LU may be at a remote node—and the NN may not even know its location.

Thus, NNs treat unsolicited BINDs from LEN Nodes as being equivalent to LOCATE/FIND GDS variables sent to it, over CP-CP sessions, by ENs. The NN will dynamically locate the required remote LU, calculate an optimum path to it based on the COS control vector that may have been included on the BIND, and then append an Route Selection CV (RSCV) to the BIND. It will then forward it to its destination. The response to the BIND will eventually get conveyed back to the LEN Node through the NN. If the response was positive the session would have been established.

Figure B.2 illustrates the various CP-CP sessions that are established within an APPN network. Table B.1 summarizes the salient characteristics of the various APPN node types and also cites representative product implementations for each type of node.

B.2 The Quintessential Characteristics of APPN

The characteristics and facilities that define APPN can be summarized as follows:

- Strictly *peer-to-peer oriented*. No reliance on a central SSCP, CP, or mainframe. APPN NN-capable mainframes will participate in APPN networks as just another peer node. APPN and HPR, as mentioned in Chapter 7, do support the notion of an optional Central Directory Server(s). Such a server is not a central point of con-

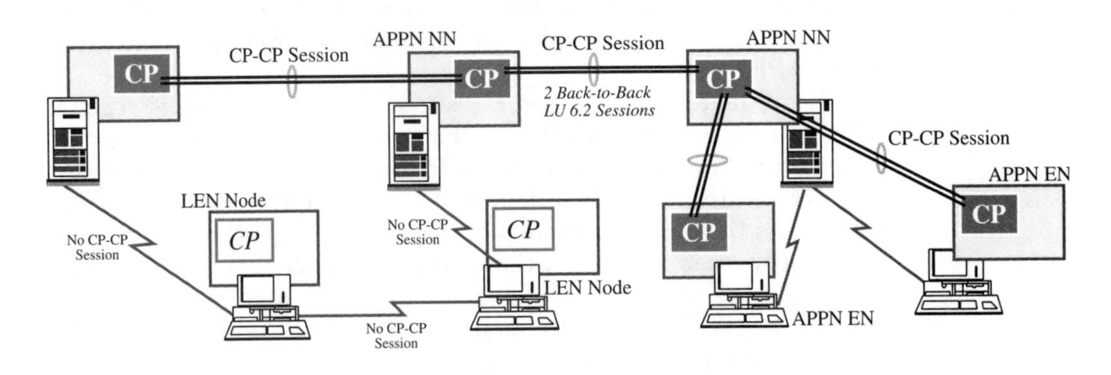

FIGURE B.2 The NN-to-NN and NN-to-EN CP-CP sessions established in APPN networks.

trol à la an SNA SSCP. Instead, it is just a means of minimizing the need for SRB-like network-wide broadcast searches.

- Full support for *Intermediate Session Routing* (ISR). This is what differentiates APPN from T2.1. In HPR, ISR is made obsolete by the new Layer 2-based ANR protocol.
- Compelling *plug-and-play networking* as is described in Chapter 7 in the context of HPR.
- *Multiple criteria-based, COS-oriented path selection* mechanism as described in Chapter 7. APPN also supports an optional spanning tree-based routing scheme that can be used to determine how to get from one NN to another without having to go through the fairly extensive processing required to calculate a COS-based optimum path.
- NNs dynamically and automatically notify each other of any changes to the network topology using a *Topology Database Update* (TDU) mechanism.
- Totally *name-based networking*. Just as with SNA, only LU names are used to establish APPN LU-LU sessions. CP-CP sessions are also established using CP Names rather than any form of numerical addresses. Unlike SNA, APPN does not even assign permanent network addresses to LUs. Instead, each session established across a given link is assigned a temporary session identification (i.e., LFSID) as was described in Appendix A.
- Ability to partition networks using *APPN border nodes*. This capability can be thought of as APPN's equivalent of SNA's SNI. Border Nodes permit transparent and seamless interoperability across autonomous APPN networks, that is, the ability to establish LU-LU sessions across independent networks. *Border nodes also sup-*

TABLE B.1 Characteristics of the APPN Node Types

	APPN End Node	*APPN Network Node*	*LEN End Node*
Derivation:	LEN EN with APPN-Capable CP	LEN EN with APPN-Capable CP	Unadulterated SNA T2.1 Peer Node
Consists of:	APPN-Capable CP; LUs –> 1 to thousands	APPN-Capable CP; LUs –> 0 to thousands	T2.1 CP; LUs –> 1 to thousands
Establishes CP-CP Session(s):	With one NN Server	With Adjacent NNs and APPN ENs that rely on it as their NN Server	No
Performs APPN Routing:	No	Yes. Intermediate Session Routing (ISR)	No
Services Provided by CP:	• Limited Local Directory Services and the ability to select a preferred link top interact with NN Server if multiple links present • Registers LUs at NN Server	• Global Directory Services: Remote LU Location. (Optional: Central Directory Server.) • Calculates optimum routes for LU-LU session • Maintains Network Topology Database and participates in TDU exchanges when necessary • Accepts LU registrations from APPN ENs • Management Focal Point	• Limited Local Directory Services • Selects link across which destination LUs can be reached.
Relies on NN for:	• Dynamic location of remote LUs • Optimum route calculation		• Does not know about NN. • NN performs dynamic location of remote LUs and route calculations.
Supports multiple links:	Yes	Yes	Yes
Quintessential implementations	CM/2, AS/400, NS DOS	AS/400, CM/2, bridge/routers	APPC/PC, S/38

port unimpeded broadcast searches across networks. However, since the networks are autonomous, a border node will not propagate TDUs from one network to another. APPN border node implementations are currently available on AS/400s and mainframes.

- Ability to assign COS-based Transmission priority (i.e., high, medium, or low) to the traffic associated with each APPN session.
- *Adaptive pacing*-based congestion control. See Appendix A.
- The ability to implement tightly interconnected *mesh networks.* APPN permits an EN to have multiple simultaneously active links to multiple NNs. However, an EN can only treat one of these NNs as its NN server. When doing an optimum route calculation for an LU-LU session the NN server is aware of the EN's other links to the other NN. In such scenarios, there is a possibility that the NN server may select a session path that bypasses it and instead uses another NN through which to route that session. Figure B.3 depicts an APPN network where ENs are simultaneously connected to multiple NNs and how some LU-LU sessions may bypass an EN's NN server.
- Supports *multiple links* between both ENs and NNs as shown in Figure B.3.

B.3 THE WEAKNESSES THAT DILUTE THE LURE OF APPN

Unfortunately, the many impressive strengths of APPN as listed above are compromised by a list of surprising, and in some instances significant, weaknesses. Some of these, such as its inexcusable inability up until quite recently to support the ubiquitous SNA/3270 sessions, explain why APPN never fulfilled its initial promise of being the successor to SNA. Today, APPN's popularity is still restricted to the AS/400 community even though full APPN NN support on mainframes has been available since 1992. APPN has now become a place-holder between SNA and HPR.

Some of APPN's major weaknesses are described below.

- Inability to natively support traditional SNA sessions. This was the major Achilles Heel of APPN. Today, the DLUS/DLUR technology, in particular the surrogate DLUR servers available on most leading bridge/routers since mid-1995, eliminates this weakness.

- Lack of dynamic alternate routing. A link failure in an APPN network will result in all the sessions routed over that link to also fail, just as in SNA, this despite all the NNs maintaining a Network Top-

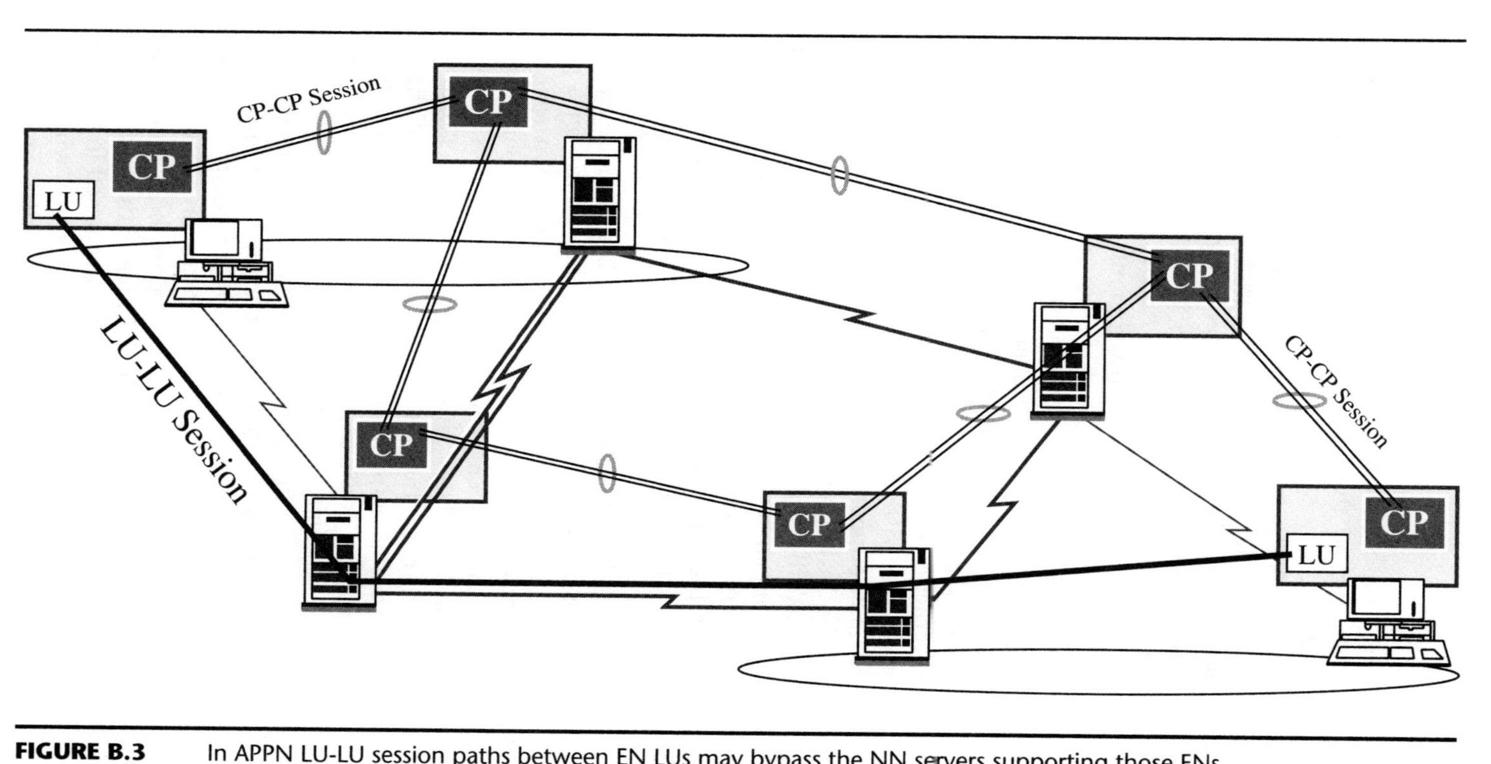

FIGURE B.3 In APPN LU-LU session paths between EN LUs may bypass the NN servers supporting those ENs.

ology Database that reflects the latest topology of the network including all the possible routes between nodes. HPR fixes this problem.

APPN's lack of dynamic alternate routing, however, can prove to be less debilitating and frustrating than is the case with SNA—particularly if all the sessions being supported are LU 6.2-based. Unlike other LUs, an LU Type 6.2 intrinsically works in store-and-forward asynchronous mode. An LU 6.2 is also capable of supporting multiple end users, and can multiplex traffic from multiple end users over the same LU-LU session. LU 6.2 thus establishes LU-LU sessions, essentially in background mode, somewhat independent of explicit end user instructions, as a common data transmission utility that may be sequentially used by multiple end users. Thus, an LU 6.2 can restart a new session and map end user interactions on to that new session without end user intervention, and even without the end users being cognizant of it. Moreover, LU 6.2s support parallel sessions that can be routed across different paths to provide a level of fault-tolerancy.

- The parallel links between nodes, even those between NNs, cannot be grouped together to form SNA-like TGs. Unfortunately, this problem is not convincingly fixed by HPR. HPR, rather than including this key facility as an architected feature, relegated it to being an implementation-dependent and implementation-specific option.
- APPN's COS-based optimum path selection scheme is not adaptive. Adaptive routing ensures that traffic on a given session is always transmitted on the most optimum path across the network. In APPN the optimum path for a given session is calculated once—when the session is being established. If a more optimum path becomes available later, as indicated by a TDU exchange, APPN NN will not attempt to reroute already established sessions. HPR also does not support adaptive routing. In mitigation, it is only fair to note that recalculating the optimum paths of previously established sessions, each time there is a topology change, can place a huge and possibly unrealistic processing burden on NNs.
- APPN due to its T2.1 node heritage, and in marked contrast to SNA, does not have a concept of an end-to-end session. Thus, an end-to-end APPN session, that traverses multiple, intermediary NN nodes, has to be constructed in the form of a series of session stages between each pair of nodes with address swapping at each hop. (Label-swapping routing is discussed in Chapter 9 in the context of NBBS.) Each APPN session stage, which is a bona fide session in its own right, has its own LFSID. Since a different LFSID

is used for each session stage a new SNA TH, with the pertinent LFSID, has to be appended to each message unit of each session as it passes through each intermediate node.

HPR uses ANRF routing as opposed to this type of hop-by-hop label-swapping routing. However, as was discussed in Chapter 7, ANR, though not swapping labels, instead peels off labels from the ANRF at each hop. This also means that the header has to be reconstituted at each hop. This is one of the few major weaknesses of HPR.

- Some implementations of APPN NNs, when calculating optimum paths for sessions, measure the congestion of other NNs not in terms of the actual traffic that may be passing through those nodes at that moment in time, but in terms of the number of sessions passing through the node. While there could be some correlation between the number of sessions passing through a node and the amount of traffic that the node is having to contend with, this measurement of node congestion is far from satisfactory.
- APPN was never diligently and lovingly architected as was the case with SNA, with overall efficiency being a paramount criteria. In general, APPN evolved out of the S/36 and AS/400 implementations. Consequently, APPN protocols and data structures, in particular the Control Vectors (CVs) liberally appended to the GDS variables, can sometimes be profligate in their use of bandwidth. HPR is much more careful and efficient, even though there is still some evidence of the lax discipline, vis-à-vis bandwidth usage, that crept in with APPN.

Glossary

ACF/NCP IBM-supplied SNA software that runs on a 37xx communications controller. Version 7 Release 1, available as of February 1994, supports SNA over frame relay à la RFC 1490.

ACF/VTAM IBM-supplied host software that is a prerequisite for implementing a host-centric SNA network.

AIW APPC/APPN Implementor's Workshop, an IBM-sponsored forum that by a curious twist of fate ended up as the de facto guardians of DLSw. AIW is not a recognized standards body.

ANR Agile Layer 2 connectionless protocol employed by HPR.

AnyNet (Gateway) Family of IBM software products that perform protocol conversion. This software can either reside within a product or be deployed within stand-alone gateway, e.g., IBM 2217. Can be used to convert SNA to TCP/IP or vice versa. Gateway can also encapsulate IPX and NetBIOS traffic within LU 6.2 message units.

APPC Marketing term for LU 6.2-based interactions.

APPN End Nodes "Smart" APPN peripheral nodes that establish a CP-to-CP session with an APPN NN. Use this session to obtain services, such as Directory Services, from the NN. Can dynamically register LUs with the NN.

APPN Network Nodes "Smart" APPN nodes responsible for intermediate node routing, directory services, route selection, and network topology management.

APPN Advanced Peer-to-Peer Networking. The born-again, peer-to-peer and plug-and-play replacement for SNA that has been around since 1986.

ARB HPR's anticipatory congestion control mechanism.

AS/400 IBM minicomputer.

ATM Strategic Layer 2, CCITT data transfer standard that combines the constant bandwidth and consistent delay characteristics of circuit switching, with the resource sharing and bursty traffic accommodating features of packet switching. The basis of ATM is the very fast switching of fixed-length, 53-byte cells.

BAN IBM's variant of RFC 1490 for encapsulating SNA/APPN traffic within frame relay. Uses the same encapsulation scheme as utilized by bridges as opposed to the SNA/APPN specific native scheme specified by RFC 1490/FRF.3.

Bandwidth The data transfer capacity of a link in terms of bits per second.

Boundary Function SNA: Set of services, including address conversion, provided by a subarea node to the peripheral nodes attached to it. HPR: Ensures interoperability between HPR and APPN nodes.

Bracket The basic unit of work within SNA. Consists of a series of requests (i.e., one or more chains) and their responses.

BSC An IBM link protocol from 1967 that is still used by devices such as automated teller machines.

Bus-and-Tag The original scheme for channel-attaching devices to an IBM mainframe. Uses two bulky cables per attachment and transfers data 8 bits at a time in parallel form. Maximum bandwidth is 4.5MBytes/sec (36Mbps).

Cache (Directory) Directory containing dynamically created entries for objects typically located through a search process. Old entries that have not been recently referenced may be overwritten with new entries.

Chain In SNA a series of related requests that are transmitted consecutively and are treated as one entity forming a complete message. This is the basic unit of error-recovery in SNA.

Channel-Attached Locally attaching a device to an IBM mainframe via a high-speed Bus-and-Tag or ESCON connection.

CP Component in T2.1 and APPN nodes that provides local directory and configuration services.

CSMA/CD Contention-based access scheme used by Ethernet.

CTS IBM-developed Layer 4-based protocol mapping and conversion scheme. Basis of AnyNet.

Cut-Through A forwarding mode in LAN switching where the switch starts to forward bits that make up a frame to its destination as soon as it as determined the relevant destination port without waiting to receive the end of the frame and checking whether it is error-free. Opposite of *store-and-forward*.

Desktop DLSw A scheme whereby SNA/APPN traffic is encapsulated within TCP/IP packets right at its source (e.g., PC/workstation) even before it reaches the LAN. Only justifiable in dial-in scenarios for mobile users.

DLSw A core set of facilities now available on most multiprotocol bridge/routers for supporting SNA/APPN and NetBIOS traffic across a TCP/IP backbone. Embraces: TCP/IP encapsulation, SDLC-to-LLC:2 conversion, Local LLC:2 acknowledgment, and so on.

DLU LU resident within a traditional SNA node that is only able to establish LU-LU sessions by depending on an SSCP to provide it with directory services.

DLUS/DLUR Technology required in order to ensure that traditional SNA traffic can be freely supported across an APPN or HPR network without any restrictions on where the SNA devices may be located.

Domain The set of SNA resources controlled by a single SSCP.

DSPU LAN-attached SNA node supported by an SNA LAN gateway.

ESCON (II) The now strategic means for mainframe channel-attachment that is based on a multimode fiber connection and serial transmission of data. Maximum bandwidth is 17Mbps (136Mbps) and channel distances can be as great as 36 miles! The latest version unveiled in September 1991 is referred to as ESCON II.

Frame Relay New-generation, high-performance, low-overhead, connection-oriented, Layer 2 *packet-switching standard* prescribed by both the CCITT and ANSI. Can be thought of as a slimmed-down and streamlined X.25. Can support speeds up to 45Mbps though only used at less than 4Mbps at present.

FRF.3 Addendum to the RFC 1490 standard that specifies how to natively (i.e., without TCP/IP or MAC/LLC headers) encapsulate SNA/APPN traffic within frame relay frames.

HPR Successor to SNA and APPN.

ILU LU 6.2s resident in T2.1 or APPN nodes that can establish LU-LU sessions without depending on an SSCP for any services.

IPX/SPX Internetwork Packet Exchange, Novell NetWare's Layer 3 protocol, and Sequenced Packet Exchange, NetWare's Layer 4 protocol.

ISDN Digital telephone service that permits multiple voice and data channels to be shared across one line.

LAN Switch Internetworking device that rapidly forwards LAN frames from one of its ports to another.

LAN-over-SNA Encapsulating TCP/IP, IPX/SPX, and NetBIOS traffic within LU 6.2 message units so that these protocols can be transported LAN-to-LAN across an SNA network.

LLC Logical Link Control—the upper portion of the OSI Layer 2, Data Link Layer, as defined by the IEEE 802.x standards. It provides Data Link Control functions such as flow control and error recovery. LLC:2 refers to a connection-oriented (i.e., session-like) version of LLC that uses a superset of the commands used by SDLC and is often used to support SNA, APPN and NetBIOS LAN traffic.

Local LLC:2 Acknowledgment Precludes LLC:2 acknowledgments having to be transported across a WAN.

LU Port through which end users gain access to an SNA network.

LU 6.2 Strategic means for program-to-program communications in SNA, APPN, or HPR networks.

Media Access Control (MAC) The lower portion of the OSI Layer 2, Data Link Layer, as defined by the IEEE 802.x standards. It provides controlled access to a LAN.

MPTN The formal architecture for CTS and AnyNet.

Multiplexing Sharing the bandwidth of a link between multiple users.

NBBS An IBM architecture for value-added ATM-centric networking.

NCIA Cisco's version of Desktop DLSw.

NetBIOS LAN protocol widely used by IBM LAN applications, most notably IBM's OS/2 LAN Server.

Networking Blueprint A framework put forward by IBM in 1992 for network-neutral applications that could be run across a variety of different network types. AnyNet is a manifestation of the CTS technology postulated by this Blueprint.

Nways IBM's brand name for strategic, multiprotocol products that perform Layer 1–3 functions.

Pacing An SNA congestion control mechanism that permits a receiver to control the rate at which data is forwarded to it.

Peripheral Node SNA Type 2 or Type 1 node.

PIR CrossComm's proprietary routing mechanism.

PIU Path Information Unit—the technical term for an SNA message unit.

PPP Point-to-Point Protocol, which is widely used to transport TCP/IP across asynchronous/synchronous links.

PU Software component within an SNA node that controls and manages the physical entities, such as links, associated with that node. Does not refer to actual devices.

PU Concentrator Type of SNA-LAN gateway that aggregates the LUs in multiple SNA nodes into a single virtual node so that a mainframe does not have to support as many DSPUs.

Remote Polling A technique for obviating the need to continually transmit BSC or SDLC polls across a backbone network.

Reverse Multiplexing Aggregating the bandwidth of multiple low-speed links to realize a high-bandwidth channel. Kind of like an SNA TG.

Reverse Protocol Conversion Permitting 3270 terminals or PCs running 3270 emulation to access non-SNA applications—in particular Unix applications

RFC 1490 A generic IETF standard for transporting alien traffic end-to-end across an FR WAN. It specifies an encapsulation scheme and a means for segmenting frames. FRF.3 extends RFC 1490 to embraces SNA/APPN traffic.

RH SNA Request/Response Header, which prefixes the data portion of an SNA message unit.

RIP One of the popular routing protocols used by IP.

Routing Information Field (RIF) A field in the Layer 2 MAC header used by Source Route Bridging to specify the LAN-to-LAN route that should be traversed by a inter-LAN packet.

RSRB Cisco's proprietary scheme for encapsulating SNA/APPN traffic within TCP/IP.

RTP Connection oriented, full-duplex end-to-end protocol used by HPR on top of ANR.

RU SNA Request/Response Unit that carries the data being transported within an SNA message unit.

S/3x S/36 or S/38 minicomputers that were the precursor to the AS/400.

SDLC-to-LLC-2 Conversion A technique used to integrate SDLC link attached SNA devices into a LAN/WAN internet where a modified remote polling process is used to make the link-attached devices appear as if they were LAN-attached.

SDLC Synchronous Data Link Control—an IBM link protocol introduced in 1974 alongside SNA, which is widely used to support link attached SNA devices.

SLIP Internet protocol used to run IP across serial links.

SMDS Switched, high-speed, connectionless public service. Will support speeds up to 155Mbps.

SNA Routing SNA's native, subarea-to-subarea routing scheme-based on network addresses and VRs.

SNA Switching TCP/IP encapsulation scheme à la DLSw that can directly route SNA traffic to the appropriate mainframe.

SNA-LAN Gateway Gateway that permits LAN-attached PCs/workstations to access mainframe or AS/400-resident SNA/APPN applications.

SNA/DS LU 6.2-based, store-and-forward mail and data delivery service.

SNMP TCP/IP's Network Management protocol.

Sockets Widely used API for TCP/IP-based interactions.

Source Route Bridging (SRB) The default technique in token-ring environments for realizing communications between devices on different LANs.

Spanning Tree Algorithm to prune a network with alternate paths so that only one active route exists between any given pair of nodes.

SRT Bridge A bridge that supports both SRB and transparent bridging.

SSCP SNA-centralized Control Point function that provides Directory, Configuration, and Management Services.

Store-and-Forward A forwarding mode in LAN switching where the switch waits until it has received a complete error-free frame before it starts forwarding it to the destination port. Opposite of *cut-through*.

Subarea(s)/Subarea Node(s) The SNA Type 4 or Type 5 nodes that support peripheral (or end) nodes and are capable of performing SNA routing. A subarea node and all the peripheral nodes attached to it form a subarea—the basis of SNA Network Addressing and Routing.

SVN IBM framework for implementing and managing switched, as opposed to routed, networks.

TCP/IP Encapsulation Encapsulating of traffic using alien protocols such as SNA/APPN or NetBIOS within TCP/IP packets.

TCP/IP Layer 3 and Layer 4 protocols in the Internet community.

TH SNA Transmission Header, which prefixes all SNA or APPN message units.

tn3270(e) Protocol conversion that permits workstation using the TCP/IP Telnet protocol to access mainframe or AS/400 resident SNA applications. Conversion may be done within the mainframe or in an external gateway. The extended version supports printing.

Transparent Bridging Bridging scheme used by Ethernet.

Transparent Switching Transparent bridging as it applies to LAN Switches.

TURBOWAYS IBM's brand name for its ATM adapters.

2217 IBM AnyNet Gateway that permits multiprotocol networking across a single-protocol APPN/HPR backbone.

3x74 3270 Control Unit that can now be used in a variety of other roles including that of an SNA LAN Gateway.

37xx IBM communications controllers (e.g., 3745, 3746, 3725).

3172 Interconnect controller whose primary role is to act as a low-cost LAN-to-mainframe gateway.

3270 Generic name for a once ubiquitous family of terminals, printers, and control units for accessing mainframe-resident applications.

5250 The equivalent of 3270s for accessing applications on AS/400s and other IBM minicomputers such as the S/36 or S/38.

List of Acronyms and Abbreviations

AAL	ATM Adaptation Layer
ACF/NCP	Advanced Communications Functions for the Network Control Program
ACF/VTAM	Advanced Communications Functions for the Virtual Telecommunications Access Method
ACP	Airline Control Program
ACTLU	Activate Logical Unit
ACTPU	Activate Physical Unit
ADPCM	Adaptive Differential Pulse Code Modulation
AIW	APPN(/APPC) Implementors' Workshop
AIX	Advanced Interactive Executive
ANR	Automatic Network Routing
ANRF	ANR Field
API	Application Program Interface
APPC	Advanced Program-to-Program Communications
APPI	Cisco's Advanced Peer-to-Peer Internetworking
APPN	Advanced Peer-to-Peer Networking
ARB	Adaptive Rate-Based Congestion Control
ASIC	Application Specific Integrated Circuit
Async	Asynchronous
ATM	Asynchronous Transfer Mode; Adobe Type Manager
B-ISDN	Broadband ISDN
BAN	Boundary Access Node
BaySIS	Bay Networks Switched Internetworking Services
BBNS	BroadBand Network Services—now called NBBS
BECN	Backward Explicit Congestion Notification
BER	Bit Error Rate
BF	Boundary Function
BGP	Border Gateway Protocol

BIU	Basic Information Unit
BNN	Boundary Network Node
BPDU	Bridge Protocol Data Unit
Bps	Bytes per second
bps	bits per second
BSC	Binary-Synchronous Communications
BTAM	Basic Telecommunications Access Method
BTU	SNA's Basic Transmission Unit
CBR	Constant Bit Rate
CBRT	Constant Bit Rate Traffic
CDCINIT	Cross-Domain Control Initiate
CDDI	Copper Distributed Data Interface
CDINIT	Cross-Domain Initiate
CDRM	Cross-Domain Resource Manager
CDRSC	Cross-Domain Resource
CDSESSST	Cross-Domain Session Started
CICS	Customer Information Control System
CIF	Cell-in-frame
CINIT	Control Initiate
CIR	Committed Information Rate
CLNP	Connectionless Network Protocol
CLP	Cell Loss Priority
CM/2	OS/2 Communications Manager/2
CMIP	OSI's Common Management Information Protocol
CMIS	Common Management Information Service
CMOS	Complimentary Metal Oxide Semiconductor
CMOT	CMIP over TCP/IP
CNM	Communications Network Management system
CNN	Composite Network Node; Cable Network News
CODEC	Coder-Decoder
COS	Class-of-Service
CP	Control Point
CPE	Customer Premise Equipment
CPI-C	Common Programming Interface for Communications
CR	Command/Response
CRC	Cyclic Redundancy Check
CSMA/CD	Carrier Sense Multiple Access with Collision Detection
CTC(A)	Channel-to-Channel (Adapter)
CTS	Common Transport Semantic
CV	Control Vector
DACTLU	Deactivate Logical Unit
DACTPU	Deactivate Physical Unit
DAF	Destination Address Field

DB/2	Database/2
DCA	Document Content Architecture
DCE	Data Circuit-Terminating Equipment; Distributed Communications Environment
DDDLU	Dynamic Definition of Dependent LU
DDLSw	Desktop DLSw
DDM	Distributed Data Management
DE	Discard Eligibility
DES	Data Encryption Standard
DFT	Distributed Function Terminal
DIA	Document Interchange Architecture
DISOSS	Distributed Office Support System
DLC	Data Link Control
DLCI	Data Link Connection Identifier
DLSw	Data Link Switching
DLU	Dependent Logical Unit
DLUS/DLUR	DLU Server/DLU Requester
DOS	Disk Operating System (PC or IBM mainframe)
DOS/VSE	Disk Operating System/Virtual Storage Extended
DSAP	Destination Service Access Point
DSPT	Display Station Pass-through
DSPU	Downstream Physical Unit
DTE	Data Terminal Equipment
E1/E2/E3	2.048Mbps link/8.448 Mbps link/34.37Mbps link
EA	Extended Addressing
EBCDIC	Extended Binary Coded Decimal Interchange Code
EFI	Expedited Flow Indicator
EISA	Extended Industry Standard Architecture
EN	End Node
ENA	SNA's Extended Network Addressing
EP	Emulation Program
ESA	SNA's Extended Subarea Addressing; IBM's Enterprise System Architecture
ESCON	Enterprise System Connectivity architecture
FCS	Frame Check Sequence
FDDI	Fiber Distributed Data Interface
FDX	Full-Duplex
FECN	Forward Explicit Congestion Notification
FEP	Front End Processor
FID	SNA's TH Format Identifier
FIFO	First-In, First-Out
FMD	SNA's Function Management Data
FR	Frame Relay

FRAD	Frame Relay Access Device
FRF	Frame Relay Forum
FRFH	Frame Relay Frame Handler
FST	Fast Sequenced Transport
FTAM	File Transfer, Access and Management
FTP	File Transfer Protocol
Gbps	Gigabits per second—Billions of bits per second
GDS	General Data Stream
GFC	Generic Flow Control
HDLC	High-Level Data Link Control
HDX	Half-Duplex
HLLAPI	High Level Language Application Program Interface
HPR	High Performance Routing
HSA	High Speed Adapter
HSS	High Speed Scanner
HSSI	High-Speed Serial Interface
ICP	Interconnect Controller Program
IDNX	IBM's Integrated Digital Network Exchange
IETF	Internet Engineering Task Force
IGRP	Cisco's Interior Gateway Routing Protocol
IHMP	Intelligent Hub Management Program
ILU	Independent Logical Unit
IMS	Information Management System
INIT-SELF	Initiate Self (i.e., Logon)
INN	Intermediate Network Node
IP	Internet Protocol
IPL	Initial Program Load
IPR	Isolated Pacing Response
IPX	Internetwork Packet Exchange
ISA	Industry Standard Architecture
ISDN	Integrated Services Digital Network
ISO	International Standards Organization
ISR	Intermediate Session Routing
J1/J2	1.586Mbps link/6.312Mbps link
JES	Job Entry System
Kbps	Kilobits per second—thousands of bits per second
LAN	Local Area Network
LEN	Low Entry Networking
LFSID	Local Form Session Identifier
LIC	Line Interface Coupler
LLC	Logical Link Control
LMI	Frame Relay's Local Management Interface
LNM	LAN Network Manager

LPDA	Link Problem Determination Application
LSA	Low Speed Adapter
LTLW	LAN-to-LAN Wide Area Network Program
LU	Logical Unit
MAC	Media Access Control
MAF	Multiple Access Face
MAP	Manufacturing Automation Protocol
MAU	Multistation Access Unit
MB	Megabytes
Mbps	Megabits per second—millions of bits per second
MCA	Micro-Channel Architecture
MIB	Management Information Base
MMMLTG	Mixed-Media, Multilink Transmission Groups
MPF	Mapping Field
MPOA	Multiprotocol Over ATM
MPTN	Multiprotocol Transport Networking
MQI	Message Queuing Interface
MSNF	Multisystem Networking Facility
MSS	Multiprotocol Switched Services
MVS	Multiple Virtual Storage
NAU	SNA's Network Addressable Unit; APPN's Network Accessible Unit
NBBS	Networking BroadBand Services
NCCF	Network Communications Control Facility
NCIA	Cisco's Native Client Interface Architecture
NC	Network Connection
NCL	Network Connection Layer
NCP	Network Control Program
NDIS	Network Driver Interface Specification
NetBEUI	NetBIOS Extended User Interface
NFS	Network File System
NHDR	Network Layer Header
NIC	Network Interface Card
NLP	Network Layer Packet
NLPID	Network Layer Protocol Identifier
NLSP	Netware Link Services Protocol
NMVT	Network Management Vector Transport
NN	Network Node
NNI	Network Node Interface
NOS	Network Operating System
NPA	Network Performance Analyzer
NPM	NetView Performance Monitor
NPSI	NCP Packet Switching Interface

NRF	Network Routing Facility
NRZ(I)	Non-Return to Zero (Inverted)
NSCP	Nways Switch Control Program
NSI	Non-SNA Interconnect
NTO	Network Terminal Option
NTRI	NCP Token-Ring Interface
OAF	Origin Address Field
OC3	155Mbps link
ODAI	OAF/DAF Assignor Indicator
ODI	Open Data-link Interface
OSA(/2)	Open Systems Adapter (Version 2)
OSF	Open Software Foundation
OSI	Open System Interconnection
OSPF	Open Shortest Path First
PAD	Packet Assembler/Disassembler
PARIS	Packetized Automatic Routing Integrated System
PBX	Private Branch Exchange
PCI	Peripheral Component Interconnect
PCM	Pulse Code Modulation
PCN	Path Control Network
PCNE	Protocol Converter for Non-SNA Equipment
PDU	Protocol Data Unit
PEP	Partitioned Emulation Program
PI	Pacing Indicator
PIR	CrossComm's Protocol Independent Routing
PIU	Path Information Unit
PLU	Primary Logical Unit
PLU	Primary LU
POWER	Performance Optimization with Enhanced RISC; Priority Output Writers, Execution Readers
PPP	Point-to-Point Protocol
PRIZMA	Packet Routing Integrated Zurich Modular Architecture
PS/2	Personal System/2
PTM	Packet Transfer Mode
PU	Physical Unit
PUCP	Physical Unit Control Point
PVC	Permanent Virtual Circuit
QOS	Quality of Service
RCP	Remote Communications Processor
RECFMS	Record Formatted Maintenance Statistics
REQMS	Request Maintenance Statistics
RFC	Request for Comment
RH	Request Header; Response Header

RIF	Routing Information Field
RIP	Routing Information Protocol
RMON	Remote Network Monitoring
RNR	Not Ready to Receive (Receive Not Ready)
RPC	Remote Procedure Call
RR	Receive Ready
RSCV	Route Selection Control Vector
RSRB	Remote Source-Route Bridging
RTM	Response Time Monitor
RTP	Rapid Transport Protocol; Research Triangle Park, N.C.
RU	Request Unit; Response Unit
SAA	Systems Applications Architecture
SABME	Set Asynchronous Balance Mode Extended
SAP	Service Access Point
SAR	Segmentation and Reassembly
SCS	SNA Character Stream
SDH	Synchronous Digital Hierarchy
SDLC	Synchronous Data Link Control
SDT	Start Data Traffic
SF	Start Field
SLIP	Serial Line Internet Protocol
SLU	Secondary LU
SMC	Standard Microsystems Corporation
SMDS	Switched Multimegabit Data Service
SNA	Systems Network Architecture
SNA/FS	SNA File Services
SNA/MS	SNA Management Services
SNA/DS	SNA Distribution Services
SNAP	Subnetwork Access Protocol; SNA Portable software
SNF	Sequence Number Field
SNI	SNA Network Interconnection; Subscriber-Network Interface
SNMP	Simple Network Management Protocol
SNRM	Set Normal Response Mode
SONET	Synchronous Optical Network
SPX	Sequenced Packet Exchange
SRB	Source-Route Bridging; Source-Route Bridge
SRT	Source-Route Transparent
SR-TB	Source-Route Translation Bridge
SSAP	Source Service Access Point
SSCP	System Services Control Point
STP	Shielded Twisted Pair; Service Transaction Program
STSN	Set and Test Sequence Numbers
SVC	Switched Virtual Circuit

SVN	Switched Virtual Networking
T1/T3	1.544Mbps link/45Mbps link
TAXI	Transparent Asynchronous Transmitter-Receiver Interface
TCAM	Telecommunications Access Method
TCP	Transmission Control Protocol; Terminal Control Program
TCP/IP	Transmission Control Protocol/Internet Protocol
TDM	Time Division Multiplexing
TDU	Topology Database Update
TG	Transmission Group
TH	Transmission Header
THDR	Transport Header
TIC	37xx Token-Ring Interface Coupler
TP	Transaction Processing
TPF	Transmission Priority Field
TPNS	Teleprocessing Network Simulator
TR	Token-Ring
TSO	Timesharing Option
UA	Unnumbered Acknowledgment
UI	Unnumbered Information
UNI	User Network Interface
UTP	Unshielded Twisted Pair
VC	Virtual Circuit or Virtual Connection; ATM Virtual Channel
VCC	Virtual Channel Connection
VCI	Virtual Channel Identifier
VCL	Virtual Channel Link
VCS	Virtual Channel Switch
VLAN	Virtual LAN
VM	Virtual Machine
VP	Virtual Path
VPI	Virtual Path Identifier
VPID	Vital Product Identification
VPL	Virtual Path Link
VR	Virtual Route
VS	Virtual Storage
VSA	Voice Server Adapter
VSE	Voice Server Extension
VTAM	Virtual Telecommunications Access Method
WAN	Wide Area Network
XI	SNA's X.25 Interconnect
XID	Exchange Identification
XNS	Xerox Network System
XRF	Extended Recovery Facility

Bibliography

SNA: Detailed Description of Traditional SNA Components, Protocols and Implementations up to LU 6.2

SNA: Theory and Practice. Anura Gurugé. New York: Pergamon, 1984. 1989.

SNA Specifications from IBM

Systems Network Architecture Formats and Protocol Reference Manual: Architectural Logic, SC30-3112. (November 1980). The bedrock SNA-4 architectural specification.

Systems Network Architecture: Introduction to Sessions Between Logical Units, GC20-1869.

Systems Network Architecture: Sessions Between Logical Units, GC20-1868.

Systems Network Architecture Formats and Protocol Reference Manual: Architecture Logic for LU Type 6.2, SC30-3269. Generic LU Type 6.2 specification for both subarea and Type 2.1 Node networks.

Systems Network Architecture LU 6.2 Reference: Peer Protocols, SC31-6808. Type 2.1 Node-specific LU Type 6.2 specification.

Systems Network Architecture Transaction Programmer's Reference Manual for LU Type 6.2. Specification of the architectural LU Type 6.2 Protocol Boundary, i.e., generalized API.

Systems Network Architecture Type 2.1 Node Reference, SC30-3422. The basic Type 2.1 node physical connection architecture. LU aspects described by the Peer Protocols manual listed above.

Systems Network Architecture Formats and Protocol Reference Manual: SNA Network Interconnection, SC30-3339.

Systems Network Architecture Management Services Reference, SC30-3346.

Systems Network Architecture File Services Reference, SC31-6807.

Systems Network Architecture Distribution Services Reference, SC30-3098.

Document Interchange Architecture: Document Distribution Services Reference, SC23-0762.

Document Interchange Architecture: Document Library Services Reference, SC23-0760.

Document Interchange Architecture: Application Processing Services Reference, SC23-0761.

Document Interchange Architecture: Interchange Document Profile Reference, SC23-0764.

Distributed Data Management Level 3.0 Architecture Reference, SC21-9526.

APPN Specifications from IBM

Systems Network Architecture: APPN Architecture Reference, SC30-3422.

HPR Specifications from IBM or the AIW

APPN High Performance Routing

Overview of ATM

Asynchronous Transfer Mode (Broadband ISDN): Technical Overview, IBM GG24-4330.

Overview of TCP/IP

The TCP/IP Companion. Martin Arick. QED, 1993.

TCP/IP Illustrated. Richard Stevens, Reading, Mass.: Addison-Wesley, 1993.

Overview of Frame Relay

Frame Relay: Principles and Applications. Philip Smith, Reading, Mass.: Addison-Wesley, 1993.

Index